Agricultural Soil Science

Agricultural Soil Science

Editor: Brian Bechdal

R CALLISTO REFERENCE

www.callistoreference.com

Callisto Reference,
118-35 Queens Blvd., Suite 400,
Forest Hills, NY 11375, USA

Visit us on the World Wide Web at:
www.callistoreference.com

ISBN: 978-1-63239-788-1(Hardback)

The publisher's policy is to use permanent paper from mills that operate a sustainable forestry policy. Furthermore, the publisher ensures that the text paper and cover boards used have met acceptable environmental accreditation standards.

Trademark Notice: Registered trademark of products or corporate names are used only for explanation and identification without intent to infringe.

Printed in the United States of America.

Cataloging-in-publication Data

Agricultural soil science / edited by Brian Bechdal.
    p. cm.
Includes bibliographical references and index.
ISBN 978-1-63239-788-1
1. Soil science. 2. Agriculture. 3. Soil fertility. 4. Soil productivity. 5. Soil management. 6. Soils and nutrition.
7. Fertilizers. I. Bechdal, Brian.
S591 .A65 2017
631.4--dc23

Table of Contents

Preface

Agricultural soil science takes into account all aspects of soil that are involved in crop production. Agricultural soil science considers soil to be a manageable resource and hence studies allied fields such as agricultural science and geology. This book aims to understand the management of cultivable soil and its characteristics such as soil texture, water content and fertility while keeping food production and economic viability as key targets. The topics covered in this extensive book deal with the core areas of agricultural soil science. It aims to present researches that have transformed this discipline and aided its advancement. It includes contributions of experts and scientists which will provide innovative insights into this field.

Various studies have approached the subject by analyzing it with a single perspective, but the present book provides diverse methodologies and techniques to address this field. This book contains theories and applications needed for understanding the subject from different perspectives. The aim is to keep the readers informed about the progress in the field; therefore, the contributions were carefully examined to compile novel researches by specialists from across the globe.

Indeed, the job of the editor is the most crucial and challenging in compiling all chapters into a single book. In the end, I would extend my sincere thanks to the chapter authors for their profound work. I am also thankful for the support provided by my family and colleagues during the compilation of this book.

Editor

Influence of Vegetation Restoration on Topsoil Organic Carbon in a Small Catchment of the Loess Hilly Region, China

Yunbin Qin, Zhongbao Xin*, Xinxiao Yu, Yuling Xiao

Institute of Soil and Water Conservation, Beijing Forestry University, Beijing, China

Abstract

Understanding effects of land-use changes driven by the implementation of the "Grain for Green" project and the corresponding changes in soil organic carbon (SOC) storage is important in evaluating the environmental benefits of this ecological restoration project. The goals of this study were to quantify the current soil organic carbon density (SOCD) in different land-use types [cultivated land, abandoned land (cessation of farming), woodland, wild grassland and orchards] in a catchment of the loess hilly and gully region of China to evaluate the benefits of SOC sequestration achieved by vegetation restoration in the past 10 years as well as to discuss uncertain factors affecting future SOC sequestration. Based on soil surveys (N = 83) and laboratory analyses, the results show that the topsoil (0–20 cm) SOCD was 20.44 Mg/ha in this catchment. Using the SOCD in cultivated lands (19.08 Mg/ha) as a reference, the SOCD in woodlands and abandoned lands was significantly higher by 33.81% and 8.49%, respectively, whereas in orchards, it was lower by 10.80%. The correlation analysis showed that SOC and total nitrogen (TN) were strongly correlated ($R^2 = 0.98$) and that the average C:N (SOC:TN) ratio was 9.69. With increasing years since planting, the SOCD in woodlands showed a tendency to increase; however, no obvious difference was observed in orchards. A high positive correlation was found between SOCD and elevation ($R^2 = 0.395$), but a low positive correlation was found between slope and SOCD ($R^2 = 0.170$, $P = 0.127$). In the past 10 years of restoration, SOC storage did not increase significantly (2.74% or 3706.46 t) in the catchment where the conversion of cultivated land to orchards was the primary restoration pattern. However, the potential contribution of vegetation restoration to SOC sequestration in the next several decades would be massive if the woodland converted from the cropland is well managed and maintained.

Editor: Manuel Reigosa, University of Vigo, Spain

Funding: This study was supported by the Fundamental Research Funds for the Central Universities (NoTD2011-2), the Open Foundation of Key Laboratory of Soil and Water Loss Process and Control on the Loess Plateau of Ministry of Water Resources (201301), the National Natural Science Foundation of China (No. 41001362) and the College Student Scientific Research Training Project of Beijing Forest University (No. 201210022013). The funders had no role in study design, data collection and analysis, decision to publish, or preparation of the manuscript.

Competing Interests: The authors have declared that no competing interests exist.

* Email: xinzhongbao@126.com

Introduction

Afforestation and other vegetation restoration techniques have been considered effective practices for the sequestration of carbon (C) to mitigate carbon dioxide (CO_2) concentrations in the atmosphere [1–4]. Soil plays an important role in the global carbon cycle. The soil C pool to a one metre depth has been estimated to sequester approximately three times the amount of carbon sequestered by the atmospheric pool and about four times that in the biotic/vegetation pool [5]. Therefore, a relatively small change in the soil C pool can significantly mitigate or enhance CO_2 concentrations in the atmosphere [6–7].

Land-use change can significantly influence the accumulation and release of SOC. When an ecosystem is disturbed by land-use change, the original equilibrium of the soil carbon pool is broken and a new equilibrium is created. During this process, soil may act as either a source or a sink of carbon depending on the ratio between inflows and outflows [8]. Many studies have reviewed the effect of land-use change on SOC. For instance, deforestation for agricultural purposes is the primary reason for the SOC loss. It

was reported that approximately 25% of SOC was lost by conversion of primary forest into cropland [9]. Houghton (1999) estimated that 105 PgC was released into the atmosphere due to the conversion of forests to agricultural lands between 1850 and 1990 [10]. However, afforestation and reforestation on agricultural lands have been cited as effective methods for increasing SOC pool and reducing the atmospheric CO_2 concentration [3,11]. Morris et al. (2007) observed that placing agricultural soils in deciduous and conifer forests resulted in soil carbon accumulations of 0.35 and 0.26 $MgCha^{-1} yr^{-1}$, respectively. Based on a meta analysis of 33 recent publications, Laganière et al. (2010) reported that afforestation increased SOC stocks by 26% for croplands. Some studies found that croplands that were converted to abandoned lands or grasslands could also increase the SOC storage [9,13–14]. Therefore, understanding the influence of land-use changes on soil organic carbon is an important step in predicting climatic change and developing potential future CO_2 mitigation strategies.

To improve the carbon sink status of afforeseation or other vegetation restoration methods on agricultural land, it is necessary to understand the control mechanisms of SOC dynamics to allow more carbon storage in soils [1]. A variety of factors will affect the quantity and quality of SOC after land-use changes. For example, climate variations have a significant effect on SOC, including temperature, precipitation, and potential evapotranspiration changes [11,15]. Laganière et al. (2010) reported that SOC restoration after afforestation was found to vary with the climate zone, with the temperate maritime zone having a higher SOC increase than others by approximately 17%. In addition, landscape and elevation also have a pronounced effect on SOC change at the catchment scale. Slope is an important topographical factor that affects soil erosion and also has an important influence on the soil nutrient loss of the slope surface [16]. Wang et al. (2012) reported that the SOCD was higher in shady slopes than in sunny slopes, and gentle slopes would generally sequester more SOC than steep slopes. Elevation differences can cause climatic and biological changes in the soil-forming environment and can influence the vertical distribution of SOC [18]. In natural ecosystems, it has been extensively documented that carbon and nitrogen are closely related [4,19]. Increased nitrogen retention may increase the carbon sequestration potential [1]. In Panama, Batterman et al. (2013) found that symbiotic N_2 fixation has potentially important implications for the ability of tropical forests to sequester CO_2 [20]. N_2-fixing tree species accumulated carbon up to nine times faster per individual than neighbouring non-fixing trees. The soil C/N ratio also has a significant effect on the rate of decomposition of organic compounds by soil microorganisms [21].

Soil erosion, as the most widespread form of soil degradation, has a large impact on the global C cycle, causing a severe depletion of SOC pools in the soil [22–24]. The total amount of C released by soil erosion each year has been estimated at approximately 4.0–6.0 Pg/year [23]. China's Loess Plateau, covering approximately 6.2×10^5 km^2, has the world's most severe soil erosion because of its unusual geographic landscape, soil and climatic conditions, and long history (over 5000 years) of human activity. Over 60–80% of the land in the Loess Plateau has been affected by soil erosion, with an average annual soil erosion of 2000 to 20000 t km^{-2} yr^{-1} [25–26]. Severe soil erosion has resulted in land degradation, which was manifested primarily in the thinning of the soil layer, nutrient loss and fertility reduction, which has directly caused decreases in the local farmers' income and has economically and socially hindered sustainable development [27–29]. Unreasonable human activities, such as deforestation and tillage on slopes, have further intensified soil erosion and land degradation in the Loess Plateau [30].

Since the 1950s, the Chinese government has launched many large-scale projects to attempt to control soil erosion and restore vegetation in the Loess Plateau, including large-scale afforestation in the 1970s and comprehensive control of soil erosion on the watershed scale in the 1980s and 1990s. Despite these efforts, there have not been significant increases in ecological benefits, which is largely due to the limitations and influences of bad natural conditions and unreasonable ecological restoration techniques. To control soil erosion and improve the quality of the local environment, "Grain for Green" was initiated by the government in the Loess Plateau in 1999. The government demanded that the agricultural lands with a slope of over 25 degrees be converted to forest, terrace orchards or grassland. To compensate famers for their economic loss, they will be given grain, cash and planting stocks by the government as subsidies and incentives for converting cultivated land back to forest, orchards or grassland. Currently, the Loess Plateau environment appears to be experiencing a recovery following more than 10 years of vegetation restoration [31–32].

Over the past decade, large-scale vegetation restoration efforts have brought obvious land-use changes to the Loess Plateau, and also have significantly influence on SOC sequestration. Therefore, understanding the effects of these dynamic changes and the corresponding changes in SOC storage caused by land-use change driven by the implementation of the Grain for Green project is important in evaluating the environmental benefits of this ecological restoration project. Recently, many studies have focused on SOC changes induced by the Grain for Green project in the Loess Plateau. These studies have mainly focused on the SOCD of different land-use types, the SOC pool, the rate of SOC change, the variation in SOC among different land-use conversions and factors which have influenced SOC sequestration in the Loess Plateau after vegetation restoration efforts [32–34]. However, for most of the researches, the study time period was shorter than 10 years as vegetation restoration was driven by the Grain for Green project. The study areas have been mostly located at the gully region of the Loess Plateau [2,17,35–36]. Little is known about the gully area of the Loess hilly region. To gain a complete understanding of the SOC change after restoration in this region, we selected the Luoyugou catchment of the Loess Plateau as the study area, which has the typical geomorphologic characteristics of the gully area of the Loess hilly region.

In this study, we hypothesised that the SOCD varied with land-use types in this catchment, and there was an increase in SOC storage of the total catchment after 10 years of restoration. Therefore, the objectives of this study were to (i) quantify the SOCD of different land-use types differ significantly in the study catchment, (ii) analyze those factors affecting SOCD, (iii) estimate the contribution of land-use conversions on SOC sequestration in the study area, in the past 10 years of restoration, and (iv) discuss the uncertainties in potential SOC sequestration during future ecological restoration efforts in the Loess Plateau.

Materials and Methods

Ethics statement

The administration of the Tianshui Experiment Station of Soil and Water Conservation, Yellow River Conservancy Commission and local farmers which are the owners of study lands gave permission for this research at each study site. We confirm that the field studies did not involve endangered or protected species.

Study area

Tianshui city (104°35′–106°44′E, 34°05′–35°10′N) of Gansu province, China, is located in the western side of the Qinling Mountains, which is the transitional zone between the Qinling Mountains and the Loess Plateau and also belongs to the third sub-region of the Loess hilly region in the middle portion of the Yellow River and the second class tributary of the Wei River. This region has the typical geomorphology of the gully area of the Loess hilly region, which includes earth-rocky mountainous areas, Loess ridges and hilly areas. In this area, the catchment landscape is an agroforestry landscape, and the climate has obvious transitional characteristics. Because of these special geomorphologic characteristics, Walter Lowermilk, deputy director of the US Department of Soil Conservation Service, chose this region in which to build an experimental station for soil and water conservation in 1941. Tianshui station, Suide station and Xifeng station now are three the best-known experimental stations for soil and water conservation in the Loess Plateau.

The Luoyugou catchment (105°30′–105°45′E, 34°34′–34°40′N), located in northern Tianshui, is a part of the observation area of the Tianshui soil and water conservation station. Its total area is 72.79 km^2, with a range from 1165–1895 m above sea level. The main topographic type is loess ridge landform, and the average slope is 19°. It has a typical continental monsoon climate with a mean annual precipitation of 548.9 mm (1986–2004). Approximately 78% of the rainfall is concentrated between July and September. The average annual temperature is 11.4°C (1986–2004),and the annual evaporation is 1293.3 mm. The main soil type in this catchment is mountain grey cinnamon soil, which, according to the Food and Agriculture Organization of the United Nations Educational, Scientific and Cultural Organization (FAO-UNESCO), belongs to the Cambisol soil group and accounts for 91.7% of the soil in the region [37–38] (Figure 1).

The main land-use types for the Luoyugou catchment include woodland, wild grassland, orchards, cultivated land, and abandoned land. In the woodland, the major tree species is black locust (*Robinia pseudoacacia L.*), most of which was artificially planted. Woodlands that are more than 30 years old have good water condition and low density because planting is located mainly beside the place beside water ditches and shady slope. They are the retained trees of the planting projects in the late 1950s to 1970s, without human management now. And most of woodland at the age of 10 to 30 years since planting is found in Fenghuang forest farm located in the gully head of this catchment, which has good management and protecting. Since 1999, the government of Tianshui city has widely implemented the "Grain for Green" project. A lot of cultivated land in the ridge top and slope top has been converted into woodland. Most of the woodland is less than 10 years since planting, and has a high density and bad management. Wild grassland is usually found on stone mountain of the northern catchment where the soil layer is thin and slope is steep. Human activities are restricted in there, only sometimes sheep may reach. The major species are dahurian bushclover (*Lespedeza dahurica (Laxm.) Schindl.*), russian wormwood (*Artemisia sacrorum Ledeb.*), digitate goldenbeard (*Bothriochloa ischaemun (L.) Keng*) and bunge needlegrass (*Stipa bungeana Trin.*), etc. The main fruit species under orchards are cherry (*Cerasus pseudocerasus (Lindl.) G. Don*), apple (*Malus pumila Mill.*) and apricot trees (*Armeniaca vulgaris Lam.*). The management system of orchards is the conventional tillage which is that the weeds beneath the trees are eliminated with herbicides and dead leaves, dried fruit and twigs are removed by manual blowers. Besides, there are not irrigation facilities and the tree water demand all comes from the rain. Fertiliser and pesticide are also applied in the orchards. This catchment is a rainfed agriculture region with a long cultivated history. The major species in the cultivated land are wheat (*Triticum aestivum L.*), maize (*Zea mays L.*) and edible rape (*Brassica campestris L.*), etc. Abandoned land is the cultivated land has been the ceased farming

activities, due to the demand of "Grain for Green" project, the bad environment conditions, the shortage of rural labour and the damage by wild animals, etc. After cultivated land abandoned, old field are spontaneously colonised by various plants, while a secondary succession process will gradually develop during different plant communities. Because of short time since abandoned, the main species in the abandoned land are annual and biennial herb and a bit of perennial herb, such as virgate wormwood (*Artemisia scoparia Waldst. et Kit.*) and green bristlegrass herb (*Utricularia australis R. Br.*).

Soil sampling and laboratory analysis

Based on the main land-use types and the percentage of each land-use-type area in the total catchment area (2008), we randomly selected 83 land-use blocks that included cultivated land (34 samples), abandoned cropland (9 samples), orchards (13 samples), wild grassland (8 samples) and woodland (19 samples) as investigation plots. Their latitudes and longitudes are shown in Table 1. Spatially separated plots were located at least 3 km from one another to help avoid pseudo-replication. In July 2012, the selected plots were surveyed and collected in the field using the Global Positioning System (GPS). The information recorded from the sampling plots in the field included geographic coordinates, slope, elevation, plant species, and water conservation measures. The years since planting or abandoned of the sampling plots were evaluated by recording tree diameters and heights, plant composition, the degree of decomposition of the topsoil crust and the crop's residual body, and by interviewing local farmers. In each sampling plot — woodland and orchards (10×10 m), cultivated land, abandoned land and wild grass land (5×5 m) — we randomly collected three soil samples using a stainless steel cutting ring 5.0 cm high and 5.0 cm in diameter to measure the soil bulk density (BD) of the topsoil layer (0–20 cm) and by taking five random soil samples between 0 and 20 cm depth with a 20 cm long soil auger. The five soil samples were manually homogenised to form a composite sample for each sampling plot, and a quarter of this sample was taken to the laboratory. These samples were air-dried and passed through a 2 mm sieve, while gravel and roots were removed from each soil sample. A quarter of each sub-sample was completely passed through a 0.25 mm sieve to determine SOC and TN. SOC was determined by the $K_2Cr_2O_7$-H_2SO_4 Walkey-Black oxidation method [39]. TN was measured using the micro-Kjeldahl procedure [40].

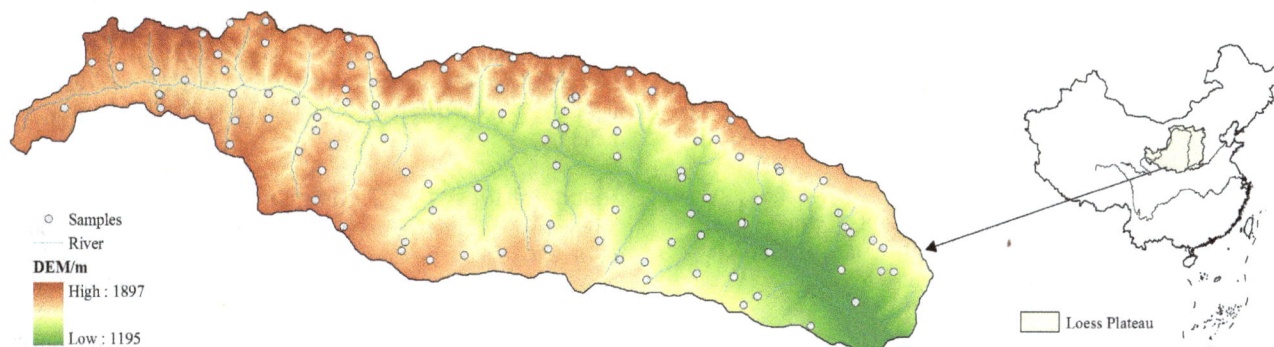

Figure 1. Location of the study area and distribution of the sampling points.

Table 1. Attributes of the studied sites.

Land-use	Sample sites	Years since planting (abandoned)/a	Elevation/m	Slope/(°)	Bulk density/(g/cm^3)	Plants
Cultivated land	34	—	1536.94	4.85	1.29	Wheat, maize, rape
Abandoned cropland	9	8–10a	1559.40	4.22	1.35	Artemisia scoparia, Leymus secalinus
Orchards	13	<5a, 5–10a, >10a	1446.45	1.92	1.32	Cherry, apple, pear
Woodland	24	<10a, 10a–30a, >30a	1639.14	12.13	1.39	Robinia
Wild grassland	8	—	1526.55	13.00	1.44	Artemisia sacrorum var. incana mattf, Stipa bungeana, Bothriochloa ischaemun

Statistical analysis

The SOCD was calculated using the following equation:

$$SOCD = SOC \times D \times BD \qquad (1)$$

where SOCD was the soil organic carbon density (Mg ha^{-1}), SOC was the soil organic carbon content (%), D was the soil layer depth (cm), BD was the soil bulk density (g cm^{-3}).

The soil organic carbon storage (SOCS) was calculated using the following equation:

$$SOCS = SOCD_i \times A_i \times 100 \qquad (2)$$

where SOCS was the soil organic carbon storage (t), $SOCD_i$ was the soil organic carbon density in the i land-use (Mg ha^{-1}), and A_i was the area of the i land-use (km^2).

The increase in SOCS caused by land-use conversion was calculated using the following equation:

$$\text{Increase of SOCD (t)} = SOCD_i \times (A_{i2}\text{-}A_{i1}) \times 100 \qquad (3)$$

$$\text{Sequestration amount of SOCS (t)} = (SOCD_i - SOCD_c) \times (A_{i2} - A_{i1}) \times 100 \qquad (4)$$

where $SOCD_i$ was the SOCD in the i land-use (Mg ha^{-1}), A_{i1} and A_{i2} were the i land-use area in 2002 and 2008, respectively (km^2), and $SOCD_c$ was the SOCD of the cultivated land (Mg ha^{-1}). All SOCD values of the different land-use types in 2002 and 2008 were the average SOCD values of the soil data collected in 2012.

All data were analysed using Excel 2007 and SPSS (Statistical Package for the Social Sciences) 18.0 software. Analysis of variance (ANOVA) was used to determine the significance of the mean difference. Fisher's LSD test was used to compare the mean values of soil variables when the results of ANOVA were significant at $p < 0.05$. Pearson's test was used to analyse the correlation between SOCD and the soil variables. For all analyses, a $p < 0.05$ was used to test statistical significance. Finally, SOCD, land-use type, elevation and slope values were used to establish the multiple regression model. SOCD was the dependent variable, and others were the independent variables.

Satellite data acquisition and processing

In our study, two types of SPOT5 multispectral images were acquired on August 25, 2002 and May 5, 2008, which were multi-spectral images (resolution: 10 m) and panchromatic images (resolution: 2.5 m). To validate the final classification result, we selected 17 points located in the Luoyugou catchment to use as the observed data based on the land-use map and some thematic maps. In July 2012, these monitoring points were collected in the field to validate the panchromatic map, while we recorded 83 points with information of land-use type and longitude and latitude to validate the accuracy of the classification results. Using ArcGIS and ERDAS software, we chose the common pre-classification, which is the supervised classification with a Maximum Likelihood Decision (MLD), to classify the preprocessed images and acquire the data of different land-use types in 2002 and 2008. The overall accuracy and Kappa coefficient of the initial classification images were 87.50% and 0.85 in 2002, respectively, and 90.10% and 0.87 in 2008, respectively. The accuracy of the 88 typical types of land-use information used to validate the classification results was 95.74%, which implied that this classification result could meet the analysis demands.

Results

Topsoil organic carbon density under different land-uses

The one-way ANOVA indicated that land-use type had a significant effect on the SOCD ($F = 8.34$, $P < 0.01$). However, only the woodland was significantly different from the other land-uses ($P < 0.05$), while no significant differences were found between other land-use types. The average SOCD of the topsoil layer (0–20 cm) in the Luoyugou catchment was 20.44 Mg/ha, and was highest in the woodland and lowest in the orchards (Table 2). Using the SOCD in the cultivated land (19.08 Mg/ha) as a reference, the SOCD in the woodland and abandoned cropland significantly increased by 33.81% and 8.49%, respectively, while the SOCD in the orchards decreased by 10.80%. The SOCD in the wild grassland was close to that of cultivated land. In this area, the coefficient of variation of the SOCD was 26.46%, and was highest (27.08%) in the woodland and smallest (16.75%) in the orchards.

Change of topsoil SOC storage in the catchment

Based on the interpretation of the remote sensing data (2000), the results showed that the total area of the four main land-use types including woodland, orchards, cultivated land and wild grassland was 67.5 km^2, accounted for 92.09% of the total region of the Luoyugou catchment. The difference of 7.02% included river, residential points, and roads that were not taken into account in the calculation of the topsoil SOC storage. From 2002 to 2008, the net decrease in the cultivated land area was

Table 2. Topsoil (0–20 cm) SOCD of different land-uses.

Land-use	Total	Woodland	Orchards	Abandoned cropland	Wild grassland	Cultivated land
Sample sites/N	83	19	13	9	8	34
SOCD Mg/ha	20.44	25.53 a	17.02 b	20.70 b	19.35 b	19.08 b
Standard deviation	5.41	6.91	2.85	4.59	3.99	3.66
Minimum	10.47	16.45	11.03	16.55	12.59	10.47
Maximum	40.71	40.71	20.72	31.14	26.02	28.86
Coefficient of variation (%)	26.46	27.08	16.75	22.17	20.62	19.16

* A different letter means a difference significant at 0.05 level.

15.11 km^2, where 7.95 km^2 of this land was converted into the woodland, and 6.51 km^2 was converted into the orchards. The increased areas of land were accounted for 52.65% and 43.12% of the total conversion area, respectively. In 2008, the SOC storage in this catchment was 1.39×10^5 t and was higher than the SOC storage in 2002 (1.35×10^5 t) by 3706.46 t. When the cultivated land was converted into the woodland, the SOC sequestration contribution was 4484.83 t; however, when the cultivated land was converted into the orchards, the contribution was -723.02 t (Table 3).

Relationship of soil organic carbon and total nitrogen content

In this catchment, soil organic carbon and total nitrogen revealed a significant positive correlation ($R^2 = 0.978$, $p < 0.01$) that increased as total nitrogen content and soil organic carbon increased (Figure 2). A one-way ANOVA indicated that land-use had significant effect on the TN ($F = 3.07$, $P = 0.021$), and that had no significant effect on the soil C/N ratio ($F = 0.86$, $P = 0.48$). The average soil C/N was 9.69, with a range of 8.73–10.77. The soil C/N ratio in the woodland was the largest at 9.80 and that in the orchards was the smallest at 9.57. Other land-uses followed the order of abandoned cropland (9.76)>cultivated land (9.66)>wild grassland (9.59).

Relationship of years since planting and topsoil SOCD

For the woodland, years since planting had a significant effect on SOCD ($F = 13.00$, $P < 0.001$). SOCD increased as years since planting increased. However, SOC was lost initially after afforestation in the cultivated land. The SOCD of the woodland that was older than 30 years was higher than that of the woodland under 10 years old (17.45 Mg/ha) and the cultivated land (19.08 Mg/ha), improving by 74.44% and 59.54% respectively (Figure 3a). However, years since planting had no significant effect on SOCD in the orchards (F = 2.01, P = 0.146), where the SOCD at >10a was 18.04 Mg/ha and was 10.13% higher at <10a. But, they all were below that of the cultivated land (Figure 3b).

Relationship of elevation and topsoil SOCD

The correlation analysis showed that there was a significant positive correlation between elevation and SOCD ($R^2 = 0.395$ $P < 0.01$), using the land-use type as the covariate. The SOCD was calculated as the average value with each 100 m used as an elevation gradient. The results showed that with increasing elevation, SOCD had an increasing trend (Figure 4). At an elevation gradient of ≥ 1700 m, the SOCD was 26.57 Mg/ha, which was higher than that at the elevation gradient of <1300 m by approximately 58.10%.

Relationship of slope and topsoil SOCD

A one-way ANOVA indicated that land-use had a significant effect on slope ($F = 2.87$, $P = 0.028$) where the average slopes of the woodland and wild grassland all were more than $11.00°$, and the average slopes of the cultivated land, abandoned cropland and orchards all were less than $5.00°$ (Table 4). The correlation analysis showed a low positive correlation between slope and SOCD ($R^2 = 0.170$, $P = 0.127$). A negative correlation was found between the slope and SOCD in the cultivated land, abandoned cropland and orchards ($R^2 = -0.210$, $F = 0.123$), and a positive correlation was found in the woodland and wild grassland ($R^2 = 0.250$, $F = 0.209$). For all correlations, the land-use type was used as the covariate.

Influence of multivariates on topsoil SOCD

Through normalisation processing of SOCD, elevation and slope and setting land-use dummy variables, the multiple regression model was established using SPSS software. The results were as follows:

$$y = 0.150 + 0.233x_1 + 0.054x_2 + 0.004s_1 - 0.025s_2 + 0.044s_3 + 0.168s_4 \left(R^2 = 0.329,\ p < 0.01 \right) \quad (5)$$

where y was the topsoil SOCD, x_1 was elevation, x_2 was slope, s_1 was wild grassland, s_2 was orchards, s_3 was abandoned land, and s_4 was woodland. The cultivated land was used as the reference when the value of s_{1-4} was 0.

According to the multiple model, elevation, slope and land-use accounted for 32.9% of the SOCD variation using the cultivated land as the reference (Eqn. 5).

Discussion

SOC sequestration in different ecological restoration types

The different types of conversion result in different trends in SOC [8]. In this study, the highest SOCD in the woodland implied that the conversion from the cultivated land to the woodland may lead to the accumulation of more SOC than when cultivated land is converted into the orchards or abandoned cropland in this area (Table 2). Some previous research has also found that the conversion from cultivated land to woodland can increase SOC levels [3,17]. This is mainly because the conversion from cultivated land to woodland can increase topsoil SOCD

Table 3. Change in SOC storage from 2002 to 2008 in the Luoyugou catchment.

Land-use	2002			2008			SOC	
	Area/km²	SOC		Area/km²	SOC			
		Storage/t	Percentage/%		Storage/t	Percentage/%	Increased/t	Sequestration amount/t
Immature woodland	8.16	19764.05	14.62	15.48	37480.99	26.98	17716.94	3759.91
Mature woodland	2.35	7145.31	5.28	2.99	9087.78	6.54	1942.48	724.92
Orchards	4.11	7377.46	5.46	10.62	19082.52	13.74	11705.06	−723.02
Wild grassland	0.71	1374.26	1.02	1.31	2536.62	1.83	1162.36	16.22
Cultivated land	52.18	99557.47	76.63	37.07	70737.09	50.92	−28820.38	–
Total	67.50	135218.55	100.00	67.46	138925.00	100.00	3706.46	3778.03

* Because the area of this catchment in 2008 was less than that in 2002, by 0.04 km², so that about 71.58 t of SOC wasn't taken into the amount of SOC storage. Immature woodland refers to trees under 30 years of age since planting (SOCD = 24.22 Mg/ha), and mature woodland refers to trees more than 30 years of age since planting (SOCD = 30.44 Mg/ha).

through increasing biomass inputs into the soil, and reducing soil erosion [9,12]. In addition, compared to frequent human activity such as tillage and grazing, in the cultivated land and abandoned cropland, less human disturbance in the woodland also enhanced SOC accumulation.

Our research found that the SOCD in the immature woodland (<10 a) was lower than that in the cultivated land. However, the SOCD of the mature woodland (>30 a) was higher than that of the cultivated land, increasing by 59.54% (Figure 3a). Therefore, in the initial conversion from the cultivated land to the woodland, SOCD may decrease, and from 10a after afforestation SOCD increase rapidly and significantly. Other studies have found similar trends [4,15]. Degryze et al. (2004) in Michigan found no difference in topsoil (0–25 cm) soil C during the first ten years after afforesting in a cropland; however, soil C of the native forest (48.6 t/ha) was significantly greater than the cropland (31.9 t/ha) [41]. Lu et al. (2013), studying afforestation in the Loess Plateau, found that the time of the SOC source to sink transition was 3 to 8 years after afforestation. The decrease of SOC in the first few years following afforestation has been attributed to low net primary productivity of plants, decreased litter inputs and increased decomposition rates [12,29].

In this study, the topsoil SOCD in the orchards was less than that in the cultivated land, which implied that a decrease in SOC may occur when the cultivated land is converted into the orchards (Figure 3b). This result is similar to the results of Yang et al. in the gully region of the Loess Plateau [42]. However, Xue et al. (2011) reported that when the slope farmland was converted into the orchards, the SOC content increased slowly with increasing years since planting and reached its peak between 20 to 30 years [43]. Thirty years later, the SOC content of the converted orchards was 4.96 g/kg and was higher than that in the slope farmland by 97%. One difference between our study and that by Xue et al. (2011) is the different reference used. The reference in their study was slope cropland, whereas we used the average value of the slope and terrace cropland as the reference. Moreover, management measures in their study were better than ours. For example, they interplanted crops into the fruit trees at the seedling stage and used more fertiliser.

Management measures are one of the main factors affecting the loss of SOC in orchards. In this catchment, fruit, trimmed branches, and litter are removed from orchards, which can cause less organic compounds to enter the soil. Clear tillage in the orchards also causes more soil erosion. Therefore, in some farmland regions that are returning to their historic uses and thus need to develop fruit trees for economic purposes, methods for reducing the loss of SOC and enhancing SOC sequestration have become important. To manage this problem, some researchers have found that orchards managed with conservation practices such as growing grass and leguminous cover crop, mulching the ground, and frequently using organic and inorganic fertilisers could reduce soil erosion and improve the SOC content [44–45]. In addition, establishing new orchard management models oriented to SOC sequestration is also very necessary.

Many studies have found that the land-use change from the cultivated land to the abandoned land can increase the SOC stored in soil [13–14]. In our study, our results also showed that the SOCD may improve when the cultivated land is converted into the abandoned land (Table 2). With the termination of human disturbance, vegetation in the abandoned land may begin the process of self-succession [14], gradually forming original vegetation communities of their region, which are secondary wild grassland or secondary forest communities. However, this process of succession is slow, as the process that turns the abandoned land into the top secondary bunge needlegrass (*Stipa bungeana Trin.*) takes 40 to 50 years [46]. Further studies are needed to understand the methods needed to accelerate community succession and better sequester SOC in the abandoned land.

The research of Li et al. (2007) in the northern region of the Loess Plateau found that the SOCD of abandoned land at approximately 10 years of age was slightly lower than that in the secondary wild land where planted grassland (*Medicago sativa*) changed in the same number of years. After 6 to 10 years of restoration, degraded artificial grassland may form secondary bunge needlegrass (*Stipa bungeana Trin.*) [47–48]. Therefore, planting artificial grass in the abandoned land can shorten the time of succession, which ensures ecological benefits and increase economic benefits at the same time. However, in this study, the SOCD of the abandoned land at 8–10 years of age was higher

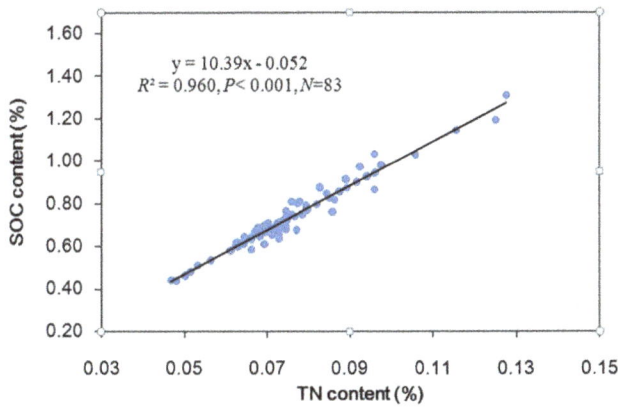

Figure 2. Relationship of topsoil soil organic carbon and total nitrogen content.

than in the wild grassland, which was different to the results of previous studies [47]. The main reason for this may be that these grasslands were distributed in the barren stony mountainous area. The average slope of these regions was 13°, and soil erosion was relatively severe. These factors led to low SOCD.

Benefits of SOC sequestration in the entire region

The SOC storage is determined by the dynamic equilibrium of SOC input and output. After approximately 10 years of ecological restoration, the area of cultivated land greatly decreased, while the area of woodland and orchards increased. From 2002 to 2008, the SOC storage of the entire area only improved by 2.74% (3706.46 t) (Table 3). Therefore, the ecological restoration has an increase in SOC storage, but not significant in the short-term. The main reason for this is that an increase in the SOC resulting from the woodland accumulation was offset by a loss in the SOC resulting from the orchards and the initial conversion stage. So, the SOC storage of the total area is in a relatively stable state. If all of the woodland values are calculated using the SOCD value of the woodland more than 30 years old (30.44 Mg/ha), then considering the loss of SOC by the orchards, the SOC storage of the total area can improve by 7.12% (9625.59 t). Therefore, SOC storage has great potential to increase in the future if the woodland converted from the cropland is well manage and maintain.

Although SOC storage has not increased in the past 10 years of restoration, with increasing ecological restoration, soil erosion has been controlled and the quality of the ecological environment, such as the air, water, vegetable coverage has gradually improved in this catchment.

Influencing factors on the SOC

Results from this study showed a strong correlation between SOC and TN (Table 2), which was consistent with earlier findings [33,36]. The trend of C:N ratio varying with the land-use types was similar to that of SOC. The woodland had a higher C:N ratio than other land-use types due to increased above and below ground biomass. The most prevalent tree in the woodland is the black locust, which has N-fixing capabilities, increased TN can promote tree growth, which results in an increase in SOC; it was found that the SOC increase was greater than the TN increase [33]. In catchments, elevation is a governing variable because of its effect on various environmental factors, such as temperature and precipitation. Our study showed a significant positive correlation between elevation and the SOCD (Figure 4). With rising elevation, topsoil SOC increased. This phenomenon was similar to results of other studies [18,49], which can be explained by the lower temperatures and increased precipitation with higher elevations. Increased precipitation can promote vegetation growth, which increases the accumulation of humus, and lower temperatures can limit the decomposition and turnover of SOC, leading to enhanced SOC storage.

The slope is one of the main topographical features affecting the soil erosion intensity as well as SOC loss and enrichment [16]. In the present study, land-use types significantly affected the correlation between the slope and SOCD, where there was a positive correlation in woodland and wild grassland and a negative correlation in cultivated land, abandoned land and orchards (Table 4). These outcomes were due mainly to the influence of vegetation type and coverage. Vegetation restoration on slopes such as planting trees and grass can increase surface vegetation coverage and effectively control soil erosion, further reducing the loss of SOC. These results have been found in other regions [17,30]. Thus, the slope land should be converted to woodland and grassland, or changed to terrace to maintain and increase SOC levels. The multiple regression analysis showed that land-use, slope and elevation accounted for 32.9% of the SOCD variation. The unexplained variation may be caused by the soil particle size,

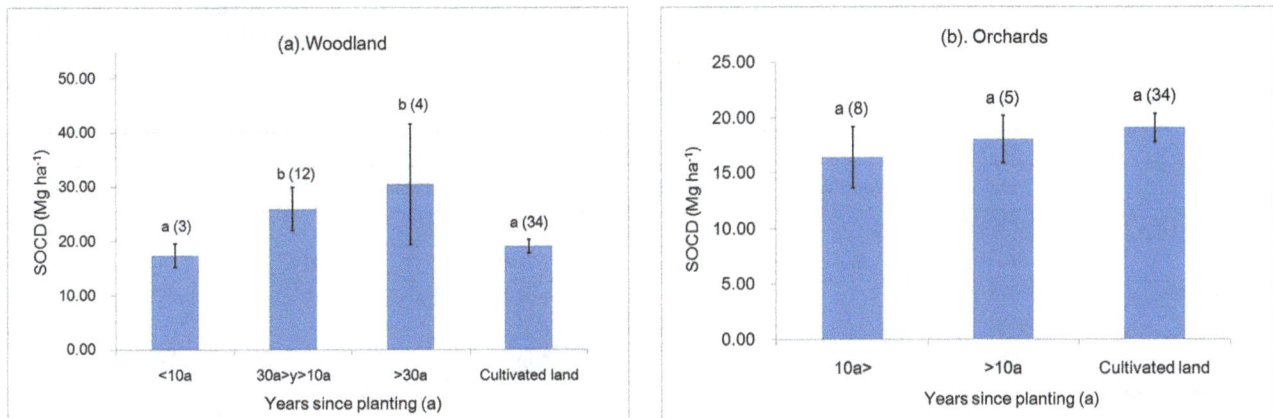

Figure 3. Relationship of years since planting and topsoil SOCD. * The error bars are the standard errors of the mean of SOCD and values above the bars is the number of observations (in parentheses). A different letter means a difference significant at $P<0.05$.

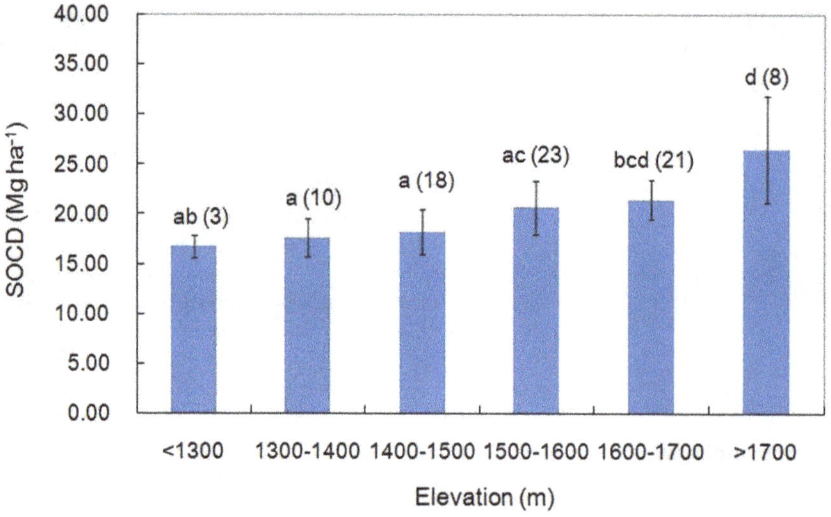

Figure 4. Relationship of elevation and topsoil SOCD. * The error bars are the standard errors of the mean of SOCD and values above the bars is the number of observations (in parentheses). A different letter means a difference significant at $P<0.05$.

soil aggregates, the aspect and the years of utilisation. Therefore, a benefits evaluation of SOC sequestration achieved through ecological restoration efforts should fully consider those influencing factors.

Uncertainties over SOC sequestration from future ecological restoration efforts

With ecological restoration efforts increasing, soil erosion in the Loess Plateau has significantly decreased, and ecological benefits such as vegetation coverage have gradually increased. However, in the long run, there are still uncertainties that will persist if we use the present SOC sequestration achievements from ecological restoration to evaluate possible future SOC sequestration benefits.

(I)   Influence of national policies: with the economy developing, the present subsidy standard for returning farmland to forests in the Loess Plateau has been increasingly unable to meet the basic living demands of local farmers. The competitiveness of this subsidy is also growing increasingly weak due to the influence of factors which are raising grain prices and creating policies which benefit farmers, including the abolition of the agricultural tax and the enhancement of subsidies for farming. Based

on the results of 2000 farmers' surveys, Cao et al. (2009) found that approximately 37.2% of farmers planned to return to cultivating forested areas and grasslands once the project's subsidies end in 2018. The smaller the subsidy amount to farmers, the fewer the farmers who are willing to participate in them. If the government is unable to improve farmers' income, much of the vegetation restored during the "Grain for Green" project is at risk of being re-converted into farmland and rangeland at the end of this project [50]. Therefore, it is very important and urgent to determine how to effectively consolidate the achievements of the "Grain for Green" project, especially once the project ends, while continually completing the task of returning farmland to forest. However, the government does not do well in this respect. Therefore, uncertainties exist concerning the benefits from SOC sequestration by ecological restoration in the future.

(II)   Influence of the environment: Most regions converted from cultivated land to forest are experiencing drought and soil impoverishment. There are no rich water resources in the Loess Plateau, and large-scale vegetation construction will certainly consume massive amounts of water. Lü et al. (2012) found that over half of the Loess

Table 4. Slope of different land-uses.

Land-use	Total	Woodland	Orchards	Abandoned cropland	Wild grassland	Cultivated land
Sample sites/N	83	19	13	9	8	32
Slope/°	6.65	11.58 a	1.92 b	4.22 ab	13.00 a	4.85 b
Standard deviation	10.85	14.76	4.80	8.39	12.10	8.93
Minimum	0	0	0	0	0	0
Maximum	45.00	45.00	15.00	20.00	30.00	30.00
Coefficient of variation %	163.16	127.46	250.00	198.82	93.08	184.12
Correlation between slope and SOCD/R^2	0.170	0.269	−0.169	−0.459	0.480	−0.229

* A different letter means a difference significant at 0.05 level. The signification of correlation between slope and SOCD all were more than 0.05 ($P>0.05$).

Plateau experienced a decrease in runoff (2–37 mm/year) with an average of 10.3 mm/year after the implementation of the "Grain for Green" project [51]. Therefore, with water consumption increases resulting from vegetation, some regions are likely to face a deficit of water resources or even a possible depletion. Drought, decreased rainfall and the excessive depletion of deep soil water by planted vegetation causes the formation of the dried soil layer [52]. The dried soil layer can lead to land degradation and heavily influences vegetation growth, which forms large-scale, low-efficiency production forests [53–54]. Once the dried soil layer appears, it is very difficult for recovery to occur. In the process of implementing "Grain for Green" some improper phenomena have occurred, including the improper selection of vegetation restoration types and excessive close planting conducted in the blind pursuit of economic benefits and political achievement [55], and this has accelerated the formation of the dried soil layer. Therefore, in the long run, ecological restoration in the Loess Plateau is likely to face water storage restrictions. Thus, the benefits of SOC sequestration may be uncertain in the future.

(III)   Influence of private and local government interests: Many regions would prefer that cultivated land be converted into orchards when the cultivated land needs to be converted due to this conversion type may generate a higher land value. However, the SOC sequestration benefits resulting from this type of conversion are lower than in the conversion from cultivated land to woodland, and this conversion type may even cause SOC loss. Over time, farmers may remove old trees in orchards converted from cultivated land, or abandoned land. In addition, according to changes in market demand, farmers may replace other tree species before the original trees have matured. The non-ecological woodland converted from cultivated land also can be harvested under reasonable conditions. However, the SOC sequestration effects of these changes have not been studied. In the future, many woodlands and orchards converted from cultivated land will almost certainly undergo cutting and regeneration. The benefits of SOC sequestration are uncertain in the future unless we can understand the SOC variation under these changes.

(IV)   The influence of urbanization and economic interests: since the reforms and open policies were established, labour exports from rural areas have increased year after year. Lack of labour has caused the abandonment of large areas of rural cultivated lands [56]. Recent research has shown that the SOC content could increase when cultivated land is converted into abandoned land. However, given the increasing speed of abandonment, it is unclear how long this phenomenon can continue. In addition, these labours may choose to go home to engage in farming in the future, affecting by the economic crisis of the original working regions and other reasons such as older age and homesickness. This transform may bring new pressure on the local ecological environment, which may cause the loss of SOC. Therefore, the future benefits of SOC sequestration are uncertain.

(V)   Influence of climate change: Here, climate change mainly refers to precipitation decline, temperature increase and elevated carbon dioxide levels. According to the results of Xin et al. (2011), the drying trend of the Loess Plateau was highly significant, and annual rainfall showed an obvious decreasing trend over the past five decades (1956–2008) by approximately −1.4 mm/a [57]. The reduced precipitation aggravated the water resource shortage situation in the Loess Plateau, which led to the poor tree growth. This outcome may affect SOC sequestration by vegetation restoration, because the temperature increased and elevated carbon dioxide levels have opposite effects on SOC sequestration. Increasing temperatures may promote SOC decomposition; however, elevated carbon dioxide levels may also improve the net primary productivity of plants and increase both the residual body of vegetation and SOC storage [58–59]. Over the last 50 years (1961–2010), the annual mean temperature has significantly increased by $1.91°C$ in the Loess Plateau [60]. The atmospheric concentration of CO_2 has increased from 280 p.p.m. to 385 p.p.m. between the pre-industrial era and 2008 [61–62]. There is no clear research concerning what functions of the increasing temperatures and elevated carbon dioxide levels for SOC storage are more pronounced. Therefore, some uncertainties exist regarding the possible benefits of SOC sequestration in the future.

Generally speaking, there are many factors that may affect SOC sequestration in the future and may increase or decrease SOC storage. Factors that may be detrimental for SOC storage include: (i) the low national subsidy standard for returning farmland into forest or grasslands and the low income of local farmers; (ii) the persistent, deteriorating environment situation and; (iii) land-use changes caused by private and local government interests. Because of these disadvantageous factors, it is very important to determine how to effectively increase SOC storage. Creating new agriculture products by the implementation of more modern agriculture techniques and offering more work opportunities in urban areas for farmers to increase their income could partially solve these problems. Additionally, the large-scale afforestation that may be exacerbating the soil water shortage in the Loess Plateau should be controlled, especially in vulnerable arid and semi-arid regions, and should fully consider the affordability of such environmental efforts before vegetation restoration is conducted.

Conclusions

In this study, land-use type had a significant influence on the topsoil SOCD. Compared with the average SOCD of cultivated land, the average SOCD in woodland was significantly larger by 33.81%, while the average SOCD in the orchards decreased by 10.80%. Therefore, the SOCD may improve when cultivated land is converted into woodland and may decline when cultivated land is converted into orchards. Over time, there was an increasing trend of SOCD in woodland, but only a small change of SOCD in the orchards. Based on the remote sensing data on land-use change, we found that the "Grain for Green" project did not significantly increase SOC storage in the Luoyugou catchment in the past 10 years of restoration. This is mainly because the increase of SOC storage by woodland was offset by a SOC storage loss in the orchards and decreases in SOC storage in the initial stage of land-use conversion. If all of the woodland area was calculated as 30.44 t/km^2 of SOCD, the potential contribution to SOC sequestration in this catchment would increase by 9625.59 t, improving the SOC storage ratio by 7.12%. Therefore, it has a huge potential for future environmental restoration efforts if we could manage and maintain the woodland converted from the cropland well. However, we cannot ignore other factors that affect SOC sequestration, including climate change and national policies.

References

1. Morris SJ, Bohm S, Haile-Mariam S, Paul EA (2007) Evaluation of carbon accrual in afforested agricultural soils. Global Change Biology, 13: 1145–1156.
2. Wang YF, Fu BJ, Lü YH, Chen LD (2011) Effects of vegetation restoration on soil organic carbon sequestration at multiple scales in semi-arid Loess Plateau, China. Catena, 85: 58–66.
3. Sauer TJ, James DE, Cambardella CA, Hernandez-Ramirez G (2012) Soil properties following reforestation or afforestation of marginal cropland. Plant Soil, 360: 375–390.
4. Li DJ, Niu SL, Luo YQ (2012) Global patterns of the dynamics of soil carbon and nitrogen stocks following afforestation: a meta-analysis. New Phytologist, 195: 172–181.
5. Lal R (2004) Agricultural activities and the global carbon cycle. Nutrient Cycling in Agroecosystems, 70: 103–116.
6. Powlson D (2005) Will soil amplify climate change? Nature, 433: 204–205.
7. Smith P, Martino D, Cai ZC, Gwary D, Janzen H, et al (2008) Greenhouse gas mitigation in agriculture. Philosophical Transactions of the Royal Society, Series B, 363: 789–813.
8. Guo LB, Gifford RM (2002) Soil carbon stocks and land-use change: a meta analysis. Global Change Biology, 8: 345–360.
9. Don A, Schumacher J, Freibauer A (2011) Impact of tropical land-use change on soil organic carbon stocks- a meta-analysis. Global Change Biology, 17: 1658–1670.
10. Houghton RA (1999) The annual net flux of carbon to the atmosphere from changes in land-use 1850–1990. Tellus, 51B: 298–313.
11. Lal R (2005) Forest soils and carbon sequestration. Forest Ecology and Management, 220: 242–258.
12. Laganière J, Angers AD, Paré D (2010) Carbon accumulation in agricultural soils after afforestation: a meta-analysis. Global Change Biology, 16: 439–453.
13. Raiesi F (2012) Soil properties and C dynamics in abandoned and cultivated farmlands in a semi-arid ecosystem. Plant Soil, 351: 161–175.
14. Novara A, Gristina L, Mantia TL, Rühl J (2013) Carbon dynamics of soil organic matter in bulk soil and aggregate fraction during secondary succession in a Mediterranean environment. Geoderma, 193–194: 213–221.
15. Paul KI, Ploglase PJ, Nyakuengama JG, Khanna PK (2002) Change in soil carbon following afforestation. Forest Ecology and Management, 168: 241–257.
16. Wang BQ, Liu GB (1999) Effects of relief on soil nutrient losses in sloping fields in hilly region of Loess Plateau. Soil Erosion and Soil and Water Conservation, 5: 18–22 (in Chinese).
17. Wang Z, Liu GB, Xu MX, Zhang J, Wang Y, et al (2012) Temporal and spatial variations in soil organic carbon sequestration following revegetation in the hilly Loess Plateau, China. Catena, 99: 26–33.
18. Zhou Y, Xu XG, Ruan HH, Wang JS, Fang YH, et al (2008) Mineralization rates of soil organic carbon along an elevation gradient in Wuyi Mountain of Southeast China. Ecology, 27: 1901–1907 (in Chinese).
19. Luo YQ, Hui DF, Zhang DQ (2006) Elevated CO$_2$ stimulates net accumulations of carbon and nitrogen in land ecosystems: a meta-analysis, Ecology, 87: 53–63.
20. Batterman SA, Hedin LO, Breugel Mv, Ransijn J, Craven DJ, et al (2013) Key role of symbiotic dinitrogen fixation in tropical forest secondary succession. Nature, 502: 224–227.
21. Huang CY (2000) Soil Science. Beijing: China Agriculture Press, 311p.
22. Gregorich EG, Greer KJ, Anderson DW, Liang BC (1998) Carbon distribution and losses: erosion and deposition effects. Soil & Tillage Research, 47: 291–302.
23. Lal R (2003) Soil erosion and the global carbon budget. Environment International, 29: 437–450.
24. Chartier MP, Rostagno CM, Videla LS (2013) Selective erosion of clay, organic carbon and total nitrogen in grazed semiarid rangelands of northeastern Patagonia, Argentina. Journal of Arid Environments, 88: 43–49.
25. Fu BJ (1989) Soil erosion and its control in the Loess Plateau of China. Soil Use and Management, 5: 76–81.
26. Shi H, Shao MA (2000) Soil and water loss from the Loess Plateau in China. Journal of Arid Environments, 45: 9–20.
27. Fu BJ, Chen LX, Qiu Y (2002) Land-use structure and ecological processes in the Loess Plateau. Beijing: The Commercial Press, pp: 1–12 (in Chinese).
28. Feng XM, Wang YF, Chen LD, Fu BJ, Bai GS (2010) Modeling soil erosion and its response to land-use change in hilly catchments of the Chinese Loess Plateau. Geomorphology, 118: 239–248.
29. Lu N, Liski J, Chang RY, Akujärvi A, Wu X, et al (2013) Soil organic carbon dynamics following afforestation in the Loess Plateau of China. Biogeosciences Discuss, 10: 11181–11211.
30. Zheng FL (2006) Effect of Vegetation Changes on soil erosion on the Loess Plateau. Pedosphere, 16: 420–427.
31. Gong J, Chen LD, Fu BJ, Huang Y, Huang Z, et al (2006) Effect of land-use on soil nutrients in the loess hilly area of the Loess Plateau, China. Land Degradation & Development, 17: 453–465.
32. Chen LD, Gong J, Fu BJ, Huang ZL, Huang YL, et al (2007) Effect of land-use conversion on soil organic carbon sequestration in the loess hilly area, loess plateau of China. Ecological Research, 22: 641–648.
33. Fu XL, Shao MA, Wei XR, Horton R (2010) Soil organic carbon and total nitrogen as affected by vegetation types in Northern Loess Plateau of China. Geoderma, 155: 31–35.
34. Chang RY, Fu BJ, Liu GH, Liu SG (2011) Soil carbon sequestration potential for "Grain for Green" project in Loess Plateau, China. Environment Management, 48: 1158–1172.
35. Wei J, Cheng JM, Li WJ, Liu WG (2012) Comparing the effect of naturally restored forest and grassland on carbon sequestration and its vertical distribution in the Chinese Loess Plateau. PLoS ONE, 7: e40123. doi:10.1371/journal.pone.0040123
36. Lei D, ShangGuan ZP, Sweeney S (2013) Changes in soil carbon and nitrogen following land abandonment of farmland on the Loess Plateau, China. PLoS ONE 8: e71923. doi:10.1371/journal.pone.0071923
37. Yu XX, Zhang XM, Niu LL, Yue YJ, Wu SH, et al (2009) Dynamic evolution and driving force analysis of land-use/cover change on loess plateau catchment. Transaction of the CSAE, 25: 219–225 (in Chinese).
38. Zhao Y, Yu XX (2013) Effects of climate variation and land-use change on runoff-sediment yield in typical watershed of loess hilly-gully region. Journal of Beijing Forestry University, 35: 39–45 (in Chinese).
39. Nelson DW, Sommers LE (1982) Total carbon, organic carbon, and organic matter. In: Page AL, Miller RH, Keeney DR (eds) Methods of soil analysis, Part 2, Chemical and microbial properties. Agronomy Society of America, Agronomy Monograph 9, Madison, Wisconsin, pp 539–552.
40. Institute of Soil Sciences, Chinese Academy of Sciences (ISSCAS) (1978) Physical and chemical analysis methods of soil. Shanghai: Shanghai Science Technology Press, pp 7–15.
41. Degryze S, Six J, Paustian K, Morris SJ, Paul EA, et al (2004) Soil organic carbon pool changes following land-use conversions. Global Change Biology, 10: 1120–1132.
42. Yang YL, Guo SL, Ma YH, Chen SG, Sun WY (2008) Changes of orchard soil carbon, nitrogen and phosphorus in gully region of Loess Plateau. Plant Nutrition and Fertilizer Science, 14: 685–691 (in Chinese).
43. Xue S, Liu GB, Zhang C, Zhang CS (2011) Analysis of effect of soil quality after orchard established in hilly Loess Plateau. Scientia Agricultura Sinica, 44: 3154–3161 (in Chinese).
44. Umali BP, Oliver DP, Forrester S, Chittleborough DJ, Hutson JL, et al (2012) The effect of terrain and management on the spatial variability of soil properties in an apple orchard. Catena, 93: 38–48.
45. Guimarães DV, Gonzaga MIS, Silva TOd, Silva TLd, Dias NdS, et al (2013) Soil organic matter pools and carbon fractions in soil under different land-uses. Soil & Tillage Research, 126: 177–182.
46. Zou HY, Cheng JM, Zhou L (1998) Natural recoverage succession and regulation of the prairie vegetation on the Loess Plateau. Research of Soil and Water Conservation, 5: 126–138 (in Chinese).
47. Li YY, Shao MA, Shang Guan ZP, Fan J, Wang LM (2006) Study on the degrading process and vegetation succession of Medicago sativa grassland in North Loess Plateau, China. Acta Prataculturae Sinica, 15: 85–92 (in Chinese).
48. Li YY, Shao MA, Zhang JY, Li QF (2007) Impact of grassland recovery and reconstruction on soil organic carbon in the northern Loess Plateau. Acta Ecologica Sinica, 27: 2279–2287 (in Chinese).
49. Leifeld J, Bassin S, Fuhrer J (2005) Carbon stocks in Swiss agricultural soils predicted by land-use, soil characteristics, and elevation. Agriculture, Ecosystems and Environment, 105: 255–266.
50. Cao SX, Xu CG, Chen L, Wang XQ (2009) Attitudes of farmers in China's northern Shaanxi Province towards the land-use changes required under the Grain for Green Project, and implications for the project's success. Land-use Policy, 26: 1182–1194.
51. Lü YH, Fu BJ, Feng XM, Zeng Y, Liu Y, et al (2012) A policy-driven large-scale ecological restoration: quantifying ecosystem services changes in the Loess Plateau of China. PLoS ONE, 7: e31782, doi:10.1371/journal.pone.0031782
52. Pan ZB, Zhang L, Yang R, Li SB, Dong LG, et al (2012) Overview on research progress of soil drought in semiarid regions of the Loess Plateau. Research of Soil and Water Conservation, 19: 287–291, 298 (in Chinese).
53. Wang L, Shao MA (2004) Soil desiccation under the returning farms to forest on the Loess Plateau. World Forestry Research, 17: 57–60 (in Chinese).
54. Chen HS, Shao MA, Li YY (2008) Soil desiccation in the Loess Plateau of China. Geoderma, 143: 91–100.
55. Niu JJ, Zhao JB, Wang SY (2007) A study on plantation soil desiccation in the upper reaches of the Fenhe River basin based on deep soil experiments. Geographical Research, 26: 773–781 (in Chinese).

Author Contributions

Conceived and designed the experiments: YBQ ZBX XXY YLX. Performed the experiments: YBQ ZBX. Analyzed the data: YBQ ZBX. Contributed reagents/materials/analysis tools: YBQ ZBX XXY YLX. Wrote the paper: YBQ ZBX XXY YLX.

56. Duan FL, Lin Z, Xiong YQ (2007) Analysis on the phenomenon of farmland abandoned by the reason of rural laborers moving out for work. Rural Economy, 16–19 (in Chinese).

57. Xin ZB, Yu XX, Li QY, Lu XX (2011) Spatiotemporal variation in rainfall erosivity on the Chinese Loess Plateau during the period 1956–2008. Regional Environmental Change, 11: 149–159.

58. William HS (1999) Carbon and agriculture: carbon sequestration in soils. Science, 284: 2095.

59. Guo GF, Zhang CY, Xu Y (2006) Effects of climate change on soil organic carbon storage in terrestrial ecosystem. Ecology, 25: 435–442 (in Chinese).

60. Wang QX, Fan XH, Qin ZD, Wang MB (2012) Change trends of temperature and precipitation in the Loess Plateau Region of China, 1961–2010. Global and Planetary Change, 92–93: 138–147.

61. Intergovernmental Panel on Climate Change (2007) Climate Change 2007: The Science of Climate Change. Cambridge University Press, Cambridge.

62. Lal R (2009) Challenges and opportunities in soil organic matter research. European Journal of Soil Science, 60: 158–169.

2

Evaluation of Three Field-Based Methods for Quantifying Soil Carbon

Roberto C. Izaurralde[1]*, Charles W. Rice[2], Lucian Wielopolski[3], Michael H. Ebinger[4], James B. Reeves, III[5], Allison M. Thomson[1], Ronny Harris[†4], Barry Francis[5], Sudeep Mitra[3], Aaron G. Rappaport[1], Jorge D. Etchevers[6], Kenneth D. Sayre[7], Bram Govaerts[8], Gregory W. McCarty[9]

1 Joint Global Change Research Institute, Pacific Northwest National Laboratory and University of Maryland, College Park, Maryland, United States of America, 2 Kansas State University, Department of Agronomy, Manhattan, Kansas, United States of America, 3 Brookhaven National Laboratory, Department of Environmental Sciences, Upton, New York, United States of America, 4 Los Alamos National Laboratory, Los Alamos, New Mexico, United States of America, 5 EMBUL, ARS, USDA, Beltsville, Maryland, United States of America, 6 Rappaport and Associates, c/o Joint Global Change Research Institute, College Park, Maryland, United States of America, 7 Soil Fertility Laboratory, Natural Resources Institute, Colegio de Postgraduados, Carretera México-Texcoco, México, 8 CIMMYT, Km. 45, Carretera México-Veracruz, Texcoco, México, México, 9 HRSL, ARS, USDA, BARC West, Beltsville, Maryland, United States of America

Abstract

Three advanced technologies to measure soil carbon (C) density (g C m^{-2}) are deployed in the field and the results compared against those obtained by the dry combustion (DC) method. The advanced methods are: a) Laser Induced Breakdown Spectroscopy (LIBS), b) Diffuse Reflectance Fourier Transform Infrared Spectroscopy (DRIFTS), and c) Inelastic Neutron Scattering (INS). The measurements and soil samples were acquired at Beltsville, MD, USA and at Centro International para el Mejoramiento del Maíz y el Trigo (CIMMYT) at El Batán, Mexico. At Beltsville, soil samples were extracted at three depth intervals (0–5, 5–15, and 15–30 cm) and processed for analysis in the field with the LIBS and DRIFTS instruments. The INS instrument determined soil C density to a depth of 30 cm via scanning and stationary measurements. Subsequently, soil core samples were analyzed in the laboratory for soil bulk density (kg m^{-3}), C concentration (g kg^{-1}) by DC, and results reported as soil C density (kg m^{-2}). Results from each technique were derived independently and contributed to a blind test against results from the reference (DC) method. A similar procedure was employed at CIMMYT in Mexico employing but only with the LIBS and DRIFTS instruments. Following conversion to common units, we found that the LIBS, DRIFTS, and INS results can be compared directly with those obtained by the DC method. The first two methods and the standard DC require soil sampling and need soil bulk density information to convert soil C concentrations to soil C densities while the INS method does not require soil sampling. We conclude that, in comparison with the DC method, the three instruments (a) showed acceptable performances although further work is needed to improve calibration techniques and (b) demonstrated their portability and their capacity to perform under field conditions.

Editor: Vishal Shah, Dowling College, United States of America

Funding: The authors acknowledge the financial support of the United States Agency for International Development, Global Climate Change Office, the United States Department of Energy's Office of Science, Biological and Environmental Research (BER) funding to the Consortium for Research on Enhancing Carbon Sequestration in Terrestrial Ecosystems (CSiTE), the USDA CSREES through its support of the Consortium of Agricultural Soils Mitigation of Greenhouse Gases (CASMGS), and the Robertson Foundation. The funders had no role in study design, data collection and analysis, decision to publish, or preparation of the manuscript.

Competing Interests: The authors have declared that no competing interests exist.

* E-mail: cesar.izaurralde@pnnl.gov

† Deceased.

Introduction

Terrestrial C sequestration through planned changes in land use and management practices has been identified as an early adoption technology to mitigate the buildup of atmospheric CO_2 [1–3]. The potential sequestration is large in agricultural soils due to the large historical losses experienced by agroecosystems [1]. Many agricultural practices (e.g., diversified crop rotations, no tillage, nutrient management, and reduced irrigation practice) alone or in combination can result in soil C sequestration [1,4]. An additional benefit of soil C sequestration is that these practices also enhance long-term soil quality and productivity [5]. For successful mitigation of atmospheric CO_2, C sequestration practices must be implemented widely, in developing as well as developed nations, and their success monitored at different scales (e.g. farm, region, and national scales) [6,7].

Changes in soil organic C stocks can be measured directly using soil sampling protocols and chemical analysis or estimated indirectly through the use of eddy covariance methods, stratified accounting procedures, or simulation models [2]. The standard protocol for measuring soil C changes involves soil sampling at the field and preparation for laboratory analysis. Soil C concentration is analyzed using dry combustion (DC); a method considered as the standard method due to the vast experience acquired using it and its precision [8]. The results are presented on a mass basis C per unit mass [g C kg^{-1}] or per unit volume [kg C m^{-3}]; alternatively they are reported as C density per unit area to a depth of 30 cm, [kg C m^{-2}]) [9,10]. Although this procedure

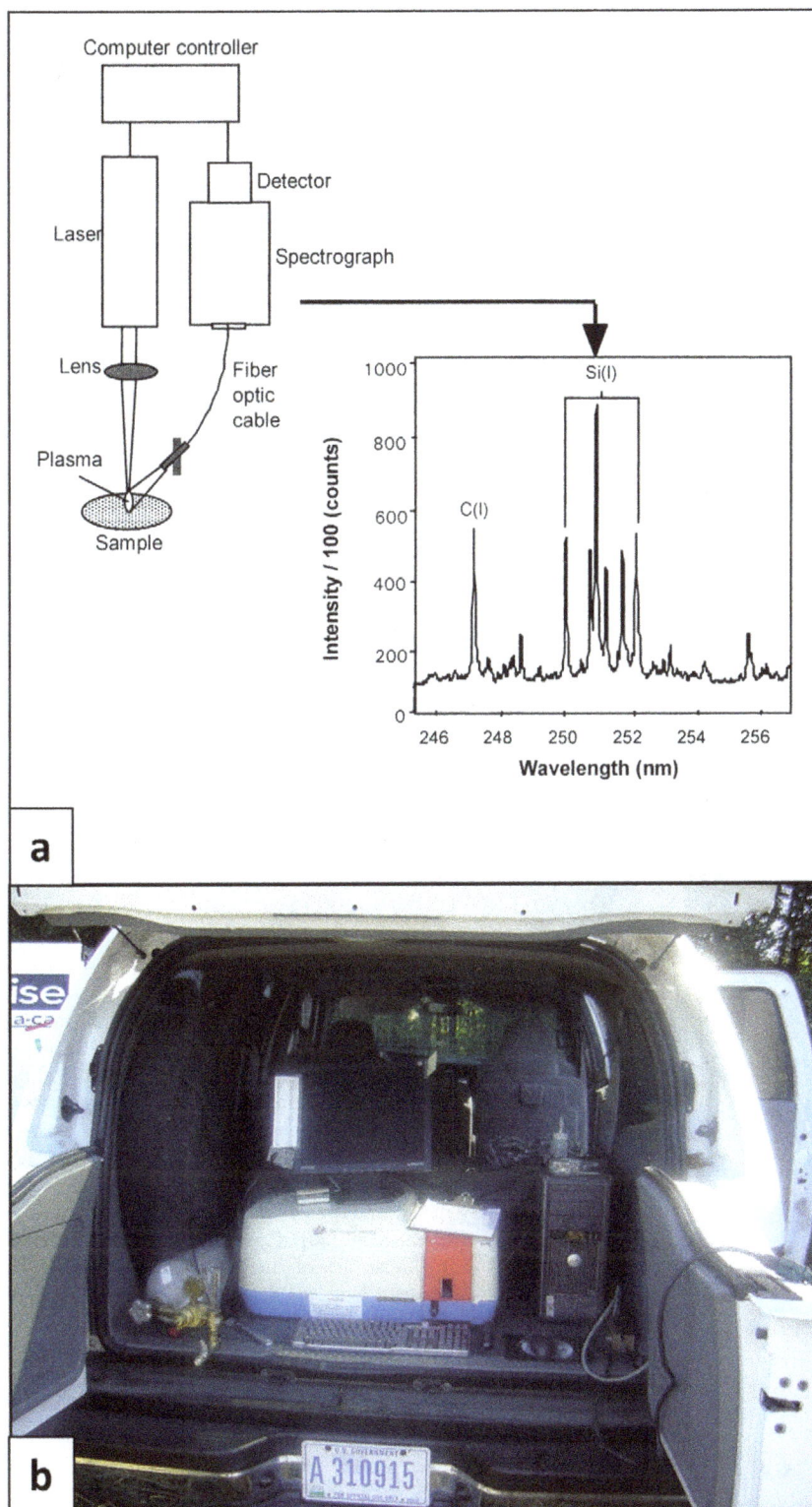

Figure 1. Schematic diagram and field setup of Laser Induced Breakdown Spectroscopy (LIBS) instrument: (a) schematic diagram and (b) picture of SUV-portable LIBS equipment used in this study.

produces excellent results, it must be done in the laboratory increasing the time, efforts and costs that restrict its routine use in agricultural C sequestration projects, particularly in developing countries. Thus, there is a need to develop portable, rapid, precise,

and cost-efficient methods for measuring soil C changes in the field [10,11].

Several technologies have been identified as potentially useful and adaptable for in-field measurement of soil C including: (a)

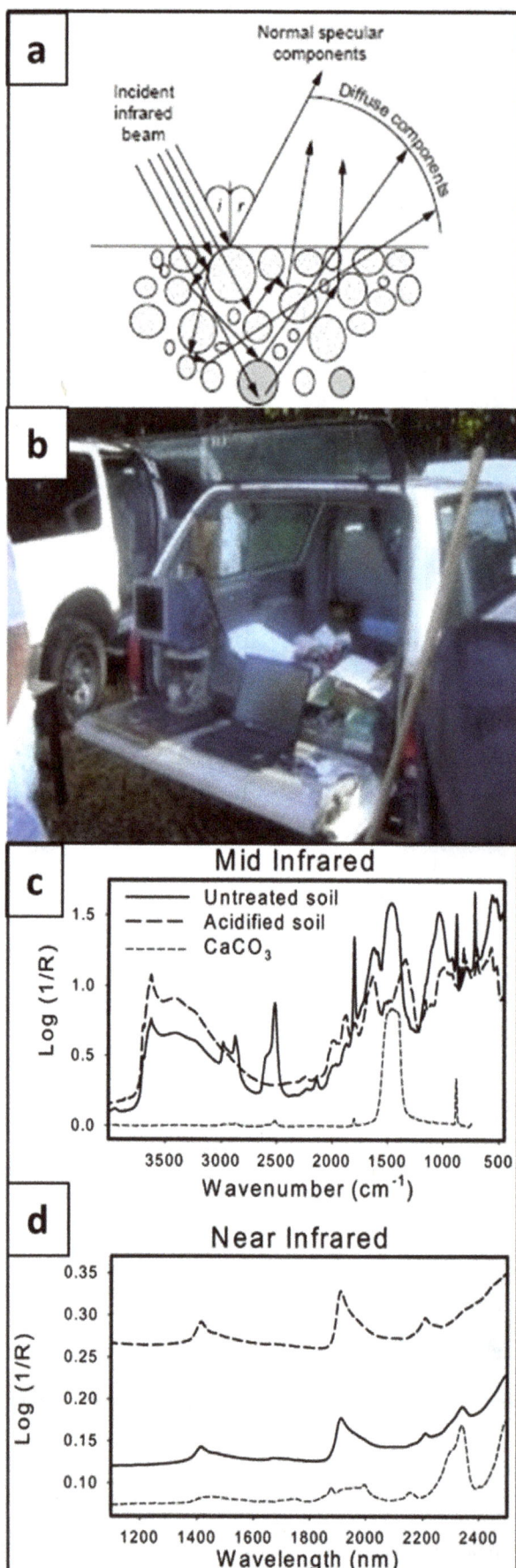

Figure 2. Spectra characteristics and field setup of Diffuse Reflectance Fourier Transform Infrared Spectroscopy (DRIFTS) instrument: (a) Diagram of diffuse reflection of IR light by soil sample, (b) SUV-portable mid-infrared (MIR) spectrometer used in this study, (c) typical mid-infrared diffuse reflectance spectra from soil and (d) near-infrared diffuse reflectance spectra from soil.

laser-induced breakdown spectroscopy (LIBS) [12], (b) diffuse reflectance mid-infrared Fourier transform spectroscopy (DRIFTS) [13], and c) inelastic neutron scattering (INS) [14–16]. These three technologies—LIBS, DRIFTS, and INS—have been used extensively for elemental and chemical analysis in other applications. For example, LIBS has been used to quantify heavy metal contamination in soils [17], DRIFTS has been applied to characterize compounds and measure their concentrations in many materials [18–20], and INS has been employed to measure whole patient elemental composition [21]. Due to their physical characteristics, soil volumes analyzed by LIBS and DRIFTS are very small in comparison to the very large volume analyzed by INS, which does not require soil sampling.

The objective of this research is to evaluate in-field measurements of soil C densities determined by: a) LIBS, b) DRIFTS, and c) INS with laboratory determinations of C densities using the DC method.

Materials and Methods

Instruments

The three novel analytical methodologies for C analysis in soil, used in this work, are based on fundamentally different physical principles with vastly different characteristics and capabilities. They are LIBS, DRIFTS and INS and are described hereafter.

Laser Induced Breakdown Spectroscopy

The LIBS technique is based on atomic emission spectroscopy (Figure 1a) [22–26]. A laser pulse is focused on a (soil) sample, creating high temperatures and electric fields that break all chemical bonds in a small volume of about 10^{-9} m^3 of the material and vaporize it into a white-hot gas of atomic ions known as microplasma [27]. The resulting emission spectrum is then analyzed using a spectrometer covering a spectral range from 190 to 1,000 nm. In order to reduce the error in the C determination it is normalized by the sum of the Al and Si intensities and taken as the standardized LIBS signal [28]. The C mass concentration was estimated from a linear fit of LIBS intensity ratio vs. C concentration on a mass basis (as determined by DC) obtained from similar soils.

Cremers et al. evaluated LIBS by measuring the total soil C of agricultural soils from Colorado and a woodland soil from New Mexico, using a subset of the Colorado samples for calibration [12]. Their tests revealed that the LIBS instrument has a detection limit of 300 mg C kg^{-1}, a precision of 4–5% and accuracy of 3–14% ($\approx$750 mg C kg^{-1}), numbers some 30–400 times higher (i.e. less precise) than the same figures from DC.

The LIBS instrument has several significant operational advantages over DC. Cremers et al. found that its throughput was less than 1 minute per sample [11]. Commercially-available portable LIBS equipment was used for the tests described in this study greatly facilitating in-field measurements (Figure 1b). The LIBS method shares with the DC method the disadvantage of requiring soil core samples that are dried, mixed, sieved, and—if carbonates are present—acid-washed. The LIBS samples must be ground and pressed into pellet form, steps that are not necessary

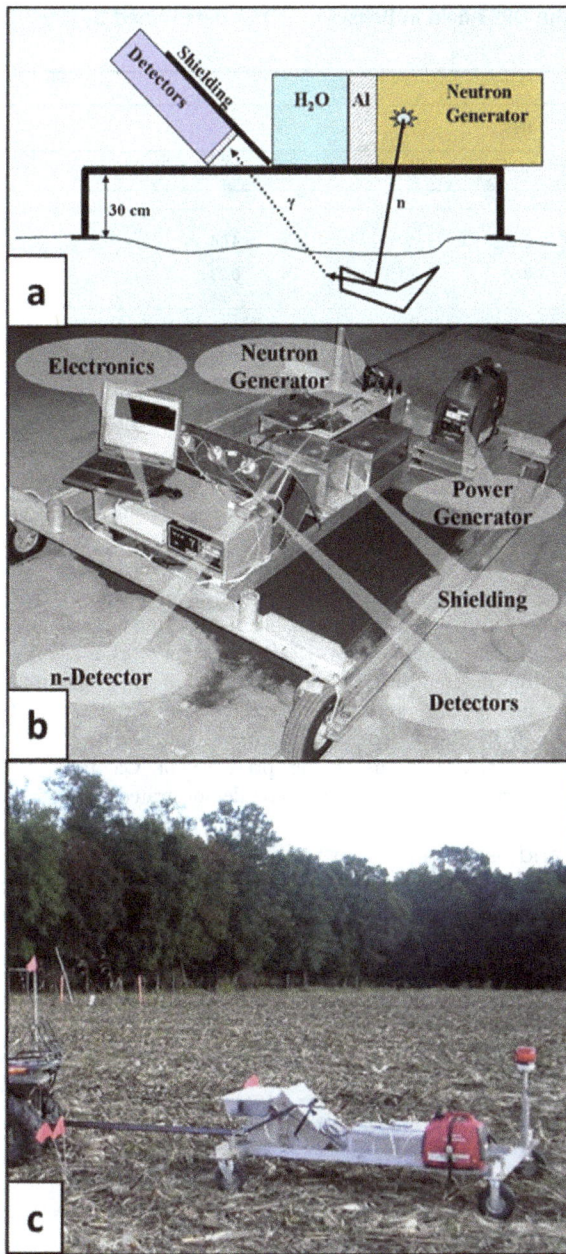

Figure 3. Deployment modes, schematic diagram, and field setup of Inelastic Neutron Scattering (INS) instrument: (a) The three-detector INS instrument in its various deployment modes: (a) Schematic diagram of a stationary INS instrument for soil studies, (b) the INS instrument used in this study, mounted on a cart for operation in field-scanning mode, and (c) INS instrument being towed behind a tractor during a field scan.

for the DC method. If maximum accuracy is not required, an un-dried whole soil core can be scanned along its length, causing the soil inside the footprint to dry, and results from different areas on the core's surface can be combined to provide statistical mixing of the sample. In this mode LIBS can provide ≈1 mm resolution (i.e., each 1 mm from top to bottom of the core).

Diffuse Reflectance Fourier Transform Infrared Spectroscopy

Unlike LIBS and INS, which probe the elemental identity of a sample's atoms, infrared (IR) spectroscopy probes the C bond identities of a sample's molecules. For soil studies, the surface of a sample is illuminated by a broadband IR source and the absorbance spectrum of the diffusely reflected component of the light is acquired by an IR spectrometer. The diffusely reflected component is light that has entered the sample and is scattered out through the same surface (Figure 2a). Like LIBS, DRIFTS and near-infrared spectroscopy (NIRS) have the operational advantages over DC of rapid throughput (1.5 to 3 minutes per sample) and portability (Figure 2b).

Many quantitative IR studies of soil – to measure component concentrations - have used the near infrared (NIR) (400–2500 nm) while qualitative investigations – to determine component chemical identities - have used the sharper absorption peaks typical of the mid- infrared (MIR) (2500–25,000 nm). The preference to use NIR over MIR is due to the weaker absorption by NIR, which would result in a longer path in the sample and hence more accurate estimates of the concentration. However, recent investigations have found that the MIR can be used for quantitative studies of low concentration components (e.g. C) in highly diverse materials such as soil [28–30]. Advanced data analyses techniques, such as partial least squares regression (PLSR), are essential for quantifying C content from the NIR or MIR spectra of undiluted samples. McCarty et al. tested the SOC-predictive performance of both the NIR and MIR wavelength ranges, on 237 soil samples taken from 14 locations in the U.S. Great Plains and demonstrated that MIR outperforms NIR by about a factor of two in precision [30].

Furthermore, because of its bond-sensitive nature, IR spectroscopy offers the possibility of directly distinguishing inorganic from organic C, thus eliminating the need of acid pretreatment to remove inorganic C (Figures 2c and 2d). However, McCarty et al. found that such direct estimation of organic C from MIR spectra produced root mean square errors (RMSE) 50% greater than those by DC [30]. Estimation of organic C by MIR spectra on soil samples that had undergone acid pretreatment had RMSE similar to those determined by DC. The field unit consisted of a Surface Optics Corporation model SOC-400 portable Fourier Transform spectrometer (SOC-400, Surface Optics, Corp., San Diego, CA). It is equipped with a non–Peltier cooled DTGS (deuterated triglycine sulfate) and KBr beam splitter by diffuse reflectance from 4000 to 400 cm^{-1} at 8 cm^{-1} resolution using a rotating sample cup (approximate path 2 mm in width around an 8 mm diam.) with KBr used for the background spectra. The samples spend about 45 s on the spectrometer, a time similar to that required in the lab for DRIFTS or NIRS.

Inelastic Neutron Scattering

The INS method is based on spectroscopy of gamma rays induced by nuclear reactions of fast neutrons with nuclei of the elements present in soil. For that purpose, a portable commercial neutron generator (NG) consisting of a small (2.5 cm diameter, 10 cm long), sealed-tube accelerator produces fast, 14 MeV, neutrons by accelerating deuterium (d) impinging on a target saturated with tritium (t) resulting in a d,t fusion reaction that emits alpha and anti-parallel neutron particles [31]. The fast neutrons interacting via INS processes induce 4.43 and 6.13 MeV gamma rays in C and O atoms, respectively. Alternatively, following elastic scatterings, some of the fast neutrons slowdown and undergo capture via thermal neutron capture (TNC) reactions; inducing a 2.2 MeV gamma ray in H. These gamma

Table 1. Soil-C density statistics to a depth of 30 cm of Plot No. 3 at the OPE3 field in Beltsville, MD as determined by dry combustion and three soil-C technologies.

| | | | | INS | |
	DC	DRIFTS	LIBS	Universal calibration	Local calibration
			------- kg C m^{-2} -------		
Mean	4.07	4.32	3.27	2.57	4.06
Std. Dev.	0.55	0.61	0.81	0.61	0.23
Dev. (%) from DC	—	6.14	−19.7	−36.9	−0.3
No. samples	9	9	9	Soil volume scanning	

rays are detected by an array of NaI detectors separated from the NG by shielding material. The entire system is mounted on a cart and non-destructively probes the soil in static and scanning modes of operation. The intensity of the measured gamma rays is proportional to the soil's C and other elements concentrations in the interrogated volume [14,22].

The high energy neutrons and gamma rays penetrate the soil quite extensively inducing detectable signal from a depth of 30 to 50 cm. The footprint of the INS system is about 1.5 m^2 resulting in a sampled volume >0.3 m^3. Schematics and an alpha prototype of the INS system are shown in Figures 3a and 3b; whereas Figure 3c shows the INS's scanning capability. A field scan represents a mean C value in the scanned area and the C signal from such large volumes averages any strong variations in the C depth profile.

The C intensity, the net area under the C gamma-ray peak, in measured spectra is calibrated against surface C density (g C m^{-2}) of synthetic soils with known amounts of C. Calibration using synthetic soils demonstrated linearity of the INS system [14].

Alternatively in the field, the INS system is calibrated directly against chemical analysis of soil samples. However, any calibration in the field has to be concerned with comparing C content in different volumes [32]. Three advantages of the INS system include: a) it is a non-destructive method, b) it does not require sample preparation, and c) it is capable of analyzing large volumes of soil and scanning large areas [33,34]. Further work is required to establish the sensitivity of the INS system to detect small changes in soil C content and its sensitivity to variations in C distribution with soil depth. Furthermore, the INS's ability to quantify inorganic C using the presence of Ca peak and stoichiometric information needs to be demonstrated.

Sites and Sampling

No specific permits were required to access and sample the described field studies described below. Access to the Beltsville Experiment was coordinated by co-authors J.B. Reeves and G.W. McCarty from USDA. Access to the El Batán Experiment was coordinated by co-authors K.D. Sayre and B. Govaerts from

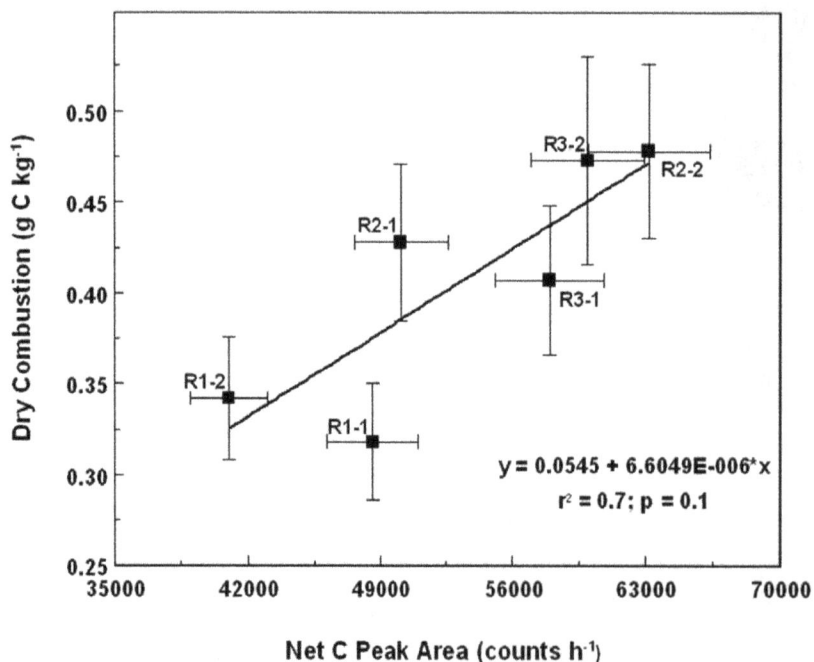

Figure 4. Correlation between INS signal and soil C density as measured by dry combustion to a depth of 30 cm in Beltsville, MD.

Table 2. Analysis of Variance of Dry Combustion, LIBS, and DRIFTS soil C density means of Plot No. 3 at the OPE3 field in Beltsville, Maryland.

	Depth interval							
	0–5		5–15		15–30		0–30	
	Mean Square	Pr>F	Mean Square	Pr>F	Mean Square	Pr>F	Mean Square	Pr>F
Treatment	0.1289	0.001	0.1825	0.051	0.7886	0.093	2.6965	0.007
Error	0.0142		0.0542		0.3005		0.4455	
R^2	0.430		0.219		0.179		0.335	
	Soil Carbon Means (kg C m^{-2})							
Dry Comb.	0.86	a†	1.76	a	1.45	ab	4.07	a
LIBS	0.68	b	1.50	b	1.09	b	3.27	b
DRIFTS	0.91	a	1.73	a	1.67	a	4.32	a

†Means followed within depth interval followed by the same letter are not significantly different at the 0.05 level of probability.

CIMMYT. Neither location was privately-owned or protected in any way. The field studies did not involve endangered or protected species.

Beltsville Experiment

The first side-by-side test was conducted on October 2–3, 2006 on a 25-ha USDA field (39°1′44.1″N; 76°50′41.7″W) in Beltsville, MD known as OPE3 (Optimizing Production Inputs for Economic and Environmental Enhancement) [35]. This field, contained within a first order agricultural catchment in the Maryland inner coastal plain, had been previously sampled using a 25-m grid pattern for the determination of soil fertility parameters including soil organic carbon using both near and MIR spectroscopy and then mapped by use of ordinary kriging [30]. At the time of the test, maize (*Zea mays* L.) had been recently harvested from the field, which showed a significant, but unmeasured amount of corn stover.

The experimental area was on a fine-loamy, mixed, semiactive, mesic Typic Fragiudult and the design consisted of three 30 m×30 m plots (R1, R2, and R3) where 9 sampling points were laid out on a square grid with each point placed 9-m apart from each other. At each grid point, six soil samples were extracted at three depth intervals (0–5, 5–15, and 15–30 cm) with a hand-held soil sampler (3.12 cm diam.) and composited into one sample per depth interval per grid point. The composite soil samples were taken to the side of the experimental area for further processing and measuring. In addition two extra soil samples per plot were taken for comparison with INS stationary measurements.

At the measuring station, the samples were processed and analyzed by the LIBS and DRIFTS instruments. Electric power was supplied from a nearby power source (120 V, 60 Hz). Alternatively, they can be successfully operated from a power inverter off the car battery.

After determining wet soil weight, each soil sample was manipulated in order to break the soil aggregates as well as to remove from the sample visible pieces of roots and crop residues. After each sample had been thoroughly mixed, a subsample was taken for subsequent laboratory determinations of soil water content, coarse fragment fraction, and soil C concentration by DC. For the DRIFTS analysis, soil samples were broken up and thoroughly mixed in a plastic Petri dish using a metal spatula (possible at BARC due to the sandy nature of the soils), mixed and

scanned. Samples for LIBS analysis were pressed in a hydraulic press to about 20 Mg force and then placed in the LIBS sample holder. The ambient air of the LIBS instrument was replaced with Ar gas to enhance the LIBS signal, and the sample was analyzed for 10 s, which provided 100 spectra per sample. The averaged spectra were collected for further quantification of soil carbon.

The INS system was operated in stationary and scanning modes and in either case the data was acquired for one hour. For the stationary measurements, the INS instrument was placed on top of the previously sampled locations (two per each plot), subsequently an area of 900 m^2 was scanned (Figure 3). Due to time constraints and technical issues, only Plot No. 3 was scanned. Correlation between INS readings in stationary mode and surface C density were used to develop a conversion factor for predicting soil C density of Plot No. 3.

All soil samples were analyzed for C at the Kansas State University (KSU) Laboratory by DC using a Carlo Erba C/N Analyzer (Carlo Erba Instruments, Milan, Italy) [35]. Carbon concentration results were converted to soil C density using soil bulk density values determined by the soil core method.

El Batán Experiment

The side-by-side test in Mexico was conducted on a 17-year old crop rotation, tillage, residue study plot located at CIMMYT, at El Batán, about 40 miles east of Mexico DF and 2240 m above sea level [36,37]. The experimental units (a total of 32) consisted of plots (22 m×7.5 m) cropped to maize (*Zea mays* L.) and wheat (*Triticum aestivum* L.), either in monoculture or in rotation, with conventional or no tillage methods, with or without crop residue removal after harvest, and with crops planted in either flat or raised beds. The plots are arranged in a randomized complete block design with two replications. All treatments planted using a flatbed system (16 treatments or 32 plots) were selected for this test.

Custom, export and import permits required by the U.S. and Mexican authorities to transfer the INS system, due to NG being a radiation producing device, precluded the participation of the INS in the Mexican tests. Thus, only the LIBS and DRIFTS techniques were tested at CIMMYT. Prior to the field test at this facility, CIMMYT researchers provided the instrument team with eight soil samples previously taken from the plots in order to facilitate the initial calibration of the LIBS and DRIFTS instruments.

Table 3. Summary of ANOVA showing mean square values for main effects, main effect means, and overall means for the 16 treatments with two replications analyzed for soil C density (kg C m^{-2}) at El Batán, Mexico in 2007.

Source	Mean Square Dry Comb.	Pr.>F	Mean Square LIBS	Pr.>F	Mean Square DRIFTS	Pr.>F
Tillage	2.3992	0.002	5.6890	0.009	3.0928	0.0003
Rotation	0.0143	0.752	1.2217	0.202	0.0151	0.772
Residue	1.7424	0.002	1.2332	0.200	0.0150	0.772
Error	0.1392		0.7085		0.1752	
R²	0.649		0.590		0.537	
Tillage						
CT	2.72	b	2.81	b	2.91	b
ZT	3.26	a	3.66	a	3.53	a
Rotation						
Monoculture	3.01	a	3.43	a	3.24	a
Rotation	2.97	a	3.04	a	3.20	a
Residue						
Retained	3.22	a	3.43	a	3.24	a
Removed	2.76	b	3.04	a	3.20	a

Treatment variables

				Soil Carbon Determination Method		
Tillage	Rotation	Crop	Residue	Dry Comb.	LIBS	MIRS
CT	Monoculture	Maize	Retained	3.39 abc†	3.83 ab	3.53 abc
		Maize	Removed	2.35 e	2.93 abcd	2.82 c
		Wheat	Retained	2.64 de	2.52 bcd	2.83 c
		Wheat	Removed	2.57 de	3.29 abcd	2.79 c
	Rotation	Maize	Retained	2.97 abcde	3.21 abcd	2.93 bc
		Maize	Removed	2.56 e	2.58 bcd	2.82 c
		Wheat	Retained	2.79 cde	2.27 cd	2.81 c
		Wheat	Removed	2.46 e	1.89 d	2.74 c
ZT	Monoculture	Maize	Retained	3.52 ab	3.91 ab	3.46 abc
		Maize	Removed	2.91 bcde	2.63 bcd	3.37 abc
		Wheat	Retained	3.45 abc	4.36 a	3.48 abc
		Wheat	Removed	3.26 abcd	3.99 ab	3.65 ab
	Rotation	Maize	Retained	3.64 a	3.97 ab	3.63 ab
		Maize	Removed	3.01 abcde	3.54 abc	3.43 abc
		Wheat	Retained	3.39 abc	3.40 abc	3.26 abc
		Wheat	Removed	2.93 bcde	3.47 abc	3.97 a

†Means within a given method followed by the same letter are not significantly different at the 0.005 level of probability.

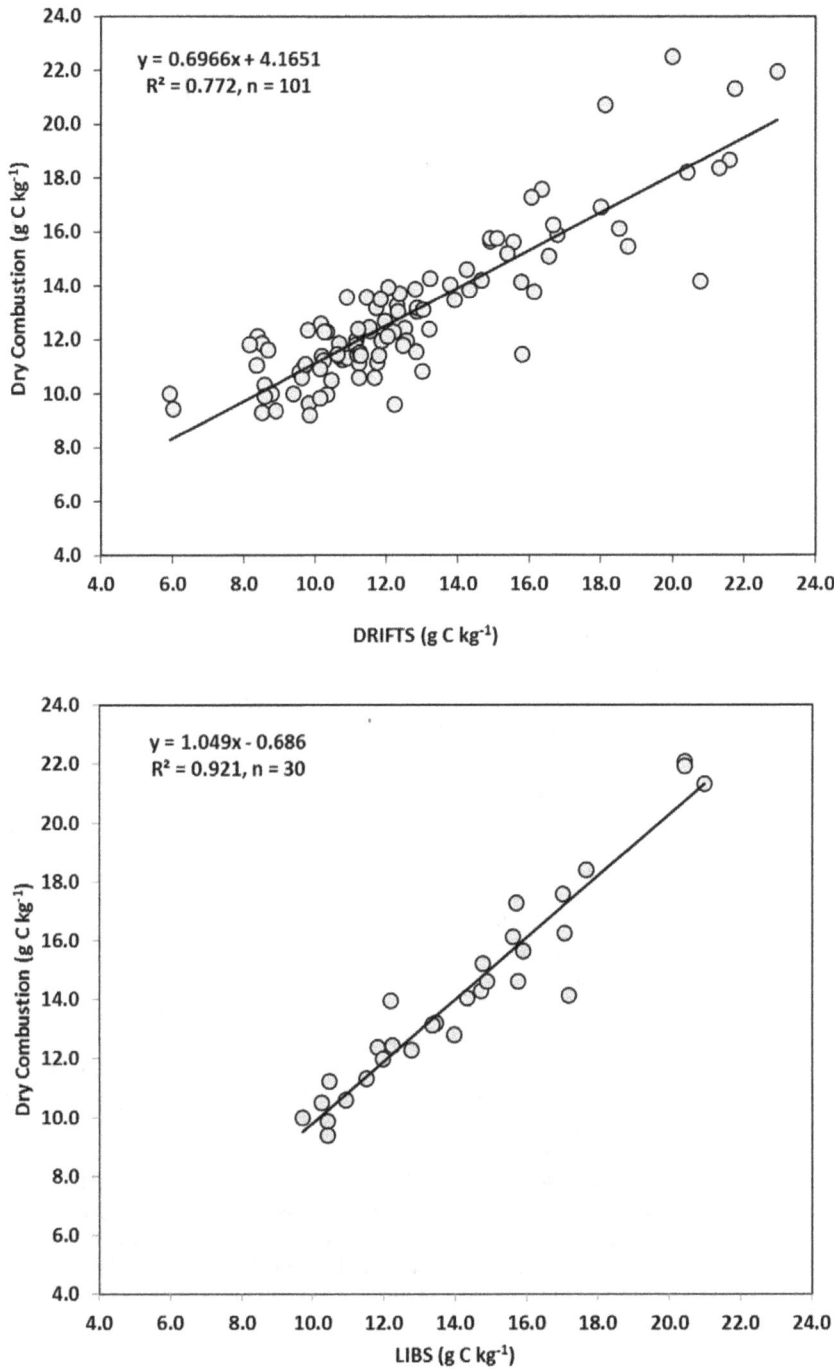

Figure 5. Comparison of calibration lines for (a) DRIFTS and (b) LIBS made by including 10% of the data in the calibration sets (see text).

A composite soil sample made of 12 subsamples per soil depth (0–5, 5–10, and 10–20 cm) was taken from each of the 32 plots. Once extracted, the soil samples were taken to an improvised processing and measuring station located about 200 m away from the plots. As in the Beltsville experiment, the composite samples were thoroughly mixed and a subsample was weighed and taken to a dry lab nearby for soil moisture determination. After separating visible pieces of crop residues and roots from the soil samples, these were set to dry, but due to the high clay content the samples dried into rock hard masses. Thus, all samples in the Mexico trials

were grounded with mortar and pestle for LIBS, DRIFTS, and DC analyses. Sample preparation for LIBS analysis was essentially the same as described above. The more intense sunlight in the CIMMYT trials, however, allowed for more thorough drying of the samples before analysis.

The rest of the samples were air dried and sent to the soils laboratory at Kansas State University for DC analysis. Soil samples were finely ground to pass 100 mesh in an agate mill. Finely ground soil samples were oven dried to 105°C for 24 h and stored. Just before analysis, the samples were oven dried again for

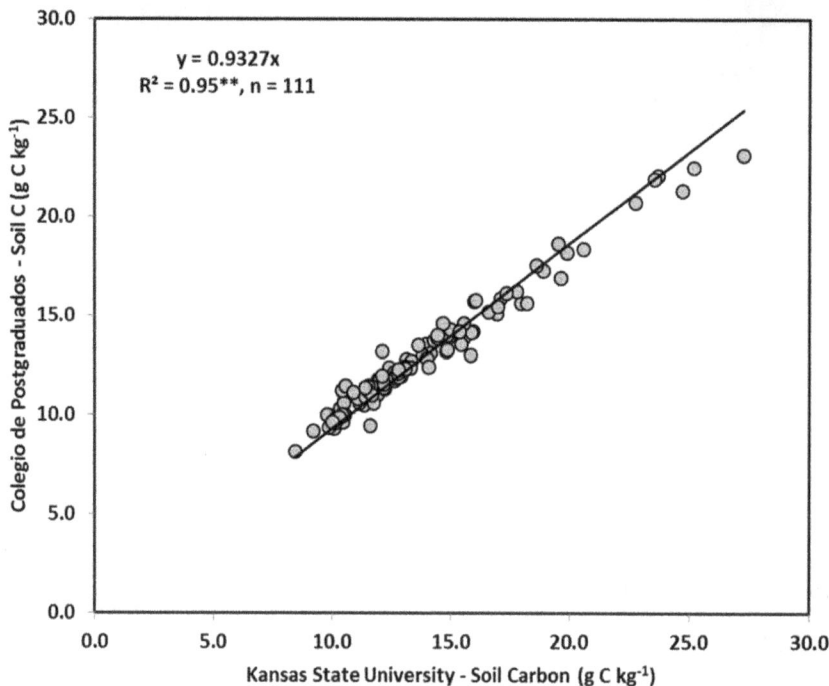

Figure 6. Comparison of dry combustion results from the two different instruments used.

2–3 h, removed from the oven and placed in a desiccator prior to analysis. The amount of total C was determined by DC (900°C) in an automatic C analyzer Shimadzu TOC 5000-A (Shimadzu Scientific, Kyoto, Japan).

Statistical Analysis

Statistical analysis consisted of regression and ANOVA analysis using SAS software (SAS Institute Inc., Cary, NC). Data for Beltsville experiment were analyzed using a Completely Randomized Design. Data for the Mexican experiment were analyzed following a Randomized Complete Block Design.

Ethics Statement

No specific permits were required to access, perform the studies, and sample the field experiments at Beltsville (Maryland) in the USA and at CIMMYT, El Batán in Mexico. At Beltsville, the technical group was allowed access to the OPE3 field by USDA – ARS scientist and team member J.B. Reeves III. At CIMMYT, the technical group was allowed access to the long-term trial by CIMMYT scientist and team member K.D. Sayre. Soil sampling conducted at both sites followed standard protocols. Soil samples taken at CIMMYT were imported to the USA by KSU professor, permit-holder, and team member C.W. Rice for analysis. Transport of equipment across interstate and international borders followed all state, federal, and international regulations.

Results

Beltsville Experiment

The mean soil C density of Plot No. 3 as determined by DC was 4.07 ± 0.55 g C m^{-2} (Table 1). A relatively low coefficient of variation of 14% suggests a rather uniform distribution of C within the estimated plot volume (270 m^3). The soil C density estimate by LIBS was about 20% lower than that by DC ($p < 0.01$) while the DRIFTS estimate was not different from that determined by DC

($p < 0.83$). In the case of INS, two values are provided (Table 1): one using a "universal" calibration and the other using a "local" calibration. Use of a "universal" calibration produced a mean estimate of soil C density about 37% lower than that obtained by the DC method. Since INS provides a single value for the entire field with a counting statistics error of about 1.5%, it requires more replications for comparing against other methods using conventional statistics. Additional "stationary" measurements at six other sites in the experimental area allowed for the development of a regression line between the DC and INS methods (Figure 4). When using the "local" calibration, there was almost a perfect agreement between the DC and INS methods.

Table 2 provides a more detailed statistical comparison of the soil C density values determined by DC, LIBS and DRIFTS for the three depths and total soil depths on Plot No. 3. Again, for the three depths, the soil C density derived from DRIFTS measurements was not significantly different from those by DC. In this case, the enhanced calibration dataset used by the DRIFTS team contributed to the close results.

El Batán Experiment

The DC measurements were repeated campus of the Colegio de Postgraduados at Montecillo, Mexico, affording a measurement of the operational accuracy (repeatability *between* instruments) of this "gold standard" technique for measuring soil carbon.

Table 3 summarizes results of the Analysis of Variance conducted on soil C density calculated to a depth of 20 cm by three methods: DC, LIBS, and DRIFTS. Here we used a Randomized Complete Block design with three main factors: tillage, rotation, and residue with two replications. Overall, R^2 was largest for the DC dataset, followed by LIBS, and DRIFTS. The DC method detected significant differences due to tillage and residue effects while the LIBS and DRIFTS methods detected significant differences only due to tillage effects. Further, a one-way ANOVA analysis (bottom half of Table 3) reveals the DC method with higher sensitivity that the other two methods to detect significant differences among means.

Subsequently, the LIBS and DRIFTS teams were provided with C concentrations for 11 (10%) of the samples to augment their respective calibration curves. The DRIFTS team used these samples from the dataset, eight samples from archived soil samples taken from the experimental site, and all of the data from OPE3 trial. Using these extra points in the calibration curve the R^2 for the DRIFTS instrument improved significantly to 0.772 (Figure 5). Using the same 11 samples to construct their new calibration curve, and using PLS methods, the LIBS team improved their instrument's R^2 to 0.919 although they were only able to use 30 of the 101 samples due to software limitations (Figure 5). These results demonstrate that there is considerable opportunity to improve the predictability of both instruments by using DC results from a small number of local samples.

The DC results from the KSU and Colegio de Postgraduados correlated rather well. Figure 6 shows linear fits of the DC values obtained at Colegio de Postgraduados to the DC values obtained at KSU. The R^2s obtained were 0.97 and 0.95, depending on whether or not a non-zero intercept was allowed between the results of the two instruments. The median value for the deviation between the two measurements, expressed as a percentage of the average of the measurements, was 5.4%, reflecting that the two medians were 3 standard deviations away from each other. These discrepancies are in accord with previous findings [6].

Discussion

Based on the results of the blind comparison in Beltsville, the DRIFTS instrument produced the closest estimates of soil C density for Plot No. 3 but required the largest amount of ancillary information to arrive at these results. In the case of the LIBS instrument, the estimates of soil C density could be improved by including more data points into the universal calibration curve. With regards to the INS instrument; since only single measurements yields the value for the entire field it needs to be reproduced to justify the complete agreement with the results by DC. It was pointed out that the INS instrument is calibrated directly in terms of soil C density, Figure 4, using large volumes without the need of knowing bulk density. The scarcity of data, only six points with two from each plot used for creating Figure 4, resulted in r^2 value of 0.7, however, in two other fields studies in mixed soils the regressions coefficients were higher than 0.9 [15,32]. Furthermore, since INS is using penetrating radiation that is exponentially attenuated it does not have a precisely defined depth of sampling. Instead, we define it as the depth from which 90% of the signal is detected. Based on Monte Carlo calculations, this effective depth is about 30 cm while for 99% it is about 50 cm. Since small signal arrives from deeper layers, variation in the depth should not play a major role in the total count [33].

The original plan was to complete the soil sampling, soil analysis in the field with the LIBS and DRIFTS instruments, as well as stationary and scanning measurements in the three plots within a period of two days (October 2–3, 2006). The activities performed during this period included: plot and sampling site demarcation, instrument setup, weighing station setup, soil sampling, prepara-

tion of samples for analysis in the field, sample analysis with LIBS and DRIFTS, stationary and scanning measurements with INS, and soil sample preparation for submission to laboratory for DC analysis. A total of 10 researchers and technicians were involved in these operations. The LIBS and DRIFTS teams were able to complete the analysis of 81 samples from the three plots and 26 samples from the stationary sites. As noted before, technical difficulties (a loose wire in the neutron generator) delayed the INS field measurements and allowed for the completion of six stationary measurements, two per plot, and two scans of plot No. 3 during the 2-day experimental period. Consequently, comparisons of the three instruments and DC were available only for one plot, Plot No. 3.

After completion of the DC analysis, some of the results were made available to the LIBS and DRIFTS teams. The DRIFTS team used results from Plot No. 1 and 2 to independently predict the values of Plot No. 3. The LIBS team did not use any of the values from Plot No. 1 and 2 to improve their "universal" calibration curve. Finally, the INS team used counts from six stationary INS measurements, two from each plot, to develop a calibration curve (soil C density (g C cm^{-2} = 54,714×INS_count − 11,026) to predict the average soil C density of Plot No. 3 to a depth of 30 cm.

In summary, this study compared the side-by-side performance of three advanced technologies to measure soil C under field conditions against standard soil carbon analysis by DC. The LIBS and DRIFTS methods and the standard DC require soil sampling and need soil bulk density information to convert soil C concentrations to soil C densities. The INS method requires some soil sampling for establishing correlations between INS and DC but no further sampling is necessary once these correlations are established. The comparative results obtained indicate an acceptable performance of the three instruments but they also show the need for improvement in terms of calibration.

In terms of transportability, the INS system is a radiation generating device and thus has to follow all transportation regulations, which, at the moment, impede international shipping. The LIBS instrument, as tested in our experiments, requires Ar gas, a press and a power source. Finally, the DRIFTS instrument was portable and can be run off a car battery or an inverter with the vehicle running. No other equipment is required other than a mortar and pestle. There are also portable MIR and NIR units available that can run off backpacks or even internal batteries.

The three instruments demonstrated their portability and their capacity to perform under field conditions. Import/export issues to developing countries (i.e., regulations, permits, and licenses) should be carefully examined in order to facilitate the smooth transport of these instruments across international borders.

Author Contributions

Conceived and designed the experiments: RCI CWRLW MHE JBR AMT. Performed the experiments: RCI CWR LW MHE JBR AMT RH BF SM AGR JDE KDS BG GWM. Analyzed the data: RCI CWR LW MHE JBR AMT RH BF SM AGR JDE GWM. Contributed reagents/materials/analysis tools: CWRLW MHE JBR RH BF SM JDE KDS BG. Wrote the paper: RCI CWRLW MHE JBR AMT JDE.

References

1. Cole CV, Duxbury J, Freney J, Heinemeyer O, Minami K, et al. (1997) Global estimates of potential mitigation of greenhouse gas emissions by agriculture. Nutrient Cycling Agroecosystems 49: 221–228.

2. Post WM, Izaurralde RC, Mann LK, Bliss N (2001) Monitoring and verifying changes of organic carbon in soil. Climatic Change 51: 73–99.

3. Smith P, Martino D, Cai Z, Gwary D, Janzen H, et al. (2008) Greenhouse gas mitigation in agriculture. Philosophical Transactions Royal Society B 363: 789–813.

4. Janzen HH, Campbell CA, Izaurralde RC, Ellert BH, Juma N, et al. (1998) Management effects on soil C storage on the Canadian Prairies. Soil Tillage Research 47: 181–195.

5. Lal R, Follett RF, Kimble J, Cole CV (1999) Managing US cropland to sequester carbon in soil. Journal Soil Water Conservation 54: 374–381.
6. Izaurralde RC, McGill WB, Bryden A, Graham S, Ward M, et al. (1998) Scientific challenges in developing a plan to predict and verify carbon storage in Canadian Prairie soils. In: Lal R, Kimble J, Follett R, Stewart BA, editors. Management of carbon sequestration in soil. Boca Raton: CRC Press. pp. 433–446.
7. Brown DJ, Hunt Jr ER, Izaurralde RC, Paustian KH, Rice CW, et al. (2010) Soil organic carbon change monitored over large areas. EOS Transactions 91: 441–442.
8. Nelson DW, Sommers LE (1996) Total carbon, organic carbon, and organic matter. In: Methods of Soil Analysis, Part 3: Chemical Methods, Book Series no. 5. Madison: Soil Science Society of America and American Society of Agronomy. pp. 961–1010.
9. Izaurralde RC (2005) Measuring and monitoring soil carbon sequestration at the project level. In: Lal R, Stewart BA, Uphoff N, Hansen DO, editors. Climate change and global food security. Boca Raton: Taylor & Francis. pp. 467
10. Izaurralde RC, Rice CW (2006) Methods and tools for designing pilot soil carbon sequestration projects. In: Lal R, Cerri CC, Bernoux, M, Etchevers J, Cerri E, editors. Carbon sequestration in Latin America. New York: The Haworth Press, Inc. pp. 457–476
11. Izaurralde RC, Rosenberg NJ, Lal R (2001) Mitigation of climatic change by soil carbon sequestration: issues of science, monitoring and degraded lands. Advances in Agronomy 70: 1–75.
12. Cremers DA, Ebinger MH, Breshears DD, Unkefer PJ, Kammerdiener SA, et al. (2001) Measuring total soil carbon with laser-induced breakdown spectroscopy (LIBS). Journal of Environmental Quality 30: 2202–2206.
13. McCarty GW, Reeves JB, Follett RF, Kimble JM (2002) Mid-infrared and near-infrared diffuse reflectance spectroscopy for soil carbon measurement. Soil Science Society of America Journal 66: 640–646.
14. Wielopolski L, Mitra S, Hendrey G, Orion I, Prior S, et al. (2004) Non-destructive Soil Carbon Analyzer (ND – SCA). Brookhaven: BNL Report No. 72200-2004, Brookhaven National Laboratory.
15. Wielopolski L, Hendrey G, Johnsen KH, Mitra S, Prior SA, et al. (2008) Nondestructive system for analyzing carbon in the soil. Soil Science Society of America Journal 72: 1269–1277.
16. Wielopolski L, Yanai RD, Levine CR, Mitra S, Vadeboncoeur MA (2010) Non-destructive carbon analysis of forest soils using neutron-induced gamma-ray spectroscopy. Forest Ecology and Management 260: 1132–1137.
17. Harmon RS, De Lucia FC, Miziolek AW, McNesby KL, Walters RA, et al. (2005) Laser-induced breakdown spectroscopy (LIBS) – an emerging field-portable sensor technology for real-time, in-situ geochemical and environmental analysis. Geochemistry: Exploration, Environment, Analysis 5: 21–28, DOI: 10.1144/1467-7873/03-059.
18. Al-Jowder O, Defernez M, Kemsley EK, Wilson RH (1999) Mid-infrared spectroscopy and chemometrics for the authentication of meat products. Journal Agriculture and Food Chemistry 47: 3210–3218.
19. Bertelli D, Plessi M, Sabatini AG, Lolli M, Grillenzoni F (2007) Classification of Italian honeys by mid-infrared diffuse reflectance spectroscopy (DRIFTS). Food Chemistry 101: 1565–1570.
20. Macauley-Patrick S, Arnold SA, McCarthy B, Harvey LM, McNeil B (2003) Attenuated total reflectance Fourier transform mid-infrared spectroscopic

21. Heymsfield SB, Wang Z, Baumgartner RN, Ross R (1997) Human body composition: Advances in models and methods. Annual Review of Nutrition 17: 527–58.
22. Chatterjee A, Lal R, Wielopolski L, Martin MZ, Ebinger MH (2009) Evaluation of different soil carbon determination methods. Critical Reviews in Plant Science 28: 164–178.
23. Moenke-Blankenburg L (1989) Laser microanalysis. New York: John Wiley & Sons.
24. Radziemski IJ, Cremers DA (1989) Spectrochemical analysis using laser plasma excitation. In Application of Laser-Induced Plasmas; Radziemski, L. J., Cremers, D. A., Eds.; Marcel Dekker: New York; p. 295.
25. Rusak DA, Castle BC, Smith BW, Winefordner JD (1997) Fundamentals and application of laser-induced breakdown spectroscopy. Critical Reviews in Analytical Chemistry 27: 257–290.
26. Ebinger MH, Norfleet ML, Breshears DD, Cremers DA, Ferris MJ, et al. (2003) Extending the applicability of laser-induced breakdown spectroscopy for total soil carbon measurement. Soil Science Society of America Journal 67: 1616–1619.
27. Ebinger MH, Harris RD (2010) United States Patent No. 7,692, 789 B1, Date of Patent: April 6, 2010.
28. Janik LJ, Merry RH, Skjemstad JO (1998) Can mid-infrared diffuse reflectance analysis replace soil extractions? Australian Journal Experimental Agriculture 38: 681–696.
29. Reeves JB, McCarty GW, Reeves VB (2001) Mid-infrared diffuse reflectance spectroscopy for the quantitative analysis of agricultural soils. Journal of Agriculture and Food Chemistry 49: 766–772.
30. McCarty GW, Reeves JB (2006) Comparison of NFAR infrared and mid infrared diffuse reflectance spectroscopy for field-scale measurement of soil fertility parameters. Soil Science 171: 94–102.
31. Csikai J (1987) Handbook of fast neutron generators. Boca Raton: CRC Press, ISBN: 0-8493-2967-1.
32. Wielopolski L, Johnsen K, Zhang Y (2010) Comparison of soil analysis methods based on samples withdrawn from different volumes: Correlation versus calibration. Soil Science Society of America Journal 74: 812–819.
33. Wielopolski L, Chatterjee A, Mitra S, Lal R (2011) In situ determination of soil carbon pool by Inelastic Neutron Scattering: Comparison with dry combustion. Geoderma 160: 394–399.
34. Wielopolski L (2011) Nuclear methodology for non-destructive multi-elemental analysis of large volumes of soil. In: Carayannis EG, editor. Planet Earth 2011 - Global warming challenges and opportunities for policy and practice. In Tech Open. pp. 467–492; ISBN: 978-953-307-733-8.
35. Gish TJ, Daughtry CST, Walthall CL, Kung K-JS (2004) Quantifying impact of hydrology on corn grain yield using ground-penetrating radar. Subsurface Sensing Technology Applications Journal 2: 493–496.
36. Govaerts B, Sayre KD, Deckers J (2006) A minimum data set for soil quality assessment of wheat and maize cropping in the highlands of Mexico. Soil & Tillage Research 87: 163–174.
37. Govaerts B, Sayre KD, Goudeseune B, De Corte P, Lichter K, et al. (2009) Conservation agriculture as a sustainable option for the central Mexican highlands. Soil & Tillage Research 103: 222–230.

3

An Improved Experimental Method for Simulating Erosion Processes by Concentrated Channel Flow

Xiao-Yan Chen[1,2]*, Yu Zhao[1], Bin Mo[1], Hong-Xing Mi[1]

1 College of Resources and Environment/Key Laboratory of Eco-environment in Three Gorges Region (Ministry of Education), Southwest University, Chongqing, China,
2 State Key Laboratory of Soil Erosion and Dryland Farming on the Loess Plateau, Institute of Soil and Water Conservation, CAS and MWR, Yangling, China

Abstract

Rill erosion is an important process that occurs on hill slopes, including sloped farmland. Laboratory simulations have been vital to understanding rill erosion. Previous experiments obtained sediment yields using rills of various lengths to get the sedimentation process, which disrupted the continuity of the rill erosion process and was time-consuming. In this study, an improved experimental method was used to measure the rill erosion processes by concentrated channel flow. By using this method, a laboratory platform, 12 m long and 3 m wide, was used to construct rills of 0.1 m wide and 12 m long for experiments under five slope gradients (5, 10, 15, 20, and 25 degrees) and three flow rates (2, 4, and 8 L min^{-1}). Sediment laden water was simultaneously sampled along the rill at locations 0.5 m, 1 m, 2 m, 3 m, 4 m, 5 m, 6 m, 7 m, 8 m, 10 m, and 12 m from the water inlet to determine the sediment concentration distribution. The rill erosion process measured by the method used in this study and that by previous experimental methods are approximately the same. The experimental data indicated that sediment concentrations increase with slope gradient and flow rate, which highlights the hydraulic impact on rill erosion. Sediment concentration increased rapidly at the initial section of the rill, and the rate of increase in sediment concentration reduced with the rill length. Overall, both experimental methods are feasible and applicable. However, the method proposed in this study is more efficient and easier to operate. This improved method will be useful in related research.

Editor: Ben Bond-Lamberty, DOE Pacific Northwest National Laboratory, United States of America

Funding: The funding which supported this work is Foundation of State Key Laboratory of Soil Erosion and Dryland Farming on the Loess Plateau under project No. K318009902-1312 (http://english.iswc.cas.cn/). The funders had no role in study design, data collection and analysis, decision to publish, or preparation of the manuscript.

Competing Interests: The authors have declared that no competing interests exist.

* Email: guangguang14@163.com

Introduction

Soil erosion is a serious environmental problem that threatens agricultural safety and sustainable development due to land degradation [1–5]. As an important component of hill slope soil erosion, rill erosion is especially dangerous on cultivated slopes and upland areas. Eroding rills are formed by concentrated surface runoff and function as sediment source areas and sediment transport vehicles [6–9]. Considering its importance, the mechanism of rill erosion has long been the focus of study.

As early as 1981, Foster et al. [10] differentiated upland soil erosion into inter-rill erosion and rill erosion. They indicated that rill erosion contributed more significantly to sediment production than inter-rill erosion. Since then, many researchers have studied rill morphology, the hydraulics of rill flow, and the rill erosion process. The results indicate that the parameters, such as sediment concentration, soil detachment rate, sediment transport capacity, soil erodibility and critical shear stress, can be used to characterize the rill erosion process. Sediment concentration and soil detachment rate are the most relevant indicators of erosion. Sediment transport capacity expresses the potential sediment entrainment ability of the concentrated flow in rills. Soil erodibility and critical shear stress provide the criteria for determining the occurrence and quantity of rill erosion [11–17].

In recent decades, various research techniques have been applied to study rill erosion. However, information regarding spatially distributed rill erosion is limited. Most existing rill erosion data are spatially integrated, as measured at the rill outlet [18–21]. Spatially integrated rill erosion data do not adequately describe the dynamics of the rill erosion process.

Traditional measurement methods cannot easily quantify the rill erosion process due to the complexity of quantifying erosion among various rill segments. Consequently, new methods that provide dynamic rill erosion process are needed.

Recently, the rare earth element (REE) was utilized to trace the temporal and spatial distribution data of rill erosion [22–24]. The method successfully quantified the rill erosion process. However, the REE method is not economical or efficient, and need special and expensive facilities to conduct the measurement. Lei et al. [13] suggested an experimental method for studying rill erosion process. In their method, the sediment concentrations produced from rills of different lengths are measured. Next, the measured sediment data are integrated to produce a spatially distributed rill erosion process. However, the approach was determined to be time consuming and disruptive to the continuity of the rill erosion process. Aksoy et al. [16,17] performed experimental analysis in a laboratory flume under simulated rainfall by pre-formed rill. They used the experimental data to relate sediment concentration to slope gradient and rainfall intensity.

Rill erosion involves such processes as infiltration of rill, winding of flow path and failure of side walls due to randomized scouring and deposition. These processes are interactive with the most important rill erosion components such as detachment rate, transport capacity, shear stress and sediment concentration along the rill to be computed or estimated.

Along a width-fluctuated rill, the sediment concentrations are much lower than those along a well-defined rill with constant water flow [25]. Naturally developed rills in laboratory experiments involve periodic change of detachment and deposition responsible for rill width fluctuation, widening where deposition occurs and narrowing when scouring takes place [26]. Randomized side wall failure during rill erosion process does cause underestimation of sediment concentration at locality. Still the underestimated sediment load in the water flow is much higher than that after deposition occurrence. Furthermore, water flow of lower sediment concentration causes higher scouring of rill bed to contribute more sediments to compensate the water flow [27,28]:

$$D_r = K_r(\tau - \tau_c)\left(1 - \frac{qc}{T_c}\right) \quad (1)$$

where, D_r (kg m^{-2} s) is the contribution of rill detachment rate to rill sediment load, K_r (s^{-1}) is the erodibility of the soil; τ (kg m^{-2}) is the shear stress of flowing water, τ_c (kg m^{-2}) is the critical shear strength of soil; c (kg m^{-1}) is the sediment concentration and q (m^2 s^{-1}) is the flow rate of a unit width, T_c (kg s^{-1} m) is the sediment transport capacity of the water flow [28].

Therefore, it seems rational to use well-defined rill for quantitative study of rill erosion.

The objectives of this study were to: 1) develop an improved experimental method for determining the rill erosion process; 2) determine the rill erosion process with experimental data under various hydraulic conditions; 3) assess the sediment concentration and erosion process with previously reported experimental data.

Methods and Materials

In regular rill erosion, the rill width changes periodically due to the erosion and deposition. And the sediment concentration could be much lower than that in well-defined rill under constant water flow. In order to overcome this problem of rill width fluctuation, well-defined rill was constructed in this experiment.

The experiments were conducted in the State Key Laboratory of Soil Erosion and Dryland Farming on the Loess Plateau, Institute of Soil and Water Conservation, Chinese Academy of Science and Ministry of Water Resources, Yangling, Shanxi Province, China. A platform that was 12 m long and 3 m wide was used as a base to construct a flume that was 12 m long, 0.6 m wide and 0.5 m deep. The flume was sub-divided into six rills that were 0.1 m wide and 12 m long using upright PVC boards to form well-defined rills (Fig. 1). The PVC board surfaces were glued with the experimental soil particles on both sides to create a roughness equivalent to the soil surface so as to minimize the boundary effect on the hydraulic and erosion processes, in case when water flow becomes contact with the boards.

The experimental soil materials were collected from the Ansai Research Station of Soil and Water Conservation on the Chinese Loess Plateau. The soil contains 15.92% clay (<0.005 mm), 63.90% silt (0.005 to 0.05 mm), and 20.18% sand (>0.05 mm), which is classified as a silt-loamy soil. The soil was air dried before being crushed and passed through an 8 mm square sieve.

The bottom 5 cm of the flume was paved with a clay loamy soil (24.9% sand particles, 43.4% silt particles and 31.8% clay

Figure 1. The experimental setup. A picture shows the experimental setup.

particles) to achieve a bulk density of approximately 1.5 g cm^{-3} (i.e., an imitation of the plow pan layer). On top of the plow pan layer, the experimental soil was packed in 5 cm thick layers to a total depth of 20 cm, at the bulk density of approximately 1.2 g cm^{-3}. The soil near the PVC boards was packed slightly higher than the middle in aim to converge the water flow to the center of the rill and to minimize the boundary effect as much as possible. The prepared rills were saturated with the rain-simulator and allowed to drain for 24 h to ensure an even and homogeneous initial soil moisture profile in an experimental run, and among different experimental runs.

A water tank was used to supply the water flow at the designed discharge rate. An additional specially designed device was used at the rill flow inlet to accelerate the water flow to a velocity level close to the rill flow. Gauze cloth approximately 0.2 m in length was placed at soil surface of the rill inlet to protect the rill surface from being directly scoured by the water flow.

Experimental runs were conducted for five slope gradients (5°, 10°, 15°, 20° and 25°) and three flow rates (2 L min^{-1}, 4 L min^{-1}, and 8 L min^{-1}) with 3 replicates.

Water flow at the designed rate was introduced into the rill at the top end after the flume was adjusted to the designed slope gradient. In Lei et al.'s [13,29] experiments, sediment-laden water samples were collected at outlets of rill of various lengths before the collected data were integrated to produce the sediment process of the entire 8 m rill. Here, after steady water flow in the rill was established, sediment-laden water samples were simultaneously taken along the rill at distances of 0.5 m, 1.0 m, 2.0 m, 3.0 m, 4.0 m, 5.0 m, 6.0 m, 7.0 m, 8.0 m, 10.0 m, and 12.0 m from the water flow inlet, which is an improvement on the previous method. Thus, the sediment delivery process was determined along the rill. Three samples were taken at 1.0-minute interval for each experimental run. Therefore each experiment lasted less than 5 minute after steady flow was established. The average value of the three samples at each location was used to determine the sediment concentration at various rill lengths. The sediment concentrations that were measured at the various locations formed the sediment delivery process along the rill.

Results and Discussion

Sediment concentration along the rill

The measured sediment concentrations along the rills under various flow rates and slope gradients are presented in Fig. 2. The experimental data exhibited a well defined trend between the sediment concentration and rill length, which can be described by:

$$S_c = A(1 - e^{-Bx}) \qquad (2)$$

where S_c (kg m^{-3}) is the sediment concentration, x (m) is the rill length, and A (kg m^{-3}) and B (m^{-1}) are regression coefficients, with A as the maximum possible sediment concentration and B as the attenuation coefficient to indicate how fast of the reduction in sediment increase.

The experimental data presented in Fig. 2 were regressed with Eq. (2). The regression parameters are listed in Table 1. The regression results indicated that the sediment concentrations for all the experimental runs increased exponentially with the rill length. Polyakov and Nearing [30] and Lei et al. [31] studied the sediment concentrations in rills and concluded that the relationship between sediment concentration and rill length is well-described by Eq. (1) too. The sediment concentration increased rapidly at the initial section of the rill, where the water contains limited amount of sediments. The increase in sediment concentration reduced rapidly along the rill. These sediment concentration distributions

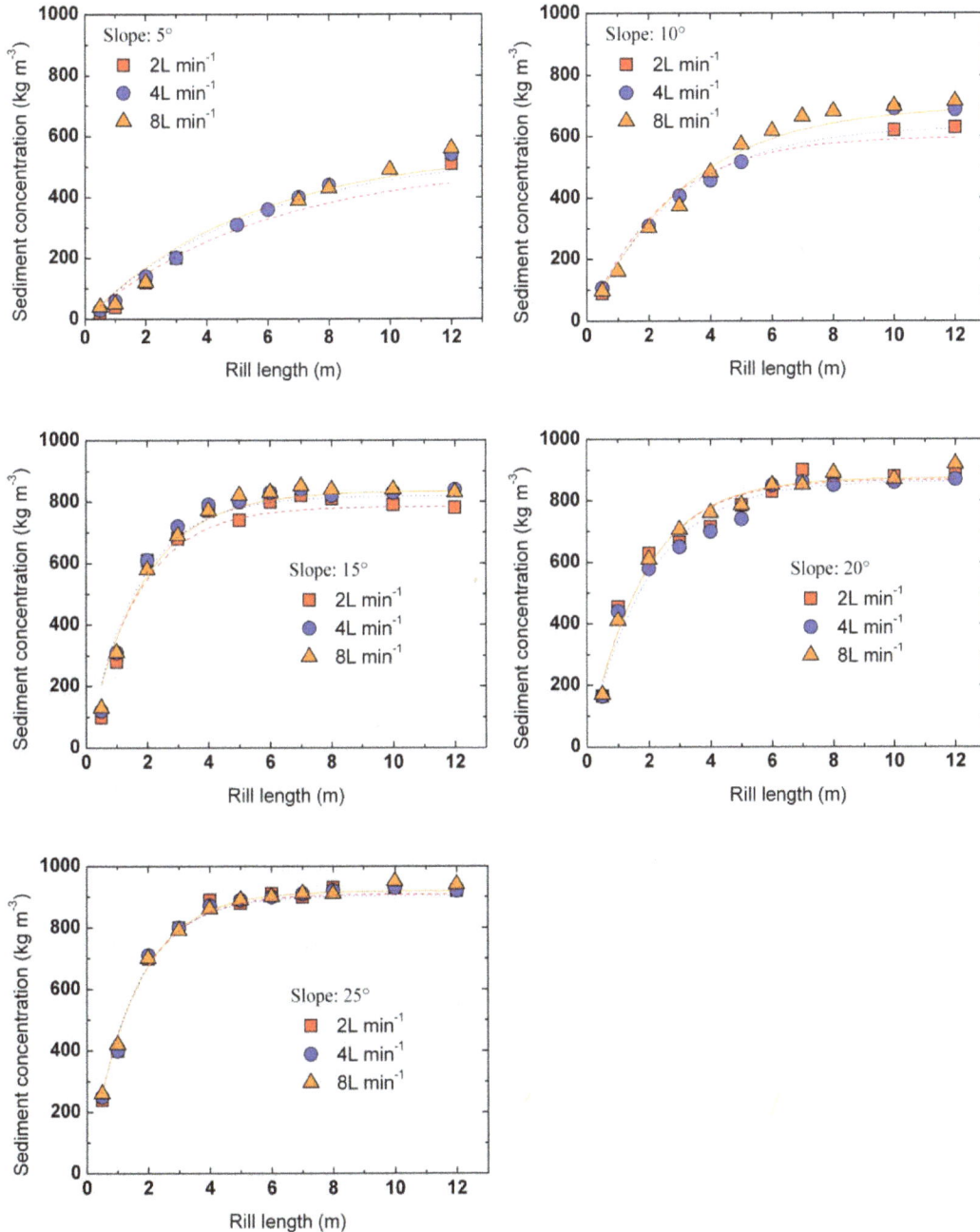

Figure 2. Sediment concentration along the rill at various slope gradients. The variation trend of the measured sediment concentrations along the rill under three flow rates and five slope gradients. The sediment concentration increased exponentially with the rill length for all of the flow rates and slope gradients. The sediment concentration also increased with the slope gradient and flow rate at lower slope gradients (5°, 10° and 15°).

were more evident at steeper slopes. It is possible that clear water (no sediment particles) that is introduced into the rill at the inlet spends limited energy to transport sediments and high energy to detach soil at the upper part of the rill. According to the Water Erosion Prediction Project (WEPP) model [28], clear water has a highest detachment potential. The increase in sediment concentration decreases to zero when the water is saturated with sediments (up to its transport capacity). Almost all the energy of the water flow is used for sediment transportation [32].

In Table 1 all the coefficients of determination (R^2) were higher than 0.95, which indicate that Eq. (2) fits well the sediment concentration along the rill. All the values of ($Prob>F$) from the F-test were rather low (i.e., less than 8.76 E^{-8}), to indicate that the results were statistically significant (F-test with $\alpha = 0.01$).

As indicated in Fig. 2, all the sediment concentrations increased with the rill length and eventually reached the maximum value of A. The value of A was the sediment concentration when the water flow was saturated with sediment particles, which represents the sediment concentration at the transport capacity of the water flow. Parameter B was a reduction coefficient that indicates the attenuation of the increase in sediment concentration. From Table 1, the value of A increases with the slope and ranges from 510 to 910. Thus, the sediment transport capacity increased rapidly with the slope gradient. A did not vary significantly with the flow rate under the same slope gradient. The value of A changed slightly between 20° and 25°, which indicates that a critical slope gradient between 20° and 25° may exist.

In summary, the sediment concentration for sediment saturated flow represents the transport capacity; thus, the sediment transport capacity and the rill length can be estimated.

Relationship between sediment concentration, slope and flow rate

The sediment concentration increased with the slope and flow rate at lower slope gradients (5°, 10° and 15°). This is consistent with the findings of other researchers [33,34]. For rill erosion, the inflow rate and slope gradient are the most influential factors. Steeper slopes and higher flow rates provided greater flow shear

force. These factors enhance soil erosion by either increasing the soil detachment rate or by weakening the protective power of the soil surface to resist erosion [35–37]. Furthermore, the curve defined by Eq. (2) fits well with the sediment concentration under the three flow rates. The sediment concentrations increased with the slope gradient, but they only increased slightly with the flow rate (Fig. 2). For steep slope gradients (20° and 25°) the sediment curves under three flow rates nearly overlapped at the same slope gradient. The slope may have a greater influence on the sediment concentration than the flow rate. Although the sediment concentrations were approximately the same, the higher flow rate delivered proportionally more sediment (proportional to the flow rate).

To better understand and clarify the relationship between sediment concentration, slope gradient and flow rate, several functions were tried to fit the relationship between the sediment concentration at transport capacity, slopes and flow rates. The following function was found a simple and appropriate fit for the experimental data:

$$S_T = k \cdot S^m \cdot q^n \qquad (3)$$

where S_T (kg m^{-3}) is the sediment concentration at transport capacity, S is the slope gradient (%), q is the flow rate (L min^{-1}) and m and n are regression coefficients. The regression results are listed in Table 2.

Regression analysis indicated that m decreased steadily with slope gradient slope and ranged from 2.45 to 1.69. This indicates that the increase in sediment concentration at transport capacity decreases with the slope gradient. However, n, decreased with slope gradient and flow rate, indicating that the increase in sediment concentration declines with the flow rate. All of the coefficients of determination (R^2) were higher than 0.84, which indicates that Eq. (3) is able to quantify the sediment concentration at transport capacity of rill erosion. Both the regression coefficients in Eq. (3), m and n, were very significant (F-test with $\alpha = 0.01$), which illustrates that the slope and flow rate were both critical parameters that influence sediment concentration at transport

Table 1. Regression parameters of Eq. (1).

Slope (°)	flow rate (L min^{-1})	A(kg m^{-3})	B (m^{-1})	R^2	F	$Prob>F$
5	2	510	0.17	0.95	172.74	5.09E-08
	4	550	0.18	0.97	1174.60	8.76E-08
	8	560	0.18	0.97	593.85	2.17E-06
10	2	625	0.41	0.99	622.72	7.85E-10
	4	635	0.35	0.96	1020.01	1.57E-13
	8	675	0.31	0.98	3072.75	1.48E-13
15	2	785	0.60	0.95	1846.31	1.99E-08
	4	820	0.60	0.97	2868.62	5.75E-09
	8	835	0.56	0.98	4788.88	1.80E-09
20	2	870	0.56	0.96	2906.29	8.80E-10
	4	865	0.52	0.96	3088.36	9.55E-10
	8	875	0.56	0.98	7395.75	7.12E-11
25	2	910	0.68	0.99	11025.60	1.35E-09
	4	905	0.69	0.99	11587.59	6.75E-10
	8	910	0.68	0.99	16567.57	8.57E-11

Table 2. Regression coefficients under various slope gradients and flow rates.

Slope (°)	flow rate (L min⁻¹)	K	m	n	R^2	Prob>F
5	2	0.83	2.45 a	1.64 a	0.95	4.59E-04
	4	0.89	2.23 b	1.13 b	0.92	4.34E-04
	8	0.87	1.99 c	1.03 bc	0.95	9.51E-04
10	2	0.80	2.12 bc	0.96 bc	0.86	7.33E-04
	4	0.74	1.94 c	0.91 c	0.89	3.11E-03
	8	0.78	1.87 cd	0.70 cd	0.88	3.64E-03
15	2	0.86	1.96 c	0.61 d	0.93	5.74E-04
	4	0.79	1.89 cd	0.57 de	0.88	1.84E-03
	8	0.83	1.76 d	0.55 de	0.84	1.15E-02
20	2	0.82	1.87 cd	0.54 de	0.95	9.73E-04
	4	0.71	1.83 cd	0.49 de	0.93	7.69E-03
	8	0.75	1.75 d	0.38 e	0.93	7.40E-03
25	2	0.78	1.79 d	0.41 de	0.95	1.20E-03
	4	0.69	1.76 d	0.39 e	0.93	7.04E-03
	8	0.74	1.69 d	0.38 e	0.90	1.67E-02

Footnote: Significant differences between the values are indicated by letters ($p<0.05$).

capacity of rill erosion process. Generally, sediment concentration at transport capacity increased with the slope gradient and flow rate (Table 2). Additionally, the values of m were all greater than n; the sediment concentration at the transport capacity increased nearly quadratically ($m \approx 2$) with the slope gradient, but increased less than linearly ($n < 1$) with the flow rate. This indicates that the slope is much more important than the flow rate with respect to the sediment transportation [38–40]. Lei et al. [31] suggested that the sediment concentration at transport capacity increased quadratically with slope gradient and exhibited an approximately linear increase with flow rate, which supported our results. Similarly, Aksoy et al. [17] concluded that slope created a linear increase in sediment concentration but the flow rate had practically no effect on sediment concentration.

According to Table 2, the m values of 5°, 10°, 15° and 20° and 25° were significantly different from each other. However, m did not display statistically significant differences between 20° and 25°. Based on the given slope and flow rate, the sediment concentration at the transport capacity increased more significantly with the slope gradients than with flow rates from 5° to 20°. However, when the slope gradient increased from 20° to 25°, the sediment concentration was more stable. The n values revealed similar results: the values at 20° and 25° were obviously different from those at the other slopes, but significant differences did not exist between the two slope gradients. That is to say, the sediment concentration had almost achieved its maximum value (i.e., a certain slope gradient between 20° and 25° was the critical erosion point of the experimental soil).

Comparison of rill erosion processes measured by various methods

In the experiments by Lei et al. [29], the soil type, designed slope gradients and flow rates were all identical to those of the present study. However, the rill length used by the previous experiment was 8.0 m, whereas the rill length used in the present study was 12.0 m. Furthermore, in the previous study, the sediment concentration process along the 8 m rill was integrated from experimental data of rills of various lengths (i.e., 0.5 m, 1 m,

2 m, etc, up to 8 m), slope gradients and flow rates. In this study, simultaneous and continuous sampling procedures were used. To compare the two data sets, only the data at the upper 8.0 m in this experiment were used (Fig. 3), with the experimental data of discontinuous rill as X-axis and the data of this study as Y-axis. The closer the dots in Fig. 3 are to the 1:1 line, the better the agreement of the two data sets. The data sets indicate that the sediment concentrations from the present experimental method were approximately 1.04, 0.88, 0.83, 0.92, and 0.87 times of the previous experimental data sets at the slope gradients of 5°, 10°, 15°, 20° and 25°, respectively. The coefficients of determination (R^2) were all greater than 0.96, which indicates that the two datasets were approximately the same and closely correlated.

The previous experimental technique produced reasonable rill erosion data, but it produced slightly higher sediment concentrations. This could have been caused by random errors in sampling, experiment preparation, etc. The method suggested in this study is more efficient. Simultaneous sampling along the rill is more rational. The relative error produced by these two methods is less than 15%, which suggests that both methods are feasible and applicable. The improved method suggested in this study can measure the rill erosion simultaneously and continuously.

Conclusions

Given the limits of the method used in previous rill erosion research, an improved and easy to use method of studying rill erosion processes by means of simultaneous sampling of sediment-laden water was suggested. The dynamic changes and distributions of sediment concentrations along the rill were measured for a silty-loam soil over a range of slope gradients and flow rates. The results indicate that sediment concentration increased exponentially with the rill length for all of the flow rates and slope gradients. The sediment concentration also increased with the slope gradient. The data proved that the slope gradient and flow rate are both important parameters for rill erosion, but the slope gradient seemed to be more influential on the sediment concentration. The data computed from the improved method were compared with

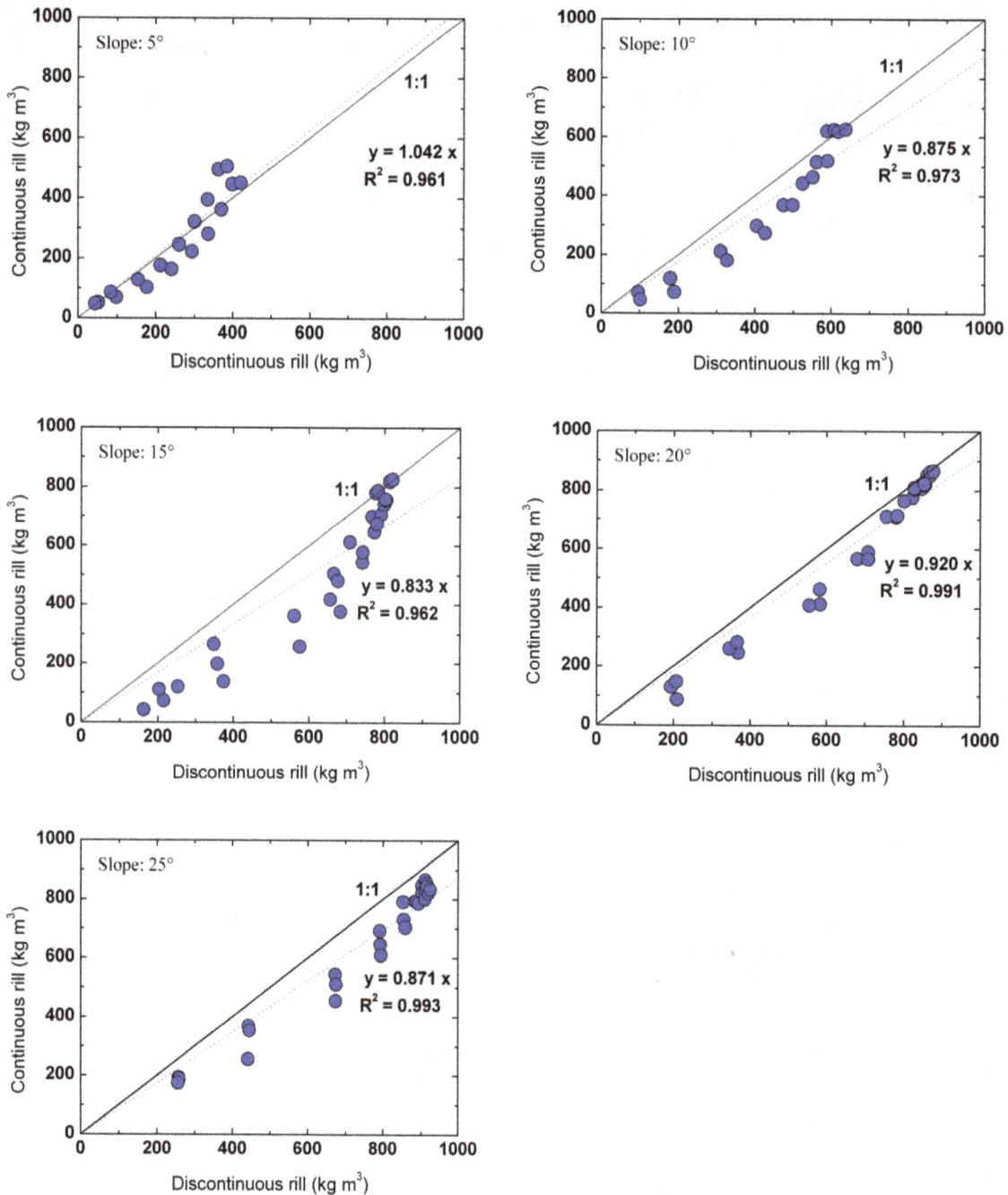

Figure 3. The sediment concentrations measured from continuous rills in this study and discontinuous rills in Lei et al. (2002). The sediment concentrations measured from continuous rills in this study and discontinuous rills in Lei et al. (2002) which used a similar methodology. The relative error produce by these two methods is less than 15%, which suggests that both of the methods are feasible.

the data from a previous study, which suggested that this study is a feasible mean to simulate rill erosion process. However, this improved experimental method needs further examination at field scale and more attention should be paid on measures to minimize the boundary effect.

Acknowledgments

Authors gratefully acknowledge the staff of State Key Laboratory of Soil Erosion and Dry Land Farming on the Loess Plateau for providing guidance of the laboratory apparatus.

Author Contributions

Conceived and designed the experiments: XYC YZ BM HXM. Performed the experiments: YZ BM HXM. Analyzed the data: YZ HXM BM. Contributed reagents/materials/analysis tools: YZ BM HXM. Wrote the paper: YZ XYC.

References

1. Wirtz S, Seeger M, Zell A, Wagner C, Wagner JF, et al. (2013) Applicability of Different Hydraulic Parameters to Describe Soil Detachment in Eroding Rills. Plos One 8, doi: 10.1371/journal.pone.0064861
2. Yang DW, Kanae S, Oki T, Koike T, Musiake K (2003) Global potential soil erosion with reference to land use and climate changes. Hydrol Process 17: 2913–2928.
3. Bhattarai R, Dutta D (2007) Estimation of soil erosion and sediment yield using GIS at catchment scale. Water Resour Manag. pp. 1635–1647.
4. Miao CY, Ni JR, Borthwick AGL (2010) Recent changes of water discharge and sediment load in the Yellow River basin, China. Prog Phys Geog 34: 541–561.
5. Miao CY, Duan QY, Yang L, Borthwick AGL (2012) On the Applicability of Temperature and Precipitation Data from CMIP3 for China. Plos One 7.
6. Ellison WD (1949) Protecting the Land against the Raindrops Blast. Sci Mon 68: 241–251.
7. Foster GR, Huggins LF, Meyer LD (1984) A Laboratory Study Of Rill Hydraulics .1. Velocity Relationships. T Asae 27: 790–796.
8. Foster GR, Huggins LF, Meyer LD (1984) A Laboratory Study Of Rill Hydraulics .2. Shear-Stress Relationships. T Asae 27: 797–804.
9. Sun LY, Fang HY, Qi DL, Li JL, Cai QG (2013) A review on rill erosion process and its influencing factors. Chinese Geogr Sci 23: 389–402.
10. Foster GR, Lane LJ, Nowlin JD, Laflen JM, Young RA (1981) Estimating Erosion And Sediment Yield on Field-Sized Areas. T Asae 24: 1253–1263.
11. Tiscarenolopez M, Lopes VL, Stone JJ, Lane LJ (1993) Sensitivity Analysis Of the Wepp Watershed Model for Rangeland Applications .1. Hillslope Processes. T Asae 36: 1659–1672.
12. Gilley JE, Elliot WJ, Laflen JM, Simanton JR (1993) Critical Shear-Stress And Critical Flow-Rates for Initiation Of Rilling. J Hydrol 142: 251–271.
13. Lei TW, Zhang Q, Zhao J, Tang Z (2001) A laboratory study of sediment transport capacity in the dynamic process of rill erosion. T Asae 44: 1537–1542.
14. Gimenez R, Govers G (2002) Flow detachment by concentrated flow on smooth and irregular beds. Soil Sci Soc Am J 66: 1475–1483.
15. Kavvas ML, Yoon J, Chen ZQ, Liang L, Dogrul EC, et al. (2006) Watershed environmental hydrology model: Environmental module and its application to a California watershed. J Hydrol Eng 11: 261–272.
16. Aksoy H, Unal NE, Cokgor S, Gedikli A, Yoon J, et al. (2012) A rainfall simulator for laboratory-scale assessment of rainfall-runoff-sediment transport processes over a two-dimensional flume. Catena 98: 63–72.
17. Aksoy H, Unal NE, Cokgor S, Gedikli A, Yoon J, et al. (2013) Laboratory experiments of sediment transport from bare soil with a rill. Hydrolog Sci J 58: 1505–1518.
18. Zhang XC, Friedrich JM, Nearing MA, Norton LD (2001) Potential use of rare earth oxides as tracers for soil erosion and aggregation studies. Soil Sci Soc Am J 65: 1508–1515.
19. Schuller P, Walling DE, Sepulveda A, Trumper RE, Rouanet JL, et al. (2004) Use of Cs-137 measurements to estimate changes in soil erosion rates associated with changes in soil management practices on cultivated land. Appl Radiat Isotopes 60: 759–766.
20. Li M, Li ZB, Liu PL, Yao WY (2005) Using Cesium-137 technique to study the characteristics of different aspect of soil erosion in the Wind-water Erosion Crisscross Region on Loess Plateau of China. Appl Radiat Isotopes 62: 109–113.
21. Porto P, Walling DE, Callegari G (2013) Using 137Cs and 210Pbex measurements to investigate the sediment budget of a small forested catchment in southern Italy. Hydrol Process 27: 795–806.
22. Lei TW, Zhang QW, Zhao J, Nearing MA (2006) Tracing sediment dynamics and sources in eroding rills with rare earth elements. Eur J Soil Sci 57: 287–294.
23. Li M, Li ZB, Ding WF, Liu PL, Yao WY (2006) Using rare earth element tracers and neutron activation analysis to study rill erosion process. Appl Radiat Isotopes 64: 402–408.
24. Zhu MY, Tan SD, Dang HS, Zhang QF (2011) Rare earth elements tracing the soil erosion processes on slope surface under natural rainfall. J Environ Radioactiv 102: 1078–1084.
25. Lei TW, Nearing MA, Haghighi K, Bralts VF (1998) Rill erosion and morphological evolution: A simulation model. Water Resour Res 34: 3157–3168.
26. Lei TW, Nearing MA (2000) Flume experiments for determining rill hydraulic characteristic erosion and rill patterns. J Hydraul Eng 11: 49–54 (in Chinese).
27. Nearing MA, Norton LD, Bulgakov DA, Larionov GA, West LT, et al. (1997) Hydraulics and erosion in eroding rills. Water Resour Res 33: 865–876.
28. Nearing MA, Foster GR, Lane LJ, Finkner SC (1989) A Process-Based Soil-Erosion Model for Usda-Water Erosion Prediction Project Technology. T Asae 32: 1587–1593.
29. Lei TW, Zhang QW, Zhao J, Xia WS, Pan YH (2002) Soil detachment rates for sediment loaded flow in rills. T Asae 45: 1897–1903.
30. Polyakov VO, Nearing MA (2003) Sediment transport in rill flow under deposition and detachment conditions. Catena 51: 33–43.
31. Lei TW, Zhang QW, Yan LJ (2009) Physically-based rill erosion model. Beijing: Science Press. 229 p.
32. Aksoy H, Kavvas ML (2005) A review of hillslope and watershed scale erosion and sediment transport models. Catena 64: 247–271.
33. Kinnell PIA (2000) The effect of slope length on sediment concentrations associated with side-slope erosion. Soil Sci Soc Am J 64: 1004–1008.
34. Sirjani E, Mahmoodabadi M (2014) Effects of sheet flow rate and slope gradient on sediment load. Arab J Geosci 7: 203–210.
35. Huang C (1998) Sediment regimes under different slope and surface hydrologic conditions. Soil Sci Soc Am J 62: 423–430.
36. Fox DM, Bryan RB (2000) The relationship of soil loss by interrill erosion to slope gradient. Catena 38: 211–222.
37. Miao CY, Ni JR, Borthwick AGL, Yang L (2011) A preliminary estimate of human and natural contributions to the changes in water discharge and sediment load in the Yellow River. Global Planet Change 76: 196–205.
38. Shi ZH, Fang NF, Wu FZ, Wang L, Yue BJ, et al. (2012) Soil erosion processes and sediment sorting associated with transport mechanisms on steep slopes. J Hydrol 454: 123–130.
39. Govers G, Rauws G (1986) Transporting Capacity Of Overland-Flow on Plane And on Irregular Beds. Earth Surf Proc Land 11: 515–524.
40. Zhang GH, Liu YM, Han YF, Zhang XC (2009) Sediment Transport and Soil Detachment on Steep Slopes: II. Sediment Feedback Relationship. Soil Sci Soc Am J 73: 1298–1304.

4

The Effects of Manure and Nitrogen Fertilizer Applications on Soil Organic Carbon and Nitrogen in a High-Input Cropping System

Tao Ren[1,2], Jingguo Wang[1], Qing Chen[1]*, Fusuo Zhang[1], Shuchang Lu[3]

1 College of Resources and Environmental Science, China Agricultural University, Beijing, China, 2 College of Resources and Environment, Huazhong Agricultural University, Wuhan, China, 3 Department of Agronomy, Tianjin Agricultural University, Tianjin, China

Abstract

With the goal of improving N fertilizer management to maximize soil organic carbon (SOC) storage and minimize N losses in high-intensity cropping system, a 6-years greenhouse vegetable experiment was conducted from 2004 to 2010 in Shouguang, northern China. Treatment tested the effects of organic manure and N fertilizer on SOC, total N (TN) pool and annual apparent N losses. The results demonstrated that SOC and TN concentrations in the 0-10cm soil layer decreased significantly without organic manure and mineral N applications, primarily because of the decomposition of stable C. Increasing C inputs through wheat straw and chicken manure incorporation couldn't increase SOC pools over the 4 year duration of the experiment. In contrast to the organic manure treatment, the SOC and TN pools were not increased with the combination of organic manure and N fertilizer. However, the soil labile carbon fractions increased significantly when both chicken manure and N fertilizer were applied together. Additionally, lower optimized N fertilizer inputs did not decrease SOC and TN accumulation compared with conventional N applications. Despite the annual apparent N losses for the optimized N treatment were significantly lower than that for the conventional N treatment, the unchanged SOC over the past 6 years might limit N storage in the soil and more surplus N were lost to the environment. Consequently, optimized N fertilizer inputs according to root-zone N management did not influence the accumulation of SOC and TN in soil; but beneficial in reducing apparent N losses. N fertilizer management in a greenhouse cropping system should not only identify how to reduce N fertilizer input but should also be more attentive to improving soil fertility with better management of organic manure.

Editor: Xiujun Wang, University of Maryland, United States of America

Funding: The authors are grateful to the National Natural Science Foundation of China (No. 31071858), Innovative Research Team of Beijing Fruit Vegetable Industry, the innovative group grant of NSFC (No. 30821003) and Basic Application and Cutting-edge Technology Research Projects of Tianjin City (09JCYBJC08600). The funders had no role in study design, data collection and analysis, decision to publish, or preparation of the manuscript.

Competing Interests: The authors have declared that no competing interests exist.

* E-mail: qchen@cau.edu.cn

Introduction

Soil organic matter plays a key role in soil biological and chemical processes, and changes in soil organic matter strongly influence soil N turnover because of the importance of available C for microbial immobilization [1-3]. Soils with higher organic matter contents may immobilize more N and reduce N loss to the environment. Otherwise, the depletion of available C will cause more rapid N turnover and losses [4]. In addition, changes in N availability can also alter soil C turnover [5]. There is no doubt that higher crop production in response to mineral N fertilizer application results in greater root exudates and more crop residues, thereby enhancing SOC sequestration in agricultural soils [6]. In addition increasing N fertilizer application can stabilize organic matter [7] and retard the mineralization of older soil organic matter [8]. N fertilization plays a positive role in enhancing the SOC [7], [9-12]. However, the addition of N fertilizer has also been reported to have a negative or no effect on SOC accumulation [13-17]. Changes in the decomposability of fresh plant litter and soil organic matter fractions, the stability of soil aggregates, and/or shifts in the microbial community can be

used to explain the decreases in SOC attributed to N fertilizer addition [13], [15]. Therefore, achieving a better understanding of the interaction between N fertilizer and SOC in agricultural soils is essential for maximizing SOC storage and minimizing potential N losses.

Intensive vegetable production systems in northern China differ from other ecosystems in which excessive nutrients and water are applied, which far exceed the resources needed for vegetable growth [18-19]. As shown in previous studies [20-21], these practices have resulted in serious N losses to the environment. Therefore, more work was done to understand how to reduce N fertilizer input with optimal N and irrigation strategies, along with catch crops in the intensive greenhouse vegetable cropping system [22-24]. Nevertheless, a recent survey of the largest greenhouse vegetable production region in northern China showed that the soil C/N ratio in greenhouse soils was lower than that of the adjacent open field soils because of the high accumulation rate of soil N as a result of an excessive N input [25]. The low soil C/N ratio implied that the C levels were insufficient for the cropping system, which would limit N immobilization by soil microorganisms [2], [26-27] and may lead to high N losses [28]. This finding

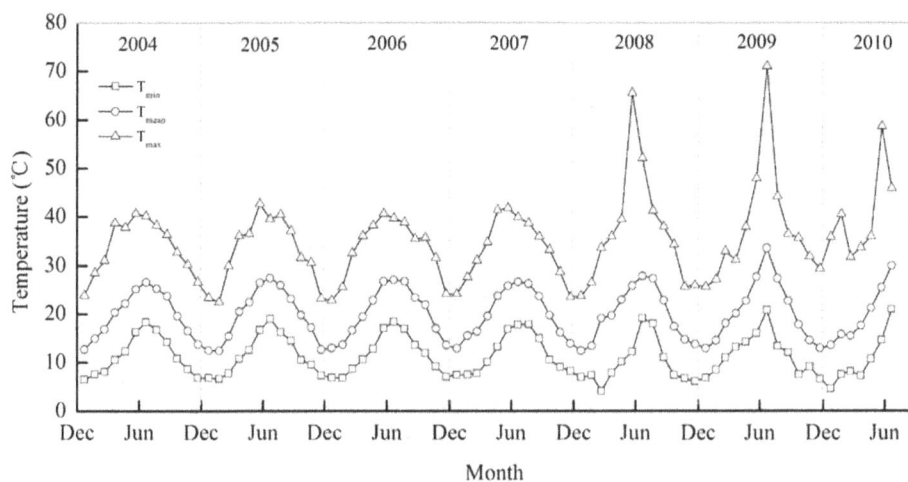

Figure 1. The monthly mean temperature, minimum and maximum temperature inside greenhouse from 2004WS to 2009AW in the year-round greenhouse tomato planting system in Shouguang, northern China.

shows that not only optimal N fertilizer management is needed to be studied, but also the soil organic matter content must be improved to enhance soil N retention capacity, which will reduce N losses to the environment and improve N use efficiency.

In a conventional greenhouse vegetable planting system, most of the plant residues are removed at harvest out of fear of infecting the next crop with fungi and other pathogens. Large amounts of organic manure with a low C/N ratio, such as poultry manure and pig manure, are the main soil carbon supplements in the greenhouse vegetable cropping system in northern China [29]. Excessive N fertilizer application is also a significant characteristic of this cropping system. However, it is unclear how the continuous application of poultry manure and excessive mineral N influences the soil organic matter and total N. Whether lower optimized N fertilizer inputs will alter the accumulation of soil organic matter and total N in the greenhouse field? A greenhouse tomato experiment into which different N management strategies were introduced was conducted from 2004 to 2010 in Shouguang county, the largest greenhouse vegetable production region in northern China. In contrast to common farming practice, optimal N management could reduce mineral N fertilizer input by 72% without decreasing the fruit yield [24]. In this experiment, different types of organic manure, including conventional dry chicken manure and wheat straw, were applied. This environment provided an opportunity to gain a better understanding of the influence of organic manure together with mineral N inputs on the SOC and total N pools, and determine if reducing mineral N fertilizer input will alter SOC and TN accumulation. Moreover, it provided an opportunity to analyse the influence of changes in the SOC pool on N losses in the greenhouse vegetable cropping system. Evaluating this system will be of great assistance in improving management practices for maintaining soil fertility and productivity while minimizing potential N losses from the high-input greenhouse vegetable cropping system.

Materials and Methods

Ethics statement

No specific permits were required for these field studies. No specific permissions were required for these locations/activities because they were not carried out on privately owned or protected areas. The field studies did not involve endangered or protected species.

Site description and crop management

The experiment was established in February 2004 in a traditional unheated commercial solar greenhouse (84×8.5 m) in Luojia (36°55′N, 118°45′E), Shouguang, northern China. The greenhouse was constructed from a vertical clay wall and covered with polyethylene film throughout the year. Thus the air temperature inside greenhouse is higher than in outside. The monthly mean temperature, minimum and maximum temperature inside greenhouse from 2004 to 2010 was showed in Figure 1. Groundwater was used for irrigation with an average of 573 mm over 11 applications for every growing season. During construction in 1999, surface soil in the greenhouse was used to build the clay wall. The silt loam that remained in the field was approximately 60 cm lower than the open field. Chicken manure was applied at 30 t DW ha^{-1}season^{-1} for the first 2 years to improve soil fertility. After 2001, chicken manure application was reduced to 15 t DW ha^{-1} season^{-1} until the experiment was established in 2004. According to FAO classification, the soil at the outset of the trial had 637 g sand kg^{-1}, 323 g silt kg^{-1} and 40 g clay kg^{-1} from 0 to 0.1m soil depth, 620 g sand kg^{-1}, 335 g silt kg^{-1} and 45 g clay kg^{-1} from 0.1 to 0.3 soil depth, 662 g sand kg^{-1}, 301 g silt kg^{-1} and 37 g clay kg^{-1} from 0.3 to 0.6 soil depth, respectively.

Tomato (*Lycopersicon esculentum* Mill.) has been the sole crop since the construction of the greenhouse in 1999. There were two cropping seasons per year, namely a winter-spring (WS) and an autumn-winter (AW) tomato crop. For the WS season, 4-week-old tomato seedlings were transplanted by hand into double rows in the middle of February, harvesting was completed in the middle of June. After a 2-month fallow period, the second (AW) crop was transplanted in early August and the final harvest was taken the following January. The tomato vines were removed from the greenhouse after each final harvest to reduce the infecting of disease carry over into the next crop.

Treatments

From February 2004 the treatments included (1) **CK**, control treatment, where neither organic manure nor mineral fertilizer N was applied. The N from irrigation water was the important source of N input in the CK treatment, ranging from 25 to 177 kg

N ha^{-1} season^{-1} with an average of 102 kg N ha^{-1} season^{-1}. The large variation in N input from irrigation water across different growing season was due to different amount of irrigation water and different concentration of NO$_3^-$N in irrigation water. (2) **MN**, organic manure treatment, only organic manure was broadcast as a basal fertilizer with no mineral N fertilizer applied. Organic manure was bought from different poultry farms every growing season and C and N content of chicken manure were different across different growing season. From 2004WS to 2006WS only dry chicken manure was used; and the application rates of dry chicken manure were 8, 11, 8, 11 and 5 t ha^{-1} season^{-1}, with an average of 271 kg N ha^{-1} season^{-1}. During the autumn-winter season in 2006 (2006AW) and onward, additional chopped wheat straw (1-5 cm long) was added to the soil together with dry chicken manure. From 2006AW to 2009AW the application rates of dry chicken manure were 8, 8, 8, 8, 8, 10 and 8 t ha^{-1} season^{-1}; and the rates of wheat straw were 2, 2.5, 4, 2, 4, 4 and 4 t ha^{-1} season^{-1}, with an average of 22 kg N ha^{-1} season^{-1}. The treatment was also labeled as "MN+S" instead of "MN". (3) **CN**, conventional N treatment, organic manure was applied as in the MN treatment (except in 2006AW). N fertilizer was applied as a side-dressing at a rate of 120 kg N ha^{-1} on 4-6 occasions based on local farmers normal management practice which depended on the weather conditions, tomato cultivar and growth stage. The average mineral N fertilizer input was 635 kg N ha^{-1} season^{-1}. From 2006AW, the CN treatment plot was split into CN and CN+S sub-treatments with plot sizes of 21.8 m^2 and 32.8 m^2, respectively. For the CN sub plot only dry chicken manure was applied; and for the CN+S treatment both wheat straw and dry chicken manure were incorporated. The chicken manure and wheat straw were applied at the same rates and timings as in the MN treatment. The mineral N fertilizer application was the same as for the original CN plot. (4) **RN**, reduced N treatment, chicken manure was applied at the same rate as in the MN and CN treatment, and mineral N fertilizer was applied as a side-dressing based on an N target value and soil mineral N content in the root zone (0-30 cm soil layer) for different growth stages from 2004WS to 2007WS. The equation used was as follows:

$$\text{Recommended fertilizer N} = \text{N target value} - NO_3^- - \text{N in the top 0.3m of the soil profile before the recommendation} - NO_3^- - \text{N from irrigation water} \quad (1)$$

N target value is calculated from crop N uptake, the necessary soil N$_{min}$ residue and soil net mineralization [30], which reflects the synchronization of crop N requirement and soil N supply. Here the initial N target values were 300 kg N ha^{-1} for the side-dressing at each stage of fruit cluster development in 2004. From 2005 the target values from transplanting to the third cluster growth stage were changed to 250 and 200 kg N ha^{-1} for the fourth cluster to the end of harvest in the WS seasons. In the AW season the N target value from transplanting to the fourth cluster growth stage was changed to 200 kg N ha^{-1}, with 250 kg N ha^{-1} for the fifth and sixth cluster growth stages [24]. Considering crop N uptake in different growth period and soil N supply, N from irrigation water, only 2-4 side-dressing events were applied according to the differences between N target values and soil N$_{min}$ content in the root zone. From 2007AW the N recommendation was simplified

based on the experiences of preceding years. Three or four side-dressing events with an interval of 7–10 days at a rate of 50 kg N ha^{-1} were required in April and October. The more detail introduction of optimized N management was reported by Ren et al [24]. Compared with the CN treatment, 71.3% of mineral N fertilizer was reduced without influencing fruit yield, with an average of 182 kg N ha^{-1} season^{-1}. From 2006AW, the RN treatment plot was split into RN and RN+S sub-treatments with the same straw amendments and plot sizes as in the CN treatment. During 2006AW and 2007WS growing season, the mineral N fertilizer application rates of RN and RN+S treatment were determined based on the N target values and soil N$_{min}$ content in the root zone before side-dressing. There were litter differences on mineral N fertilizer application rates between RN and RN+S treatment. Since 2007AW, the mineral N fertilizer application was the same for RN and RN+S treatment.

All treatments were set up in a randomized block design with three replicates. Plots were separated for each other by plastic film at a depth of 30 cm. The exogenous C and N inputs are shown in Table 1. The average C input from chicken manure was 2779 kg C ha^{-1} season^{-1}, with a range of 1114 to 5445 kg C ha^{-1} season^{-1}. Beginning in 2006AW, an average of 1119 kg wheat straw C ha^{-1} season^{-1} was supplied, and the total C input was as high as 3483 kg C ha^{-1} season^{-1}. The N sources included mineral N fertilizer, organic manure and irrigation water. In the CK treatment, the N from irrigation water was the only source of N input, with an average of 102 kg N ha^{-1} season^{-1}. The average exogenous N inputs for the MN, RN and CN treatments were 354, 527 and 984 kg N ha^{-1} season^{-1}, respectively.

Urea was the main mineral N fertilizer. All plots received P$_2$O$_5$ as calcium monophosphate (12% P$_2$O$_5$) and K$_2$O as potassium sulfate (50% K$_2$O) during each growing season. The average input during the past 12 growing seasons were 350 kg P$_2$O$_5$ ha^{-1} and 563 kg K$_2$O ha^{-1} per season.

Soil sampling and analysis

Soil organic carbon and total N concentrations. Three soil cores (3.5 cm in diameter) were taken from each plot to a depth of 0.6 m and subdivided into 0-0.1, 0.1-0.3 and 0.3-0.6 m increments on April 18, 2004, and January 20, 2010. Fresh soil cores were taken to the lab immediately, mixed thoroughly to provide a composite sample from each plot, sieved through a 2 mm mesh and then air-dried and stored in plastic bottles. After removing any carbonates with 1 M HCl, the SOC and total N concentrations were determined using a C and N analyzer (vario MACRO CN, Elementar, Germany).

At the same time, soil bulk density in different soil layers was measured by the cutting ring method. In 2004, three samples were collected from the whole greenhouse and the average soil bulk densities were 1420 kg m^{-3}, 1450 kg m^{-3} and 1480 kg m^{-3} for the 0-10 cm, 10-30 cm and 30-60 cm soil layers, respectively. The bulk density of each plot monitored in 2010 is shown in Table 2 and no significant differences were observed among the different treatments.

Soil organic carbon fraction. Changes in the quantity and quality of the SOM pool are generally difficult to detect in the short term following agricultural management. Labile and recalcitrant SOM separated by different methods provide a more sensitive indicator for evaluating the effect of different management strategies on SOM dynamics [31-33]. Here, soil organic carbon fractionation procedures were carried out as described by Blair et al [31]. Air-dried soil samples containing approximately 15 mg of C were oxidized with 25 mL of 333 mmol L^{-1} KMnO$_4$ for 1 h at 25°C on a shaker at 180 rpm. The samples were then

Table 1. Exogenous C and N inputs in the greenhouse tomato production system in Shouguang, northern China (kg ha^{-1} season^{-1}).

Growing season	C¹						N¹					
	CK	MN(MN+S)³	RN	RN+S	CN	CN+S	CK	MN(MN+S)	RN	RN+S	CN	CN+S
2004WS²	0	2860	2860	-	2860	-	56	316	644	-	1186	-
2004AW²	0	5445	5445	-	5445	-	118	478	638	-	1198	-
2005WS	0	3476	3476	-	3476	-	54	370	497	-	1000	-
2005AW	0	3902	3902	-	3902	-	177	435	636	-	1155	-
2006WS	0	1114	1114	-	1114	-	165	327	465	-	927	-
2006AW	0	4095	3423	4095	5135	5807	133	456	651	671	1200	1212
2007WS	0	2258	1698	2258	1698	2258	144	308	509	486	838	848
2007AW	0	4806	3406	4806	3406	4806	82	417	492	517	872	897
2008WS	0	2110	1606	2110	1606	2110	25	180	321	330	771	780
2008AW	0	3052	1606	3052	1606	3052	39	202	382	402	782	802
2009WS	0	4205	2592	4205	2592	4205	135	441	605	641	1105	1141
2009AW	0	3857	2214	3857	2214	3857	99	322	489	523	769	802
Average	0	3432	2779	3483	2921	3728	102	354	527	510	984	926

¹C input from chicken manure and wheat straw; N input from mineral fertilizer, organic manure and irrigation water;

²WS: winter-spring growing season, AW: autumn-winter growing season;

³Since 2006AW chopped wheat straw and dry chicken manure was broadcast as a basal fertilizer and the treatment was labeled as "MN+S" instead of "MN";

Table 2. Soil bulk density in the soil profile in Jan. 2010 in a year-round greenhouse tomato planting system in Shouguang, northern China (kg m^{-3}).

Treatment	Soil layer (cm)		
	0-10	**10-30**	**30-60**
CK	1333±41	1545±70	1589±63
MN(MN+S)[1]	1296±118	1548±79	1586±56
RN	1301±82	1502±30	1460±42
RN+S	1285±35	1508±126	1546±87
CN	1227±213	1584±129	1512±127
CN+S	1283±43	1493±114	1542±132
Average	1288	1530	1539

[1]Since 2006AW chopped wheat straw and dry chicken manure was broadcast as a basal fertilizer and the treatment was labeled as "MN+S" instead of "MN";

centrifuged, diluted and spectrophotometrically measured at 565 nm. The oxidized carbon was considered labile C, and the remainder represented the non-labile C.

Apparent N losses. The nutrient balance is often used to estimate potential environmental risks [34-35]. The apparent N losses were calculated according to Equation (2), as described by Ren et al [24], as follows:

$$N_{loss} = N_{min\ initial} + N_{manure} + N_{fert} + N_{irri} - N_{crop} - N_{min\ harvest} \quad (2)$$

where N_{loss} = apparent N loss, $N_{min\ initial}$ = soil N_{min} content at 0-0.6 m before transplanting, N_{manure} = total N input from organic manure, N_{fert} = N from mineral fertilizer, N_{irri} = NO_3^--N from irrigation water, N_{crop} = total N uptake by tomato aboveground parts, and $N_{min\ harvest}$ = soil N_{min} content at 0-0.6 m at the end of the harvest.

The parameters were calculated as follows:

Three soil cores (3.5 cm in diameter) were collected from each plot at a depth of 0.6 m and then subdivided into 0-0.3 m and 0.3-0.6 m increments before transplanting and at the end of the harvest for each season. Fresh soil cores were mixed thoroughly to give a composite sample from each plot and then passed through a 2 mm sieve. Next, 12 g subsamples were weighed and extracted by shaking with 100 mL of 1 mol L^{-1} KCl for 1 h. The extract was stored at -18°C until an analysis of the NO_3^--N and NH_4^+-N concentrations could be carried out with a continuous flow analyzer (TRAACS Model 2000). The water content of the soil samples was also gravimetrically determined to calculate the soil N_{min} (NO_3^--N+NH_4^+-N) content on a dry matter basis. The soil bulk density was used to convert the mineral N in mg per kg of soil to kg per hectare.

The input of N from irrigation water was determined by recording the amounts applied, and water samples were collected during each irrigation event over the entire growing season. The samples were stored frozen until NO_3^--N and NH_4^+-N analysis.

Plant samples were collected from each plot at the end of the harvest; divided into leaves, fruit, stems and roots; and weighed before and after drying at 70°C for 48 h. The dried shoots were ground before determining the total N, which was conducted using a modified Kjeldahl method with salicylic acid. N uptake was calculated as the product of dry matter and total N concentration in different parts.

Data analysis. Analysis of variance (*ANOVA*) was used to determine the significance of treatment effects based on a randomized complete block design. Multiple comparisons of mean values were performed using either Duncan's multiple range tests or Fisher's protected least significant difference (*LSD*) test at the 0.05 level of probability. Statistical analysis was performed using version 6.12 of the SAS software package (SAS Institute Inc., Cary, NC).

Results

Soil organic carbon and total N concentrations

Soil organic carbon (SOC) and total soil N (TN) concentrations in soil profiles to a depth of 60 cm are shown in Figure 2. The highest SOC concentration was seen in the 0-10 cm layer, below which it decreased for all treatments. In April of 2004, the SOC of the soil profile was similar across all N treatments, except there were some variations in the 30-60 cm layer associating with uneven soil fertility. After 6 years, SOC in the CK treatment was significantly lower than that of the other treatments in the 0-10 cm soil layer. However, there were no significant differences among the other treatments. In the 10-30 cm layer, SOC differed significantly between the RN and CN treatments. In comparison to the initial value, SOC decreased significantly in all CK treatment layers after 6 years of cultivation, but the only significant change in SOC for the CN treatment was an increase in the 0-10 cm layer.

Patterns in total N in the soil profile followed a similar pattern to that of the SOC (Figure 1). In January 2010, soil TN in the CK treatment was significantly lower than that of the other treatments in the 0-10 cm layer. In addition, no significant differences were observed among the MN, RN and CN treatments. The TN values did not show significant differences across N treatments in the 10-60 cm layer in 2010.

Soil organic carbon and total N pool

Total SOC and TN pool in the profile above 60 cm were 46.6-55.1 t C ha^{-1} and 6.8-8.3 t N ha^{-1}, respectively (Table 3). After 6 years of cultivation, approximately 13.1 t C ha^{-1} and 2.2 t C ha^{-1} were lost from the CK and MN treatments, respectively. The decreased bulk density was mainly attributed to the decreased SOC in the MN treatment (Table 2). The SOC pool increased to 0.8 t C ha^{-1} and 1.7 t C ha^{-1} in response to the RN and CN treatments, respectively. For the N pool, the only reduction occurred in the CK treatment. Approximately 0.63 t N ha^{-1} was lost over the last 6 years. Average accumulation rates of SOC and TN in the 0-60 cm layer were -2.17, -0.37, 0.14 and 0.28 t C ha^{-1} a^{-1} and -0.10, 0.14, 0.00 and 0.06 t N ha^{-1} a^{-1} for the CK, MN,

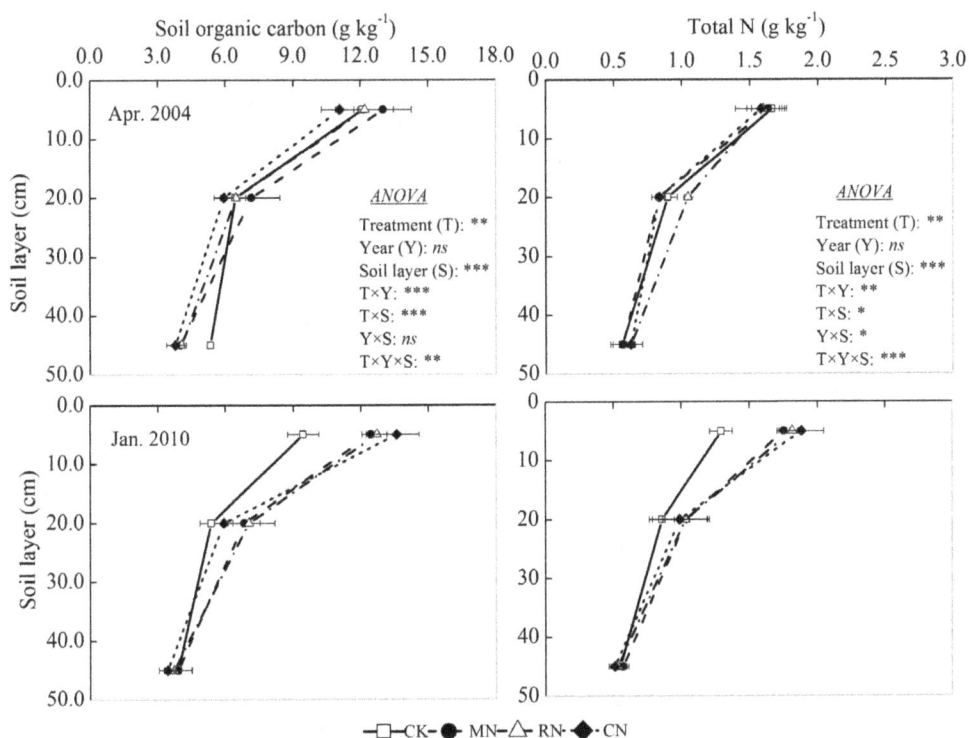

Figure 2. The distribution of soil organic carbon and total N concentrations in the soil profile with different N treatments in the year-round greenhouse tomato planting system in Shouguang, northern China. Note: *, ** and *** indicate significant differences at $P<$ 0.05, $P<0.01$ and $P<0.001$, ns denotes no significant difference.

RN and CN treatments, respectively. The addition of straw over 4 years made little difference to SOC and TN concentration

Soil organic carbon fractions

Figure 3 shows the distribution of labile and non-labile carbon in different soil layers. In April 2004, the labile C concentrations were similar across all N treatments in the same layer. However, in January 2010, the soil labile C concentration in the CK treatment was significantly lower than that of the other treatments in the 0-30 cm layer. Compared with the initial values in 2004, the soil labile C concentration in the CK treatment did not decrease significantly according to paired-sample T test; however, labile C increased significantly in the RN and CN treatments for the 0-10 cm layer.

Changes in the concentration of the non-labile C fraction in the soil profile were similar to those of the SOC. In January 2010, the soil non-labile C concentration in the 0-10 cm layer increased as the N application increased. When compared with April 2004, it decreased significantly in the CK treatment. Nevertheless, it increased significantly in the CN treatment. For all other treatments, no significant changes were observed in the soil profile.

Apparent N losses

Figure 4 shows the annual apparent N losses from 2004 to 2009 as obtained estimated from the nitrogen balance. For the CN treatment, the average annual mean apparent N losses were as high as 1529 kg N ha^{-1} a^{-1}, accounting for 77% of the annual exogenous N input. A significant decrease in the annual apparent N losses of 633 kg N ha^{-1} a^{-1} occurred in the RN treatment because the rate of N fertilization had been reduced to less than one-third of that in the CN treatment. For the CK treatment, soil

N and nitrate in the irrigation water were the major sources of N, which were less input than N removed by plant uptake, resulting in a negative N balance. These changes could indirectly explain the decreased TN in the CK treatment. Although the straw treatments led to the addition of C, the apparent N losses from treatments with and without straw amendments did not differ significantly. The N surpluses were 490 and 1285 kg N ha^{-1} a^{-1} N for RN+S and CN+S, respectively.

Interactions between C and N in soil

Figure 5 shows the relationship between the changes in SOC and TN concentration after 6 years of cultivation, as well as annual apparent N losses and average N inputs. The SOC and TN concentrations increased in response to N addition but showed a negative response to excessive N application (Figure 5a, b). The total N in the soil did not increase linearly with the increase in applied mineral N, which might be explained by changes in the SOC and its fractions (Figure 5c). The minor changes in SOC in response to the current type and amount of organic manure application, limited N storage in soil and N was close to saturation. With the increase in fertilizer N application, the apparent N losses linearly increased and conventional N fertilizer management caused the highest apparent N losses (Figure 5d).

Discussion

Effect of N fertilizer on SOC

Greater root exudates and more crop residues in response to mineral N fertilizer application were the dominant reasons why N fertilizer application improved the SOC [6]. In this experiment, although 71.3% of mineral N fertilizer was cut down in the RN treatment compared with the CN treatment, there were no

Table 3. Distribution of the soil organic C and N pools in the soil profile of a greenhouse tomato production system in Shouguang, northern China.

		SOC pool (t ha⁻¹)						N pool (t ha⁻¹)					
		CK	MN (MN+S)²	RN	RN+S	CN	CN+S	CK	MN (MN+S)²	RN	RN+S	CN	CN+S
Apr-04	0-10 cm	17.2±0.2	18.5±1.8	17.3±1.8	-	15.7±1.2	-	2.36±0.11	2.32±0.17	2.27±0.17	-	2.25±0.27	-
	10-30 cm	18.8±1.3	20.8±3.7	18.8±2.0	-	17.3±1.3	-	2.61±0.21	2.41±0.15	3.04±0.07	-	2.43±0.05	-
	30-60 cm	23.7±0.6	17.8±0.5	18.1±0.9	-	16.8±1.8	-	2.50±0.39	2.52±0.33	2.79±0.37	-	2.78±0.14	-
Jan-10	0-10 cm	12.2±0.9	16.0±0.5	16.4±0.6	17.9±0.9	17.5±1.3	17.6±1.0	1.66±0.11	2.26±0.06	2.34±0.05	2.42±0.09	2.4±0.21	2.54±0.12
	10-30 cm	16.5±1.5	20.9±2.2	21.6±3.5	17.2±1.2	18.2±1.3	18.6±2.1	2.63±0.28	3.17±0.52	3.15±0.50	2.70±0.23	3.02±0.61	2.95±0.18
	30-60 cm	18.0±2.8	18.0±2.9	17.1±0.1	17.2±1.5	15.8±1.8	18.7±0.7	2.55±0.22	2.64±0.20	2.62±0.14	2.68±0.06	2.35±0.19	2.86±0.41
Δ(0-60 cm)¹		-13.1±4.5	-2.2±1.0	0.8±6.1	-	1.7±3.6	-	-0.63±0.50	0.82±0.27	0.01±0.74	-	0.34±0.47	-
Accumulation rate (t ha⁻¹ a⁻¹)		-2.17	-0.37	0.14	-	0.28	-	-0.10	0.14	0.00	-	0.06	-

^{1}Δ(0-60 cm) = SOC (N) pool$_{2004}$ - SOC (N) pool$_{2010}$;
^{2}Since 2006AW chopped wheat straw and dry chicken manure was broadcast as a basal fertilizer and the treatment was labeled as "MN+S" instead of "MN";

significant differences on fruit yields and plant biomass between the CN and RN treatment [24]; root exudates were presumed to be similar between these two treatments. Besides, in our greenhouse system crop residues were removed from the greenhouse at harvest because of the risks of disease carryover, so organic manure was the major C supplement. The C input from organic manure was the same for the CN and RN treatments in the same growing season (Table 1). Moreover, no significant changes in the root C/N ratio (data not shown), soil microbial community [36] and soil organic matter fractions (Figure 3) between the RN and CN treatments were found. Thus, no significant differences in SOC pool between the CN and the RN treatment were observed in this experiment, which was similar to other work [14], [17]. Apparently, N was not limiting factor in this cropping system and reduced mineral N input would not alter SOC and TN accumulation. All these findings indicate that when organic manure is used, optimizing N fertilizer input over a continuous 6-year period would not affect the SOC or TN contents in the greenhouse vegetable cropping system; but it was helpful in reducing the environmental risks without influencing fruit yields (Figure 4).

Effect of organic manure on SOC

Organic manure application brought lots of N, with the average of 252 kg N ha⁻¹. In contrast to the CK treatment, significant increments on fruit yields were achieved; yet there were no differences on fruit yields and plant biomass between MN+S treatment and RN, CN treatment in most growing seasons [24], demonstrating that N from organic manure was important N source for crop growth and excessive mineral N fertilizer application was wasteful without considering N from organic manure. In addition, Organic manure application is considered to be a consistent method for maintaining soil fertility over the long-term [10], [37]. In this unique production system, the major source of organic carbon is organic manure. If the input of organic manure is excluded, only 44-146 kg C ha⁻¹ season⁻¹ from residual roots is incorporated [38]. If no organic manure is applied, SOC concentration, especially for stable organic carbon, will decline significantly (Figure 2, 3). These results are similar to those of another long-term greenhouse tomato experiment [39].

Before the start of this experiment, about 210 t ha⁻¹ chicken manure had been applied across the whole greenhouse during 1999-2003; and the high organic manure application might build higher soil organic carbon pool in a short time. Therefore, in contrast to the treatments with organic manure application, there was the lack of a big difference on SOC in the CK treatment for the next 6 years. As well, no observable changes in soil organic C was found over 6 years of successive chicken manure applications. Whether it implied the soil organic carbon pool in our study was saturated? Although the coarse-textured soil has lower capacity for C and N stabilization, the saturated soil organic carbon pool in 0-30cm soil layer could be high to 75.6-96.8 t ha⁻¹ according to Hassink [40] and Six's C-saturation model [41]. These values were higher than it reported in our study, indicating that soil organic carbon could be improved with optimum management. Quantity and quality of input organic manure significantly influenced soil organic C dynamics. In contrast to farmers' normal manure application, the application rate of chicken manure in the experiment was lower, ranging from 8 t ha⁻¹ season⁻¹ to 12 t ha⁻¹ season⁻¹. Whether SOC content would be enhanced with the organic manure application rate increase? Indeed, high rates of organic manure application were conducive to enhanced SOC and SOC fractions [42-43]. Ge et al [39] demonstrated that it would take 10-15y with 75 t ha⁻¹ a⁻¹ of horse manure to increase

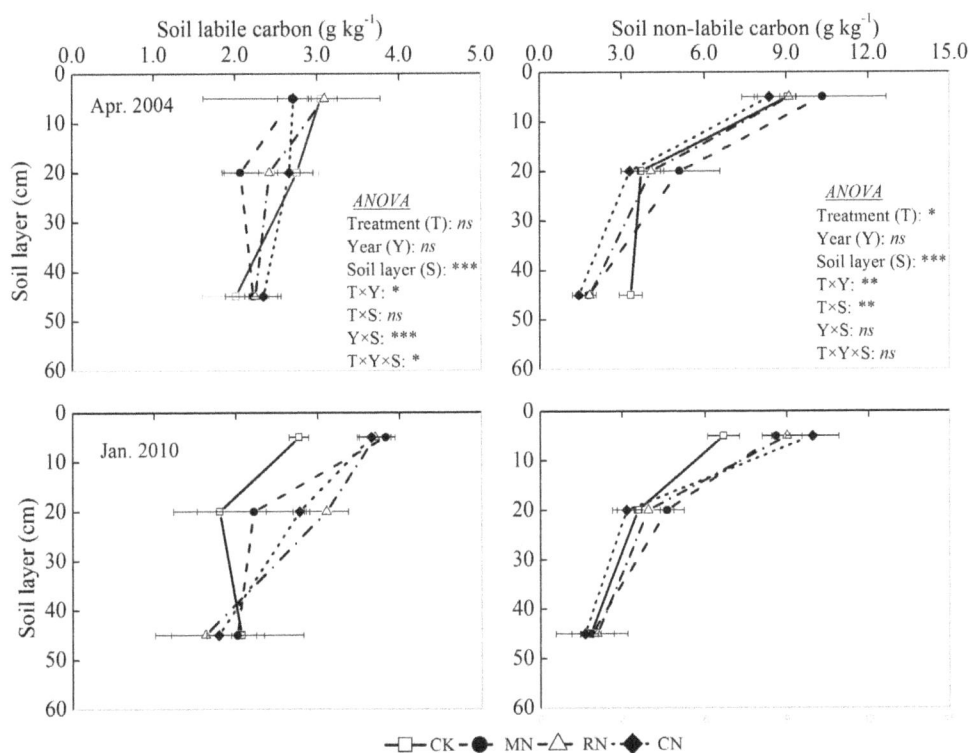

Figure 3. The distribution of the soil labile and non-labile carbon concentrations in the soil profile with different N treatments in the year-round greenhouse tomato planting system in Shouguang, northern China. Note: Note: *, ** and *** indicate significant differences at $P<0.05$, $P<0.01$ and $P<0.001$, ns denotes no significant difference.

the soil organic matter content from 24 g kg^{-1} to 30-40 g kg^{-1} in the greenhouse vegetable soil. However, environmental pressures associating with excessive application of organic manure were serious [44]; and in Europe the applications of organic manure are restricted to not exceed 170 kg N ha^{-1} y^{-1} by the legislation. In our experiment N input from organic manure averaged 252 kg N ha^{-1} season^{-1}, with 335 kg N ha^{-1} y^{-1} apparent N loss. Obviously it will not be an effective way to enhance SOC relying on the increments of organic manure application rate in greenhouse vegetable cropping system. Changing the type of organic manure input might be an important way to heighten SOC. For chicken and pig manure with high proportions of water-soluble C and easily biodegradable organic compounds, approximately 45-62% of C is evolved as CO_2-C within 30 days [45-46]. Plaza et al [47] reported a significant decrease in the total organic C in soils amended with pig manure slurry. However, manure with a greater ratio of C/N, or high content of recalcitrant C could reduce mineralization of bio-labile compounds, thereby enhancing soil organic matter [48-49]. Long term field experiments showed that the benefits in SOC content were higher from application of rice straw and compost than that from pig manure [42], [50].

To improve soil organic carbon content, wheat straw was added from in 2006AW. However, no significant increase in the SOC or the SOC fraction was observed after 4 years of cultivation. This result was similar to that of Antil's study [51], which found that the SOC in bulk soil decreased or was not affected by a slurry + straw treatment in both fallow and cropped plots, even after 28 and 38 yrs. According to the mechanisms of real and apparent priming effects [52], it is assumed that when chicken manure and wheat straw are applied together, microorganisms may first use the C from chicken manure to activate the microbial community;

however, when the easily decomposed organic carbon is consumed, activated microorganisms will use the wheat straw carbon. Most of added straw carbon is then utilized by microorganisms, perhaps explaining why there was no effect on SOC when chicken manure and wheat straw were incorporated together. The short duration might be another important reason why no differences were seen. Additional long-term studies should be conducted to determine if the application of a mixture of wheat straw and chicken manure is an effective method of enhancing soil organic carbon in the greenhouse vegetable cropping systems over the long run. Overall, developing an optimum organic manure management system to enhance soil fertility is now one of the important issues in greenhouse vegetable cropping systems in China. In comparison to increasing the application rate, shifting the type of organic manure from pig and chicken manure to manure with a wider ratio of C/N or high content of recalcitrant C may be more practical for enhancing soil organic matter in greenhouse vegetable cropping system.

Effect of soil organic carbon content on the fate of N

Similar to the results from 2004 to 2007 [24], approximately 77% of the exogenous N input was surplus in the CN treatment. With the exception of immobilized N in soil clays, N leaching [53] and N_2O emissions [21] were the major N loss processes in these vegetable cropping systems. Furthermore, warm and moist conditions with sufficient available nitrate and labile carbon can lead to denitrification loss, which might also be an important N loss process [54]. In any case, an excessive N surplus would lead to high potentially N losses. Therefore, most studies on the prevention of N losses have focused on reducing the N input and on irrigation strategies [24], [55] and catch crops [56]. In the

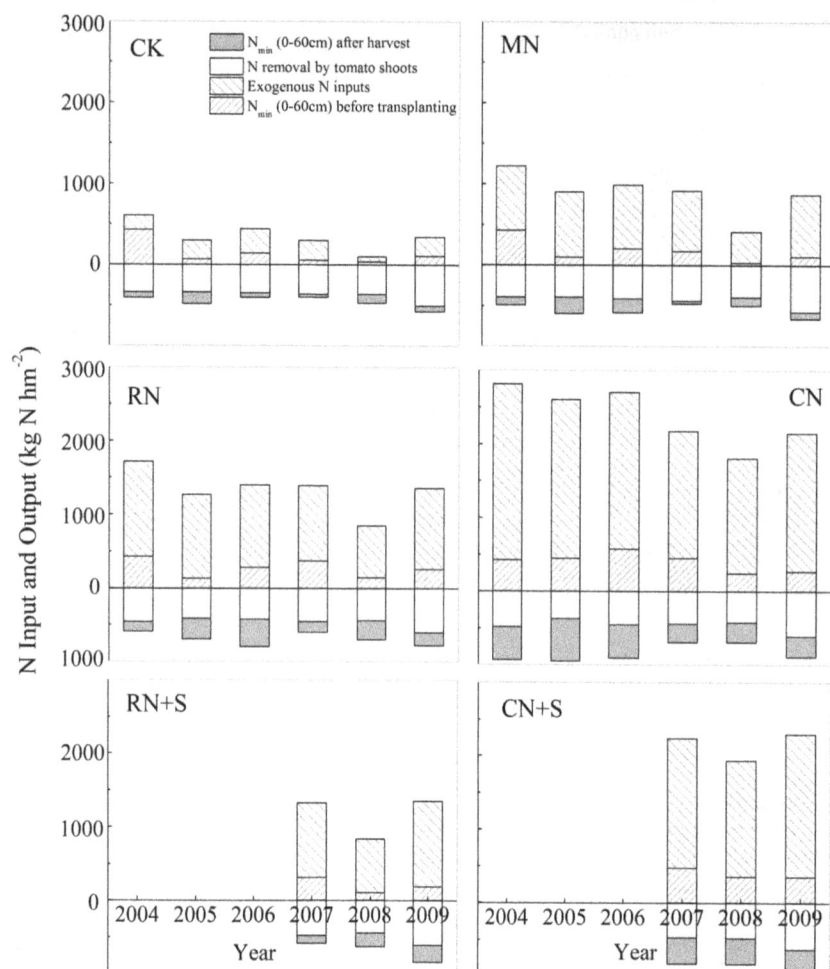

Figure 4. Apparent N loss with different treatments in the year-round greenhouse tomato planting system from 2004 to 2009 in Shouguang, northern China.

experiment, the rate of N fertilization in the RN treatment had been reduced to less than one-third of that in the CN treatment and apparent N losses were also decreased. Other than that, SOC plays an important role in regulating soil N turnover and the improvement of SOC is beneficial to increase potential rates of N immobilization and reduce N losses [57]. Yang et al [58] showed that the rate of absolute N change increased linearly with changes in the size of the C pool change and organic N capital was determined by long-term carbon sequestration. A similar tendency was observed in the present study (Figure 3c). In our experiment, SOC concentration was not increased even though high amounts of organic manure were applied during a 6 year period; Changes in organic N were similarly limited. Moreover, the mineralization rate of total organic N might be greater than the retention of exogenous N to soil organic N pool because of the great amount of organic manure applied within several years before the experiment started. Thus more exogenous N was lost to the environment. The low soil C/N ratio indirectly revealed that there was insufficient C in the cropping system would limiting N immobilization by soil microorganisms [2], [26] and leading to high N losses [28]. Therefore, improving soil organic matter content and enhancing potential rates of soil N immobilization according to optimal organic manure management was crucial, as well reducing N

fertilizer input, to lower N losses in greenhouse vegetable cropping system.

Conclusion

Organic manure represents a major organic C source in conventional greenhouse vegetable cropping systems in China. Without additions of organic manure, SOC, particularly stable C is likely to decline. However, no significant increment in SOC was observed with the addition of high amounts of low C/N ratio organic manure or plant residues. Shifting the type of manure from chicken manure to manures with a wider ratio of C/N, or high content of recalcitrant may be more effective in enhancing soil fertility in greenhouse vegetable production. On the basis of organic manure application, optimized N fertilizer inputs according to root-zone N management did not influence the accumulation of SOC and TN in soil; but beneficial in reducing apparent N losses.

The SOC concentration was a dominant limiting factor for soil total N enhancement. Given the current type and quantity of organic manure application, the SOC concentration was unchanged and most applied N was in excess and was lost to the environment. Therefore, integrating nutrient management, including optimized N fertilizer input, as well as enhanced the soil

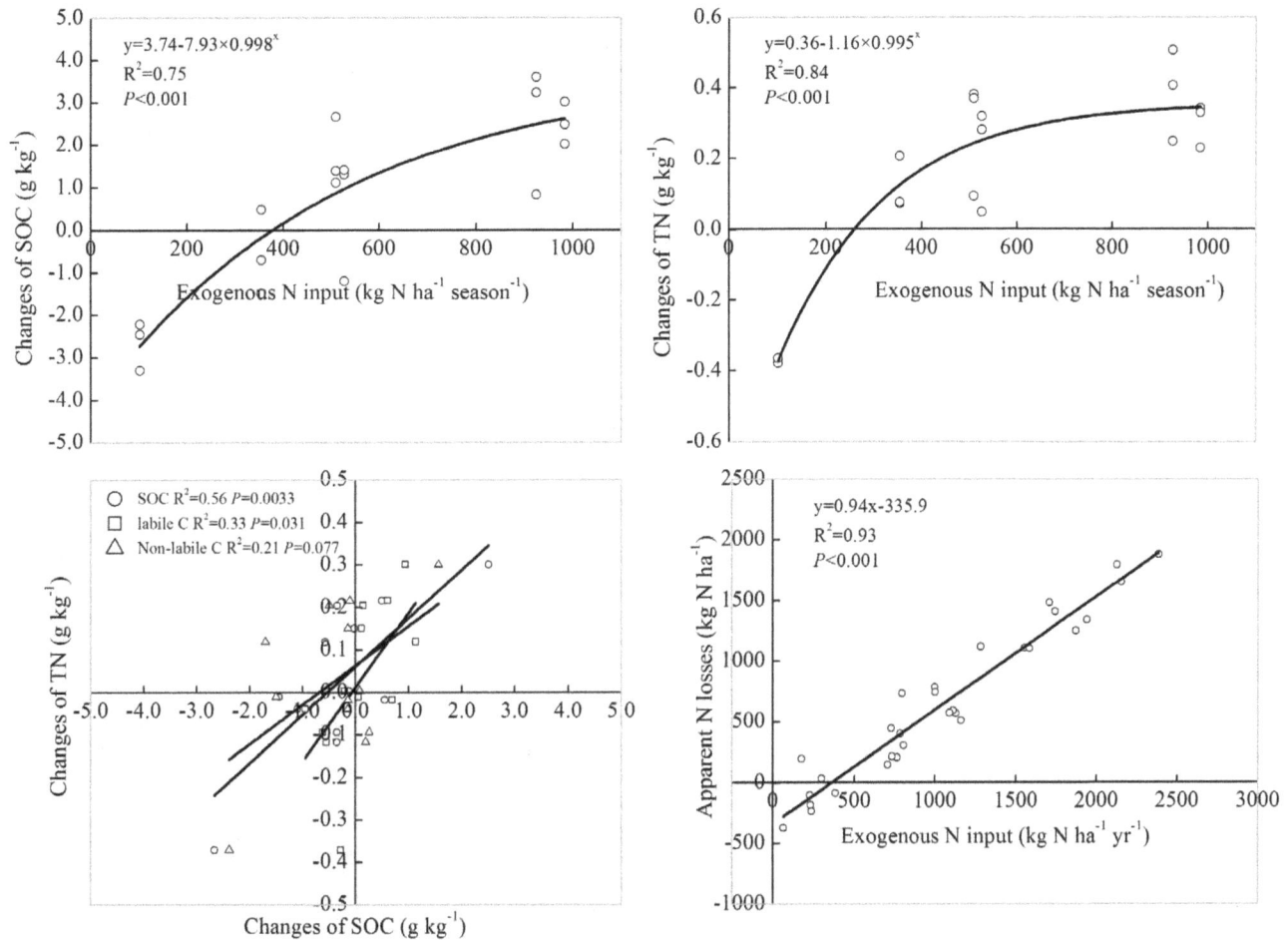

Figure 5. The relationship between the changes in SOC and TN concentration after 6 years of cultivation, as well as annual apparent N losses and the average N inputs in the year-round greenhouse tomato planting system in Shouguang, northern China.

organic matter content, should be considered to maintain soil fertility and productivity, minimize potential N losses and achieve sustainable development in greenhouse vegetable cropping systems.

Author Contributions

Conceived and designed the experiments: QC JW FZ. Performed the experiments: TR. Analyzed the data: TR SL. Contributed reagents/materials/analysis tools: QC JW. Wrote the paper: TR QC.

References

1. Bird JA, van Kessel C, Horwath WR (2002) Nitrogen dynamics in humic fractions under alternative straw management in temperate rice. Soil Sci Soc Am J 66: 478-488.
2. Accoe F, Boeckx P, Busschaert J, Hofman G, Van Cleemput O (2004) Gross N transformation rates and net N mineralization rates related to the C and N contents of soil organic matter fractions in grassland soils of different age. Soil Biol Biochem 36: 2075-2087.
3. Paré MC, Bedard-Haughn A (2013) Soil organic matter quality influences mineralization and GHG emissions in cryosols: a field-based study of sub-to high Arctic. Global Change Biol 19: 1126-1140.
4. Compton JE, Boone RD (2002) Soil nitrogen transformation and the role of light fraction organic matter in forest soils. Soil Biol Biochem 34: 933-943.
5. Neff JC, Townsend AR, Gleixner G, Lehman SJ, Turnbull JT, et al. (2002) Variable effects of nitrogen additions on the stability and turnover of soil carbon. Nature 419: 915-917.
6. Christopher SF, Lal R (2007) Nitrogen management affects carbon sequestration in North American Cropland soils. Crit Rev Plant Sci 26: 45-64.
7. Swanston C, Homann PS, Caldwell BA, Myrold DD, Ganio L, et al. (2004) Long-term effects of elevated nitrogen on forest soil organic matter stability. Biogeochemistry 70: 227-250.
8. Hagedorn F, Spinnler D, Siegwolf R (2003) Increased N deposition retards mineralization of old soil organic matter. Soil Biol Biochem 35: 1683-1692.
9. Malhi SS, Harapiak JT, Nyborg M, Gill KS, Monreal CM, et al. (2003) Total and light fraction organic C in a thin Black Chernozemic grassland soil as

affected by 27 annual application of six rates of fertilizer N. Nutr Cycl Agroecosyst 66: 33-41.
10. Blair N, Faulkner RD, Till AR, Poulton PR (2006) Long-term management impacts on soil C, N and physical fertility part I: Broadbalk experiment. Soil Till Res 91: 30-38.
11. Jagadamma S, Lal R, Hoeft RG, Nafziger ED, Adee EA (2007) Nitrogen fertilization and cropping systems effects on soil organic carbon and total nitrogen pools under chisel-plow tillage in Illinois. Soil Till Res 95: 348-356.
12. Lemke RL, VandenBygaart AJ, Campbell CA, Lafond GP, Grant B (2010) Crop residue removal and fertilizer N: effects on soil organic carbon in a long-term crop rotation experiment on a Udic Boroll. Agr Ecosyst Environ 135: 42-51.
13. Mack MC, Schuur EAG, Bret-Harte MS, Shaver GR, Chapin III FS (2004) Ecosystem carbon storage in arctic tundra reduced by long-term nutrient fertilization. Nature 431: 440-443.
14. Dolan MS, Clapp CE, Allmaras RR, Baker JM, Molina JAE (2006) Soil organic carbon and nitrogen in a Minnesoota soil as related to tillage, residue and nitrogen management. Soil Till Res 89: 221-231.
15. Fonte SJ, Yeboah E, Ofori P, Quansah GW, Vanlauwe B, et al. (2009) Fertilizer and residue quality effects on organic matter stabilization in soil aggregates. Soil Biol Biochem 73: 961-966.
16. Liu LL, Greaver TL (2010) A global perspective on belowground carbon dynamics under nitrogen enrichment. Ecol Lett 13: 819-828.

17. Lu M, Zhou XH, Luo YQ, Yang YH, Fang CM, et al. (2011) Minor stimulation of soil carbon storage by nitrogen addition: A meta-analysis. Agr Ecosys Environ 140: 234-244.

18. Chen Q, Zhang XS, Zhang HY, Christie P, Li XL (2004) Evaluation of current fertilizer practice and soil fertility in vegetable production in the Beijing region. Nutr Cycl Agroecosyst 69: 51-58.

19. He FF, Chen Q, Jiang RF, Chen XP, Zhang FS (2007) Yield and nitrogen balance of greenhouse tomato (*Lycopersicum esculentum* Mill.) with conventional and site-specific nitrogen management in Northern China. Nutr Cycl Agroecosyst 77: 1-14.

20. Song XZ, Zhao CX, Wang XL, Li J (2009) Study of nitrate leaching nitrogen fate under intensive vegetable production pattern in northern China. CR Biol 332: 385-392.

21. He FF, Jiang RF, Chen Q, Zhang FS, Su F (2009) Nitrous oxide emissions from an intensively managed greenhouse vegetable cropping system in Northern China. Environ Pollut 157(5): 1666-1672.

22. Mao XS, Liu MY, Wang XY, Liu CM, Hou ZM, et al. (2003) Effects of deficit irrigation on yield and water use of greenhouse grown cucumber in the North China Plain. Agr Water Manage 61(3): 219-228.

23. Guo RY, Li XL, Christie P, Chen Q, Jiang RF, et al. (2008) Influence of root zone nitrogen management and a summer catch crop on cucumber yield and soil mineral nitrogen dynamics in intensive production systems. Plant Soil 313: 55-70.

24. Ren T, Christie P, Wang JG, Chen Q, Zhang FS (2010) Root zone soil nitrogen management to maintain high tomato yields and minimum nitrogen losses to the environment. Sci Hortic 125: 25-33.

25. Lei BK, Fan MS, Chen Q, Six J, Zhang FS (2010) Conversion of wheat-maize to vegetable cropping systems changes soil organic matter characteristics. Soil Sci Soc Am J 74(4): 1320-1326.

26. Degens BP, Schipper LA, Sparling GP, Vojvodic-Vukovic M (2000) Decreased in organic C reserves in soils can reduce the catabolic diversity of soil microbial communities. Soil Biol Biochem 32: 189-196.

27. Cookson WR, Abaye DA, Marschner P, Murphy DV, Stockdale EA, et al. (2005) The contribution of soil organic matter fractions to carbon and nitrogen mineralization and microbial community size and structure. Soil Biol Biochem 37: 1726-1737.

28. Gundersen P, Callesen I, de Vries W (1998) Nitrate leaching in forest ecosystems is related to forest floor C/N ratios. Environ Pollut 102: 403-407.

29. Zeng XB, Bai LY, Li LF, Su SM (2009) The status and changes of organic matter, nitrogen, phosphorus and potassium under different soil using styles of Shouguang of Shangdong Province. Acta Ecol Sin 29(7): 3737-3746 (in Chinese).

30. Feller C, Fink M (2002) N_{min} target values for field vegetables. Acta Hort 571: 195-201.

31. Blair GJ, Lefroy RDB, Lisle L (1995) Soil carbon fractions based on their degree of oxidation, and the development of a carbon management index for agricultural systems. Aust J Agric Res 46: 1459–1466.

32. Six J, Paustian K, Elliott ET, Combrink C (2000) Soil structure and organic matter: I. Distribution of aggregate-size classes and aggregate-associated carbon. Soil Sci Soc Am J 64: 681-689.

33. McLauchlan KK, Hobbie S (2004) Comparison of labile soil organic matter fractionation techniques. Soil Sci Soc Am J 68: 1616-1625.

34. Öborn I, Edwards AC, Witter E, Oenema O, Ivarsson K, et al. (2003) Element balances as a tool for sustainable nutrient management: a critical appraisal of their merits and limitations within an agronomic and environmental context. Eur J Agron 20: 211-225.

35. Sieling K, Kage H (2006) N balance as an indicator of N leaching in an oilseed rape-winter wheat-winter barley rotation. Agric Ecosyst Environ 15: 261-269.

36. Zhao XC (2011) Effects of fertilization and crop rotation on soil microbial community structure of greenhouse tomato. Master Thesis, China Agricultural University, Beijing China (In Chinese).

37. Edmeades DC (2003) The long-term effects of manures and fertilizers on soil productivity and quality: a review. Nutr Cycl Agroecosyst 66: 165-180.

38. Lei BK, Chen Q, Fan MS, Zhang FS, Gan YD (2008) Changes of soil carbon and nitrogen in Shouguang intensive vegetable production fields and their impacts on soil properties. Plant Nutr Fert Sci 14(5): 914-922 (in Chinese).

39. Ge XG, Zhang EP, Zhang X, Wang XX, Gao H (2004) Studies on changes of filed-vegetable ecosystem under long-term fixed fertilizer experiment (I) changes of soil organic matter. Acta Hortic Sin 31(1): 34-38 (in Chinese).

40. Hassink J (1997) The capacity of soils to preserve organic C and N by their association with clay and silt particles. Plant Soil 191: 77-87.

41. Six J, Conant RT, Paul EA, Paustian K (2002) Stabilization mechanisms of soil organic matter: Implications for C-saturation of soils. Plant Soil 241: 155-176.

42. Liu J, Schulz H, Brandl S, Miehtke H, Huwe B, et al. (2012) Short-term effect of biochar and compost on soil fertility and water status of a Dystric Cambisol in NE Germany under field conditions. J Plant Nutr Soil Sci 175: 698-707.

43. Wang XJ, Jia ZK, Liang LY, Han QF, Ding RX, et al. (2012) Effects of organic manure application on dry land soil organic matter and water stable aggregates. Chin J Appl Ecol 23(1): 159-165 (in Chinese).

44. Ju XT, Kou CL, Zhang FS, Christie P (2006) Nitrogen balance and groundwater nitrate contamination: comparison among three intensive cropping systems on the North China Plain. Environ Pollut 143: 117-125.

45. Ajwa HA, Tabatabai MA (1994) Decomposition of different organic materials in soils. Biol Fert Soils 18: 175-182.

46. Cayuela ML, Velthof GL, Mondini C, Sinicco T, van Groenigen JW (2010) Nitrous oxide and carbon dioxide emissions during initial decomposition of animal by-products applied as fertilizers to soils. Geoderma 157: 235-242.

47. Plaza C, Garcia-Gil JC, Polo A (2005) Effects of pig slurry application on soil chemical properties under semiarid conditions. Agrochimica 49: 87-92.

48. Piccolo A, Spaccini R, Nieder R, Richter J (2004) Sequestration of a biologically labile organic carbon in soils by humified organic matter. Climatic Change 67: 329-343.

49. Adani F, Genevini P, Ricca G, Tambone F, Montoneri E (2007) Modification of soil humic matter after 4 years of compost application. Waste Manage 27: 319-324.

50. Li ZP, Liu M, Wu XC, Han FX, Zhang TL (2010) Effects of long-term chemical fertilization and organic amendments on dynamics of soil organic C and total N in paddy soil derived from barren land in subtropical China. Soil Till Res 106: 268-274.

51. Antil RS, Gerzabek MH, Haberhauer G, Eder G (2005) Long-term effects of cropped vs. fallow and fertilizer amendments on soil organic matter I. organic carbon. J Plant Nutr Soil Sci 168: 108-116.

52. Blagodatskaya E, Kuzyakov Y (2008) Mechanisms of real and apparent priming effects and their dependence on soil microbial biomass and community structure: critical review. Biol Fert Soils 45:115-131.

53. Lin Y (2010) Solute transportation and soil H^+ production budgets in a greenhouse vegetable production system. Master Thesis, China Agricultural University, Beijing China (In Chinese).

54. Ryden JC, Lund JL (1980) Nature and extent of directly measured denitrification losses from some irrigated vegetable crop production units. Soil Sci Soc Am J 44:505-511.

55. Zotareli L, Scholberg JM, Dukes MD, Mu~noz-Carpena R, Icerman J (2009) Tomato yield, biomass accumulation, root distribution and irrigation water use efficiency on a sandy soil, as affected by nitrogen rate and irrigation scheduling. Agr Water Manage 96: 23-34.

56. Constantin J, Mary B, Laurent F, Aubrion G, Fontaine A, et al. (2010) Effects of catch crops, no till and reduced nitrogen fertilization on nitrogen leaching and balance in three long-term experiments. Agr Ecosyst Environ 135: 268-278.

57. Schimel DS (1986) Carbon and nitrogen turnover in adjacent grassland and cropland ecosystems. Biogeochemistry 2: 345-357.

58. Yang YH, Luo YQ, Finzi AC (2011) Carbon and nitrogen dynamics during forest stand development: a global synthesis. New Phytol 190(4): 977-989.

5

Impact of Transgenic Wheat with *wheat yellow mosaic virus* Resistance on Microbial Community Diversity and Enzyme Activity in Rhizosphere Soil

Jirong Wu[1,2,3❾], Mingzheng Yu[1,2,3❾], Jianhong Xu[1,2,3], Juan Du[1,2,3], Fang Ji[1,2,3], Fei Dong[1,2,3], Xinhai Li[4*], Jianrong Shi[1,2,3*]

1 Institute of Food Safety and Detection, Jiangsu Academy of Agricultural Sciences, Nanjing, China, 2 Key Lab of Food Quality and Safety of Jiangsu Province—State Key Laboratory Breeding Base, Nanjing, China, 3 Jiangsu Center for GMO evaluation and detection, Nanjing, China, 4 Institute of Crop Sciences, Chinese Academy of Agricultural Sciences, Beijing, China

Abstract

The transgenic wheat line N12-1 containing the *WYMV-Nib8* gene was obtained previously through particle bombardment, and it can effectively control the wheat yellow mosaic virus (WYMV) disease transmitted by *Polymyxa graminis* at turngreen stage. Due to insertion of an exogenous gene, the transcriptome of wheat may be altered and affect root exudates. Thus, it is important to investigate the potential environmental risk of transgenic wheat before commercial release because of potential undesirable ecological side effects. Our 2-year study at two different experimental locations was performed to analyze the impact of transgenic wheat N12-1 on bacterial and fungal community diversity in rhizosphere soil using polymerase chain reaction-denaturing gel gradient electrophoresis (PCR-DGGE) at four growth stages (seeding stage, turngreen stage, grain-filling stage, and maturing stage). We also explored the activities of urease, sucrase and dehydrogenase in rhizosphere soil. The results showed that there was little difference in bacterial and fungal community diversity in rhizosphere soil between N12-1 and its recipient Y158 by comparing Shannon's, Simpson's diversity index and evenness (except at one or two growth stages). Regarding enzyme activity, only one significant difference was found during the maturing stage at Xinxiang in 2011 for dehydrogenase. Significant growth stage variation was observed during 2 years at two experimental locations for both soil microbial community diversity and enzyme activity. Analysis of bands from the gel for fungal community diversity showed that the majority of fungi were uncultured. The results of this study suggested that virus-resistant transgenic wheat had no adverse impact on microbial community diversity and enzyme activity in rhizosphere soil during 2 continuous years at two different experimental locations. This study provides a theoretical basis for environmental impact monitoring of transgenic wheat when the introduced gene is derived from a virus.

Editor: Newton C M Gomes, University of Aveiro, Portugal

Funding: This work was supported by the National Special Transgenic Project (2014ZX08011-003), Natural Science Foundation of Jiangsu Province, China (BK20130721) and Jiangsu Agriculture Science and Technology Innovation Fund [cx(11)4064]. The funders had no role in study design, data collection and analysis, decision to publish, or preparation of the manuscript.

Competing Interests: The authors have declared that no competing interests exist.

* E-mail: lixinhai@caas.cn (XL); shiji@jaas.ac.cn (JS)

❾ These authors contributed equally to this work.

Introduction

Since the first successful genetically engineered (GE) plant was reported in 1983 [1], the planting area of transgenic crops has increased rapidly [2]. The global area cultivated commercially with transgenic crops has increased from 1.7 million ha in 1996 to 170.3 million ha in 2012 [3]. With the continued release and use of transgenic crops, there is a growing concern about their impact on the biota and soil microbial processes, such as nutrient cycling, and the potential risk of gene transfer from transgenic crops to indigenous soil microbes [4–5]. The microbes in rhizosphere soil play an important role in plant growth and development [6–7]. Transgenic crops planted in soil will inevitably interact with microorganisms such as bacteria, fungi, and actinomycetes [8–10]. Thus, transgenic crops may affect soil microbial population structure and quantity [11–13]. Additionally, root exudates have marked effects on soil microbial diversity and spatial distribution

[14–15]. At this time, most studies of environmental risk assessment focused on transgenic Bt crops such as transgenic cotton, rice and maize containing the *Bt* gene [16–18]; these studies provided basic methods for environmental risk assessment for other crops.

Enzymes in the rhizosphere soil derived from animal, plant roots and soil microbial cell secretion and decomposition of residues are an important component of the soil ecosystem [19]. They play an important role in soil biochemical processes and directly affect soil fertility [19]. Urease is associated with nitrogen transformation in the soil, while sucrase is associated with soil organic matter, nitrogen and phosphorus contents, and dehydrogenase is associated with the redox ability of the soil [19]. Previous studies showed that transgenic plants might affect enzyme activities in rhizosphere soil [11,20–21]. Therefore, it is important to investigate the impact of transgenic crops on rhizosphere soil enzyme activity when performing environmental safety risk assessments.

The first report of transgenic plants with virus resistance, expressing the coat protein of the *tobacco mosaic virus* (TMV) and delaying the development of disease, appeared in 1986 [22]. The same strategy was subsequently used to create resistance to a range of other viruses [23–24]. The exogenous genes of the transgenic virus-resistant crops are generally derived from the virus itself, including genes encoding coat protein and replicase [22–24]. Sequences derived from the genomes of plant viruses have been used to generate viral resistance in transgenic crop plants, but potential safety issues have been raised due to the environmental risks of transgenic plants with virus resistance, including hetero-encapsidation, virus recombination, gene flow, synergism and effects on non-target organisms [25,26].

Wheat yellow mosaic disease, caused by the wheat yellow mosaic virus (WYMV) at turngreen stage, is a serious illness affecting wheat in the middle and lower reaches of the Yangtze River region in China [27–28]. Disease-resistant variety breeding is one of the most cost-effective ways to control this disease through conventional wheat breeding. In recent years, conventional wheat breeding in combination with genetic engineering techniques has been applied to address wheat yellow mosaic disease, and some disease-resistant wheat lines have been cultivated. Using the particle bombardment method, genes from WYMV encoding replicase WYMV-Nib8 were transferred to the disease-sensitive variety Yangmai158 (Y158), and the disease-resistant transgenic wheat line named N12-1 was obtained by successive backcross with Y158 [29]. N12-1 showed stable and effective resistance to wheat yellow mosaic disease in a previous study [30].

Considering the above risks, transgenic virus-resistant wheat may affect the microbial community diversity in rhizosphere soil and change the population structure. Exogenous insertion of genes may also cause changes in the metabolic pathways of genetically modified crops and alter the composition of root exudates, resulting in changes in soil enzyme activity [31]. Thus, further studies on the impact on soil microbial community diversity and enzyme activities should be performed. In this study, environmental risk assessment of N12-1 was performed during 2 consecutive years of wheat cultivation under field conditions at two different experimental stations. The research involved primarily: (i) differences in soil microbial (bacterial and fungal) diversity in rhizosphere soil between N12-1 and Y158 using polymerase chain reaction–denaturing gradient gel electrophoresis (PCR-DGGE) and (ii) the activity of enzymes (urease, sucrase and dehydrogenase) in rhizosphere soil. In this report, we provide a theoretical basis for environmental transgenic wheat monitoring.

Materials and Methods

Ethics statement

In our study, the research samples were rhizosphere soils in the presence of transgenic and non-transgenic wheat. This presented no ethical issue.

Plant materials and field trial

Transgenic wheat line N12-1 and its recipient Yangmai158 (Y158) provided by the Chinese Academy of Agricultural Sciences (CAAS) were applied in this study. N12-1, which contains the *WYMV-Nib8* gene from wheat yellow mosaic virus, can effectively control the WYMV disease transmitted by *Polymyxa graminis* at turngreen stage. Y158 was one of the most popular varieties in the middle and lower reaches of the Yangtze River region in China.

However, it is sensitive to WYMV disease and the yield decreased significantly due to effects of this severe disease.

This study was performed at Luhe experimental station for transgenic crop, Jiangsu Academy of Agricultural Sciences (Luhe) and Xinxiang experimental station for transgenic crop, Henan Academy of Agricultural Sciences (Xinxiang). The physical and chemical properties of the soil are provided in Table 1. pH value, water content, available nitrogen, phosphorus potassium and organic matter content were determined by potentiometry method, alkali solution diffusion method, sodium bicarbonate method, ammonium acetate extraction method, potassium dichromate method, respectively [32]. The experiment was conducted in two successive growth seasons of wheat (October 2010-June 2011 and October 2011-June 2012) in the same field in which transgenic crops had never been planted. Each variety (line) had four blocks, each of which was 10×6 m. The materials were planted in a row with a row length of 6 m and row spacing of 0.3 m. Distance between plants was 3 cm within a row. Completely random design was applied to arrange the experiment performed in the field, and the wheat was subjected to conventional field management, that was 375 kg/ha of compound fertilizer ($N:P_2O_5:K_2O = 1:0.4:1$) as base fertilizer and 225 kg/ha of urea as topdressing at seedling stage.

Soil sampling

Rhizosphere soil samples were collected in both years at Luhe and Xinxiang at four growth stages [seeding stage (SS), turngreen stage (TS), grainfilling stage (GS), maturing stage (MS)]. Rhizosphere soil was defined as the soil still attached to the roots after the roots were shaken by hand. For each sampling site, five wheat plants were selected to collect rhizosphere soil and each block contains five sampling site. Rhizosphere soil from the five sampling sites per block was mixed as a composite rhizosphere soil sample. The soil samples were sieved using a 20-mesh sieve and then stored at 4°C until further use, usually within one month before DNA extraction.

Soil DNA extraction

Total community DNA was extracted from 0.5 g of rhizosphere soil using an UltraClean Soil DNA Isolation Kit (MoBio Lab, USA). DNA extraction was performed according to the manufacturer's protocol.

PCR amplification of 16S and18S rDNA fragments for DGGE analysis

The 16S rDNA fragments of bacteria were amplified by using the primer pair GC338f (5′-CGCCGCGCGCGGCGGGGCG-GGGCGGGGGCACGGGGGGGACTCCTACGGGAGGCAGC-AG-3′, the sequence underlined was the GC clamp) and 518r (5′-ATTACCGCGGCTGCTGG -3′) as described by Bakke et al. [33]. High fidelity polymerase of KOD-Plus-Neo (Toyobo, Japan) was applied to perform PCR amplification and avoid mutations in the PCR product. Briefly, the reaction mixture consisted of 1 μl of template DNA (1–5 ng), 5 μl 10×PCR Buffer, 5 μl of 2 mM dNTPs, 3 μl of 25 mM $MgSO_4$, 0.5 μl of 10 μM forward primer, 0.5 μl of 10 μM reverse primer, and 1 U of DNA polymerase, after which ddH_2O was added to a final volume of 50 μl. The thermal cycling program was performed with an initial denaturation at 94°C for 5 min, followed by 35 cycles at 95°C for 15 sec, 58°C for 15 sec, and 68°C for 30 sec before the final extension at 68°C for 10 min. Products were checked by electrophoresis in 1% (wt/vol) agarose gels followed by ethidium bromide staining.

Table 1. Main physical and chemical properties of the soil from two experiment locations before planting.

Experiment station	Physical and chemical properties					
	pH	water (%)	available nitrogen (mg/kg)	available phosphorus (mg/kg)	available potassium (mg/kg)	organic matter (%)
Luhe	5.8	20.55	110.16	90.81	857.99	1.44
Xinxiang	8.5	4.92	70.39	28.26	863.69	0.68

The 18S rDNA fragments of fungi were amplified by using the primer pair (GC-Fungi: 5′-CGCCCGCCGCGCCCCGCGCCC-GGCCCGCCGCCCCCGCCCCATTCCCCGTTACCCGTT-G-3′; NS1: 5′- GTAGTCATATGCTTGTCTC -3′, the sequence underlined was the GC clamp) as described by Das et al. [34]. The protocol for PCR amplification was similar as above. All products were purified before electrophoresis using a Cycle Pure Kit (Omega, USA).

PCR-DGGE

DGGE analysis for 16S rDNA and 18S rDNA products was performed with the DCode System (Bio-Rad, USA). Polyacrylamide gels were composed of a denaturing gradient of 50–65% (bacteria) and 30–38% (fungi) urea, 0.17% (vol/vol) TEMED, 0.047% (wt/vol) ammonium persulfate, 6% acrylamide-N,N_-methylenebisacrylamide (37.5:1) and 1×TAE. PCR products (up to 50 μl) were applied to the gel. DGGE was performed at 50 V in 1×TAE at 60°C for 12 h (bacteria) and at 50 V in 1×TAE at 60°C for 20 h (fungi), respectively. A silver staining method was used for the detection of DNA in DGGE gels.

Migration and intensity of DGGE bands were analyzed using Quantity One according to the manual. The bands that shared identical migration positions were considered to be the same species. Shannon's diversity index (H) of bacterial and fungal DGGE profiles was calculated with the following formula [35]:

$$H = -\sum_{i=1}^{S} \frac{n_i}{N} \ln \frac{n_i}{N}$$

Simpson's diversity index (D) was calculated with the following formula:

$$D = \sum_{i=1}^{S} \left[\frac{n_i}{N}\right]^2$$

Evenness (E) was calculated with the following formula:

$$E = \frac{H}{\ln S}$$

Figure 2. Shannon's index of bacterial communities at different growth stages. Error bars indicate standard errors (n = 4). Different letters above bars denote a statistically significant difference between the means of the fields. A: Luhe; B: Xinxiang. SS: seeding stage; TS: turngreen stage; GS: grainfilling stage; MS: maturing stage.

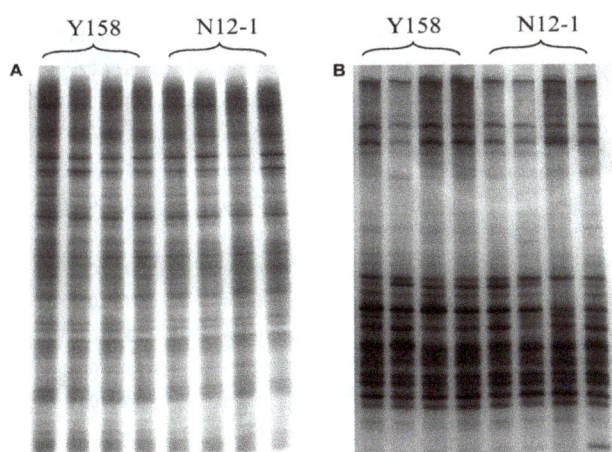

Figure 1. DGGE profiles of 16S rDNA and 18S rDNA fragments amplified from DNA extracted from rhizosphere soil of N12-1 and Y 158 at turngreen stage from Luhe experiment station in 2011. A: bacteria; B: fungus.

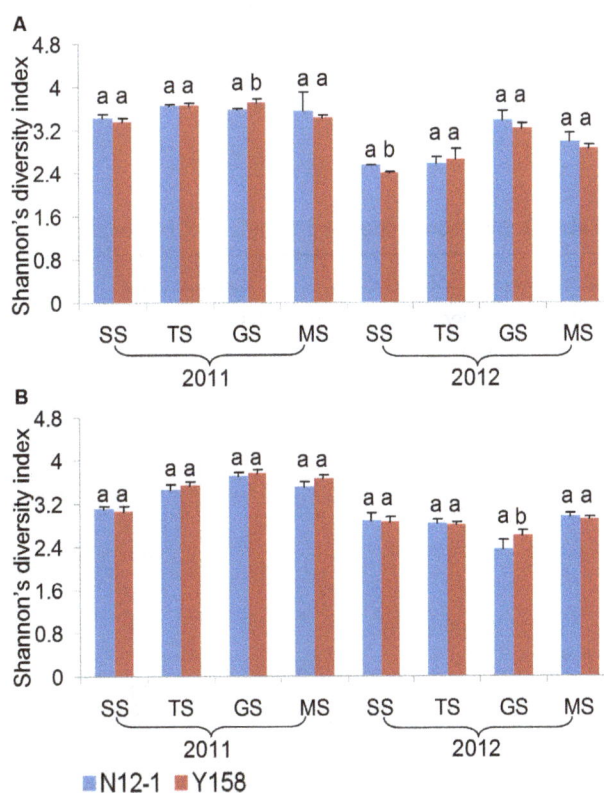

A

B

■ N12-1 ■ Y158

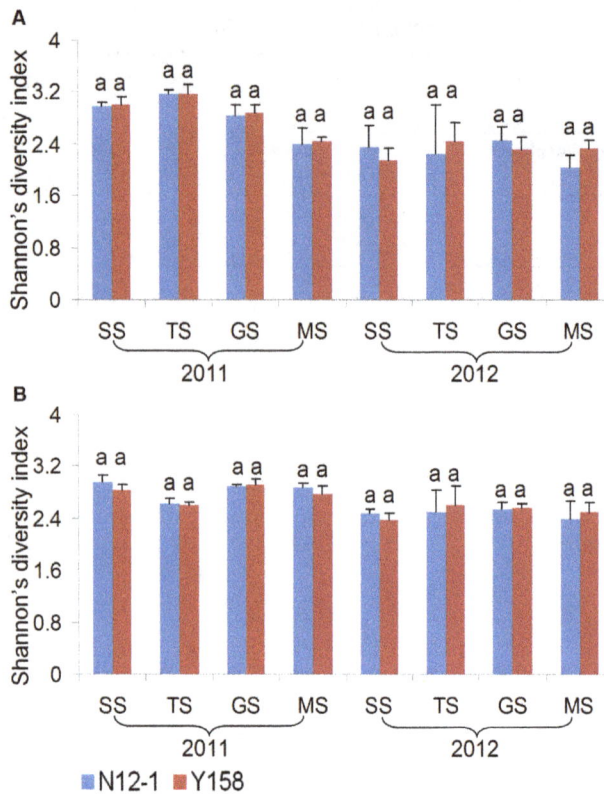

Figure 3. Shannon's index of fungi communities at different growth stages. Error bars indicate standard errors (n = 4). Different letters above bars denote a statistically significant difference between the means of the fields. A: Luhe; B: Xinxiang. SS: seeding stage; TS: turngreen stage; GS: grainfilling stage; MS: maturing stage.

n_i represented the square of individual peaks detected by Quantity One; N represented the square of all peaks in the same lane; S represented the number of bands in the same lane.

Band sequencing

Visible bands in the fungi DGGE gel were picked with sterile tips and transferred into a 200 µl tube. Sterile ddH$_2$O (50 µl) was added to the tube and the gel was pounded to pieces. The tubes with broken gels were incubated at room temperature overnight and then centrifuged at 12000 rpm for 5 min. The supernatant solution was used as the template for PCR, which was performed as described above or for fungi using primers without GC-clamps. The PCR products were purified (Omega, USA), ligated into pM19-T vector (Takara, Japan) and transformed into competent cells (*E coli* DH5α, Takara, Japan) according to the instructions of the manufactures and plated on LB solid medium with ampicillin. Positive clones were selected by PCR with primer pair of NS1 and Fungi (GC-Fungi without GC-clamps) and plasmids were extracted for sequencing (Invitrogen, Shanghai). All the sequences that have been sequenced successfully were submitted to GenBank (Accession numbers: KJ755390-KJ755404).

Enzyme activity analysis

Activities of urease, sucrose and dehydrogenase were analyzed in this study. Urease and sucrose activities in soil were assayed using the method of Guan [36]: urease activity was determined by measuring the release of NH$_3$ as mg.(g.d)$^{-1}$, and sucrose activity was determined based on 3,5-dinitrosalicylic acid colorimetry as mg.(g.d)$^{-1}$. Dehydrogenase activity was determined based on the reduction of triphenyltetrazolium chloride (TTC) to triphenylformazan (TPF), as described by Serra-Wittling et al. [37] with minor modifications, which was expressed as µg.(g.d)$^{-1}$. The data were subjected to analysis of variance, and the means and standard deviations of four replicates were calculated.

Table 2. Simpson's index and Evenness of bacterial community.

Experiment station	Growth stage	Variety(line)	Simpson's index		Evenness	
			2011	2012	2011	2012
Luhe	SS	N12-1	0.04±0.00a	0.10±0.00a	0.95±0.01a	0.88±0.01a
		Y158	0.04±0.00a	0.12±0.00b	0.95±0.01a	0.86±0.02a
	TS	N12-1	0.03±0.00a	0.09±0.01a	0.84±0.01a	0.92±0.03a
		Y158	0.03±0.00a	0.08±0.02a	0.95±0.01a	0.93±0.01a
	GS	N12-1	0.03±0.00a	0.04±0.01a	0.95±0.00a	0.93±0.01a
		Y158	0.03±0.00b	0.05±0.01a	0.96±0.01a	0.93±0.02a
	MS	N12-1	0.04±0.00a	0.06±0.02a	0.99±0.10a	0.93±0.03a
		Y158	0.04±0.00a	0.07±0.00a	0.95±0.01a	0.90±0.02a
Xinxiang	SS	N12-1	0.05±0.01a	0.06±0.01a	0.93±0.03a	0.98±0.01a
		Y158	0.05±0.01a	0.06±0.01a	0.95±0.01a	1.00±0.01a
	TS	N12-1	0.04±0.01a	0.07±0.01a	0.93±0.02a	0.93±0.02a
		Y158	0.04±0.00a	0.07±0.01a	0.93±0.00a	0.94±0.02a
	GS	N12-1	0.03±0.00a	0.11±0.02a	0.95±0.01a	0.92±0.03a
		Y158	0.03±0.00a	0.08±0.01b	0.95±0.01a	0.96±0.01a
	MS	N12-1	0.04±0.01a	0.07±0.00a	0.93±0.02a	0.90±0.01a
		Y158	0.03±0.00a	0.07±0.01a	0.96±0.01b	0.89±0.02a

SS: seeding stage; TS: turngreen stage; GS: grainfilling stage; MS: maturing stage. The alphabets after the value represented the significance level of the index.

Table 3. Simpson's index and Evenness of fungi community.

Experiment station	Growth stage	variety(line)	Simpson's index		Evenness	
			2011	2012	2011	2012
Luhe	SS	N12-1	0.06±0.01a	0.11±0.03a	0.92±0.02a	0.94±0.02a
		Y158	0.06±0.01a	0.14±0.02a	0.94±0.01a	0.91±0.03a
	TS	N12-1	0.05±0.00a	0.04±0.06a	0.92±0.00a	0.78±0.23a
		Y158	0.05±0.01a	0.11±0.05a	0.91±0.01a	0.97±0.08a
	GS	N12-1	0.07±0.01a	0.10±0.02a	0.93±0.01a	0.95±0.02a
		Y158	0.07±0.01a	0.11±0.03a	0.93±0.01a	0.93±0.02a
	MS	N12-1	0.12±0.02a	0.16±0.03a	0.88±0.03a	0.86±0.04a
		Y158	0.11±0.01a	0.12±0.02b	0.89±0.04a	0.92±0.04a
Xinxiang	SS	N12-1	0.06±0.01a	0.10±0.01a	0.93±0.03a	0.87±0.02a
		Y158	0.07±0.00b	0.10±0.01a	0.93±0.01a	0.86±0.03a
	TS	N12-1	0.08±0.01a	0.09±0.03a	0.95±0.01a	0.97±0.02a
		Y158	0.08±0.00a	0.08±0.02a	0.95±0.01a	0.97±0.00a
	GS	N12-1	0.07±0.00a	0.08±0.01a	0.94±0.02a	0.96±0.01a
		Y158	0.07±0.01a	0.08±0.01a	0.93±0.01a	0.96±0.01a
	MS	N12-1	0.07±0.01a	0.12±0.04a	0.89±0.03a	0.88±0.04a
		Y158	0.07±0.01a	0.10±0.02a	0.91±0.02a	0.89±0.03a

SS: seeding stage; TS: turngreen stage; GS: grainfilling stage; MS: maturing stage. The alphabets after the value represented the significance level of the index.

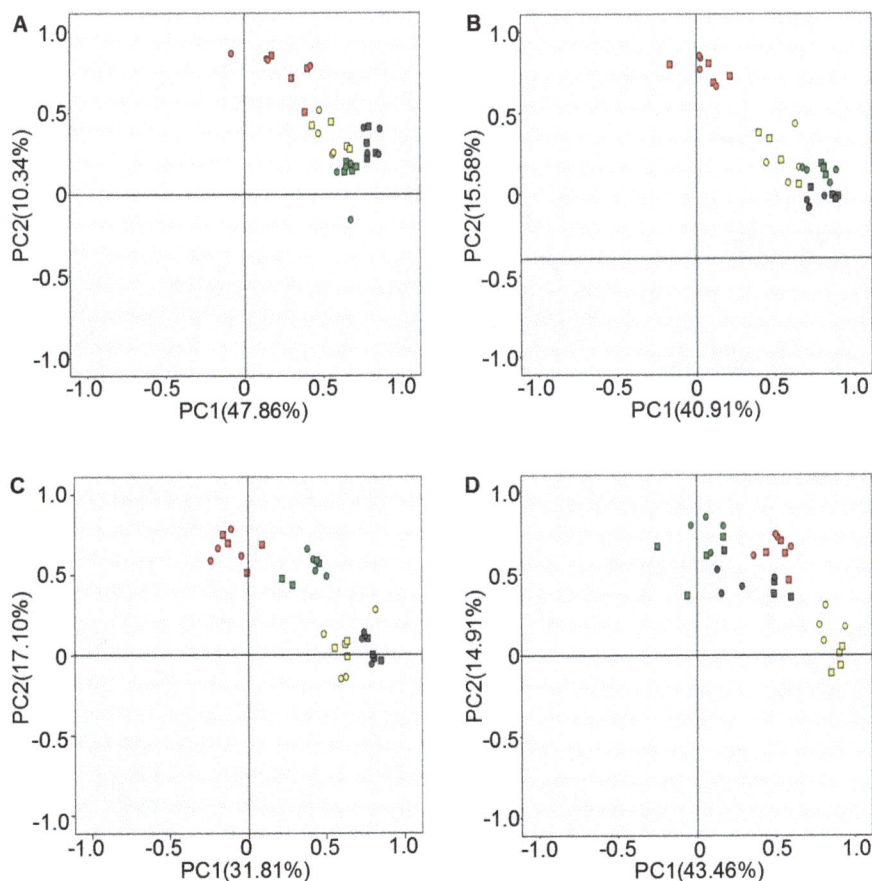

Figure 4. Principal component analysis of bacterial community diversities in rhizosphere soil. A: Luhe in 2011; B: Luhe in 2012; C: Xinxiang in 2011; D: Xinxiang in 2012. Square: N12-1; Round: Y158. Gray: seeding stage; Green: turngreen stage; Red: grainfilling stage; Yellow: maturing stage. Band position and presence (presence/absence) were used to carry out PCA analyses.

Statistical analysis

SPSS 16.0 was applied to determine whether the indices and enzyme activities differed between years, varieties and growth stages by ANOVA. PCA analyses were carried out based on band position and presence (presence/absence), and then the correlation matrix principal component analysis was performed by SPSS 16.0 [35]. Microsoft Excel 2003 was used to construct column diagrams.

Results

Impact of transgenic wheat on bacterial and fungal community diversity

One of the DGGE profiles of 16S rDNA and 18S rDNA fragments amplified from DNA extracted from rhizosphere soil was presented as figure 1. Three diversity indices (Shannon's, Simpson's, evenness) were used to analyze the bacterial and fungal DGGE profiles of the soil samples from Luhe and Xinxiang at four different growth stages in 2011 and 2012. For bacteria, the effect of wheat line on DGGE diversity indices was insignificant, except GS stage in 2011, SS in 2012 at Luhe and GS stage in 2012 at Xinxiang for Shannon's diversity index (Fig. 2). The Simpson's diversity index showed the same results as Shannon's diversity index (Table 2). For evenness, only one difference was found at the MS stage at Xinxiang in 2011 (Table 2). For fungi, the effect of wheat line on DGGE diversity indices was insignificant, except for

SS in 2011 at Xinxiang and MS in 2012 at Luhe for Simpson's index (Fig. 3; Table 3).

Principal component analysis of bacterial community diversity

Principal components analysis (PCA) using both band position and presence/absence as parameters were performed to further analyze DGGE fingerprint profiles. For experiments conducted at Luhe, the contribution rates of the two principal components were 47.86% and 10.34% in 2011 (Fig. 4A) and 40.91% and 15.58% in 2012 (Fig. 4B), respectively. Different growth stages showed a distinct separation along the principal components axes, whereas different replications of experimental materials formed a cluster at the same growth stage. This was consistent with the result of Shannon's diversity analysis. In 2011, the first principal component axis clearly separated the GS and SS stage (Fig. 4A), but separated the GS and MS stage in 2012 (Fig. 4B). The second principal component axis clearly distinguished the GS stage in 2011 (Fig. 4A) and the GS stage in 2012 (Fig. 4B).

For experiments conducted at Xinxiang, the contribution rates of the two principal components were 31.81% and 17.10% in 2011 (Fig. 4C) and 43.46% and 14.91% in 2012 (Fig. 4D), respectively. Different growth stages also showed a distinct separation along the principal components axes, whereas different replications of experimental materials clustered together at the same growth stage. In 2011, the first principal component axis

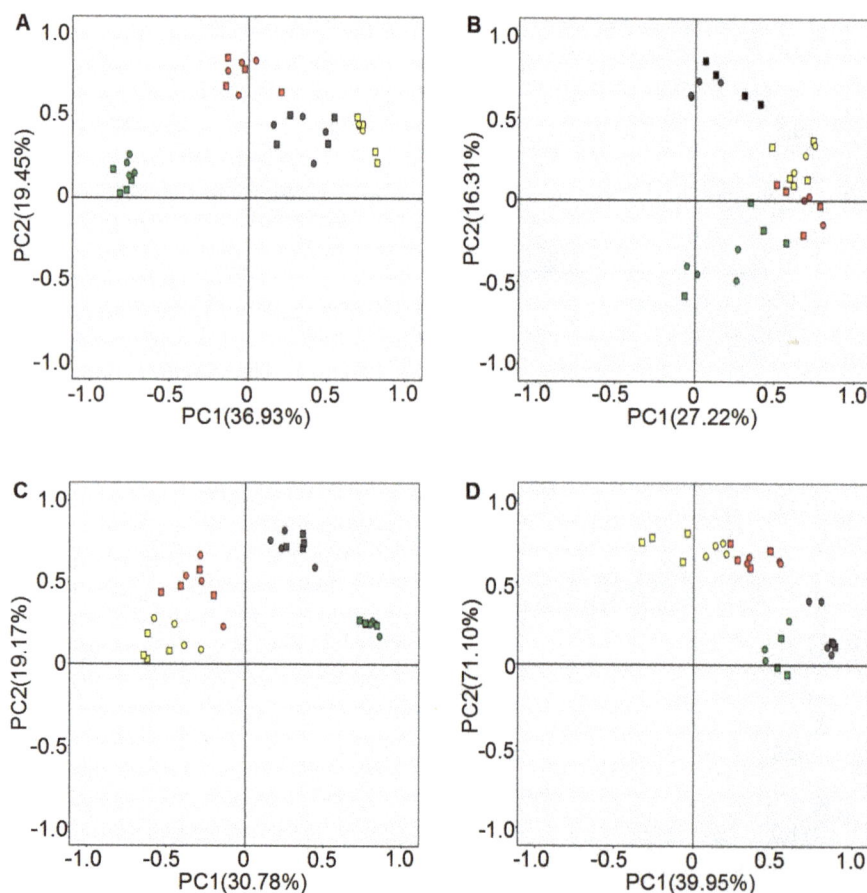

Figure 5. Principal component analysis of fungi communities diversity in rhizosphere soil. A: Luhe in 2011; B: Luhe in 2012; C: Xinxiang in 2011; D: Xinxiang in 2012. Square: N12-1; Round: Y158. Gray: seeding stage; Green: turngreen stage; Red: grainfilling stage; Yellow: maturing stage. Band position and presence (presence/absence) were used to carry out PCA analyses.

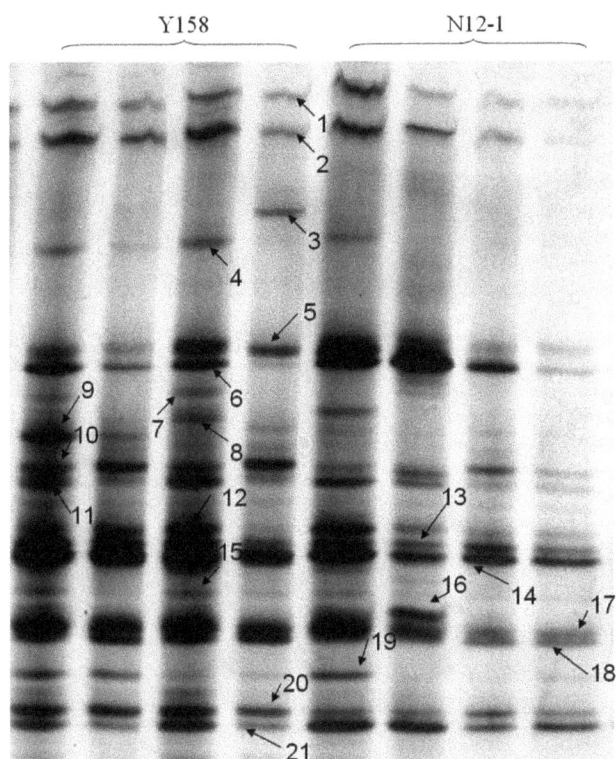

Figure 6. PCR-DGGE gel profile of fungi communities used for band sequencing. The numbers means different bands picked for sequencing.

clearly separated the four growth stages (Fig. 4C), but separated the MS stage from the other three stages in 2012 (Fig. 4D). The second principal component axis clearly distinguished the TS and GS stages from the SS and MS stages in 2011 (Fig. 4C), and distinguished the MS stage in 2012 (Fig. 4D).

These PCA analysis results showed that growth stage played an important role in bacterial community diversity, rather than the presence of transgenic and non-transgenic wheat.

Principal component analysis of fungal community diversity

For experiments at Luhe, the contribution rates of the two principal components were 36.93% and 19.45% in 2011 (Fig. 5A) and 27.22% and 16.31% in 2012 (Fig. 5B), respectively. Different sampling times showed a distinct separation along the principal components axes, whereas different replications of experiment materials formed a cluster at the same sampling time. In 2011, the first principal component axis clearly separated the four growth stages (Fig. 5A), but separated SS and TS from GS and MS in 2012 (Fig. 5B). The second principal component axis clearly distinguished the TS stage in 2011 (Fig. 5A), and the SS and TS stage in 2012 (Fig. 5B).

For experiments at Xinxiang, the contribution rates of the two principal components were 30.78% and 19.17% in 2011 (Fig. 5C) and 39.95% and 17.10% in 2012 (Fig. 5D). Different sampling times also showed a distinct separation along the principal components axes, whereas different replications of experimental materials formed a cluster at the same sampling time. In 2011, the first principal component axis clearly separated the SS and TS stages (Fig. 5C), but separated the SS and MS stages in 2012 (Fig. 5D). The second principal component axis could not clearly

Table 4. Blast results of the bands from the DGGE gels of fungl community analysis.

No. of bands	Accession No.	Blast result	identity
1	GU214699.1	*Septoria dysentericae* strain CPC 12328 18S ribosomal RNA gene	100%
2	GQ330624.1	Uncultured *Mucorales* clone PR3 4E 28 18S ribosomal RNA gene	95%
3		Cannot be amplified	
4	AJ515922.1	Uncultured soil ascomycete partial 18S rDNA gene	100%
5	EU120944.1	Uncultured *Cystofilobasidiales* (aff. Guehomyces) clone Y9 18S ribosomal RNA gene	100%
6	AJ515941.1	Uncultured soil ascomycete partial 18S rDNA gene	99%
7		Cannot be amplified	
8	AY789390.1	*Peziza varia* strain ZW-Geo94-Clark 18S small subunit ribosomal RNA gene	99%
9	FJ176814.1	*Saccobolus dilutellus* isolate AFTOL-ID 1299 18S small subunit ribosomal RNA gene	97%
10	FO181499.1	Balen uncultured eukaryote partial 18S ribosomal RNA	80%
11	AY771600.1	Polyozellus multiplex isolate AFTOL-ID 677 18S small subunit ribosomal RNA gene	99%
12	GU190186.1	*Cochliobolus* sp. Enrichment culture clone NJ-F5 18S small subunit ribosomal RNA gene	100%
13		Cannot be amplified	
14	AJ515948.1	Uncultured soil ascomycete partial 18S rDNA gene	99%
15		Cannot be amplified	
16	AJ301992.1	*Myrothecium leucotrichym* 18S RNA gene	99%
17	JX159444.1	Uncultured *Filobasidium* clone Cegs 957 18S ribosomal RNA gene	99%
18	KC171701.1	Uncultured fungus isolate DGGE gel band f10 18S ribosomal RNA gene	100%
19		Cannot be amplified	
20		Cannot be amplified	
21	EU120947.1	Uncultured *Ascobolus* clone Y12 18S ribosomal RNA gene	99%

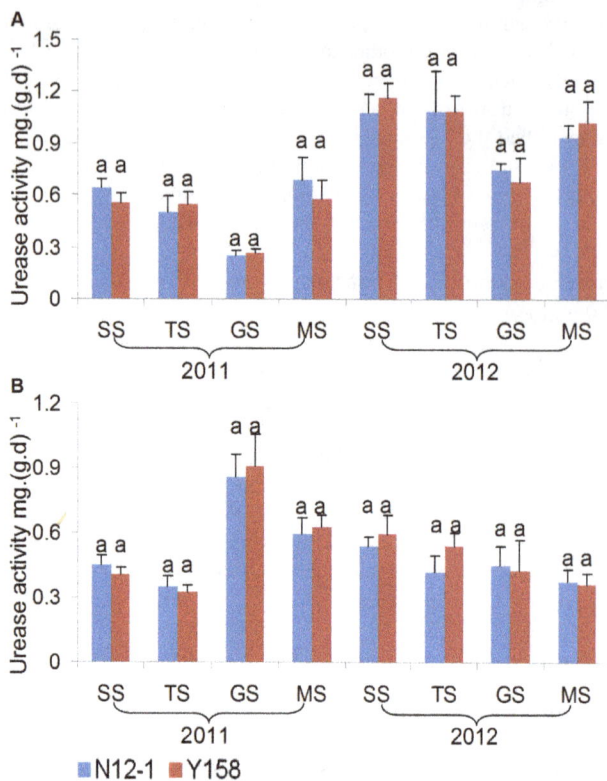

Figure 7. Urease activity in rhizosphere soil at different growth stages. Error bars indicate standard errors (n = 4). Different letters above bars denote a statistically significant difference between the means of the fields. A: Luhe; B: Xinxiang. SS: seeding stage; TS: turngreen stage; GS: grainfilling stage; MS: maturing stage.

Figure 8. Sucrase activity in rhizosphere soil at different growth stages. Error bars indicate standard errors (n = 4). Different letters above bars denote a statistically significant difference between the means of the fields. A: Luhe; B: Xinxiang. SS: seeding stage; TS: turngreen stage; GS: grainfilling stage; MS: maturing stage.

distinguish any growth stage in 2011 (Fig. 5C), but could distinguish the MS and GS stages from the SS and TS stages in 2012 (Fig. 5D).

These PCA analysis results showed that fungal communities exhibited marked diversity at different growth stages, rather than between the transgenic line and non-transgenic wheat recipient.

Band sequencing

A total of 21 visible bands from the DGGE gel of fungi from Luhe in 2011 were subjected to sequencing (Fig. 6), and 15 were sequenced successfully. Using NCBI BLAST, we found that most of the sequenced bands represented uncultured fungi. Others were partial 18S rRNA sequences of *Septoria dysentericae*, *Peziza varia*, *Saccobolus dilutellus*, *Polyozellus*, *Cochliobolus*, and *Myrothecium leucotrichym* (Table 4).

Enzyme activity analysis

Urease, sucrase, and dehydrogenase activities in rhizosphere soil were applied as indicators for environmental risk assessment of transgenic wheat N12-1 in this study.

In general, there was no consistent significant difference in the enzyme activity between soils of transgenic wheat N12-1 and its recipient Y158 within the same growth stage during the 2 years. Only one significant difference in activity was observed; for dehydrogenase at the MS stage at Xinxiang in 2011. In 2011, the dehydrogenase activity in soil of N12-1 was significantly ($p<0.05$) higher than in soil of its recipient Y158 (Figs. 7–9). Significant differences were observed between years ($p<0.01$) and among

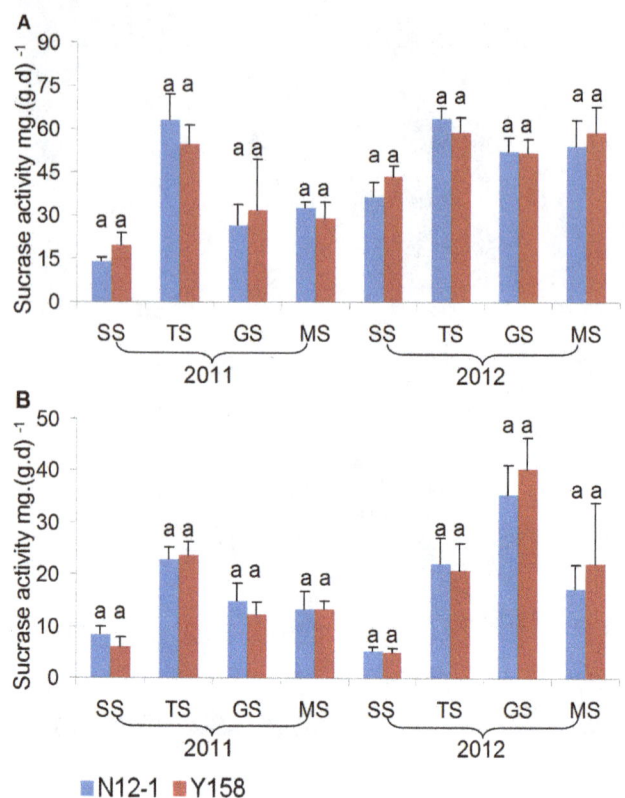

growth stages ($p<0.001$) at both Luhe and Xinxiang, with the exception of dehydrogenase among growth stages at Xinxiang ($p<0.25$) (Table 5). These results showed that N12-1 had a minor impact on soil enzyme activities.

Discussion

With the cultivation of more varieties of virus-resistant transgenic plants and large-scale planting, environmental impact monitoring after commercial release has attracted increasing attention from the scientific community and public [38,39]. In soil, there are high microbial population densities and large numbers of microbial species that interact with the plants and surrounding environment and have an effect on the function of the soil ecosystem, such as the enzyme activity and physicochemical properties.

Soil microbial analysis has been used widely to evaluate the impact of various exogenous chemical or environmental pollutants (such as herbicides, fertilizers, heavy metals, et al.) on soil fertility and crop yields [7]. Therefore, monitoring changes in soil microbial populations will increase our understanding of the potential risks of introduction of exogenous genes to soil [4,7]. In our study, two years and two locations of field research was performed to compare the impact of transgenic wheat with genes encoding replicase from WYMV on microbial population diversity in agricultural systems. One of the major outcomes was that transgenic insertion did not significantly alter bacterial or fungal population diversity at each growth stage; however, growth stage and planting year had important effects on microbial diversity.

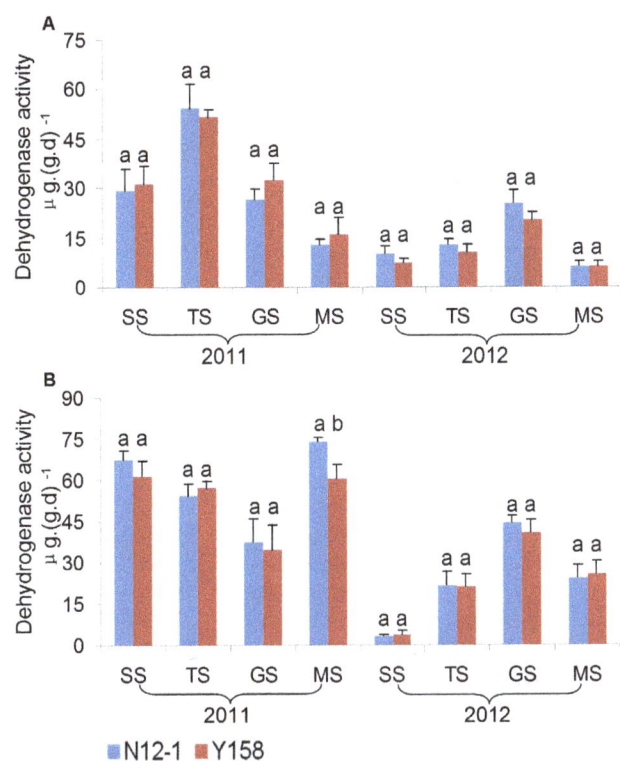

Figure 9. Dehydrogenase activity in rhizosphere soil at different growth stages. Error bars indicate standard errors (n = 4). Different letters above bars denote a statistically significant difference between the means of the fields. A: Luhe; B: Xinxiang. SS: seeding stage; TS: turngreen stage; GS: grainfilling stage; MS: maturing stage.

This result was similar to the findings of Meyer et al. [40]. In that study, the authors found that the effects of GM wheat on plant-beneficial root-colonizing microorganisms are minor and not of ecological importance. Lupwayi et al. reported that glyphosate-resistant wheat–canola rotations under low-disturbance direct seeding and conventional tillage did not affect the functional diversity of rhizosphere soil bacteria in 18 of 20 site-years [20]. The observation that certain growth stages (mainly SS and GS) showed differences between transgenic and non-transgenic wheat may be due to inconsistencies in the soil at seeding time, and at later growth stages the temperature and humidity increased rapidly. The differences between years and growth stages indicated that the diversity of bacteria and fungi might be affected by various environmental factors, such as temperature, humidity, and light. In studies of GM crops against virus, there was no significant difference between microbial communities with transgenic or non-transgenic watermelon resistant or cucumber green mottle mosaic virus (CGMMV), but significant changes in the microbial community were observed during the growing season [41]. Transgenic tomato resistant to cucumber mosaic virus (CMV) had no effect on the variation of soil microbial communities, in which soil position and environmental factors played more dominant roles [42]. Also, non-environmental factor, such as root exudates, may also play an important role for diversity changes of bacteria and fungi between years and growth stages [14–15]. Fang et al. thought that bacterial communities differed due to changes in root exudates quantity and composition by developing corn plant, which select different bacterial groups during root colonization [43]; Donegan et al. have speculated that the reason for the different in the communities of genetically modified plants is due to differences in the root exudates patterns of these plants [44]. However, Wei et al. reported the opposite result [45]. In a transgenic alfalfa study performed using the cultivation-dependent plating method, statistically significant differences in densities of rhizospheric bacteria between transgenic and non-transgenic

Table 5. Generalized Linear Mixed Model results for overall effects on enzyme activity.

Location	Enzyme	Effect	F Value	p Value
Luhe	Urease	Years	282.21	0.00
		Growth stage	36.89	0.00
		Variety (Line)	0.01	0.91
	Sucrase	Years	74.66	0.00
		Growth stage	33.54	0.00
		Variety (Line)	0.05	0.83
	Dehydrogenase	Years	82.20	0.00
		Growth stage	16.23	0.00
		Variety (Line)	0.03	0.87
Xinxiang	Urease	Years	6.38	0.01
		Growth stage	7.97	0.00
		Variety (Line)	0.103	0.75
	Sucrase	Years	11.45	0.00
		Growth stage	23.67	0.00
		Variety (Line)	0.10	0.75
	Dehydrogenase	Years	73.88	0.00
		Growth stage	1.40	0.25
		Variety (Line)	0.11	0.74

alfalfa clones were observed for ammonifying bacteria, cellulolytic bacteria, rhizobial bacteria, denitrifying bacteria and *Azotobacter* spp. [46]. These results indicated that transgenic crops containing a viral gene conferring resistance to viral disease had little effect on soil microbial diversity (excluding a small number of studies) compared with non-transgenic crops. Transgenic wheat also had no adverse effects on soil biological indicators, such as *Folsomia candida* [47] and earthworm [48]. Duc et al. found that GM wheat with race-specific antifungal resistance against powdery mildew (Pm3b), and two with nonspecific antifungal resistance, had no impact on the soil fauna community (mites, springtails, annelids, and diptera). However, sampling date and location significantly influenced the soil fauna community and decomposition processes [49].

Soil enzymes in the soil nutrient cycle and energy transfer play an important role in soil ecology, and are derived mainly from soil microbial populations. Many studies have used soil enzymes as indicators of soil microbial activity and fertility [50–52]. In our study, urease, sucrase and dehydrogenase were used as indicators of the impact of transgenic wheat on soil quality. The results showed no significant difference in enzyme activity in rhizosphere soil between transgenic and non-transgenic wheat at each growth stage at two locations in 2 years, excluding dehydrogenase during the maturing stage at Xinxiang in 2011. In other studies of transgenic crops, there was no consistent significant difference in soil enzymes between transgenic and non-transgenic plants, but there were differences among seasons and crop varieties [53–54]. These results are consistent with our study. In other studies, some enzymes showed significant differences between transgenic and non-transgenic plants [11,20–21]. There have been no previous studies of soil enzyme activities of transgenic wheat. Thus, our results should be confirmed in future studies and at more experimental locations. Additionally, other types of transgenic wheat, such as insect-resistant and stress-tolerant varieties, should be used to perform risk assessments.

Due to the complexity of DGGE profiles, several bands can be difficult to identify visually, and different bands represent different microbes. These issues make it difficult to compare varieties. Thus, the combination of DGGE and cloning sequencing methods is often used to investigate the impact of transgenic plants on microorganisms in rhizosphere soil [55,56]. In our study, most of the bands from fungi DGGE gels represented uncultured fungal taxa. This is in agreement with the fact that only ~1% of microbes in soil can be artificially cultured and identified [57].

With the development of sequencing technology, the way we study microbial communities has been changed. Traditionally, the study of genes from natural environments included cloning DNA into a vector, inserting that vector into a host, screening, and Sanger sequencing. Sequence-by-synthesis methods provide faster, cheaper, and simpler methods for (meta)genome sequence that bypass the PCR amplification bias, cloning bias and labor-intensive Sanger method [58]. Currently, massively parallel high-throughput pyrosequencing methods can process hundreds of thousands of sequences simultaneously [59]. Fierer et al. used metagenomic and small subunit rRNA analyses to study the genetic diversity of bacteria, archaea, fungi, and viruses in soil [60]; Uroz et al. used functional assays and metagenomic analyses to reveal difference between the microbial communities [61]. Li et al. analyzed the impact on bacterial community in midguts of the asian corn borer larvae by transgenic *Trichoderma* strain overexpressing a heterologous *chit42* gene with chitin-binding domain by using 16s rRNA library. All above studies have used the next generation sequencing technology [62]. Now, this technology is being adopted to study the microbial community in rhizosphere soil of transgenic plants gradually [62].

In conclusion, our study has produced weak evidence for the effect of virus-resistant transgenic wheat on soil microbial community diversity and enzyme activities. The community structure was markedly affected by natural variations in the environment related to wheat growth stage and planting year. Little difference was observed in bacterial and fungal communities in the presence of the wild-type Y158 or the transgenic line N12-1. This requires further investigation using extended field observations involving more varieties for more years. Based on this information, we can determine whether the altered composition is attributable to the presence of transgenic crops, or is simply part of the variation driven by the presence of different genotypes [63]. These studies should also involve more soil types and longer-term monitoring to account for the variability of the natural environment.

Acknowledgments

We thank everyone that helped with the fieldwork and Dr. YZ from CAAS for providing seeds of transgenic and non-transgenic wheat.

Author Contributions

Conceived and designed the experiments: JW MY JX XL JS. Performed the experiments: JW MY JX JD FJ FD. Analyzed the data: JW MY. Wrote the paper: JW MY JS.

References

1. Horsch RB, Fraley RT, Rogers SG, Sanders PR, Lloyd A, et al. (1984) Inheritance of functional foreign genes in plants. Science 223: 496–498
2. Vauramo S, Pasonen HL, Pappinen A, Setälä H (2006) Decomposition of leaf litter from chitinase transgenic birch (*Betula pendula*) and effects on decomposer populations in a field trial. Appl Soil Ecol 32: 338–349.
3. James C (2013) Global Status of Commercialized Biotech/GM Crops: 2012. ISAAA Brief No. 44. ISAAA, Ithaca, New York.
4. McGregor AN, Turner MA (2000) Soil effects of transgenic agriculture: biological processes and ecological consequences. NZ Soil News 48(6):166–169
5. Sengeløv G, Kristensen KJ, Sørensen AH, Kroer N, Sørensen SJ (2001) Effect of genomic location on horizontal transfer of a recombinant gene cassette between Pseudomonas strains in the rhizosphere and spermosphere of barley seedlings. Cur Microbiol 42:160–167
6. Gyaneshwar P, Naresh Kumar G, Parekh LJ, Poole PS (2002) Role of soil microorganisms in improving P nutrition of plants. Plant Soil 245: 83–93
7. Kent AD, Triplett EW (2002) Microbial communities and their interactions in soil and rhizosphere ecosystems. Annu Rev Microbiol 56:211–236
8. O'Callaghan M, Glare TR. (2001) Impacts of transgenic plants and microorganisms on soil biota. New Zeal Plant Prot 54: 105–110.
9. Andow DA, Zwahlen C (2006) Assessing environmental risks of transgenic plants. Ecology Letters 9:196–214
10. Liu B, Zeng Q, Yan F, Xu H, Xu C (2005) Effects of transgenic plants on soil microorganisms. Plant Soil 271:1–13
11. Chen ZH, Chen LJ, Zhang YL, Wu ZJ (2011) Microbial properties, enzyme activities and the persistence of exogenous proteins in soil under consecutive cultivation of transgenic cottons (*Gossypium hirsutum* L.). Plant Soil Environ, 57: 67–74
12. Blackwood CB, Buyer JS (2004) Soil microbial communities associated with Bt and non-Bt corn in three soils. J Environ Qual 33:832–836
13. Donegan KK, Palm CJ, Fieland VJ, Porteous LA, Ganio LM, et al. (1995) Changes in levels, species and DNA fingerprints of soil microorganisms associated with cotton expressing the *Bacillus thuringiensis* var. *kurstaki* endotoxin. Appl Soil Ecol 2:111–124
14. Bais HP, Weir TL, Perry LG, Gilroy S, Vivanco JM (2006) The role of root exudates in rhizosphere interactions with plants and other organisms. Annu Rev Plant Biol 57:233–266
15. Saxena D, Flores S, Stotzky G (1999) Insecticidal toxin in root exudates from Bt corn. Nature 402: 480–481

16. Raybould A, Higgins LS, Horak MJ, Layton RJ, Storer NP, et al. (2012) Assessing the ecological risks from the persistence and spread of feral populations of insect-resistant transgenic maize. Transgenic Res 21:655–664

17. Devarc MH, Jones CM, Thies JE (2004) Effect of Cry3Bb transgenic corn and tefluthrin on the soil microbial community: biomass, activity, and diversity. J Environ Qual 33: 837–843

18. Lu H, Wu W, Chen Y, Zhang X, Devare M, et al. (2010) Decomposition of *Bt* transgenic rice residues and response of soil microbial community in rapeseed–rice cropping system. Plant Soil 336: 279–290

19. Burns RG (1982) Enzyme activity in soil: Location and a possible role in microbial ecology. Soil Biol Biochem 14: 423–427

20. Lupwayi NZ, Hanson KG, Harker KN, Clayton GW, Blackshaw RE, et al. (2007) Soil microbial biomass, functional diversity and enzyme activity in glyphosate-resistant wheat–canola rotations under low-disturbance direct seeding and conventional tillage. Soil Biol Biochem 39: 1418–1427

21. Sun CX, Chen LJ, Wu ZJ, Zhou LK, Shimizu H (2007) Soil persistence of *Bacillus thuringiensis* (Bt) toxin from transgenic Bt cotton tissues and its effect on soil enzyme activities. Biol Fertil Soils 43: 617–620

22. Abel PP, Nelson RS, De B, Hoffman N, Rogers SG, et al. (1986) Delay of disease development in transgenic plants that express the tobacco mosaic virus coat protein. Science 232, 738–43.

23. Beachy RN, Loesch-Fries S, Tumer NE (1990) Coat-protein mediated resistance against virus infection. Annu Rev Phytopathol 28, 451–474.

24. Fuchs M, Gonsalves D (2007) Safety of virus-resistant transgenic plants two decades after their introduction: lessons from realistic field risk assessment studies. Annu Rev Phytopathol 45: 173–202

25. Tepfer M (2002) Risk assessment of virus-resistant transgenic plants. Annu Rev Phytopathol 40: 467–491

26. Robinson DJ (1996) Environmental risk assessment of release of transgenic plants containing virus-derived inserts. Transgenic Res 5: 359–362

27. Chen J (1993) Occurrence of fungally transmitted wheat mosaic viruses in China. Ann Appl Biol 123: 55–61

28. Han C, Li D, Xing Y, Zhu K, Tian Z, et al. (2000) *Wheat yellow mosaic virus* widely occurring in wheat (*Triticum aestivum*) in China. Plant Dis 84: 627–630

29. Xu H, Pang J, Ye X, Du L, Li L, et al. (2001) Study on the gene transferring of Nib8 into wheat for It's resistance to the yellow mosaic virus by bombardment. Acta Agronomica Sinica 27: 688–693 (in Chinese with English abstract)

30. Wu H, Zhang B, Gao D, Xu H, Cheng S (2006) Disease resistance test of transgenic wheat lines with *WYMV-Nib8* gene and their application in breeding. J Triticeae Crops 26: 11–14 (in Chinese with English abstract)

31. Conner AJ, Glare TR, Nap J (2003) The release of genetically modified crops into the environment. Part II. Overview of ecological risk assessment. The Plant J 33: 19–46

32. Lu RK (1999) Methods of Soil Agrochemical analysis. Beijing: China Agricultural Science and Technology Press

33. Bakke I, Schryver PD, Boon N, Vadstein O (2011) PCR-based community structure studies of bacteria associated with eukaryotic organisms: a simple PCR strategy to avoid co-amplification of eukaryotic DNA. J Microbiol Meth 84: 349–351

34. Das M, Royer TV, Leff LG (2007) Diversity of fungi, bacteria, and actinomycetes on Leaves decomposing in a stream. Appl Environ Microbiol 73: 756–767

35. Liu W, Lu HH, Wu W, Wei QK, Chen YX, et al. (2008) Transgenic Bt rice does not affect enzyme activities and microbial composition in the rhizosphere during crop development. Soil Biol Biochem 40: 475–486

36. Guan SY (1986) Soil Enzyme and Its Research Methods. Beijing: China Agricultural Press, 62–142

37. Serra-Wittling C, Houot S, Barriuso E (1995) Soil enzymatic response to addition of municipal solid-waste compost. Biol Fert Soils 20: 226–236

38. Graef F, Züghart W, Hommel B, Heinrich U, Stachow U, et al. (2005) Methodological scheme for designing the monitoring of genetically modified crops at the regional scale. Environ Monit Assess 111: 1–26

39. Züghart W, Benzler A, Berhorn F, Sukopp U, Graef F (2008) Determining indicators, methods and sites for monitoring potential adverse effects of genetically modified plants to the environment: the legal and conceptual framework for implementation. Euphytica 164: 845–852.

40. Meyer JB, Song-Wilson Y, Foetzki A, Luginbühl C, Winzeler M, et al. (2013) Does wheat genetically modified for disease resistance affect root-colonizing pseudomonads and arbuscular mycorrhizal fungi? PLoS ONE, 8(1): e53825. doi:10.1371/journal.pone.0053825

41. Yi H, Kin H, Kim C, Harn CH, Kim HM, et al. (2009) Using T-RFLP to assess the impact on soil microbial communities by transgenic lines of watermelon rootstock resistant to cucumber green mottle mosaic virus (CGMMV). J Plant Biol 52: 577–584

42. Lin C, Pan T (2010) PCR-denaturing gradient gel electrophoresis analysis to assess the effects of a genetically modified cucumber mosaic virus-resistant tomato plant on soil microbial communities. Applied Environ Microbiol 76: 3370–3373

43. Fang M, Kremer RJ, Motavalli PP, Davis G (2005) Bacterial diversity in rhizospheres of nontransgenic and transgenic corn. Appl Environ Microb 71:4132–4136

44. Donegan KK, Seidler RJ, Doyle JD, Porteous LA, Digiovanni G, et al. (1999) A field study with genetically engineered alfalfa inoculated with recombinant *Sinorhizobium meliloti*: effects on the soil ecosystem. J Appl Ecol 36:920–936

45. Wei XD, Zou HL, Chu LM, Liao B, Ye CM, et al. (2006) Field released transgenic papaya affects microbial communities and enzyme activities in soil. Plant Soil 285:347–358

46. Faragova N, Gottwaldova K, Farago J (2011) Effect of transgenic alfalfa plants with introduced gene for alfalfa mosaic virus coat protein on rhizosphere microbial community composition and physiological profile. Biologia 66: 768–777

47. Romeis J, Battini M, Bigler F (2003) Transgenic wheat with enhanced fungal resistance causes no effects on *Folsomia candida* (Collembola: Isotomidae). Pedobiologia 47: 141–147

48. Lindfield A, Nentwig W (2012) Genetically engineered antifungal wheat has no detrimental effects on the key soil species *Lumbricus terrestris*. The Open Ecology J 5:45–52

49. Duc C, Nentwig W, Lindfeld A (2011) No adverse effect of genetically modified antifungal wheat on decomposition dynamics and the soil fauna community – a field study. PLoS ONE 6(10): e25014. doi:10.1371/journal.pone.0025014

50. Weaver RW, Angle JS, Bottomiey PS (1994) Methods of soil analysis. Part 2. Microbiological and Biochemical properties, No. 5. Soil Sci Soc Am, Madison

51. Alef K, Nannipieri P (1995) Methods in Applied Soil Microbiology and Biochemistry. San Diego, CA: Academic Press

52. Dick RP, Breakwel DP, Tureo RF (1996) Soil enzyme activities and biodiversity measurements as integrating microbiological indicators. In: Doran JW, Jones AJ. ed. Methods for Assessing Soil Quality. Soil Sci Soc Am, Madison, WI, 247–272

53. Icoz I, Saxena D, Andow DA, Zwahlen C, Stotzky G (2008) Microbial populations and enzyme activities in soil In Situ under transgenic corn expressing Cry proteins from *Bacillus thuringiensis*. J. Environ. Qual 37: 647–662

54. Shen RF, Cai H, Gong WH (2006) Transgenic Bt cotton has no apparent effect on enzymatic activities of functional diversity of microbial communities in rhizosphere soil. Plant Soil 285:149–159

55. Tan F, Wang J, Feng Y, Chi G, Kong H, et al. (2010) Bt corn plants and their straw have no apparent impact on soil microbial communities. Plant Soil 329: 349–364

56. Weiner N, Meincke R, Gottwald C, Radl V, Dong X, Schloter M, et al. (2010) Effects of genetically modified potatoes with increased zeaxanthin content on the abundance and diversity of rhizobacteria with in vitro antagonistic activity do not exceed natural variability among cultivars. Plant Soil 326:437–452

57. Kowalchuk GA, Bruinsma M, Van Veen JA (2003) Assessing responses of soil microorganisms to GM plants. Trends Ecol Evol 18: 403–410

58. Cardenas E, Tiedje JM (2008) New tools for discovering and characterizing microbial diversity. Curr Opin Biotech 19: 544–549

59. Hirsch PR, Mauchline TH, Clark IM (2010) Culture-independent molecular techniques for soil microbial ecology. Soil Bio Biochem 42: 878–887

60. Fierer N, Breitbart M, Nulton J, Peter S, Lozupone C, et al. (2007) Metagenomic and small-subunit rRNA analyses reveal the genetic diversity of bacteria, archaea, fungi, and viruses in soil. Appl Environ Microbiol 73: 7059–7066

61. Uroz S, Ioannidis P, Lengelle J, Cébron A, Morin E, et al. (2013) Functional assays and metagenomic analyses reveals differences between the microbial communities inhabiting the soil horizons of a Norway spruce plantation. PLoS ONE 8(2): e55929. doi:10.1371/journal.pone.0055929

62. Li Y, Fu K, Gao S, Wu Q, Fan L, et al. (2013) Impact on bacterial community in midguts of the asian corn borer larvae by transgenic *Trichoderma* Strain overexpressing a heterologous chit42 gene with chitin-binding domain. PLoS ONE 8(2): e55555. doi:10.1371/journal.pone.0055555

63. Chun YJ, Kim DY, Kim H, Park KW, Jeong S, et al. (2011) Do transgenic chili pepper plants producing viral coat protein affect the structure of a soil microbial community? Applied Soil Ecol 51: 130–1

6

The Impact of Using Alternative Forages on the Nutrient Value within Slurry and Its Implications for Forage Productivity in Agricultural Systems

Felicity V. Crotty[1], Rhun Fychan[1], Vince J. Theobald[1], Ruth Sanderson[1], David R. Chadwick[2], Christina L. Marley[1]*

[1] Institute of Biological, Environmental and Rural Sciences, Aberystwyth University, Gogerddan, Aberystwyth, United Kingdom, [2] Environment Centre Wales, School of Environment, Natural Resources and Geography, Bangor University, Bangor, Gwynedd, United Kingdom

Abstract

Alternative forages can be used to provide valuable home-grown feed for ruminant livestock. Utilising these different forages could affect the manure value and the implications of incorporating these forages into farming systems, needs to be better understood. An experiment tested the hypothesis that applying slurries from ruminants, fed ensiled red clover (*Trifolium pratense*), lucerne (*Medicago sativa*) or kale (*Brassica oleracea*) would improve the yield of hybrid ryegrass (*Lolium hybridicum*), compared with applying slurries from ruminants fed ensiled hybrid ryegrass, or applying inorganic N alone. Slurries from sheep offered one of four silages were applied to ryegrass plots (at 35 t ha^{-1}) with 100 kg N ha^{-1} inorganic fertiliser; dry matter (DM) yield was compared to plots only receiving ammonium nitrate at rates of 0, 100 and 250 kg N ha^{-1} year^{-1}. The DM yield of plots treated with 250 kg N, lucerne or red clover slurry was significantly higher than other treatments (P<0.001). The estimated relative fertiliser N equivalence (FNE) (fertiliser-N needed to produce same yield as slurry N), was greatest for lucerne (114 kg) >red clover (81 kg) >kale (44 kg) >ryegrass (26 kg ha^{-1} yr^{-1}). These FNE values represent relative efficiencies of 22% (ryegrass), 52% (kale), 47% (red clover) and 60% for lucerne slurry, with the ryegrass slurry efficiency being lowest (P = 0.005). Soil magnesium levels in plots treated with legume slurry were higher than other treatments (P<0.001). Overall, slurries from ruminants fed alternative ensiled forages increased soil nutrient status, forage productivity and better N efficiency than slurries from ruminants fed ryegrass silage. The efficiency of fertiliser use is one of the major factors influencing the sustainability of farming systems, these findings highlight the cascade in benefits from feeding ruminants alternative forages, and the need to ensure their value is effectively captured to reduce environmental risks.

Editor: Marinus F.W. te Pas, Wageningen UR Livestock Research, Netherlands

Funding: The authors acknowledge the Department for the Environment, Food and Rural Affairs, (DEFRA) United Kingdom, (LS3642) for the financial support of this work. The funders had no role in study design, data collection and analysis, decision to publish, or preparation of the manuscript.

Competing Interests: The authors have declared that no competing interests exist.

* E-mail: cvm@aber.ac.uk

Introduction

Managing nutrients on farms is essential to ensure agroecosystem sustainability, often through the use of nutrient budgeting. Balancing the input and output of nutrients within the farm system is critical to ensuring both short-term productivity and long-term sustainability [1]. The efficient use of feed and fertiliser is central to the sustainability of farming systems. There is a strong impetus that considers animal manure as a source of essential plant nutrients and as a means to improve soil quality [2–4], rather than considering it a waste product. Globally, since 2007, agriculture commodity prices rose to historically high levels, leading to concerns about global food availability and food security [5]. Maximising the efficiency of use of nutrients within a system is the key to reducing bought-in fertiliser inputs, which are costly in both economic and environmental terms. Integrating fertiliser use with slurry supply has been known for ~30 years to be a key way of mitigating and minimising the impact of grazing animals [6]. Life cycle assessment (LCA) studies have suggested a more holistic approach to reduce environmental impacts of farming, improving manure storage, reducing inorganic fertilisers and increasing the use of leguminous forage [7] to reduce the carbon footprint.

Up to 95% of the feed nutrients consumed by ruminant livestock may be excreted in faeces and urine [8]. Therefore, managed correctly, farmyard manure and slurry offer great potential as valuable nutrient balancers, building soil fertility and reducing the need for expensive inorganic fertilisers. The value of fertiliser utilised in the UK is estimated at £1,621 million in 2011, with the value of fertiliser consumed doubling since 2006 [9]. Regular applications of organic fertilisers can also improve both soil structure and condition by increasing water holding capacity, drought resistance, structural stability and biological activity [10]. Using farmyard manure and slurry, provides additional environmental benefits, for example greenhouse gas abatement, and increasing the organic matter content of soils. Adjusting for the fertiliser requirement with manure and slurry, will potentially lead to reductions in inorganic fertiliser application rates [11].

Globally, the increasing demand for animal protein is focusing attention on the source of feed, its suitability, quality and the safety of future supply. It has been estimated that about 1000 million tonnes of animal feed is produced worldwide per annum, and 60% of the world total is from 10 countries [12]. The agricultural feed industry continues to rely heavily on imports of protein for livestock, for example in the UK, the total cost of animal feed rose to £4.4 billion in 2011 [9]. Fluctuations in world feed prices and increasing consumer concerns regarding traceability following numerous crisis's, has led to an upsurge in further demand for home-grown sources of high-quality feed.

The feeding of ryegrass silage often requires the addition of concentrate feed to achieve commercially-viable productivity in ruminants. Advances in silage technology have improved the possibility to ensile alternative forages as high protein winter forage for livestock, giving farmers another option which may reduce their reliance on bought-in concentrate feed. A study comparing consumption of grass and legume silages with concentrates on milk production in dairy cows found higher DM intake and milk yield with the legume silage compared to the grass [13]. In an experiment comparing ryegrass silage to alternatives, lambs offered alternative ensiled forages, notably lucerne and red clover, had a higher dry matter (DM) intake and live-weight gain than lambs offered ryegrass silage [14]. Furthermore, the food conversion and nitrogen (N) use efficiency was higher in lambs offered alternative silages compared with those offered ensiled ryegrass, with lambs offered kale silage having the most efficient use of N. These findings demonstrate the potential for using ensiled alternative forages compared with ensiled ryegrass to improve nutrient use efficiency, and thus, the sustainability of ruminant systems.

In order to determine the effects of incorporating different forages into a livestock system, understanding of the total loss of nutrients by the animal is needed to determine the full economic and environmental impact within a farm nutrient budget plan. Results from earlier experiments with ensiled forages indicate that at least 60% of the N in the forage will be excreted by the ruminant animal [15] resulting in a valuable high N source. This has the potential to replace inorganic N within a farm nutrient plan, reducing the reliance on inorganic N inputs (if correctly stored and applied). Consequently, there is a need to establish the benefits and limitations of integrating different forages into ruminant livestock systems in order to balance efficient production with environmental impact.

Whilst much is known about factors influencing N availability to crops following the application of typical manure types [4], [16], little attention has been paid to the efficiency with which crops can utilise the nutrients from slurries and manures derived from livestock fed alternative forages and their impact on soil nutrient status. Sheep were used as an example of a ruminant system, in an experiment conducted to test the hypothesis that applying slurries from ruminants fed ensiled red clover (*Trifolium pratense*), lucerne (*Medicago sativa*) or kale (*Brassica oleracea*) would alter the yield of swards of hybrid ryegrass (*Lolium hybridicum*) compared with applying slurries from ruminants fed ensiled hybrid ryegrass, or just applying inorganic N alone.

Materials and Methods

Ethics statement

The Institute had an ethics committee who meet at regular intervals throughout the year as part of an Ethical Review Process, as required by the Home Office (UK). The experiment reported here did not involve any regulated procedures bound by the

Animals (Scientific Procedures) Act 1986 (ASPA) (UK) and did not require Home Office approval. No specific permits were required for this study, because the performance of this study was in accordance with guidelines set by the Institute. No specific permits were required for the described field studies, because the field was owned/managed by the Institute. No specific permits were required for these locations/activities, because the location is not privately-owned or protected in any way and the field studies did not involve endangered or protected species. Ethical considerations made during experiments, related to the nutritional welfare of the sheep kept to obtain slurry for this study.

Experimental site, plot establishment and maintenance

Twenty-eight field plots (12×2.5 m) of hybrid ryegrass (cv. AberExcel) were sown at the rate of 36 kg ha^{-1} in early September in four replicate blocks, in a randomised complete block design. The plots were sited on an area of stony, well-drained loam of the Rheidol series at the Institute of Biological, Environmental and Rural Sciences (IBERS) site, University of Aberystwyth, Wales (52° 26' 55" N, 4° 1' 27" W) (Table 1 for full details of site characteristics). To achieve an optimal soil pH (of 6.0), ground limestone was applied at the rate of 5 t ha^{-1}. Compound fertiliser was applied to achieve phosphate and potash indices of 2+ to 3 [4], muriate of potash at the rate of 140 kg K$_2$O ha^{-1} and triple super phosphate at a rate of 100 kg P$_2$O$_5$ ha^{-1}. Plots were treated with the insecticide Dursban 4 (chlorpyrifos 480g l^{-1}; Dow Agrosciences, Hitchin, Herts.) applied at 1.5 litres ha^{-1} as a preventative measure as there was likely to be an established population of wireworms (Elateridae) and leatherjackets (Tipulidae) present, prior to sowing. Lupus slug pellets (3% methiocarb; Bayer plc, Bury St Edmunds, Suffolk) were applied at 5 kg ha^{-1} to aid establishment of the ryegrass, as due to the temperate climate (mild and wet) slugs are constantly prevalent (Table 1 for meteorological information). Plots were also treated with a herbicide (UPL Grassland Herbicide, dicamba, 25 g L^{-1}; MCPA, 200 g L^{-1}; mecoprop-P 200 g L^{-1}; United Phosphorus, Warrington, UK) at 5 L ha^{-1}. Ryegrass plots were cut in December to a height of 6 cm.

During the following establishment year, the plots were maintained by cutting to a height of 6 cm on 12 March, 13 May, 25 June, 8 August, 24 September and 10 December and the harvested material removed. Artificial N fertiliser was added to all plots, as 34.5% ammonium nitrate, on 5 occasions: 11 March, 28 March and prior to cuts 2, 3 and 4 to provide a total of 200 kg N ha^{-1} annum^{-1}. Potassium and phosphate fertiliser were added as previously, to maintain indices of 2+ to 3.

Animals and slurry collection

Lambs were used as a ruminant model organism for slurry production, due to size, replication and cost considerations. Slurries were collected from 80 Suffolk-cross finishing lambs fed on ensiled red clover (cv. Merviot), lucerne (cv. Vertus), ryegrass (cv. AberExcel) or kale (Kaleage, a hybrid combining cv. Pinfold and cv. Keeper) during an eight-week period. A description of the feeding experiment during which slurries were obtained was provided in Marley et al., [14]. Prior to slurry collection, the lambs were grouped within gender and according to live weight (mean 30.9 kg (± 2.29)) for a six week standardisation period and then adapted to their respective silage treatment over a 14 day period, where the first seven days the alternative forage was introduced as a proportion of the diet (i.e. 0.75:0.25, 0.5:0.5, 0.25:0.75 and 1:0 of treatment and ryegrass silage offered). A further seven day period with *ad libitum* access to their allocated silage as their sole diet, was permitted for full dietary adaptation. After which, the slurry

Table 1. Site characteristics, previous cropping and initial soil analysis (mean ± standard error).

Location characteristics	
UK Ordinance Survey Grid ref	52° 26' 55" N, 4° 1' 27" W
Altitude (a.s.l.)	30 m
Soil series	Rheidol
Soil type	stony, loam
Annual rainfall (10 year average)	1094 (±54) mm/yr
Drainage status	well-drained
Site history	Grass/Barley
Initial soil analysis	
pH (H_2O)	5.75 (±0.036)
Ammonium-N (mg kg^{-1} DM)	10.1 (±1.15)
Nitrate N (mg kg^{-1} DM)	15.1 (±1.99)
Extractable Phosphorus (ppm)	15 (±2.0)
Potassium (ppm)	90 (±5.7)
Calcium (ppm)	1186 (±35.1)
Magnesium (ppm)	157 (±6.0)
Weather conditions over two harvest years	
Average temperature (°C; two year average)	10.6 (±0.83)
Maximum temperature (°C; two year average)	14.0 (±0.86)
Minimum temperature (°C; two year average)	7.1 (±0.82)
Solar radiation (MJ/m^2/day; two year average)	9.7 (±1.21)
Number of days above 5 °C first harvest year	316
Number of days above 5 °C second harvest year	320
Total rainfall (mm; total first harvest year)	843.2
Total rainfall (mm; total second harvest year)	1101.2
Monthly rainfall (mm; two year average)	81.0 (±7.86)

collection period began, with the lambs continuing to be fed the alternative forage *ad libitum* as the sole diet for eight weeks, whilst slurry was obtained from beneath all 20 lambs within each treatment. Lambs were housed as four replicate groups of five lambs for each treatment (n = 20 per forage treatment) and placed in a sheep housing facility that was arbitrarily divided into four blocks with five pens for each of the replicate groups within each treatment. Mesh flooring placed over plastic trays where used in one of the four blocks and the lambs were rotationally moved every 14 days, in their respective replicated blocks, so that faeces and urine were collected from beneath all 20 lambs within each treatment during the 8 week experiment. Slurry obtained from the different lambs within each treatment was bulked and mixed; however each slurry was kept separate between the individual forage treatments. Each pen of lambs was offered forage *ad libitum*, with feeding levels designed to ensure a refusal margin of 10% each day. Fresh water was available to the lambs at all times. Lambs on red clover, lucerne and ryegrass silage were fed first-cut silage during weeks 1–4 and second-cut silage during weeks 5–8.

Preparation, storage and the application of slurries and inorganic fertilisers

The faeces and urine collected were diluted initially 1:1 with water (except kale-fed excreta which was sufficiently dilute) and mixed thoroughly using a 'Hilta Drysite' diaphragm pump (Morris Site machinery, Wolverhampton, UK) to form slurries. Slurries were collected over an 8 week period from January – March and

stored until required for land spreading. Storage was at 4°C in 1 m^3 plastic vessels, with a narrow opening at the top and a tap at the base. The vessels were loosely sealed to reduce losses of ammonia nitrogen.

Slurry from animals fed on the four different silages were applied (in addition to 100 kg N ha^{-1} inorganic fertiliser N) to field plots of ryegrass (12×2.5 m per plot) and compared with plots receiving ammonium nitrate at the rate of 0, 100 and 250 kg N ha^{-1} year^{-1}, in a randomised block design with a total of 7 treatments in 4 replicate randomised blocks. Slurries were applied manually using calibrated watering cans with a spoon attachment to simulate a splash-plate (surface broadcast) application. At application, the slurries were all diluted so that all slurries were of the same dilution ratio, and were applied at a ratio of 1:2.5 with water to allow the material to be applied evenly to the plot surface. All slurries were kept well mixed and were the same volume across plots at application; slurry was randomly applied within a set time on the same day to avoid any effects of weather conditions or time of day at application.

Slurries were applied at the rate of 35 t ha^{-1} as a split dressing, with half applied on 26 March and the remainder applied on 20 May, the year following plot establishment. All plots treated with slurry also received ammonium nitrate at 100 kg N ha^{-1} year^{-1} applied as a base application at the rate of 25 kg N ha^{-1} on four occasions (on 18 March, and also immediately after first, second and third cut), using a Gandy plot fertiliser (BLEC Landscaping Equipment Ltd., Spalding, Lincolnshire). Control plots, comprised of plots receiving ammonium nitrate at the rate of 0, 100 and 250 kg N ha^{-1} year^{-1} (to be referred to as 0N, 100N and 250N onwards), ammonium nitrate was applied on the same dates on the solely inorganic N plots as it was applied to slurry-treated plots. Water was applied to all control plots at a rate of 35 t ha^{-1} annum^{-1} on the same dates as slurry was applied, to control variability between treatments. Potassium and phosphate fertiliser were applied as a compound of muriate of potash and triple super phosphate at the rate of 154 kg K$_2$O ha^{-1} and 100 kg P$_2$O$_5$ ha^{-1}, to all experimental plots, to ensure neither element was limited during the harvest years.

Soil and slurry analysis

Preliminary soil samples were taken 15 months after sowing the ryegrass, in the first harvest year prior to slurry application, from a W-formation across each replicate block of each set of plots and bulking each replicate block together (n = 4). Extra samples were taken from the experimental site at each depth to calculate bulk density and water content to allow for the calculation of nutrients per ha.

Experimental soil samples were taken at 0–7.5 cm for mineral analysis, and 0–30 cm and 30–60 cm for N analysis (at some sites bedrock was less than 60 cm from the soil surface, thus less than 30–60 cm depth was taken). Soil analysis was carried out on samples obtained immediately prior to the first slurry application, from cores taken in a W-formation as described above. Further soil analysis was determined from samples obtained six months after the first slurry application and 18 months after the first slurry application, from 6 replicate samples (cores 0–7.5 cm) taken per plot, bulked to form one sample per plot for mineral analysis. Soil samples of 0–30 cm and 30–60 cm were taken for soil N analysis and processed immediately, with soil N being determined as nitrate (NO$_3$-N) and ammonium-N (NH$_4$-N). Soil mineral analysis (0–7.5 cm cores) was determined for calcium (Ca), magnesium (Mg), potassium (K); and phosphorus (P). Soil P was determined as bicarbonate extractable (Olsen) P and 0.01 M CaCl$_2$ extractable P (a measure of potentially mobile P) whilst the other minerals were extracted from soil using acetic acid and measured by inductively coupled plasma (ICP). Soil pH was determined as 1:1 (soil:water) mixture, shaken for 30 min before the pH was measured.

Sub-samples of each slurry type were collected at the time of spreading and analysed for pH, dry matter (DM) content, total N, nitrate-N and ammonium-N. Ammonium-N and nitrate-N were extracted from slurry using a 2 M KCl solution (10 g slurry in 50 ml KCl shaken for 1 h then filtered). Nitrate was determined by reduction of nitrate to nitrite using a cadmium column followed by colorimetric measurement at 520 nm. Ammonium-N was determined colorimetrically at 660 nm. Total N in slurry samples was determined using a Kjeldahl method (Tecator Kjeltec Auto 1030, Tecator, Höganäs, Sweden). The two-step process involved digesting the sample using sulphuric acid and a digestion catalyst which converts the organic N content to the ammonium form. The sample digest was then analysed for ammonium-N by distillation and titration. DM was determined by drying a known amount at 105 °C for at least 24 hours. The pH was determined after mixing 10 g of slurry with 50 ml deionised water. The solution was allowed to settle for 30 min before the pH was measured.

Sward density, herbage yield, nitrogen offtake and sward composition

Plant population densities were monitored during the spring and autumn of each year. The mean ryegrass tiller count m^{-2} was determined from eight randomly-placed 12×18.75 cm quadrats

per replicate block, in the autumn and spring, post slurry application.

During the first year after slurry application (first harvest year), plots were cut on 18 May, 30 June, 19 August, 12 October and 10 December. In the second harvest year, plots were cut on two occasions – 16 May and 6 July to measure any residual carry-over effects. Plots were harvested using a Haldrup 1500 plot harvester (J. Haldrup a/s, Løgstør, Denmark), and cut to a height of 6 cm. Yield was determined by weighing the material cut from an area of 12 m×1.5 m within each plot. Sub-samples of forage, as harvested, were taken to determine dry matter (DM) yield, N offtake and the botanical composition of the sward. All sample material was stored at −20°C prior to subsequent chemical analysis. The DM contents of the herbage was determined by drying to constant weight at 80 °C in a forced-draught oven, and the DM content of the samples taken for chemical analysis after freeze-drying. Total N of the herbage cut was determined using a Leco FP 428 nitrogen analyser (Leco Corporation, St. Joseph, MI, US).

Statistical analysis

Effects of fertiliser treatment on plot yields, N balance and recovery were assessed by analysis of variance according to the randomised block design. Differences in the composition of slurry applied to the slurry plots were assessed similarly on the relevant subset of the design. Soil mineral composition on two sampling dates and N content at two depths were compared by split plot analysis of variance with fertiliser treatment effects assessed at the whole plot level and effects of sampling date and/or depth and their interaction with fertiliser assessed at the sub-plot level. Where applicable, multiple comparisons within tables of means were made using the Student Newman Keuls test [17] with the experiment-wise type I error rate set at 5%. The total inorganic fertiliser N equivalence (FNE) of each slurry was estimated by a within-block reverse interpolation assuming a quadratic diminishing response in DM yield across the three inorganic N treatments (including 0 N) (N = 4 per treatment). Slurry N efficiency in terms of DM yield was estimated as total inorganic N equivalence less 100 kg (applied as ammonium nitrate) relative to slurry N applied. To understand the difference in N utilisation for each treatment, the apparent N recovery (ANR) was calculated according to the method of Kanneganti et al., [18]. The N offtake relative to 0N or 100N, was calculated; ANR = ((NTRT-NCON)/NTOT)*100 where NTRT is N offtake, NCON is N offtake from control and NTOT is total N applied, all measured in kg ha^{-1} yr^{-1} and expressed as a percentage of the difference in total N applied. All data were analysed using GenStat (14th Edition, [19]) and are presented as mean and S.E.D (standard error of the difference), unless otherwise stated.

Results

Slurry

Lambs fed on kale silage produced a higher amount of excreta than lambs on other silages (P<0.001), the total dry matter (DM) from lambs fed on kale silage was lower than lambs fed the legume silages, and it also had a significantly lower N content (Table 2). Kale slurry had less than a third of the DM content of all the other slurries applied (Table 2). Lambs fed lucerne and red clover produced an intermediate amount of slurry compared to kale and ryegrass, however these two alternatives had the highest dry matter and N content (both P<0.001) compared to the other slurries. Hybrid ryegrass fed lambs produced the least amount of slurry per day, although ryegrass had lower dry matter and N content. There also were significant differences in composition between the slurries applied (P<0.001) for pH, nitrate and ammonium and total N contents (Table 2). In terms of pH all slurries were significantly different from each other (P<0.05), with ryegrass having the lowest pH and lucerne the highest (Table 2). Lucerne and red clover slurry both had high total-N content, whilst kale had the lowest total-N followed by ryegrass which was intermediate (Table 2). Nitrate N concentration was higher in kale slurry (1.07 mg kg^{-1}) than in the remaining slurries (P<0.05) which showed levels <1 mg kg^{-1}. The ammonium-N content of the ryegrass and kale slurries were similar and significantly lower than the red clover slurry which in turn was lower than the lucerne slurry. Lucerne slurry showed the highest percentage concentration of ammonium-N compared to all other slurries (P<0.05). Overall, kale slurry was the most different to the legume slurries, with the hybrid ryegrass slurry as an intermediate. All environmental variables were considered to be the same for each treatment, as the replicated plots were all located within the same 100 m^2 area (Table 1).

Soil

Looking at the composition of soils after slurry application there was no evidence of interaction between effects of treatment and sampling date for any of the analytes measured (Table 3). This lack of interaction significance was because the general trend appears to be the same across all treatments; between autumn and spring pH, K, and P contents decreased (P<0.01), and Ca and Mg levels increased (P<0.01), suggesting differences in the release rates of essential nutrients over time. The Ca and P contents were not significantly affected by treatment (P=0.322 and P=0.333 respectively), however, there were significant differences over time (Table 3). Using an analysis of variance, near significant differences were also noticeable in the pH level of the soil between treatments (P=0.054). There was a significant difference over time, with all pH's decreasing; due to this trend across treatments the interaction was not significant. The level of soil K was lower with the 250N treatment than with the other treatments (P<0.05). Mg levels were higher in soils treated with legume slurry than the other treatments and were highest for red clover slurry treated soil (P<0.05).

The ammonium, nitrate or total N content of soil at both the 0–30 cm or 30–60 cm depth was assessed at the beginning and end of the growing season, however no differences were found between treatments (P>0.05) (Table 4). However, there were significant differences found between depths for nitrate and total N (both P<0.001) in the autumn, with lower levels in the 30–60 cm sample than the 0–30 cm sample (Table 4). Significant differences were also found between the two different depths after the growing season had finished (P<0.001) for nitrate, ammonium and total N (Table 4).There were significant changes in the soil mineral N content over the growing season, particularly for nitrate and ammonium (P<0.001), with a reduction in nitrate in the top 0–30 cm of soil over the season and an increase in the 30–60 cm layer. Whilst for ammonium there was an increase in both depths over the season. There were no significant changes found depending on treatment and time of sampling, nor was the interaction between treatment, depth and time of sampling significant (Table 4).

Dry matter yields

Overall DM yields were significantly different between treatments (P<0.001), with all slurry treatments and 250N inorganic fertiliser treatment having significantly greater yields than the 100N and 0N treatments (Figure 1, dotted line representing the

Table 2. Mean composition of slurries (fresh weight) as applied to plots of hybrid ryegrass.

Forage Fed	Undiluted slurry (per lamb per day) (kg)	N content (%) undiluted slurry	Dry Matter (g kg^{-1})	pH	NO$_3$-N (mg/kg)	NH$_4$-N (mg/kg)	NH$_4$-N (% Total N)	Total N (g/kg)
H. Ryegrass	0.92[a]	0.517[b]	68.1[b]	7.2[a]	0.14[a]	300[a]	9.1[a]	3.35[b]
Kale	2.73[c]	0.404[a]	20.9[a]	8.2[c]	1.07[b]	290[a]	12.1[a]	2.43[a]
Lucerne	1.69[b]	0.882[d]	72.5[c]	8.4[d]	0.29[a]	1688[c]	31.3[c]	5.46[c]
Red clover	1.34[b]	0.740[c]	72.7[c]	8.0[b]	0.19[a]	989[b]	20.4[b]	4.87[c]
s.e.d	0.185	0.0418	0.673	0.036	0.090	69.95	2.19	0.306
Probability	<0.001	<0.001	<0.001	<0.001	<0.001	<0.001	<0.001	<0.001

Analysis of variance was used to assess differences between composition of slurry for all organic fertiliser treatments. Treatment effects were apportioned using a Student Newman Keuls test (different superscripts following mean indicating significant differences (P<0.05) between treatments), N=4.

Table 3. Mean mineral composition (g kg^{-1}) and pH of soils (0–7.5 cm cores) from plots of hybrid ryegrass in the autumn and following spring after application of inorganic N or slurry from lambs offered different silages.

	Sampling	Treatment							Mean	Treatment (T)		Sampling (S)		T.S
		0N	100N	250N	HRG	Kale	Lucerne	Red Clover		s.e.d.	Prob	s.e.d.	Prob	Prob
K	Autumn	248	201	142	241	209	207	206	208^b	15.5	<0.001	5.9	<0.001	0.422
	Spring	208	158	137	191	172	166	189	174^a					
	Mean	228^b	179^b	140^a	216^b	190^b	187^b	198^b						
Ca	Autumn	1638	1378	1458	1555	1480	1592	1335	1491^a	83.7	0.322	49.3	<0.001	0.547
	Spring	2028	1959	2148	2093	2101	2088	2081	2071^b					
	Mean	1833	1668	1803	1824	1791	1840	1708						
Mg	Autumn	89	90	88	95	95	101	105	94^a	1.8	<0.001	1.1	0.003	0.327
	Spring	93	97	92	94	95	103	112	98^b					
	Mean	91^a	93^a	90^a	94^a	95^{ca}	102^b	108^c						
P	Autumn	45	38	41	42	38	47	43	42^b	2.4	0.333	1.5	<0.001	0.507
	Spring	23	24	27	25	21	24	25	24^a					
	Mean	34	31	34	34	30	35	34						
pH	Autumn	6.72	6.61	6.65	6.74	6.74	6.85	6.72	6.72^b	0.054	0.054	0.024	0.003	0.657
	Spring	6.64	6.56	6.64	6.65	6.58	6.72	6.67	6.64^a					
	Mean	6.68^{ab}	6.59^a	6.64^{ab}	6.70^{ab}	6.66^{ab}	6.79^b	6.70^{ab}						

Analysis of variance was used to assess differences between composition of soils for all treatments (T), sampling time (S) and the interaction between treatment and sampling time (T.S). Effects were apportioned using a Student Newman Keuls test (different superscripts following mean indicating significant differences (P<0.05) between treatments). N = 4.

Table 4. Mean N content of soil (mg kg DM^{-1}) from plots (0–30 cm and 30–60 cm) of hybrid ryegrass in the autumn and following spring after application of inorganic N or slurry from lambs offered different silages.

Sampling	Nitrogen	Depth	Treatment							Mean	Treatment (T)		Depth (D)		T.D	Sampling (S)		T.S	D.S	T.D.S
			0N	100N	250N	HRG	Kale	Lucerne	Red Clover		s.e.d.	Prob	s.e.d.	Prob	Prob	s.e.d.	Prob	Prob	Prob	Prob
Autumn	NO$_3$-N	0–30 cm	4.9	5.6	6.8	5.9	6.6	6.3	6.1	6.0^b	0.40	0.232	0.18	<0.001	0.275	0.18	<0.001	0.655	<0.001	0.326
		30–60 cm	0.4	0.3	0.6	0.4	0.4	0.4	0.4	0.4^a										
	NH$_4$-N	0–30 cm	5.3	4.8	4.8	5.2	5.8	5.0	4.9	5.1	0.80	0.720	0.52	0.129	0.669	0.22	<0.001	0.828	0.003	0.274
		30–60 cm	3.1	3.7	4.3	5.4	3.4	4.6	5.5	4.3										
	Total N	0–30 cm	10.2	10.4	11.6	11.1	12.3	11.3	11.1	11.1^b	0.87	0.392	0.62	<0.001	0.831	0.30	0.929	0.791	<0.001	0.372
		30–60 cm	3.5	4.0	4.9	5.9	3.8	5.0	5.9	4.7^b										
Spring	NO$_3$-N	0–30 cm	3.0	2.9	2.6	2.7	2.7	2.3	2.9	2.7^b	0.28	0.669	0.41	<0.001	0.894					
		30–60 cm	0.6	0.5	1.0	1.1	0.3	1.7	1.0	0.9^a										
	NH$_4$-N	0–30 cm	8.0	6.6	7.2	6.9	6.6	7.4	7.9	7.2^b	0.78	0.840	0.44	<0.001	0.743					
		30–60 cm	4.3	4.3	5.4	5.5	5.2	5.8	4.9	5.0^a										
	Total N	0–30 cm	10.9	9.5	9.8	9.6	9.3	9.7	10.8	10.0^b	0.91	0.724	0.76	<0.001	0.864					
		30–60 cm	4.9	4.8	6.3	6.6	5.5	7.5	5.9	5.9^a										

Analysis of variance was used to assess differences between N content of soils for all treatments (T), depth (D), sampling time (S) and the interactions between treatment and depth (T.D), treatment and sampling time (T.S), depth and sampling time (D.S), and treatment, depth and sampling time (T.D.S). Effects were apportioned using a Student Newman Keuls test (different superscripts following mean indicating significant differences (P<0.05) between treatments). N = 4.

"control" 100N yield). Of the different treatments the DM yield increased significantly from ryegrass<kale<red clover<lucerne< 250N. Treating plots with slurries from animals fed on different forages or with different levels of inorganic N did not alter the yield of unsown (weed) species (P = 0.121). However the percentage of unsown species (by mass) in total yield was significant (P = 0.001), with 100N and 0N inorganic fertiliser treatments having a significantly greater proportion of unsown species in comparison to lucerne and 250N fertiliser, which had the lowest unsown species proportion. Total ryegrass (sown species) DM yield was significantly different between treatments (P<0.001; Figure 1). All treatments had significantly greater yields than the control (100N); ryegrass and kale slurry had similar yields, which were significantly lower than red clover yields, which was significantly lower than lucerne and 250N sown species yield. There were no significant differences in ryegrass tiller counts between treatments (P = 0.246), or over time (P = 0.569). Nor was there any effect of slurry or fertiliser applications on ryegrass tiller counts taken in two or three years post-sowing.

The DM yield was positively correlated with the amount of N the crops received (Figure 2). Considering that different amounts of N were applied for each treatment (Table 5), this is probably the main factor that contributed to the differences in yield. Overall, when considering the estimated relative inorganic fertiliser N equivalence (FNE) of each slurry and the efficiency of N from the different slurries used to produce DM yield, there were significant differences between treatments. After subtraction of 100 kg inorganic N ha^{-1}, the N applied as slurry was equivalent to 114 kg for lucerne, 81 kg for red clover, 44 kg for kale and 26 kg inorganic N ha^{-1} yr^{-1} for ryegrass slurries. Given slurry N application rates of 117, 85, 170 and 191 kg N ha^{-1} in terms of

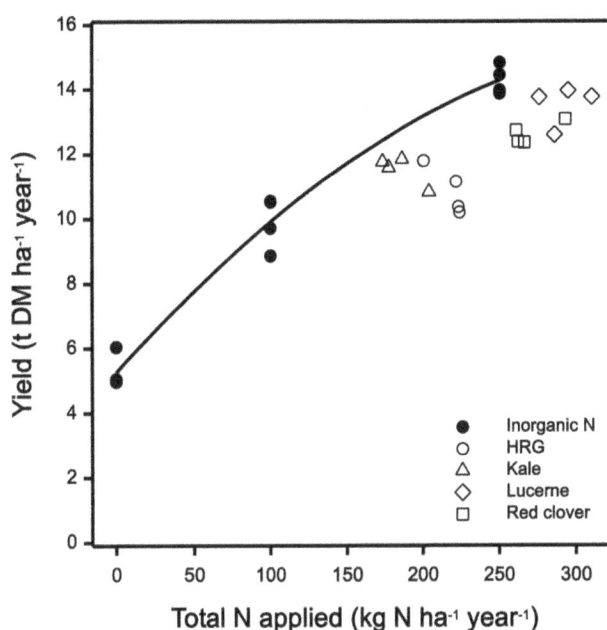

Figure 2. Total annual yield (kg DM ha^{-1}) compared to the total N applied (kg N ha^{-1}). Plots of hybrid ryegrass treated with slurries from sheep offered four different forage diets (H. ryegrass (HRG), kale, lucerne or red clover) or with inorganic nitrogen at a rate of 0, 100 and 250 kg N ha^{-1} $year^{-1}$, (N = 4). Estimated relative fertiliser N equivalence is indicated by the quadratic regression line.

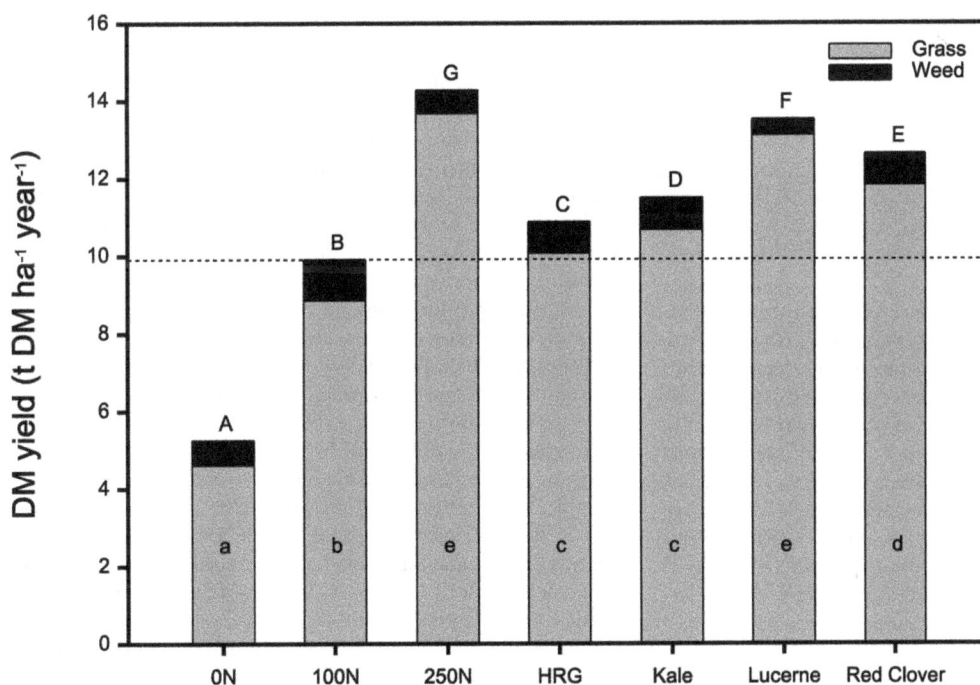

Figure 1. Total annual dry matter yield (t DM ha^{-1} $year^{-1}$) of sown and unsown (weed) species. Plots of hybrid ryegrass treated with slurries from sheep offered four different forage diets (H. ryegrass (HRG), kale, lucerne or red clover) or with inorganic nitrogen at the rate of 0, 100 and 250 kg N ha^{-1} $year^{-1}$, (N = 4). Dotted line indicates yield obtained for the control (100N). There were significant differences between treatments for total yield and sown yield (P<0.001). Treatment effects were apportioned using a Student Newman Keuls test looking at total yield (capital letters) and total sown species yield (lowercase letters) indicate significant differences (P<0.05) between treatments. There were no significant differences found between unsown species yield.

Table 5. Mean total N input, offtake and N balance (kg ha^{-1} year^{-1}), apparent N recovery (%) and estimated relative fertiliser N equivalence (FNE) for slurry N efficiency, for plots of hybrid ryegrass treated with inorganic N or slurry from lambs offered different silages, (N = 4).

	Total N input[1]	N offtake	N balance[2]	Apparent N Recovery (%)		FNE[5]
				Applied N[3]	Slurry N[4]	(%)
0N	25	99^a	−74^a			
100N	125	165^b	−40^b	65^b		
250N	275	268^f	7^c	68^b		
H. Ryegrass	242^b	197^c	45^d	45^a	29	23^a
Kale	210^a	196^c	14^c	52^a	37	52^b
Lucerne	317^c	246^e	71^d	50^a	43	60^b
Red Clover	296^c	223^d	73^d	46^a	35	47^b
s.e.d.	10.9$^\#$	8.4	12.9	5.5	8.5	7.7
Prob	<0.001$^\#$	<0.001	<0.001	0.003	0.462	0.005

^{a,b}Different superscript letters denote significant differences between means (P<0.05).
$^\#$relates to means for slurry treatments only
[1]Sum of inorganic N, slurry N plus atmospheric N deposition at a rate of 25 kg ha^{-1} year^{-1} [52].
[2]The N balance was calculated by subtracting offtakes, summed over the entire period (five cuts) from total N input [53].
[3, 4]Apparent N recovery (ANR) was calculated for each plot within each replicate block according to the method of Kanneganti et al., [18] as N offtake relative to 0N (3) or 100N (4), expressed as a percentage of the difference in total N applied. ANR = ((NTRT-NCON)/NTOT)*100 where NTRT is N offtake, NCON is N offtake from control and NTOT is total N applied, all measured in kg ha^{-1} yr^{-1}.
[5]Estimated by reverse interpolation assuming a quadratic diminishing response in DM yield across the three inorganic N treatments (including 0 N treatment).

DM yield, these FNE values where significantly greater for all of the alternative forages compared to ryegrass (Table 5). This showed that efficiency of use of ryegrass slurry N for DM yield relative to fertiliser N is lower than that of the other alternative slurries (P = 0.005).

In terms of N yield, offtakes harvested from the different treatments were found to be significantly different between all treatments (P<0.001). The majority of individual treatments had significantly different N offtakes apart from kale and ryegrass which were similar (Table 5). All slurry treatments had greater N offtakes than the 0N and 100N treatments. However, the greatest offtake was the 250N treatment. The nitrogen balance was significantly different between treatments, with the greatest deficit being the 0N input, however 100N input, also had a deficit and both were significantly different to the other N balances. The 250N and all slurry treatments, had a positive N balance; with the ryegrass, lucerne and red clover treatments having the greatest surplus. Apparent N recovery represents the amount recovered by the crop from the fertiliser/slurry. Relative to the 0N treatment, 100N and 250N showed total N recoveries of 65% and 68% while the slurries all showed significantly lower (P<0.05) values, but there was no significant difference between the slurry recoveries, ranging from 45% to 52% (Table 5). Apparent recovery of slurry N calculated by reference to the 100N treatment did not differ significantly (P = 0.462) between the slurries but with the lowest value associated with slurry derived from a ryegrass silage diet (28%) and the highest associated with slurry from lucerne silage diet (43%) (Table 5).

Discussion

The aim of this experiment was to improve our understanding of the plant-animal-soil nitrogen cycle [20] within livestock production systems. Optimising nutrient supply has the greatest potential to balance intensive livestock production, by converting the detrimental increases in N from animal excreta into a benefit,

via the utilisation of slurry, simultaneously, reducing chemical costs and decreasing the environmental impact of farming. Our study illustrates how the use of home-grown alternative forages could reduce the input and output of nutrients within farming systems, thus ensuring both short-term productivity and long-term sustainability. A study looking at the economics of storage, transporting and spreading slurry found that despite high energy costs, it was actually a much lower cost per kg of available N compared to inorganic fertiliser [21]. Previous research has tended to focus on comparison between ranges of fertilisers (form of fertiliser) [22], how they are applied (surface application versus shallow injection) [23] or from which species of livestock they originate [24] but few studies have examined the effects of different forage diets within the same livestock species, or the nutrient value of this as a farm resource.

Understanding the nutrient budgets of farming systems at different scales is central for the efficient use of the available nutrients, to effectively improve the long-term sustainability and environmental impact of farming systems [25]. The utilisation of slurry rather than inorganic fertiliser has the potential to impart large economic value, directly by the reduction in expenditure on inorganic fertilisers and exploiting a natural farm resource. For example, an investigation of the profitability and performance of grazing steers on ryegrass with inorganic fertiliser compared to a ryegrass and legume mix, found no difference in performance but an increased cost of US$19 ha^{-1} for the ryegrass with inorganic fertiliser [26]. Whilst within Europe, it is thought the introduction of legume and grass-legume silages (compared to grass silage) has the potential economic gain of €137 ha^{-1}, corresponding to a gain of as much as €1300 million to the European livestock farming sector [27]. Indirectly the use of slurry will also provide a number of ecosystem services, through the changes in soil structure, the direct addition of organic matter and the favouring of different soil food webs [28].

This study used sheep as an example of a ruminant organism for slurry production, due to their size, ease of replication and total cost considerations for the overall experiment involving several treatments. Ruminant research is known to focus on sheep, particularly when using specialised feed to produce slurry e.g. [29–30]. However, it is recognised that it is difficult to draw full comparisons between cattle and sheep, given species differences in grazing habits, digestive efficiencies, and intakes [31], however research focusing on the slurry component has found less differences than may normally be expected [32]. One of the main differences between slurries used in the current experiment, before application was the dry matter content and the amount of manure produced. These differences have a long-term management impact; the amount of dilution needed before application, as well as the potential storage issues if these slurries were used in normal farming practice. Although in practice, farmers do not dilute slurries to produce spreadable material, water is added through the washing of housing units and drainage. In the European nitrates directive, 58% of England has been classified as a nitrate vulnerable zone [33], leading to protection measures and stricter control of fertiliser application. The differences in N content in the slurries, could potentially lead to different measures being needed. It should be noted that the ryegrass silage offered to produce the slurries in this experiment had a crude protein content that was 1.6% below the average ryegrass silage produced in the UK in the same season due to weather conditions delaying the silage harvest (see [14]). Therefore, the proportion of ammonium-N to total-N in the ryegrass slurry treatment may have been correspondingly lower than a typical ryegrass slurry treatment.

The chemical composition of the slurries in this study before application, were significantly different (pH, NO_3-N, NH_4-N and Total N), however this didn't lead to significantly different N levels within the soil, suggesting the differences were ameliorated by the uptake of the growing crop or lost to the environment. Slurry with a lower pH has a reduced risk of ammonia volatilisation after application, compared to those with a higher pH [34]. In this study, ryegrass slurry was significantly lower than the other treatments, with lucerne having a significantly higher pH than the other slurries. Higher DM content within slurry also poses a greater risk of methane emissions during storage [35], and ammonia volatilisation after application as the slurry does not infiltrate the soil as quickly. Kale slurry had the lowest DM content, thereby posing the least risk compared to the other treatments in this respect. A study investigating the impacts of different slurries on gaseous emissions after spreading, however found kale slurry to have the largest N_2O emissions (compared to lucerne and ryegrass) [36]. Plant available N varied between the different slurries, with kale slurry having the greatest NO_3 levels, which is the form most plants absorb N through the root system. However, the lucerne and red clover slurry had the highest NH_4 levels, which can be readily converted to nitrate in the soil [37]. As nitrate-N is the most susceptible to leaching, these slurries could potentially cause a problem in nitrate vulnerable zones if not correctly managed.

The compositional differences in the slurries likely led to the different mineral levels (e.g. K and Mg) in the soil after slurry application. However, soil mineral levels varied over time, suggesting that there may have been differences in release rate or mobilisation. Significant differences were found in the Mg level in the soil after slurry application, with the greatest found in soil where red clover slurry had been applied, also leading to potential carry-over effects. There was no significant difference between the N content of the soil between treatments after slurry application, reducing the potential for variation in future crops. The overall mineral N content of the soil showed significant differences in depth and over time, these changes were unlikely to be due to the different slurry applications, as N content is known to change with depth [38].

A key finding in this experiment was that of the DM yield of the ryegrass after slurry application. Treating plots of hybrid ryegrass with lucerne slurry (plus 100 kg N ha^{-1}) had similar DM yield (sown species) compared with plots receiving 250 kg N ha^{-1} of inorganic N alone. The 250N and lucerne treatments had the greatest positive N balances; however kale slurry had the greatest apparent N recovery, followed by lucerne. Suggesting lucerne slurry could be comparable for ryegrass growth, without loss of yield, to inorganic fertiliser treatments. This is likely to be due to the N in these slurries being more efficiently used. The utilisation of slurry as fertiliser is a common practice but, it should be noted that it is not usually slurry that has been produced from animals fed only on a single forage diet, which was the approach taken here for experimental purposes. Our results show the effect a change of diet can have on slurry and the cascade in effects this could have on production. The estimated relative fertiliser N equivalence (FNE), was greatest for lucerne; although all three alternative forages were greater than ryegrass. The estimated FNE values represented efficiencies of 47–60% for the alternative forages compared to only 22% for ryegrass. It should be noted that the FNE of the slurry is only an estimate based on an assumed diminishing N response curve produced from the three inorganic fertiliser treatments (replicated four times at the same experimental site). Although this does not provide an absolute FNE value for these slurry treatments, it does provide a valid relative value when comparing treatments within the context of this experiment.

Not all N applied to crops is taken up by the plant, some is lost to the environment as ammonia volatilisation or denitrification. However, some N will remain in the soil, in crop residues (roots and non-cut grass) and assimilated into soil microbial biomass. Sampling the N composition of the soils in the spring after slurry application, shows there is N remaining within our soils, with significant differences in the interaction between depth and sampling time. All of the deeper soil samples (30–60 cm) taken in the spring having greater amounts of N (Total N, NO_3-N, NH_4-N) generally across treatments, then they did the previous autumn, suggesting the transfer of N further down the soil column.

Applying slurries from ruminants fed on different forages also did not significantly alter the DM yield of unsown (weed) species present in ryegrass swards. In fact, the 100N and 0N inorganic fertiliser treatments had a significantly greater proportion of unsown (weed) species in comparison to the other treatments; it is likely that this is because the ryegrass in these low N treatments could not out-compete the weed species to the same extent, it could in the forage slurry and 250N treatments.

Previous studies have focused on the amount of N excreted following consumption of different forage compositions [39]. Investigations of N uptake and yield of corn amended with slurry of different forage-fed cattle, has also shown variation between different forage slurries [40]. These studies concentrated on cattle and forages commonly fed in the USA like soybean and corn, and found differential effects on soil N mineralisation and plant N uptake after application to soil [41]. Dairy diets are often formulated so that crude protein (CP) levels remain similar, independent of feed; these calculations are based on a total CP value (N×6.25). This approach does not account for differences in the N use efficiency (NUE) of the total CP, and there are differences in NUE occurring among these alternative silages [14]. The higher intakes recorded for ruminants offered legume silages relative to grass silage was attributed to legume silages having a higher passage rate due to higher rumen outflow rate [42]. Using proximate analyses for in vitro digestibility values to predict the nutritive value of legume forages may not be accurate, e.g. [43], who showed the degradability of the CP in vivo was the only reliable method to determine N utilisation efficiencies between forages with similar CP values.

Feeding these alternative forage diets resulted in a higher NUE, whilst being produced from legumes which were grown with minimal N fertiliser addition. This subsequently leads to agricultural benefits which are two-fold, with slurries replacing inorganic N for crops that need it, whilst being produced without any inputs. The value of fertiliser utilised in the UK is estimated at £1,621 million in 2011 [9], if by modifying feeding regime slightly the utilisability of slurry can increase, this would reduce costs to the farming industry, making the farming system more sustainable. Research has shown that there are various factors that can influence the efficient transfer of nitrogen from organic manures to plants. Factors include the total N, readily available N, dry matter content and C:N ratio of the manure [44]; the amount applied, timing of application, application method, rainfall and soil type in the field [4], [45]. In grassland soils, organic manures compared to inorganic fertiliser, are known to increase the organic C, the total N, the activity of decomposers, and the supply of nutrients via the soil food web [46]. Manure slurry has also been found to promote a higher bacterial activity and provide greater mineralisable N compared to inorganic fertiliser [47]. Thus as well as being comparable in yield to inorganic fertiliser, using manure slurry has greater value through the provision of more ecosystem services.

The nutrient composition of manures from different livestock and guidelines on their expected values are available [4]. However, it is recommended that farmers analyse their own manure nutrient compositions, as depending on the forages fed to these animals, they may vary. Our results highlight that slurry from sheep can differ significantly in nutrient value depending on food source and this should be considered as part of routine farm management when slurries are used as fertiliser. However, through Defra's "Farm Practice Survey's" it was found that only 23% of UK farmers tested the composition of their slurry [48]. Future work should focus on encouraging the use of different management systems by farmers (e.g. MANNER and MANNER-NPK [15]) to effectively fertilise crops through slurry spreading.

Modelling the profitability of whole farming systems found variation in fertiliser prices to have a relatively small effect on net margins, largely because this cost comprised a small proportion of total costs [49]. However, it is still a significant amount to be considered, particularly when margins are already low. The agricultural industry continues to rely heavily on imports of protein for livestock production, however the effect of feeding concentrates on the nutrient value of livestock slurry was not the focus of this experiment. This study has highlighted the importance of understanding the nutrient content of manures, and how a change in food source can impact yields of future crops. Farmers need to consider how these differences in slurry could affect plant growth and not base application rates on fixed values per ha, as there will be different N loading rates, and therefore different yield responses. We still need to understand the effect variable nutrients of forage provided to all livestock have on slurry composition and spreading guidelines. The efficiency of fertiliser use is the key to the sustainability of farming systems.

The results of this study has shown that slurry derived from ensiled alternative forages is comparable to inorganic fertiliser, when considering DM yield of a future forage crop (hybrid ryegrass). The use of high-protein alternative forages can reduce the need for expensive amendments, building soil fertility, and improving nutrient efficiency in ruminant livestock systems. Thus, optimising nutrient requirements and maximising nutrient capture and retention within the farming system; resulting in a more beneficial and sustainable scenario for production and the environment, than currently exists. Optimisation of the entire manure management continuum [50]; is key to the development of sustainable livestock production systems [51]. Our data could be used for this purpose – to inform sustainability indices and farm nutrient budgets, including carbon foot-printing on livestock farms aiming to reduce reliance on imported feeds and fertilisers

Conclusions

Overall, the findings have shown the potential to use slurry from ruminants fed home-grown alternative forages as a valuable fertiliser within livestock systems, and the impact of that at a farm nutrient level on the subsequent use of nutrients within slurries produced – improved N fertiliser equivalence compared to ryegrass only slurries. The utilisation of slurry rather than inorganic fertiliser has the potential to impart large economic value, directly by the reduction in expenditure on inorganic fertilisers and exploiting a natural farm resource. However, the value of these slurries will depend on farmers having suitable storage and spreading facilities, to reduce any potential environmental risks from these higher N-slurries and further highlights a requirement for farmers to implement industry guidelines to regularly measure the N value of their slurry. There is a need to identify and develop strategies that will allow the use of these alternative forage crops to further mitigate the impact of livestock systems on nutrients and carbon cycling at a UK and global scale.

Acknowledgments

The authors would like to thank John Roberts, Gareth Lewis and Rob Davies for their assistance with the field work. A special acknowledgement is given to the late Raymond Jones for his contribution to this research. Meteorological data was supplied by the Met Office, UK. We also acknowledge the Department for the Environment, Food and Rural Affairs, (DEFRA) UK, (LS3642) for the financial support of this work.

Author Contributions

Conceived and designed the experiments: CLM RF DRC. Performed the experiments: RF VJT. Analyzed the data: FVC RS. Wrote the paper: FVC CLM DRC.

References

1. Fortune S, Conway JS, Philipps L, Robinson JS, Stockdale EA, et al. (2000) N, P and K for some UK organic farming systems – implications for sustainability. Soil Organic Matter and Sustainability. Wallingford, UK.: CABI. pp. 286–293.
2. Begum F, Bajracharya RM, Sitaula BK, Sharma S (2013) Seasonal dynamics, slope aspect and land use effects on soil mesofauna density in the mid-hills of Nepal. Int J Biodiv Sci, Ecosyst Serv Manage 9: 1–8.
3. Shrestha RK, Lal R, Rimal B (2013) Soil carbon fluxes and balances and soil properties of organically amended no-till corn production systems. Geoderma 197–198: 177–185.
4. DEFRA (2010) Fertiliser Manual (RB209). In: DEFRA (Ed.), Fertiliser Manual. The Stationery Office (TSO): Norwich, p. 252.
5. FAO (2010) Current world fertilizer trends and outlook to 2014. In: Food & Agriculture Organisation of the United Nations (Ed.), Current world fertiliser trends and outlook to 2014. FAO: Rome
6. Unwin RJ, Pain BF, Whinham WN (1986) The effect of rate and time of application of nitrogen in cow slurry on grass cut for silage. Agric Wastes 15: 253–268.
7. O'Brien D, Shalloo L, Patton J, Buckley F, Grainger C, et al. (2012) A life cycle assessment of seasonal grass-based and confinement dairy farms. Agric Syst 107: 33–46.

8. Powell JM, Ikpe FN, Somda ZC (1999) Crop yield and the fate of nitrogen and phosphorus following application of plant material and feces to soil. Nutr. Cycl. Agroecosyst. 54: 215–226.

9. DEFRA (2011) Agriculture in the United Kingdom. In: DEFRA (Ed.), Agriculture in the United Kingdom. DEFRA: London, p. 134.

10. Mader P, Fliessbach A, Dubois D, Gunst L, Fried P, et al. (2002) Soil fertility and biodiversity in organic farming. Science 296: 1694–1697.

11. MacLeod M, Moran D, Eory V, Rees RM, Barnes A, et al. (2010) Developing greenhouse gas marginal abatement cost curves for agricultural emissions from crops and soils in the UK. Agric Syst 103: 198–209.

12. FAO (2004) Protein sources for the animal feed industry - Expert Consultation and Workshop. In: Food & Agriculture Organisation of the United Nations (Ed.), Animal production and health proceedings. FAO: Rome

13. Dewhurst RJ, Fisher WJ, Tweed JKS, Wilkins RJ (2003) Comparison of Grass and Legume Silages for Milk Production. 1. Production Responses with Different Levels of Concentrate. Journal of Dairy Science 86: 2598–2611.

14. Marley CL, Fychan R, Fraser MD, Sanderson R, Jones R (2007) Effects of feeding different ensiled forages on the productivity and nutrient-use efficiency of finishing lambs. Grass Forage Sci 62: 1–12.

15. Marley CL, Fychan R, Fraser MD, Sanderson R, Jones R (2009) The effects of feeding ensiled alternative forages compared with ensiled ryegrass on excreta losses from growing lambs; In: Advances in Animal Bioscience. Southport: UK. Cambridge University Press. pp. 162.

16. Chambers BJ, Lord EI, Nicholson FA, Smith KA (1999) Predicting nitrogen availability and losses following application of organic manures to arable land: MANNER. Soil Use Manage 15: 137–143.

17. Sokal RR, Rohlf FJ (1995) Biometry - The principles and practice of statistics in biological research. New York, USA: Freeman and Company. pp. 240.

18. Kanneganti VR, Klausner SD, Kaffka SR (1996) Nitrogen recovery by Orchardgrass from dairy manure. In: 1996 Research Summaries. U.S. Dairy Forage Research Center, Washington, USA., pp 26–28.

19. Payne RW, Murray DA, Harding SA, Baird DB, Soutar DM (2011) Introduction to GenStat for Windows 14th Edition. Hemel Hempstead, UK: VSN International. pp.150.

20. Wardle DA, Bardgett RD, Klironomos JN, Setala H, van der Putten WH, et al. (2004) Ecological linkages between aboveground and belowground biota. Science 304: 1629–1633.

21. Wiens MJ, Entz MH, Wilson C, Ominski KH (2008) Energy requirements for transport and surface application of liquid pig manure in Manitoba, Canada. Agric Syst 98: 74–81.

22. Fortune S, Robinson JS, Watson CA, Philipps L, Conway JS, et al. (2005) Response of organically managed grassland to available phosphorus and potassium in the soil and supplementary fertilization: field trials using grass-clover leys cut for silage. Soil Use Manage 21: 370–376.

23. Groot JCJ, Van Der Ploeg JD, Verhoeven FPM, Lantinga EA (2007) Interpretation of results from on-farm experiments: manure-nitrogen recovery on grassland as affected by manure quality and application technique. 1. An agronomic analysis. NJAS – Wagen J Life Sci 54: 235–254.

24. Komiyama T, Kobayashi A, Yahagi M (2013) The chemical characteristics of ashes from cattle, swine and poultry manure. J Mater Cycles Waste Manag 15: 106–110.

25. Berry PM, Stockdale EA, Sylvester-Bradley R, Philipps L, Smith KA, et al. (2003) N, P and K budgets for crop rotations on nine organic farms in the UK. Soil Use Manage 19: 112–118.

26. Butler TJ, Biermacher JT, Kering MK, Interrante SM (2012) Production and Economics of Grazing Steers on Rye-Annual Ryegrass with Legumes or Fertilized with Nitrogen. Crop Sci 52: 1931–1939.

27. Rochon JJ, Doyle CJ, Greef JM, Hopkins A, Molle G, et al. (2004) Grazing legumes in Europe: a review of their status, management, benefits, research needs and future prospects. Grass Forage Sci 59: 197–214.

28. Zhang W, Ricketts TH, Kremen C, Carney K, Swinton SM (2007) Ecosystem services and dis-services to agriculture. Ecol Econ 64: 253–260.

29. Sorensen P, Jensen ES (1998) The use of N-15 labelling to study the turnover and utilization of ruminant manure N. Biol Fert Soils 28: 56–63.

30. Thomsen IK (2001) Recovery of nitrogen from composted and anaerobically stored manure labelled with N-15. Eur J Agron 15: 31–41.

31. Cushnahan A, Gordon FJ, Ferris CPW, Chestnut DM, Mayne CS (1994) The use of sheep as a model to predict the relative intakes of silages by dairy cattle. Anim Prod 59: 415–420.

32. Kyvsgaard P, Sorensen P, Moller E, Magid J (2000) Nitrogen mineralization from sheep faeces can be predicted from the apparent digestibility of the feed. Nutrient Cycling in Agroecosystems 57: 207–214.

33. DEFRA (2013) Nitrate vulnerable zones. Available: http://www.defra.gov.uk/food-farm/land-manage/nitrates-watercourses/nitrates/. Accessed 25.06.13.

34. Misselbrook TH, Chadwick DR, Pain BF, Headon DM (1998) Dietary manipulation as a means of decreasing N losses and methane emissions and improving herbage N uptake following application of pig slurry to grassland. J Agric Sci 130: 183–191.

35. Velthof GL, Nelemans JA, Oenema O, Kuikman PI (2005) Gaseous nitrogen and carbon losses from pig manure derived from different diets. J Environ Qual 34: 698–706.

36. Cardenas LM, Chadwick D, Scholefield D, Fychan R, Marley CL, et al. (2007) The effect of diet manipulation on nitrous oxide and methane emissions from manure application to incubated grassland soils. Atmos Environ 41: 7096–7107.

37. Jarvis SC, Stockdale EA, Shepherd MA, Powlson DS (1996) Nitrogen mineralization in temperate agricultural soils: Processes and measurement. Advances in Agronomy 57: 187–235.

38. DuPont ST, Culman SW, Ferris H, Buckley DH, Glover JD, (2010) No-tillage conversion of harvested perennial grassland to annual cropland reduces root biomass, decreases active carbon stocks, and impacts soil biota. Agric Ecosyst Environ 137: 25–32.

39. Powell JM, Broderick GA, Grabber JH, Hymes-Fecht UC (2009) Effects of forage protein-binding polyphenols on chemistry of dairy excreta. J Dairy Sci 92: 1765–1769.

40. Powell JM, Grabber JH (2009) Dietary Forage Impacts on Dairy Slurry Nitrogen Availability to Corn. Agron J 101: 747–753.

41. Powell JM, Wattiaux MA, Broderick GA, Moreira VR, Casler MD (2006) Dairy diet impacts on fecal chemical properties and nitrogen cycling in soils. Soil Sci Soc Am J 70: 786–794.

42. Dewhurst RJ, Evans RT, Scollan ND, Moorby JM, Merry RJ, et al. (2003) Comparison of grass and legume silages for milk production. 2. In vivo and in sacco evaluations of rumen function. J Dairy Sci 86: 2612–2621.

43. Fraser MD, Fychan R, Jones R (2001) The effect of harvest date and inoculation on the yield, fermentation characteristics and feeding value of forage pea and field bean silages. Grass Forage Sci 56, 218–230.

44. Chadwick DR, John F, Pain BF, Chambers BJ, Williams JR (2000) Plant uptake of nitrogen from the organic nitrogen fraction of animal manures: a laboratory experiment. J Agric Sci 134: 159–168.

45. Nicholson FA, Chambers BJ, Dampney PMR (2003) Nitrogen value of poultry litter applications to root crops and following cereal crops. J Agric Sci 140: 53–64.

46. Murray PJ, Crotty FV, Van Eekeren N (2012) Management of Grassland Systems, and Soil and Ecosystem Services. In: Wall DH, Bardgett RD, Behan-Pelletier V, Herrick JE, H JT et al., editors. Soil Ecology and Ecosystem Services. Oxford, UK: Oxford University Press. pp. 424

47. van Eekeren N, de Boer H, Bloem J, Schouten T, Rutgers M, et al. (2009) Soil biological quality of grassland fertilized with adjusted cattle manure slurries in comparison with organic and inorganic fertilizers. Biol Fert Soils 45: 595–608.

48. DEFRA (2013) Good agricultural practice, nutrients and fertilisers. Available: http://www.defra.gov.uk/food-farm/land-manage/nutrients/. Accessed 25.06.13.

49. Ashfield A, Crosson P, Wallace M (2013) Simulation modelling of temperate grassland based dairy calf to beef production systems. Agric Syst 115: 41–50.

50. Chadwick D, Sommer S, Thorman R, Fangueiro D, Cardenas L, et al. (2011) Manure management: implications for greenhouse gas emissions. Anim Feed Sci Technol 166–167: 514–531.

51. Velthof GL, Bannink A, Oenema O, Van Der Meer HG, Spoelstra SF (2000) Relationships between animal nutrition and manure quality: a literature review on C, N, P and S compounds. In: Green World Research. Wageningen, The Netherlands: Alterra. pp. 42.

52. Kirkham FW (2001) Nitrogen uptake and nutrient limitation in six hill moorland species in relation to atmospheric nitrogen deposition in England and Wales. J Ecol 89: 1041–1053.

53. Vos J, van der Putten PEL (2000) Nutrient cycling in a cropping system with potato, spring wheat, sugar beet, oats and nitrogen catch crops. I. Input and offtake of nitrogen, phosphorus and potassium. Nutr Cycl Agroecosyst 56: 87–97.

7

Epidemiological Study of Hazelnut Bacterial Blight in Central Italy by Using Laboratory Analysis and Geostatistics

Jay Ram Lamichhane[1,2], **Alfredo Fabi**[1,2], **Roberto Ridolfi**[1], **Leonardo Varvaro**[1,2]*

1 Department of Science and Technology for Agriculture, Forestry, Nature and Energy (DAFNE), Tuscia University, Viterbo, Italy, **2** Hazelnut Research Center, Viterbo, Italy

Abstract

Incidence of *Xanthomonas arboricola* pv. *corylina*, the causal agent of hazelnut bacterial blight, was analyzed spatially in relation to the pedoclimatic factors. Hazelnut grown in twelve municipalities situated in the province of Viterbo, central Italy was studied. A consistent number of bacterial isolates were obtained from the infected tissues of hazelnut collected in three years (2010–2012). The isolates, characterized by phenotypic tests, did not show any difference among them. Spatial patterns of pedoclimatic data, analyzed by geostatistics showed a strong positive correlation of disease incidence with higher values of rainfall, thermal shock and soil nitrogen; a weak positive correlation with soil aluminium content and a strong negative correlation with the values of Mg/K ratio. No correlation of the disease incidence was found with soil pH. Disease incidence ranged from very low (<1%) to very high (almost 75%) across the orchards. Young plants (4-year old) were the most affected by the disease confirming a weak negative correlation of the disease incidence with plant age. Plant cultivars did not show any difference in susceptibility to the pathogen. Possible role of climate change on the epidemiology of the disease is discussed. Improved management practices are recommended for effective control of the disease.

Editor: Anna-Liisa Laine, University of Helsinki, Finland

Funding: The authors have no support or funding to report.

Competing Interests: The authors have declared that no competing interests exist.

* E-mail: varvaro@unitus.it

Introduction

Hazelnut (*Corylus avellana* L.) represents an economically important nut crop of Italy. The country is the second largest producer worldwide after Turkey [1]. In Italy, the production of this crop is concentrated in Campania, Latium, Sicily and Piedmont regions in order of importance [2]. The province of Viterbo has 90% of the hazelnut cultivations of Latium region where Tonda Gentile Romana is the predominant cultivar in over 85% of the orchards [3].

Bacterial blight of hazelnut is caused by *Xanthomonas arboricola* pv. *corylina* (hereafter Xac) [4]. The disease first occurred in the U.S.A. on *Corylus maxima* [5] and further spread in other continents [6,7,8,9,10]. Recent reports of bacterial blight disease on hazelnut regard the countries like Iran [11], Germany [12], Poland [13] and Chile [14] explaining the movement of the pathogen between the countries *via* propagation materials. However, this disease is not widespread in Europe and as such the European and Mediterranean Plant Protection Organizations included this pathogen in the A2 list of quarantine microorganism [9,10]. The damage caused by Xac regards mainly young hazelnut plants (1–4 year old) in orchards killing up to 10% [9,10]. The losses are even more severe in nurseries, where suckering is widely practiced on the mother plants. However, devastating damage can occur also on older (7–8 years) plants [10].

In Italy, Xac was first reported in Latium [15,16] and successively in Campania regions [17]. During the early nineties, endemic presence of the pathogen was described in central Italy [18]. Recently, hazelnut plants infected by Xac were noticed also from the Italian islands [19–21]. Nonetheless, no economically important loss, associated to hazelnut bacterial blight, was reported previously, from central Italy [22].

The current status of the bacterial blight has been changing drastically, for some years now. Frequent occurrence of this disease with severe damage, found across the orchards, was a serious matter of concern by growers. A prime example could be the severe canker symptoms caused by bacterial blight on cv. Tonda di Giffoni [23], the only Italian cultivar that did not bear the canker symptoms in the field for a century [24].

The need to carry out a detailed epidemiological study raised following the outbreaks of this disease across central Italy. Apparently, distribution and incidence of the disease were heterogeneous across the Viterbo province which suggested the possible role of pedoclimatic factors in disease occurrence. Recent finding of the disease in other European countries [12,13] and the outbreaks in central Italy [23] could in part be associated to possible effect of climate change. The role of the latter on crop-disease interaction has been a serious matter of concern by many authors [25–28]. Regarding the soil, its physical and chemical properties play an important role in the plant health and the consequent disease occurrence and spread [29,30]. In addition, crop management practices influence significantly the occurrence and control of hazelnut bacterial blight [31].

Spatial patterns of plant pathogens and diseased plants can facilitate the determination of relationships between inoculum density and disease incidence; optimal sampling parameters; the influence of cultural, biological and environmental factors on population dynamics; and the risk assessment of genetically altered microorganisms [32–36]. Different methods of spatial pattern analysis are used to characterize the spatial position of plant pathogens and diseased plants [37–44]. Spatial autocorrelation functions use the linear correlation between the spatial series and the same series at a further distance interval to detect spatial dependence and have been applied to studies in plant pathology [41,45,46].

Geostatistics represent one of the most applied techniques to study spatially related data [47]. Their application has undergone a rapid expansion in plant disease epidemiology and management [48–52]. More specifically, spatial pattern analysis has been used to investigate factors affecting plant diseases [53–56]. The spread of plant diseases can be estimated by spatial interpolation methods, due to its relation with geographical variables such as soil and climatic characteristics [56]. These techniques were successfully applied to analyze plant disease epidemics even at plot or field scales [33,57–59]. Modelling of the spatial autocorrelation represents a crucial point of geostatistical analysis. This can be performed by examining the variogram estimation. The model variogram is incorporated into a procedure for surface interpolation known as "kriging". The advantage of the latter is of having two outcomes, a surface map of the variable and a surface map of the kriging standard deviation (KSD). The second provides a relative measure of confidence in the estimates [60–63]. Pedoclimatic factors can be considered as regionalized variables and as such they can be investigated by means of geo-statistics and kriging. Furthermore, regression analysis can be applied on frequency distribution of disease incidence in relation to the several spatialized pedoclimatic parameters investigated.

The aims of this study were a) to investigate the current status of the bacterial blight disease and its incidence across the main hazelnut cultivated areas of central Italy and b) to analyze the possible correlation between the disease incidence and the spatial distribution of pedoclimatic factors that can contribute in occurring and spreading of the disease, by means of geostatistics.

Materials and Methods

Ethics Statement

No specific permits were required for the described field studies. At each study site, the landowner granted us permission to collect hazelnut samples. The studies carried out during the consecutive years 2010–2012 did not involve endangered or protected species. Field surveys were made across the hazelnut orchards, in the Province of Viterbo, central Italy while laboratory and green house studies were carried out at Tuscia University.

Description of the study area

Viterbo province is located in Latium region, central-western part of Italy between 42.15° and 42.74° north latitude and 11.60° and 12.44° east longitude (Figure 1). Large-scale mechanized production of hazelnut occurs in the province, which is situated in the inner part of the coastal area adjacent to the Tyrrhenian Sea. The topography of the study sites is hilly, and the elevation ranges from 265 to 520 m above sea level (asl). The hill areas where main hazelnut orchards are located are known as "Cimini Hills".

Field surveys and data collection

The need to investigate on the current status of the bacterial blight disease, across the hazelnut orchards of central Italy, was based in part on growers report of disease outbreaks in hazelnut fields throughout the province during early summer of 2010 [23].

Twelve municipalities, known for the cultivation of hazelnut, within the province of Viterbo were surveyed (Figure 1). The observations were made over the time period between March and September for all the three years. Each municipality had different number of sites, ranging from a minimum of 1 to a maximum of 6, depending on their extent of hazelnut cultivated area. The size of site varied from 5 to 25 ha and, the total number of sites was 30. From each site, 900 hazelnut trees were randomly surveyed (300/ year). Data related to the surveyed municipalities, average age of plant from each site and the cultivars are reported in Table S1. Each site surveyed for bacterial blight infection, was geographically referenced by the Universal Transverse Mercator (UTM) coordinate system [64] with a handheld Global Positioning system (GPS) instrument. The instrument used was a Garmin III Plus (Garmin International Inc., Olathe, KS).

Sample collection and isolation

Samples were collected over the time period between March and September for all the three years. Plant parts showing characteristic symptoms of bacterial blight (Figures 2, 3 and 4), as those reported in literature [6,9,17,18,22,23,31,65,66], were cut, separately put into sterile plastic lab bags and brought to the laboratory. The pruning shears, knife and hand-saw used to cut the samples were sterilized, each time the sample was taken, by immersing them in copper solution for 2 min followed by two rinses in sterile distilled water (SDW), each for 1 min. The number of the samples taken from each site ranged from 15 to 60, depending on the presence and incidence of bacterial blight. In fields where the disease symptoms were sporadic, only few samples were collected within an approximate radius of 100 m. In fields where the disease incidence was high, samples were collected from several neighboring plants. This sampling scheme was adopted to obtain more bacterial isolates possible from different diseased tissues of a given site in order to further investigate the possible phenotypic differences among them.

Immediate sample processing was made, for the isolation of the causal agent. The samples were surface sterilized in 1% NaOCl for 1 min followed by two rinses in SDW (each for 1 min), excised and each section was then crushed in SDW, left for 5 minutes, in order to allow bacterial streaming [23]. Streaking was made by taking a loopful of the resulting suspension onto the surface of the following media: nutrient agar supplemented with 5% sucrose (NAS), yeast extract-dextrose-calcium carbonate agar (YDCA), glucose yeast extract calcium carbonate agar (GYCA) and yeast extract bacto peptone glucose agar (YPGA) media [10,14,23]. The plates were incubated at $26 \pm 1°C$ and examined daily up to three days for bacterial growth.

Disease incidence

One thousand hazelnut trees/site were randomly analyzed each year (total of 3000 trees/site) for the presence of bacterial blight on leaves, stems, fruits, twigs and branches. Any symptoms suggestive to bacterial blight, based on the literature review, were taken into consideration. The incidence of the disease was calculated by the proportion of diseased plants within the total [67]. Statistical analysis were performed using software package Statistica ® (StatSoft, Inc., Tulsa, USA). Student's t test, Pearson correlation and regression analysis were performed to determine the effect of plant age and pedo-climatic factors on disease incidence (DI). However, the influence of plant cultivars on DI was not evaluated for the fact that over 85% of the cultivation in Viterbo is represented by cultivar Tonda Gentile Romana and the remaining are grown in different proportions from one site to another. Moreover, the aforementioned cultivars are the clonal populations

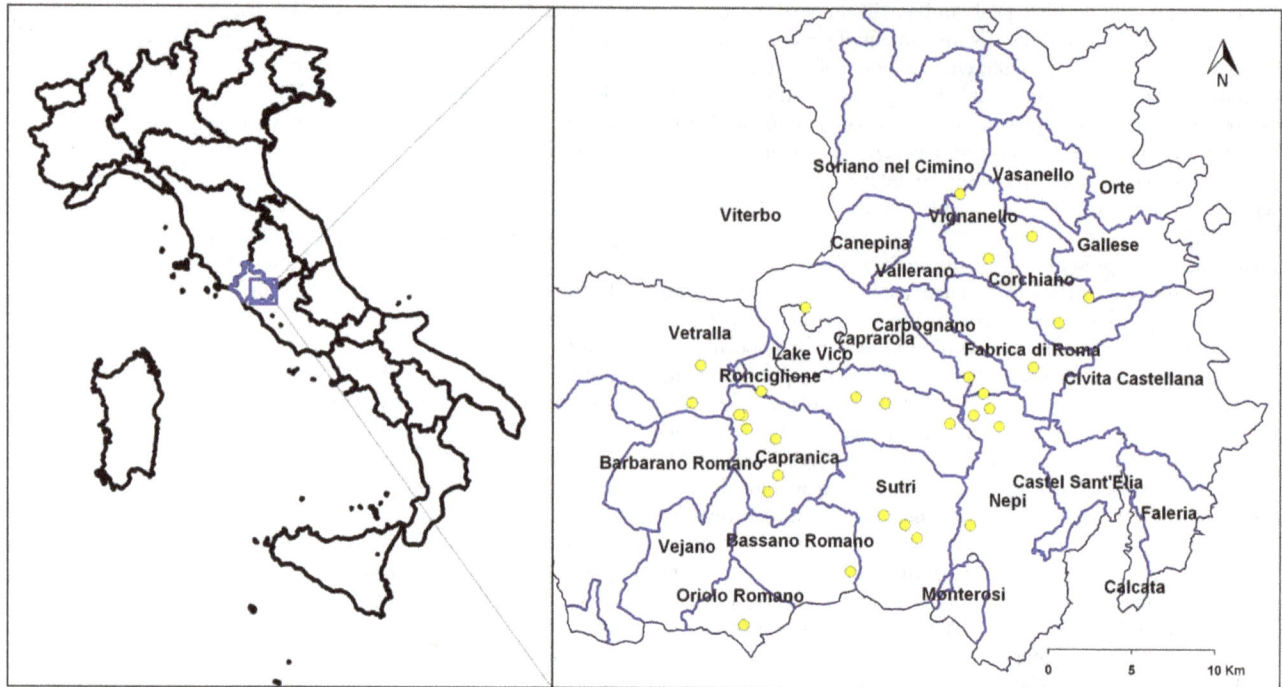

Figure 1. Feature map of the municipalities monitored for bacterial blight. The yellow balls within each municipality represent the number of investigated sites for each municipality, within the province of Viterbo, central Italy.

derived from the asexually multiplied local hazelnut accessions and as such may not present genetic differences.

Characterization of the isolates

A consistent number of bacterial isolates were obtained from the diseased tissues sampled across the sites. The colony morphology was evaluated on the media described above. The shape, size, color, margin and pigment were considered. In addition, the

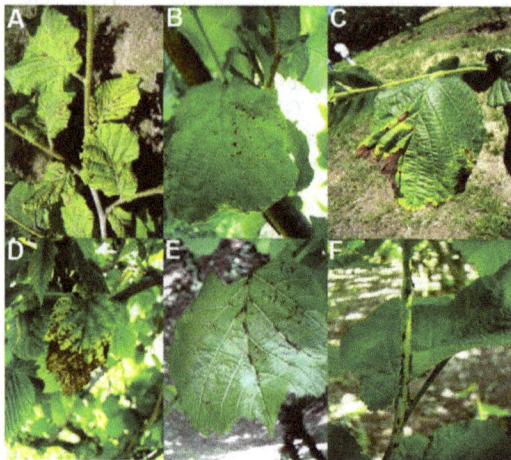

Figure 2. Characteristic leaf symptoms of bacterial blight observed across the hazelnut field in the spring. Water-soaked necrotic spots developed on the leaves at the beginning of the infection (A and B), the presence of extended browning area on the leaves, due to the merging of the necrotic spots, in the late stage of infection (C and D), presence of longitudinal necrosis on the mid-ribs and secondary veins of the lower leaf surface (E) and on young stem (F).

ability of the isolates to metabolize succinate-quinate was tested on SQ medium [68]. Ten isolates from each site were used for the biochemical and nutritional tests prior to their identification. One recommended strain, NCPPB 2896 [10] and strains (Xaco1, Xaco2) isolated recently from central Italy, were used as reference [69]. Successively, the isolates were tested for the production of levan, oxidase, pectinolytic enzyme, urease, indole, catalase, lecithinase, tobacco hypersensitivity, metabolism test, H_2S from peptone and cysteine; hydrolysis of gelatin, aesculin, arbutin and starch; lypolisis of tween 80; tyrosinase activity; tolerance to 0.1% 2,3,5-triphenyltetrazolium chloride (TTC); maximum growth temperature; growth in 2% and 5% NaCl and litmus milk reaction [70,71]. In addition, growth ability by using different carbohydrates as sole carbon source was tested [72].

Lesion tests

The ability of the isolates to induce the hypersensitive defense reaction (HR) was tested on bean pods [73]. The hypersensitive reaction to infection of bean pods by the plant pathogenic

Figure 3. Characteristic symptoms of bacterial blight observed on hazelnut fruit in the field at the beginning of the summer. Oily lesions developed on fruit involucres (A) and fruit shell (B).

Figure 4. Characteristic symptoms of bacterial blight observed on woody parts across the hazelnut field during late summer. Canker symptoms and cracking of bark tissues, caused by *Xatnthomonas arboricola* pv. *corylina*, on the twigs and branches of hazelnut observed in the field (A, B and C), browning of the sub-cortical tissues (D), wilting of shoots (E) and dieback of branches (F). Figures E and F report the symptoms caused by Xac, rarely found in the field and very similar to those caused by *Pseudomonas syringae* pv. *avellanae*.

Xanthomonads, except those pathogenic to bean, was considered as an indicator of bacterial plant pathogenesis [74]. *Pantoea agglomerans* (strain Pag1) was used as negative control [69].

Pathogenicity tests

Two-year-old, healthy potted hazelnut plants (cvs. Tonda Gentile Romana and Tonda di Giffoni) were inoculated with 30 isolates (one from each site) obtained during our study. The inoculum containing 10^6 cfu·mL^{-1} was prepared as described previously [23]. Two different inoculation techniques were used on different plants. In the first case, leaves were spray inoculated (10 mL per plant) with a sprayer containing the bacterial suspension. All the plants were covered with plastic bags from 2 hrs before until 2 hrs after spraying to maintain a high relative humidity (90 to 100%). In the second case, wounds made on the woody parts of the plant (twig and branch) were drop inoculated (0.10 µL per wound). The wounds were then sealed with parafilm (Pechiney Plastic Packaging, Chicago, IL, USA) for three days. For each isolate, eight plants were used (4 plants for each cultivar, 2 plants each inoculation technique). Aforementioned reference strains and the SDW were used respectively as positive and negative controls on the same number of plants. The plants were maintained in the greenhouse. The inoculated plants were inspected weekly until the symptoms appeared.

16S rDNA gene sequencing

Bacterial cultures for DNA extraction were grown on NA medium for 24 h at 26±1°C. DNA was extracted with Pure-Link^{tm}kit (Invitrogen, Carlsbad, CA, USA) following manufacture's instruction. The purity and integrity of DNA was determined by 1% agarose gel electrophoresis. The concentration of DNA was measured by Qubit^{tm} fluorometer (Invitrogen, Carlsbad, CA, USA), and the final concentration was adjusted to 60 ng/µL.

Molecular identification was achieved by sequencing the complete 16S rRNA gene. To this purpose we used the universal primers NOC1F (AGA GTT TGA TCA TGG CTC AG), NOC1R (GTA TTA CCG CGG CTG CTG GCA C), NOC3F

(GCA TGG CTG TCG TCA GCT CGT G) and NOC3R (ACG GTT ACC TTG TTA CGA CTT). The PCR reaction was prepared by mixing 1 µL of 60 ng concentrated template DNA, 1 µL of each primer at 10 µM and 12 µL of DNA Polymerase master mix (GoTaq®Flexi, Promega). The final volume was adjusted at 25 µL by adding SDW. The PCR reactions were performed according to the following program: 1 cycle at 95°C for 5 min., 35 cycles consisting of 30 sec. at 95°C, 1 min. at 55°C and 1 min. at 72°C and 1 cycle (final elongation) at 72°C for 4 min.

The PCR product was loaded on 1% agarose gel electrophoresis to determine the presence and size of the bands. The bands at the expected size were sequenced, analyzed with chromas version 1.45 (32-bit) and aligned with other sequences present in the Genebank by using CLUSTAL W software available online. The similarity of our sequences was compared with those of reference strains present on NCBI by blasting and the sequences were deposited in GeneBank.

Climatic and environmental data

The meteorological data recorded hourly by different stations, located around the Cimini Hills, were collected. Temperature data (°C), only of the last ten years, were recorded by 19 stations across the province. Regarding the rainfall (mm), data collected from 1970 to 2012 from several meteorological stations, most of them not active any longer, were considered. In addition, a database of soil chemical-physical analysis of more than 180 geo-referred sites in the area was set up (Figure 5). Besides meteorological data, some essential soil parameters contributing to the presence and spread of the disease were investigated. In particular, the effects of rainfall [75], total Nitrogen amount (%) in the soil [76,77], soil Mg/K ratio were evaluated. The latter is of considerable importance that explains certain specific Mg deficiency *in planta*, especially on acidic soils [76]. Our study sites, characterized by the volcanic soils, have this deficiency which might increase plant susceptibility to various diseases [78–80]. The influence of lower pH associated to soil aluminium (meq/hg) was investigated given its influence in plant-pathogen interaction [81–83]. Finally, the average thermal shock (Δ), given by the difference among the day/night temperature, associated to every frost event recorded was calculated. The latter was found strictly associated to dieback of hazelnut, caused by *Pseudomonas syringae* pv. *avellanae*, in the same area [84]. All the values were spatialized according the geostatistical methods as described above. Kriging was applied to the real or log-transformed data. An exception was made for the average rainfall data given that they were collected in a period of about forty years since 1970 from several meteorological stations, most of them not active any longer. In particular, rainfall map consisted in a vectorial filled contour layer made up of interpolated level curves with an interval of 1 mm of rainfall. The level curves traced were at a distance of 100 mm for simplicity although the actual resolution was of 1 mm. Estimated meteorological and pedo-chemical parameters, together with their associated standard error (SE), were then correlated with the average DI for each site by means of Regression Analysis (RA). In all the correlation graphs the real DI is expressed in logarithmic scale, thus the correlations between the incidence itself and the various parameters are straight-line logarithmic correlations.

The analyses were performed on the incidence averaged over the three years rather than on each temporal observation independently. Previous preliminary tests conducted on the seasonal epidemiological data, of bacterial canker and bacterial blight of hazelnut, in relation to the seasonal rainfall trend in Viterbo showed no significant correlation at year level. However, a significant correlation was found between the average rainfall of at

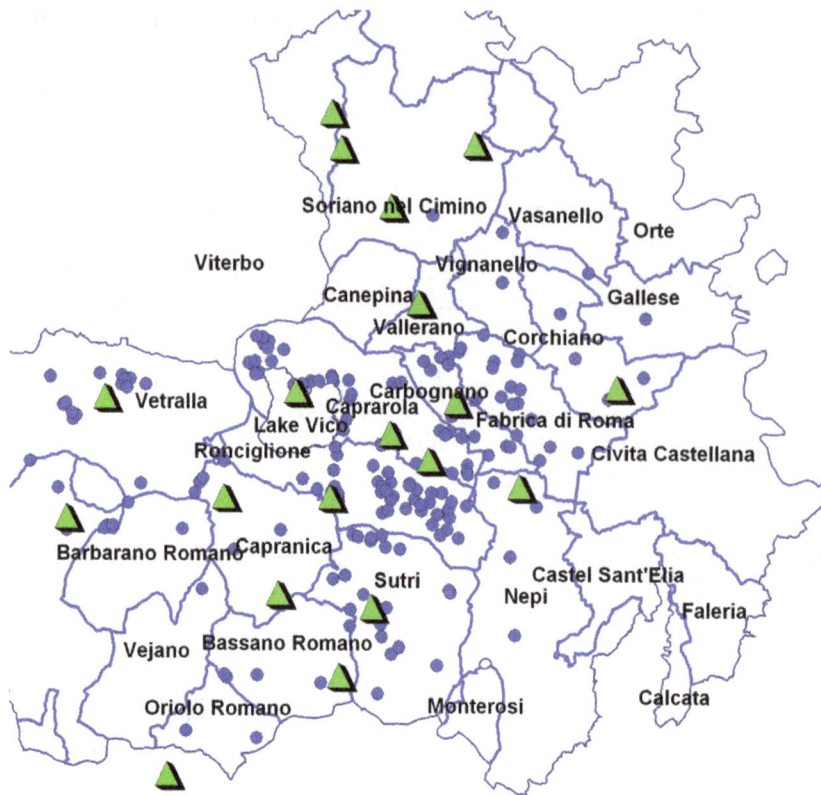

Figure 5. Feature map showing the locations from where pedoclimatic data were collected. Soil analyzed areas (blue beads) and meteorological stations (green triangles) across the hazelnut sites of Viterbo province.

least 10 years and the incidence of bacterial canker caused by *P. avellanae* [84]. Additionally, the spread of the bacterial diseases seemed very less correlated to specific rainfall events which may occur throughout the territory as a rainstorm over the years. Indeed, the disease does not spread with a particular way across the territory, within the single year characterized by particular meteorological events or intense periods of drought, remaining confined to areas more or less circumscribed [85]. All these data suggest that incidence averaged over the three years is much more reliable compared to the single temporal observation, especially when the range of the sites is wide.

Geostatistical analysis and kriging of the meteorological and pedological data

In order to obtain a basic knowledge of the data set, conventional statistical analysis of the pedoclimatic data was prior conducted using software package Statistica ® (StatSoft, Inc., Tulsa, USA). Student's t test, Pearson correlation and regression analysis were performed to determine the correlation of pedoclimatic factors with DI. Prior to conduct geostatistical analyses, on the spatial patterns of data collected from the meteorological stations and sites of Cimini Hills, the latitude and longitude coordinates of the points were first transformed into plane coordinates with a Universal Transverse Mercator (UTM) projection, Zone 33N, European Datum 50, in ArcGIS (version 8.0, ESRI, Redlands, CA). The UTM coordinates were used in the subsequent statistical analyses. The UTM system gives the positions in meters and is preferred over latitude-longitude in decimal degrees. The position coordinates in meters facilitate the correct computation of distances between sample locations in a

plane (two dimensions), essential in a geostatistical analysis. Geostatistical results are presented as semivariograms, which represent the average of squared differences in values between pairs of samples separated by a given distance [86]. Semivariograms are statistical measures that assume normally distributed input sample data where local neighbourhood means and standard deviations show no trends [42,43]. Statistically, abnormal distribution of pedoclimatic data can have an adverse impact on semivariogram analysis and further interpolation of data sets. To moderate this effect, the logarithmic transformation was also applied to measured pedoclimatic data before geostatistical analysis [60,61]. Coefficient of skewness (Cs) and coefficient of kurtosis (Ck) are calculated for pedoclimatic data, before and after the logarithmic transformation. Although transformed data cannot pass relative test for normal distribution, they obviously obtained lower Cs and Ck in comparison with those untransformed ones. Semivariogram calculation was then conducted with log-transformed data.

The experimental variograms of the data were fitted by a spherical model with different ranges and sill variances. Eight lags, with a lag distance of 1600 m, were used for the semivariogram calculation, then the model fitting give us key parameters describing spatial structure of data. These parameters include the Nugget value (N), Partial Sill (pS), Sill (S = N+pS), and Range value. Sill is an estimated semivariance that marks where a plateau begins. The Nugget value presents the variability at zero distance (spatial random variance of regional variable). The pS value presents the sill proportion of explainable semivariance. The Range is defined as the separation distance corresponding at about 95% of the Sill [60,61]. The N/S ratio (NSR) was selected to express short-distance autocorrelation of regionalized variables [87]. Low NSR indicates high spatial autocorrelation or spatial continuity over short distances. The spatial dependence was defined using the nugget to sill ratio convention [88], whereby nugget/sill <0.25, >0.25–<0.75 and >0.75 corresponds to strong, moderate and weak spatial dependence, respectively.

The theory of regionalized variables [89] was applied. The objective was to model and identify the spatial structure of the variable and to estimate its values across the studied area. Linear and nonlinear models were fitted to semivariograms by least squares regression using Geostatistical Analyst module of ArcGIS (ESRI). The pedoclimatic variables modelled were regressed on distance with or without logarithmic transformation. Successively, the Gaussian, spherical, linear, linear to sill, and exponential models, were used to describe semivariograms [90]. The coefficient of determination (r^2), structural variance (proportion of spatial structure), and mean square error (MSE) were used to evaluate the goodness-of-fit of data and to choose the best regression model [91]. Anisotropy was determined by comparison of semivariogram characteristics. Oriented semivariograms that displayed differences among semi-variogram characteristics for different directional orientations indicated anisotropy or directionality in the degree of spatial dependence [86]. Ordinary kriging was applied to estimate logarithmic values of pedoclimatic data for each of 19 meteorological stations or 180 soil analysis sites. In addition, vectorial maps were gained using anti-logging calculation. The whole application, including semi-variogram calculation and kriging (ArcGIS version 8.0, ESRI, Redlands, CA, USA), was applied to log-transformed data. The reliability of the applied model was tested by cross validation between measured and interpolated values. Interpolated data have been depicted as colored raster maps, limiting the prevision to the areas where data were collected.

Results

Disease symptoms in the fields

Initial symptoms began to appear on leaves, during the early spring, when the phase of leaf development was terminated and the temperature was more favorable to leaf infection. Also the stem and new shoots were affected over time. The woody parts of the plants were asymptomatic until the late spring but the symptoms began to appear afterwards.

Water-soaked necrotic spots (Figure 2 A and B) appeared on the leaf surfaces. Initially, only few spots were seen but their number increased from March to September and over the sampling years 2010–2012. The type, the number and the size of the spots varied significantly within and among the orchards (Figure 2). Often, leaves showed numerous oily polygonal lesions which merged together causing a general chlorosis of the lamina. Infected leaves sometimes showed browning of the leaf margins (Figure 2 C and D). Longitudinal browning and necrosis along the mid-ribs and secondary veins on the lower leaf surface were frequently found (Figure 2 E). In some cases, even the lesions on the new stems can be seen (Figure 2 F). The dieback of new lateral shoots was common. During the summer, black heel symptoms with the browning of the shell and corresponding part of the involucres were observed on the fruits. In addition, oily lesions of different size (2–6 mm long) can be seen, on the involucres and shell before lignifications (Figure 3 A and B). The formations of 10–20 cm long cankers, with longitudinal cracking of the bark, were seen along the branches and main trunks, especially on young plants (Figure 4 A and B). Moreover, the appearance of brownish and/or black necrosis was observed over time on the convex side of the layered

branches and shoots (Figure 4 C). When the bark of the infected branch was excised, the necrosis was visible also on the cortical tissues (Figure 4 D). Interestingly, affected twigs were often bent and the swelling was very common on the bent parts. In the late summer, shoot wiltings and necrosis (Figure 4 E) was spread to the trees stump causing the complete dieback of one or more branches (Figure 4 F).

Leaf spots and new shoots dieback were equally present both on the plants younger than 6-year old and those having older age. The presence of canker and the dieback of one or more branches can be seen only on the plants younger than 6 years. All the examined cultivars were infected.

Disease incidence

The incidence of the disease ranged from <1% to almost 75% across the study sites (Figure 6, Table S1). All the surveyed sites presented the characteristic symptoms on the herbaceous parts of the plant whereas the canker symptoms and the consequent dieback of one or more branches were present only on young plants. Equal to the plant age, disease incidence varied from site to site (Table S1). Disease incidence and plant age ahowed a weak negative correlatin (t = −3.74; P<0.001; r^2 = 0.33)(Appendix S1).

Identification of the isolates

Yellow-mucoid, shiny and rounded bacterial colonies were observed on NAS, YDCA, GYCA and YPGA media after 72 hrs of incubation. The colonies were of approximately 2.5 to 3 mm in diameter on NAS but they were slightly smaller on the other media. All of them showed oxidative metabolism and produced the HR on tobacco leaves and bean pods. The isolates were negative for gram, oxidase, indole, lecithinase, urease, tyrosinase, nitrate reduction, growth in 5% NaCl, lypolisis of tween 80 and positive for catalase, growth in 2% NaCl, hydrolysis of gelatin, esculin, arbutin and starch, growth at 35°C and production of H_2S from peptone and cysteine. In addition, all of them gave alkaline reaction to litmus milk, tolerated to 0.1% TTC and produced dark green and yellow pigments on SQ and YDCA media respectively. The isolates used glycerol, D-mannitol, fructose, galactose, trealose, sucrose, glucose, maltose, L-asparagine, D-sorbitol, L-arginine, salicin and inulin as sole carbon sources and caused their acidification except of the last four. All of them were negative for the utilization of L(+) tartrate.

On the basis of the results obtained from the morphological, physiological, biochemical and nutritional tests, the isolates causing the aforementioned symptoms were identified as *Xanthomonas arboricola* pv. *corylina*. The comparison of our sequences with those present in the database further confirmed the results regarding the belonging of our isolates. Our sequences (Accession numbers JQ861273, JQ861274 and JQ861275) shared 99% identity with the analogous sequences of Xac available in the NCBI database.

Lesion tests

All Xac isolates produced water-soaked lesions at the site of inoculation of bean pods, after 4 days. No symptoms were observed on those inoculated with *P. agglomerans* (Figure S1).

Pathogenicity tests

All the isolates reproduced the symptoms of the disease on the plants artificially inoculated. Water-soaked necrotic spots (Figure S2 A) followed by shoot witlings (Figure S2 B) were observed on leaves within a month after the artificial inoculation. By contrast, canker symptoms began to appear only after a month (approx. 5

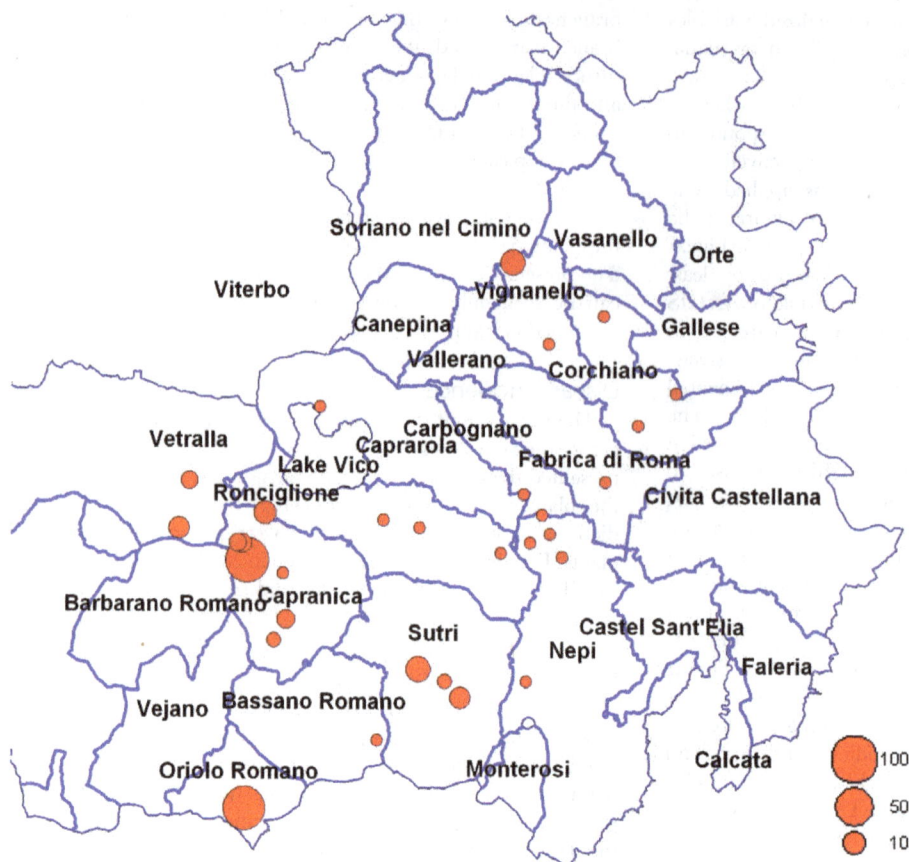

Figure 6. Incidence of the bacterial blight (%) across the study sites in the province of Viterbo. The red circle size inside the map indicates the different disease incidence expressed in logarithmic scale.

weeks) (Figure S2 C). The canker observed on the artificially inoculated plants were smaller (2–4 cm) than those observed in the field. Results showed that both of the cultivars are similarly susceptible (t = 0.59; P = 0.64; r^2 = 0.23) to the pathogen. No significant difference (t = 1.38; P = 0.43; r^2 = 0.17) in time of symptom appearance was found among the isolates and reference strains. Plants inoculated with the reference strains produced the same symptoms whereas control plants inoculated with SDW remained healthy. Bacteria re-isolated from the diseased parts of the plants had the same characteristics of the inoculated strains.

Statistical analysis and kriging of the meteorological and pedoclimatic data

Details on the type of distribution (leptokurtic and platykurtic) of pedoclimatic data and the difference between normal and log-transformed data are provided in Appendix S2.

Table 1 shows the semivariogram parameters of the spherical model applied to the pedoclimatic data. The spatial dependence ranged from moderate (for the total nitrogen and Mg/K ratio) to high (for the thermal shock). In particular, total nitrogen data had an N/S ratio of 0.636, inferring moderate spatial dependence. This means that 36.4% of the total variation in total nitrogen present can be explained by spatial variations while the remaining 63.6% was attributable to unexplained sources of variations. For thermal shock, an N/S ratio of 0.153 was indicative that 84.7% of total variation was spatial variation while only 15.3% was due to other sources of variation. The spatial dependence was moderate also for soil aluminium and pH (Table 1).

Summary statistics of cross validation prediction errors (CVPE) applied to Log data are shown in table 2. Here the term "prediction error" was used for the difference between the prediction and the actual measured value. Cross-validation provides indexes useful to determine the goodness of the model used in the study. For a model that provides accurate predictions, the mean prediction error (MPE) should be close to 0 if the predictions are unbiased, the standardized root-mean-square prediction error should be close to 1 if the standard errors are accurate, and the root-mean-square prediction error should be small if the predictions are close to the measured values. The goal should be to obtain the standardized mean prediction errors (SMPE) close to 0, the small root-mean-square prediction errors (RMSPE) and the average standard error (ASE) near root mean

Table 1. Semivariogram parameters object of study.

Parameter	N	pS	N/S ratio	Range (m)
Total Nitrogen	0.116	0.066	0.636	12649
Mg/K ratio	0.082	0.102	0.446	5635
Δ	0.005	0.026	0.153	10335
Aluminium	1.222	1.592	0.434	12648
Soil pH	0.008	0.005	0.611	12650

N: nugget; pS: partial Sill; S: sill; N/S ratio = [N/(N+pS)]; Δ: thermal shock.

Table 2. Summary statistics of cross validation prediction errors applied to log-transformed data.

Parameter	Mean	RMSPE	ASE	SMPE	SRMS
Total Nitrogen	0.000089	0.052660	0.053480	−0.002782	0.975300
Mg/K ratio	−0.012700	0.349200	0.292900	−0.059400	1.235000
Δ	−0.134000	1.553000	1.883000	−0.051540	0.825100
Aluminium	0.106900	0.908300	4.045000	0.020600	0.283700
Soil pH	−0.007366	0.561900	0.537800	−0.011860	1.040000

RMSPE: root mean square prediction error; ASE: average standard error; SMPE: standardized mean prediction error; SRMS: standardized root mean square; Δ: thermal shock.

square prediction error (RMSPE); and the standardized root-mean-square prediction errors (SRMSPE) close to 1.Among the investigated parameters, the best fitting model was found for total nitrogen followed by soil pH, Mg/K ratio, thermal shock and then for aluminium (Table 2).

Appendix S1 reports the correlation results of statistical tests, p-values and Pearson correlation coefficient matrixes of the logarithmic disease incidence (Log DI) with pedoclimatic variables and plant age. The degree of freedom for the correlation analysis (Appendix S1; Table 1), Pearson correlation analysis (Appendix S1; Table 2) and regression analysis (Appendix S1; Table 1) are 28, 22 and 7.22, respectively. Values showed a strong positive correlation of DI with average rainfall (t = 11.54; P<0.001; $r^2 = 0.82$), thermal shock (t = 11.65; P<0.001; $r^2 = 0.82$) and average soil nitrogen (t = 17.34; P<0.001; $r^2 = 0.91$). Additionally, the correlation was strong negative for Mg/K ratio (t = −10.64; P<0.001; $r^2 = 0.80$) and weak positive for aluminium (t = 3.99; P<0.001; $r^2 = 0.36$). No correlation of DI was found with soil pH (t = −1.10; P = 0.28; $r^2 = 0.04$).

The incidence of bacterial blight, expressed in logarithmic scale, in relation to the average rainfall data of the last 40 years, are reported in Figure 7A. The incidence appeared well correlated with the higher average annual rainfall (1000–1300 mm rainfall classes of rain per year). The correlation was further confirmed by linear regression analysis ($r^2 = 0.82$) (Figure 8A, Appendix S1).

The thermal shock values (in °C) in correspondence to the disease incidence are shown in figure 7B. The incidence increased across the areas where these values ranged between 13 and 15°C. The linear regression between the data of the heat shock, estimated with the logarithm of the real values of the disease incidence confirmed the correlation. The significance of the correlation itself was confirmed by the value of r^2 equal to 0.82 (Figure 8B; Appendix S1). The change in temperature was divided into eight classes covering a distance of values between 8.5 and 14.5°C, which are also associated with a kriging standard error (from 1.784 to 2.128).

The estimated total amount of nitrogen, divided into eight classes, and the incidence of bacterial blight are demonstrated in Figure 7C. The lowest class corresponded to an estimated content of nitrogen that varied from 0.015 to 0.075% up to the highest class, where the content of this element was between 0.21 and 0.375%. A kriging standard error (ranging from 0.047 to 0.06) was associated to the prediction of the amount of total nitrogen in the soil. A direct linear relationship between the nitrogen contents and the logarithm of the disease incidence was confirmed by the linear regression value ($r^2 = 0.91$) (Figure 8C; Appendix S1). A low global standard error was observed for this correlation which is able to

better discriminate the DI per site in relation to the nitrogen content. The Mg/K ratio in the soils, divided into eight classes, are reported in Figure 7D. The ratio ranged from the minimum of 0.1 up to the maximum of 2.7. For each of these values a standard error (between 0.297 to 0.412) was associated. At low Mg/K ratios the disease incidence was greater showing an inverse correlation. The correlation was confirmed by analysis of linear regression with an r^2 value of 0.80 (Figure 8D; Appendix S1).

Pearson correlation values among each pair of variables are reported in Appendix S1. Results showed that nitrogen was negatively correlated with Mg/K ratio (P<0.001; r = −0.95) and plant age (P<0.001; r = −0.54) while it was positively correlated with thermal shock (P<0.001; r = 0.95), rainfall (P<0.001; r = 0.95) and aluminium (P<0.001; r = 0.68). Likewise, Mg/K ratio was negatively correlated with thermal shock (P<0.001; r = −0.92), rainfall (P<0.001; r = −0.96), soil aluminium (P<0.001; r = −0.68) and positively correlated with plant age (P<0.001; r = 0.62). Equally, a negative correlation of thermal shockwith plant age (P<0.001; r = −0.48) and a positive one with rainfall (P<0.001; r = 0.93) and soil aluminium (P<0.001; r = 0.64) were observed. Regarding the rainfall, its correlation was negative with plant age (P<0.001; r = −0.59) and positive with soil aluminium (P<0.001; r = 0.65). Finally, a moderate negative correlation of soil aluminium (P<0.01; r = −0.49) was observed with plant age whereas there was a weak positive correlation (P<0.05; r = 0.38) among soil pH and plant age. It is important to note that soil pH did not show any correlation with the pedoclimatic factors analyzed except the weak positive correlation found with plant age as described above.

Results of multiple regression analysis are shown in Appendix 1. Only the regression coefficients for nitrogen were highly significant (β = 1.20; t = 4.83; P<0.001). It means that the null hypothesis (H₀: $β_i = 0$ for all i, where $β_i$ = regression coefficient for every factor) can be rejected for this value. By contrast, the regression coefficient for Mg/K ratio (β = 0.36; t = 1.55; P = 0.13), thermal shock (β = 0.03; t = 0.16; P = 0.87), rainfall (β = 0.06; t = 0.29; P = 0.77), aluminium (β = −0.10; t = −1.34; P = 0.19), soil pH (β = −0.03; t = −0.49; P = 0.62) and plant age (β = −0.13; t = −1.75; P = 0.09) were not significant and as such the null hypothesis cannot be rejected.

The correlation maps of DI with soil aluminium content and pH, reported into eight classes, are shown in Appendix S3. The values of linear regression analysis demonstrated a weak positive correlation between the DI and soil aluminium content ($r^2 = 0.36$) whereas no correlation of DI was found with soil pH (Appendix S1; Appendix S3). . Moreover, in both cases, the associated standard error of the two pedological parameters was found to be generally higher for all the sites if compared to the other cases described above (Appendix S3).

Discussion

Our study reports some important results which are significant to explain the role of pedoclimatic factors in the occurrence and spread of bacterial blight disease. We did not observe phenotypic and pathogenic differences among Xac isolates, obtained during the years 2010–2012. This indicates that different DI across our study sites might not be related to the different degrees of virulence of the isolates. In addition, equal susceptibility of hazelnut cultivars to artificial inoculations further confirms this hypothesis. However, It should be emphasized that hazelnut grown in the Latium region (cvs. Tonda Gentile Romana, Tonda di Giffoni and Nocchione) are the clones derived from local cultivars which are even similar to Spanish ones [92]. The aforementioned cultivars are the result

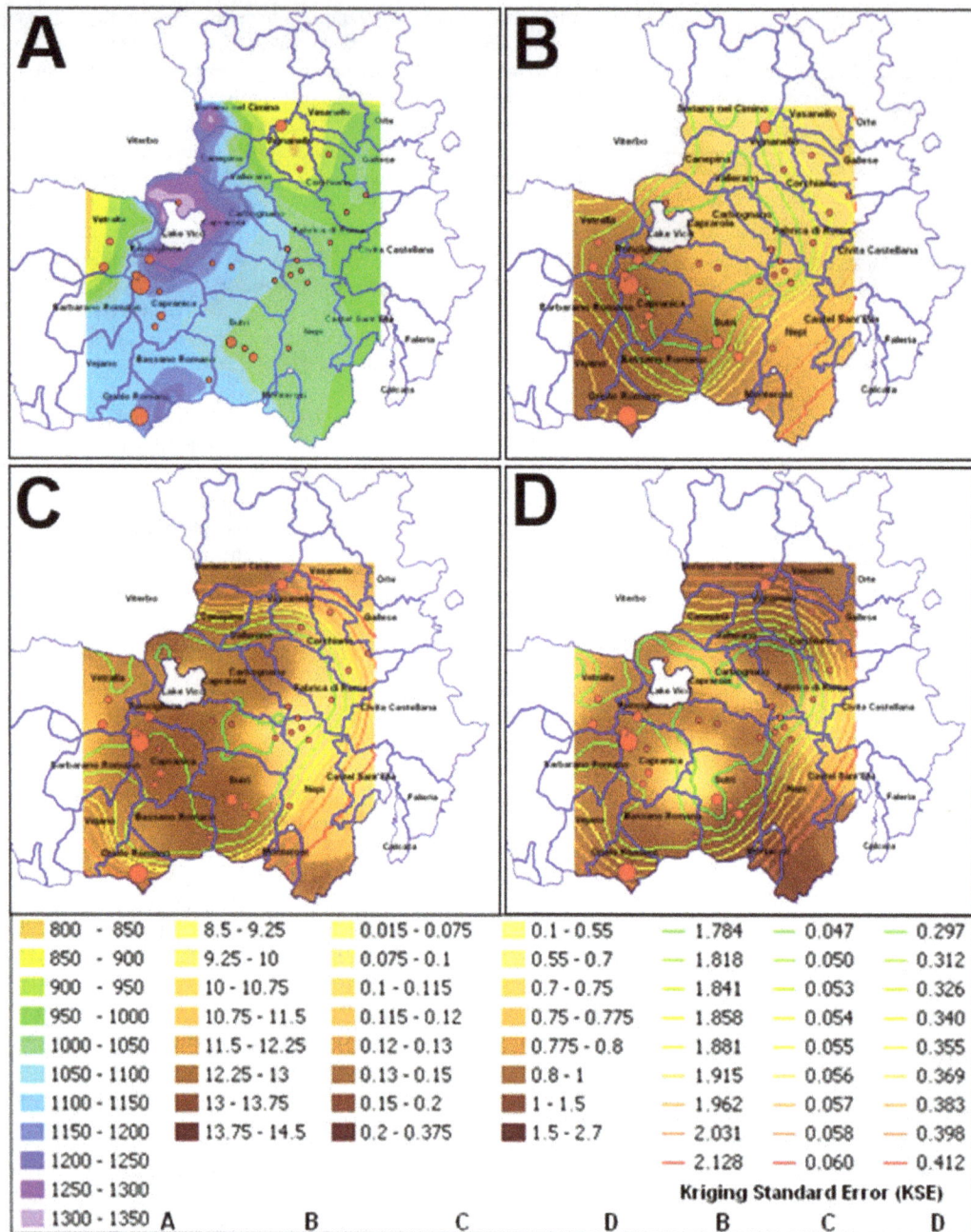

Figure 7. Map illustrating the correlation of average disease incidence (%) with different pedoclimatic factors. Average rainfall values (A), kriging of thermal shock associated with frost (B), kriging of nitrogen content in the soil (C) and kriging of Mg/K ratio in the soil (D). The values are expressed in the following units: rainfall (mm), thermal shock (Δ = °C) obtained from the difference between maximum and minimum daily temperature, content of nitrogen in the soil (%) and the last is a ratio between magnesium (Mg) and potassium (K) content in the soil. The red circle size inside the maps indicates the different disease incidence (<1% to 75%) expressed in logarithmic scale.

of asexual multiplication carried out by farmers, in practicing traditional agriculture system, in the years sixties-eighties [3,93]. The lack of genetic breeding programs of this crop resulted in the diffusion of the same clones over the years across Italy by increasing crop vulnerability to biotic and abiotic stresses [94]. The true variability in infectivity profiles can be detected only by using different host cultivars, not existing yet in Italy. A previous detailed study on the genetic and phenotypic characterization demonstrated that Xac populations of different geographic origin,

including that of central Italy, are homogeneous [95] which is in full agreement with the results of our phenotypic tests.

Recent disease outbreaks on cv. Tonda di Giffoni in the province of Viterbo with the presence of longitudinal canker might have an explanation. "Tonda di Giffoni" is a typical cultivar grown in Campania region, which is located to the southern part of Latium region having a mild winter temperature and rare spring frosts [24]. Indeed, the presence of bacterial blight symptoms on leaves and sprouts was described from Campania region but canker formation did not occur [24]. The cultivar has

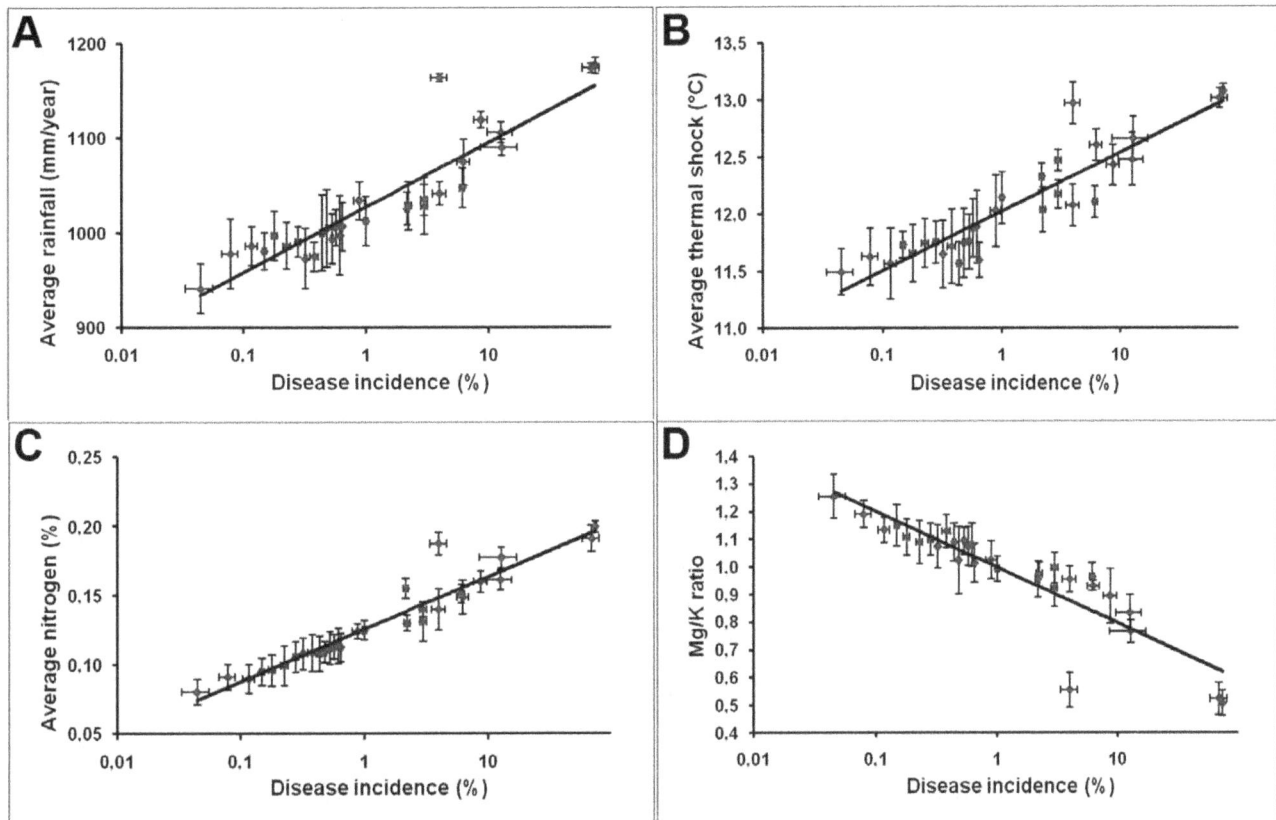

Figure 8. Correlation of the average disease incidence (%) of each site with different pedoclimatic factors. (A) average rainfall values (t = 11.54; P<0.001; r^2 = 0.82), (B) thermal shock associated with frost (t = 11.65; P<0.001; r^2 = 0.82), (C) nitrogen content in the soil (t = 17.34; P<0.001; r^2 = 0.91) and (D) Mg/K ratio in the soil (t = −10.64; P<0.001; r^2 = 0.80). The horizontal and vertical bars represent the standard error of disease incidence and pedoclimatic variables, respectively. Six were the average number of replicate in each site (n = 6).

been introduced very recently by growers in Latium, a region with cold winter temperatures and frequent spring frosts [84,96], because this cultivar is characterized by a vigorous growth and an early production. Introduction of this cultivar to areas of Latium region may have predisposed trees to infection by Xac which has an epiphytic phase and enters and cause the canker through wounds [7–10].

Concerning the relationship of plant age with the DI caused by Xac, no strong correlation was found among these parameters. However, the DI was weak and negatively correlated to the plant age confirming that young plant are the most affected by the disease as previously reported [9,10]. Geostatistical analysis showed different levels of spatial autocorrelation for spatialized parameters. The best autocorrelation was found for the thermal shock (N/S ratio<0.25). The other two parameters showed less spatial autocorrelation (N/S ratio between 0.25 and 0.75). In all cases, the spherical model adopted results to producing good CVPE [97–99], especially for total nitrogen.

In recent years, a large number of studies were made on plant disease epidemics using spatial pattern analysis in order to relate the observed characteristics of epidemics to the underlying ecological and pathological processes [100–102]. However, specific studies on the spatial pattern analysis of the pedoclimatic factors affecting plant diseases are lacking in the literature. Often, the role of pedoclimatic factors on the occurrence and spread of the diseases is ignored by the researchers. The lack of information on the correlation between the plant disease incidence and pedoclimatic factors hindered our ability to develop more

sustainable management strategies for many plant diseases [103]. The latter continue to cause large losses on a wide range of important crops because management practices, including the use of genetic resistance, are not complete and in many cases, sole reliance on chemical products has led to pathogen resistance in the field for many species in the genus [104–110]. This study demonstrates how the knowledge of pedoclimatic factors could clearly lead to better understand the role of different factors on plant disease occurrence and spread.

Geostatistic analysis showed significant correlation among the DI and investigated pedoclimatic parameters, from geostatistical point of view. Disease incidence appeared correlated to higher values of annual rainfall, a higher content of nitrogen in the soil, higher thermal shock values and a lower Mg/K ratio in the soil. Additionally, a weak positive and negative correlations of soil aluminium and plant age, respectively, were detected with DI. However, The last two factors have less impact on disease occurrence and development compared to other pedoclimatic variables investigated. The significance of these correlations is likely the consequence on the infection process and spread of the disease.

Higher nitrogen level in contributing disease development [77,111] is of vital concern. However, it must be emphasized that the higher nitrogen levels found across our study sites, are not high, in absolute terms and might not be due to the natural presence of this trophic element. Especially, large amounts of mineral nitrogen, indiscriminately used in the last five decades, in the name of intensive agriculture might have unbalanced significantly the nitrogen content in the soil. The return to balanced fertilization practices with regard to nitrogen may well be in favor of a restoration of balance that was lacking for long time. Previously, a possible problem caused by the unbalanced fertilizations, as a cause of increased occurrence and spread of hazelnut diseases in the studied area, was hypothesized [112].

High coefficient of determination value confirmed a strong correlation among the average rainfall and DI. Rainfall is considered one of the most influencing factors on phytobacterial disease occurrence [113]. Since Xac has the epiphytic phase on hazelnut plants [9,10], probably, it becomes airborne in splash droplets during the rainfall, as other epiphytic bacteria [114]. In addition, the importance of rainfall in triggering the multiplication of the epiphytes and the consequence for the epidemiology of bacterial diseases of plants have been demonstrated [115]. Once bacteria multiply on the phylloplane high levels of bacterial populations can be reached. As a consequence, during the rainfall period (common across our study sites), the bacterial pathogen spreads at a rate that might cause severe infection, as bacteria within lesions are released very readily from wet leaves [116–119]. Moreover, the potential number of cells available for dispersal from a single spot on a single infected leaf may be greater than the entire epiphytic population on the surfaces of several thousand leaves or plants [78]. Previous reports suggested that the presence of leaf symptoms, caused by Xac, are rare in orchards [9,10]. In contrast to that report, we found very frequent presence of leaf symptoms across the study sites, especially in certain areas where the DI was almost 75%. Since hazelnut cultivation is practiced within a very concentrated area of Viterbo Province, the pathogen can easily disperse through the rainfall rapidly across the entire cultivation sites. These considerations might turn very useful in order to take preventive measures in controlling disease occurrence.

In addition to the key role, rainfall might play a secondary role in causing disease occurrence across our study sites. The presence of compact soil is common in the area due to the lack of soil breaking up practices [120]. As the consequence, soil aeration, drainage and the consequent root elongation did not occur. Root asphyxia phenomenon is very frequent during the major raining seasons, causing plants general suffering. All these stress conditions predispose plants to bacterial attacks [121].

The relationship between the DI and lower values of Mg/K ratio is more difficult to interpret. No study in literature is available in this regard. However, this might be partly explained by the fact that magnesium is an antagonist of potassium or *vice versa* because of the competition between the ions [122–125]. Since Mg^{2+} and K^+ as Ca^{2+} are similar in size and charge, exchange sites cannot distinguish the difference between them. An indiscriminate acceptance of either of these ions (the most abundant), at the expense of other, at the exchange sites, is the consequence. Moreover, the binding strengths of K^+ is much stronger than Mg^{2+} and for this the first easily out-competes the second at exchange sites [122]. The presence at very high concentration of one of these ions results in a complete suppression in uptake mechanism of the other [124–131]. It is well established that most of our study sites has the soils of volcanic origin, rich of potassium [132], confirming the reason of lower Mg/K ratio. Magnesium is one of the essential elements for plant growth and production [133,134] and its deficiency renders plants susceptible predisposing them to the disease.

A strong correlation is likely to show the role of thermal shock on the DI. The higher thermal shock values, especially those referred to the critical periods (presence of night frost), damage the plant tissues and create the micro-lesions. The latter represent an ideal route of entry for bacteria and the consequent infection begin process. Recent studies showed higher thermal shock values registered across Viterbo province [84,96] with consequent effects on the health status of hazelnut plants.

In addition to the parameters described above, the role of soil aluminium (Al) content associated to lower pH values might play an important role in disease occurrence. Previous studies showed the presence of acidic soil (pH<5) across some of our study sites [120,135] which are much lower than the optimal pH values (5.5–7.8), for hazelnut crop [136]. Soils with lower pH values can increase the susceptibility of fruit tree species to bacterial diseases, especially to those caused by pseudomonads [81–83]. In addition, lower pH values result fatal in association to higher Al in soil. This metal can be present in different forms in the soil. The trivalent Al form, (Al^{3+}), dominates in acid condition (pH<5) which is toxic to plants [137]. Poor root elongation, growth reduction and premature aging of the plants are the consequence [138,139]. The phytopathogenic bacteria, in these circumstances, easily explicate their virulence causing the disease. However, we did not find any significant correlation among the low pH, high aluminium soil values and the incidence of bacterial blight. Pearson correlation coefficient matrix, among soil pH and aluminium, further confirmed no significant correlation with the DI. This is probably because acidic soils do not increase host susceptibility to diseases caused by Xanthomonads, unlike those described for pseudomonads. Nonetheless, the poor correlation of the DI to soil aluminium values is also affected by the relatively high CVPE values of the model applied.

Factors that depend strictly by agronomic practices (total nitrogen and Mg/K ratio of soil), can be improved on behalf of crop requirement. Nevertheless the negative effect of rainfall and thermal shock are not easily manageable. In this context, the potential effects (direct and indirect) of climate change should not be overlooked. Increased atmospheric CO_2, heavy and unseasonal rains, increased humidity, drought, cyclones and hurricanes and warmer winter temperature are the major climate change factors influencing disease occurrence, severity and spread [25,140–144]. Changes to any one of these climatic factors can affect the distribution and biology of plant pathogens with very serious economic consequences [145,146]. Several studies hypothesized the increasing CO_2 as the main cause of changes in crop architecture, leading to increased humidity within the canopy and more suitable condition for pathogen survival [28,147]. In addition, increased photosynthetic rate, under elevated CO_2 levels [146] might lead to the availability of new growth flushes earlier in the season for pathogen to colonize and multiply in [27]. Changes in rainfall pattern and temperature might affect the epidemiology of plant diseases including the survival of primary inoculums [148], the rate of disease progress and even the duration of epidemics [27]. These phenomena might be favorable for growth, multiplication and spread also of hazelnut bacterial blight pathogen.

The results presented here confirm the role of some pedoclimatic factors in the occurrence and spread of hazelnut bacterial blight disease. This is the first epidemiological study, based on detailed data, on this disease from central Italy. Improved crop management, through adequate agronomic techniques, is possible across our study sites, especially in the areas where the risk of bacterial blight is higher. Detailed information on biology, epidemiology and control of Xac are described [7,8,9,10]. This information is essential for the effective control of the pathogen. However, besides the general control measures, the disease strictly related to pedoclimatic conditions, at local level requires specific control measures. Referring to our study sites, improvement of soil drainage system, through soil breaking up practices might be necessary. The latter might be sufficient once every five years, provided that it is done at greater depth, without compromising mechanical harvest. The application of copper-based compounds, the only chemicals allowed in Italy, should be done following the phenomena that cause lesions on plants (pruning, spring frost and hail). Balanced nitrogen fertilization is essential across our study sites. Increased use of mineral fertilizers, in the last five decades, caused serious problems in soil mineral equilibrium. Lower soil pH values in the area are attributed to the indiscriminate use of soil acidifying nitrogen fertilizers [135]. Reduction in use of mineral nitrogen and increased use of organic substances must be done. Soil liming is advisable in areas where pH values are particularly lower to avoid plant stress [136]. An increase in pH values in these soils can result in increased absorption of calcium, magnesium and potassium [138]. Finally, the problem of lower Mg/K ratio can be addressed by the application of foliar fertilization of magnesium salts.

Supporting Information

Appendix S1 Statistical tests, Pearson correlation table and multiple regression analysis of the disease incidence with pedoclimatic factors and plant age.

Appendix S2 Transformation of pedoclimatic variables.

Appendix S3 Maps and correlation graphs of disease incidence with soil aluminium, pH and plant age.

References

1. FAO statistics (2010) Available: http://www.faostat.fao.org/site/339/default.aspx. Accessed 2012 Feb 9.
2. Me G, Valentini N (2006). La corilicoltura in Italia e nel mondo. Petria 16: 7–18
3. Pedica A, Vittori D, Ciofo A, De Pace C, Bizzarri S, et al. (1997) Evaluation and utilization of *C. avellana*. Genetic resources to select clones for hazelnut varietal turnover in the Latium region (Italy). Acta Hort 445: 123–134.
4. Vauterin L, Hoste B, Kersters K, Swings J (1995) Reclassification of *Xanthomonas*. Int J Syst Bacteriol 45: 472–489.
5. Barss HP (1913) A new filbert disease in Oregon. Oregon Agricultural Experiment Station Biennial Crop Pest and Horticulture. Rep 14: 213–223.
6. Bradbury JF (1987) *Xanthomonas campestris* pv. *corylina*. CMI Descriptions of Pathogenic Fungi and Bacteria No. 896. CAB International, Wallingford, UK.
7. OEPP/EPPO (1986) Data sheet on quarantine organisms, 134: *Xanthomonas campestris* pv. *corylina* (Miller et al. 1940) Dye 1978. OEPP/EPPO Bull 16: 13–16.
8. OEPP/EPPO (1990) Specific quarantine requirements. EPPO Technical Documents No. 1008.
9. OEPP/EPPO (2004) Diagnosis protocols for regulated pests *Xanthomonas arboricola* pv. *corylina*. OEPP/EPPO Bull 179: 179–181.
10. OEPP/EPPO (2004) Diagnosis protocols for regulated pests *Xanthomonas arboricola* pv. *corylina*. OEPP/EPPO Bull 34: 155–157.
11. Kazempour MN, Ali B, Elahinia SA (2006) First report of bacterial blight of hazelnut caused by *Xanthomonas arboricola* pv. *corylina* in Iran. J Plant Pathol 88: 341.
12. Poschenrieder G, Czech I, Friedrich-Zorn M, Huber B, Theil S, et al. (2006) Ester nachweiss von *Pseudomonas syringae* pv. *coryli* (pv. nov.) and *Xanthomonas arboricola* pv. *corylina* an *Corylus avellana* (Haselnuss) in Deutschland. Bayerische Landesanstalt fur Landwirtschaft-Insitut fur Pflanzenschutz. Jahrb pp. 32–33.
13. Pulawska J, Kaluzna M, Kolodziejska A, Sobiczewski P (2010) Identification and characterization of *Xanthomonas arboricola* pv. *corylina* causing bacterial blight of hazelnut: a new disease in Poland. J Plant Pathol 92: 803–806.
14. Lamichhane JR, Grau P, Varvaro L (2012) Emerging hazelnut cultivation and the severe threat of bacterial blight in Chile. J Phytopathol. doi:10.1111/jph.12004
15. Petri L (1932) Rassegna dei casi fitopatologici osservati nel 1931. Bollettino R. stazione Pat. Veg Roma 12: 1–64.
16. Petri L (1933) Rassegna dei casi fitopatologici osservati nel 1931. Bollettino R. stazione Pat. Veg Roma 13: 1–73.
17. Noviello C (1968) Osservazioni sulle malattie parassitarie del nocciolo con particolare riferimento alla Campania. Annali della Facoltà di Scienze Agrarie dell'Università di Napoli, Portici. Ann 4: 3–31.
18. Scortichini M, Rossi MP (1991) Presenza endemica di *Xanthomonas campestris* pv. *corylina* in noccioleti del Lazio. Inf Fitopatol 41: 251–256.
19. Fiori M, Loru L, Marras PM, Virdis S (2006) Le principali avversità del nocciolo in Sardegna. Petria 16: 71–88.
20. Siscaro G, Longo S, Catara V, Cirvilleri G (2006) Le principali avversità del nocciolo in Campania. Petria 16: 59–70.
21. Virdis S (2008) Studio delle principali malattie del nocciolo in Sardegna. Phd Thesis, University of Sassari, Sardegna, Italy. 72 p.
22. Varvaro L (1993) Le fitopatie del nocciolo nell'alto Lazio: un triennio di osservazioni e di strategie di lotta. Inf Fitopatol 2: 54–58.
23. Lamichhane JR, Fabi A, Varvaro L (2012) Severe outbreak of bacterial blight caused by *Xanthomonas arboricola* pv. *corylina* on hazelnut, cv. Tonda di Giffoni, in central Italy. Plant Dis 96: 1577.
24. Mazzone P, Ragozzino A (2006) Le principali avversità del nocciolo in Campania. Petria 16: 19–30.
25. Chakraborty S, Newton AC (2011) Climate change, plant diseases and food security: an overview. Plant Pathol 60: 2–14.
26. Fitt BDL, Fraaije BA, Chandramohan P, Shaw MW (2011) Impacts of changing air composition on severity of arable crop disease epidemics. Plant Pathol 60: 44–53.
27. Luck J, Spackman M, Freeman A, Trebicki P, Griffiths W, et al. (2011) Climate change and diseases of food crops. Plant Pathol 60: 113–121.
28. Pangga R, Hanan J, Chakraborty S (2011) Pathogen dynamics in a crop canopy and their evolution under changing climate. Plant Pathol 60: 70–81.
29. Broders KD, Wallhead MW, Austin GD, Lipps PE, Paul PA, et al. (2009) Association of soil chemical and physical properties with *Pythium* species diversity, community composition, and disease incidence. Phytopathology 99: 957–967.
30. Duffy BK, Ownley BH, Weller DM (1997) Soil Chemical and Physical Properties Associated with Suppression of Take-all of Wheat by *Trichoderma koningii*. Phytopathology 87: 1118–1124.
31. Miller PW (1949) Filbert bacteriosis and its control. Oregon Agricultural Experiment Station Technical Bulletin No. 6.
32. Campbell CL, Noe JP (1985) The spatial analysis of soil-borne pathogens and root diseases. Annu Rev Phytopathol 23: 129–148.

Figure S1 Reaction observed on the bean pods inoculated with bacteria. Pods inoculated with strains of *Pantoea agglomerans* (A) and *Xanthomonas arboricola* pv. *corylina* (B). In the first case, the tissues of the pods inoculated with *P. agglomerans* did not collapse (C) whereas tissue collapsing was observed in the second case (D). Figures A and B are referred to the naked-eye observation whereas stereomicroscope (Stemi DV4) observation was made for C and D (5 X).

Figure S2 Characteristic symptoms of bacterial blight developed on the artificially inoculated hazelnut plant. Water-soaked necrotic spots on the leaves (A), shoot dieback (B) and canker formation (C) observed respectively at 3, 4 and 5 weeks after inoculation. Figures are referred to 2-year old potted plants (cv. Tonda Gentile Romana).

Table S1 Study areas, hazelnut cultivars, plant age and bacterial blight incidence across the Viterbo province.

Acknowledgments

The authors would like to thank the farmer's Associations: ASSOFRUTTI, APRONVIT and APNAL for their kind availability during the field visit and data collection.

Author Contributions

Conceived and designed the experiments: JRL AF LV. Performed the experiments: JRL AF RR. Analyzed the data: JRL AF. Contributed reagents/materials/analysis tools: LV. Wrote the paper: JRL AF LV.

33. Chellemi DO, Rohrbach KJ, Yost RS, Sonoda RM (1988) Analysis of the spatial pattern of plant pathogens and diseased plants using geostatistics. Phytopathology 78: 221–226.

34. Gent DH, Farnsworth JL, Johnson DA (2011) Spatial analysis and incidence-density relationships for downy mildew on hop. Plant Pathol 61: 37–47.

35. Henne DC, Workneh F, Rush CM (2012) Spatial patterns and spread of potato zebra chip disease in the Texas Panhandle. Plant Dis 96: 948–956.

36. Orum TV, Bigelow DM, Nelson MR, Howell DR, Cotty PJ (1997) Spatial and temporal patterns of *Aspergillus flavus* strain composition and propagule density in Yuma County, Arizona, soils. Plant Dis 81: 911–916.

37. Ferrin DM, Mitchell DJ (1986) Influence of initial density and distribution of inoculums on the epidemiology of tobacco black shank. Phytopathology 76: 1153–1158.

38. Gray SM, Moyer JW, Bloomfield P (1986) Two dimensional distance class model for quantitative description of virus-infected plant distribution lattices. Phytopathology 76: 243–248.

39. Madden LV, Louie R, Abt JJ, Knoke JK (1982) Evaluation of tests for randomness of infected plants. Phytopathology 72: 195–198.

40. Martins L, Castro J, Macedo W, Marques C, Abreu C (2007) Assessment of the spread of chestnut ink disease using remote sensing and geostatistical methods. Eur J Plant Pathol 119: 159–164.

41. Nicot PC, Rouse DI, Yandell BS (1984) Comparision of statistical methods for studying spatial patterns of soilborne plant pathogens in the field. Phytopathology 74: 1399–1402.

42. Noe JP, Campbell CL (1985) Spatial pattern analysis of plant parasite nematodes. J Nematol 17: 86–93.

43. Proctor CH (1984) On the detection of clustering and anisotrophy using binary data from a lattice patch. Commun Stat Theor Meth 13: 617–638.

44. Ramirez BN, Michell DJ (1975). Relationship of density of chlamydospores and zoospores of *Phytopthora palmivora* in soil to infection of papaya. Phytopathology 65: 780–785.

45. Clark I (1979) Pratical Geostatistics. Elsever Applied Science Publishers, Essex, England. 129 p.

46. Cliff AD, Ord JK (1977) Spatial Autocorrelation. Pion, London, 178 p.

47. Goodchild MF (1993) The state of GIS for environmental problem solving. In: Goodchild M F, Parks BO, Steyaert LT, editors. Environmental Modeling with GIS. Oxford University Press, London. pp. 8–15.

48. Nelson MR, Felix-Gastelum R, Orum TV, Stowell LJ, Myers DE (1994) Geographic information systems and geostatistics in the design and validation of regional plant virus management programs. Phytopathology 84: 898–905.

49. Nelson MR, Orum TV (1997) Geographic information systems and geostatistics in the design of regional plant disease and insect pest management programs. AAAS Annu Meet Sci Innov Expo 163: A22.

50. Nelson MR, Orum TV, Jaime-Garcia R, Nadeem A (1999) Applications of geographic information systems and geostatistics in plant disease epidemiology and management. Plant Dis 83: 308–319.

51. Orum TV, Bigelow DM, Nelson MR, Howell DR, Cotty PJ (1997) Spatial and temporal patterns of *Aspergillus flavus* strain composition and propagule density in Yuma County, Arizona, soils. Plant Dis 81: 911–916.

52. Wu BM, Van Bruggen AHC, Subbarao KV, Pennings GGH (2001) Spatial analysis of lettuce downy mildew using geostatistics and geographic information systems. Phytopathology 91: 134–142.

53. Broders KD, Wallhead MW, Austin GD, Lipps PE, Paul PA, et al. (2009) Association of soil chemical and physical properties with *Pythium* species diversity, community composition, and disease incidence. Phytopathology 99: 957–967.

54. Martins LM, Oliveira MT, Abreu CG (1999) Soils and climatic characteristic of chestnut stands that differ on the presence of the Ink Disease. Acta Hort 494: 447–449.

55. Martins LM, Lufinha MI, Marques CP, Abreu CG (2001) Small format aerial photography to assess Chestnut Ink Disease. For Snow Landsc Res 73: 357–360.

56. Martins LM, Macedo FW, Marques CP, Abreu CG (2005) Assessment of Chestnut Ink Disease spread by geostatistical methods. Acta Hort 693: 621–625

57. Gottwald TR, Avinent L, Llácer G, de Mendoza Hermoso A, Cambra M (1995) Analysis of the spatial spread of sharka (Plumb pox virus) in apricot and peach orchards in eastern Spain. Plant Dis 79: 266–278.

58. Larkin RP, Gumpertz ML, Ristaino JB (1995) Geostatistical analysis of *Phytophthora* epidemic development in commercial bell pepper fields. Phytopathology 85: 191–203.

59. Stein A, Kocks CG, Zadoks JC, Frinking HD, Ruissen MA, et al. (1994) A geostatistical analysis of the spatio-temporal development of the downy mildew epidemics in cabbage. Phytopathology 84: 1227–1239.

60. Eastman JR (2001) Guide to GIS and Image Processing. Vol I, Clark University Worcester, MA USA. 171 p.

61. Eastman JR (2001) Guide to GIS and Image Processing. Vol II, Idrisi Release Worcester, MA:Clark University. 144 p.

62. Deutsch CV, Journel AG (1992) GSLIB, Geostatistical Software Library and Users Guide. New York: Oxford Press. 340 p.

63. Yamamoto JK (1999) Quantification of Uncertainty in Ore-Reserve Estimation: Applications to Chapada Copper Deposit, State of Goias, Brazil. Natural Res J 8:153–163.

64. Star J, Estes JE (1990) Geographic Information Systems: An Introduction. Prentice Hall, Englewood.Cliffs, New Jersey. 303 p.

65. Gardan L (1986) *Xanthomonas campestris* pv. *corylina*. EPPO Data sheets on quarantine organisms. EPPO/EPO Bull 16: 13–16.

66. Gardan L, Devaux N (1983) Bacterial blight of hazelnut caused by *Xanthomonas corylina*. Proc. International congress on hazelnut, Avellino, Italy pp. 443–450.

67. Seem RC (1984) Disease incidence and severity relationships. Ann Rev Phytopathol 22: 133–150.

68. Lee YA, Hildebrand DC, Schroth MN (1992) Use of quinate metabolism as a phenotypic property identify members of *Xanthomonas campestris* dna homology group 6. Phytopathology 82: 971–973.

69. Lamichhane JR, Fabi A, Varvaro L (2012) Bacterial species associated to brown spots of hazelnut in central Italy: Survey, isolation and characterization. Acta Hort, In press.

70. Lelliot RA, Stead DE (1987) Methods for the diagnosis of bacterial diseases of plants. In: Preece TF editor. Methods in Plant Pathology, Volume 2. Blackwell Scientific Press, London, UK. 216 p.

71. Schaad NW, Jones JB, Chun W (2001) Laboratory Guide for the Identification of Plant Pathogenic Bacteria. Third edition, 373 p.

72. Ayers SH, Rupp P, Johnson WT (1919) A study of the alkali forming bacteria in milk. USDA Bull. 882 p.

73. Klement Z, Lovrekovich L (1961) Defence Reactions Induced by Phytopathogenic Bacteria in Bean Pods. J Phytopathol 41: 217–227.

74. Klement Z, Goodman RN (1967) The hypersensitive reaction to infection by bacterial plant pathogens. Ann Rev Phytopathol 5: 17–44.

75. Roberts SJ (1997) Effect of weather conditions on local spread and infection by pea bacterial blight (*Pseudomonas syringae* pv. *pisi*). Eur J Plant Pathol 103: 711–719.

76. Agrios GN (2005) Plant Pathology. Amsterdam: Elsevier-Academic Press. 948 p.

77. Balestra GM, Varvaro L (1997) Influence of nitrogen fertilization on the colonization of olive phylloplane by *Pseudomonas syringae* subsp. *savastanoi*. In: Rudolph K, Burr TJ, Mansfield JW, Stead D, Vivian A, et al. editors. *Pseudomonas syringae* Pathovars and Related Pathogens. Kluwer Academic Publishers, Dordrecht, The Netherlands. pp. 88–92.

78. Shear GM, Wingard SA (1944) Some ways by which nutrition may affect severity of disease in plants. Phytopathology 34: 603–605.

79. Barnett HL (1959) Plant disease resistance. Ann Rev Microbiol 13: 191–210.

80. Dordas C (2008) Role of nutrients in controlling plant diseases in sustainable agriculture. A review. Agron Sustain Dev 28: 33–46.

81. Melakeberhan H, Jones AL, Hanson E, Bird GW (1995) Effect of low soil pH on aluminium availability and on mortality of cherry seedlings. Plant Dis 79: 886–892.

82. Vigoroux A, Bussi C (1993) Influence of water availability and soil calcic amendment on susceptibility of apricot to bacterial canker. Acta Hort 384: 607–611.

83. Weaver DJ, Wehunt EJ (1975) Effect of soil pH on susceptibility of peach to *Pseudomonas syringae*. Phytopathology 65: 984–989.

84. Fabi A, Varvaro L (2009) Application of geostatistics in studying epidemiology of hazelnut diseases: a case study. Acta Hort 845: 507–514.

85. Fabi A, Varvaro L (2006) Spatial and temporal distribution of dieback of hazelnut on Cimini hills (Central Italy) by use of Geographic Information System and Geostatistics. Proc. 12th Congr. Medit. Phytopath. Union, June 11–15, Rhodes Island (Greece), 217–219.

86. Tangmar BB, Yost RS, Uehara G (1985) Application of geostatistics to spatial studies of soil properties. Adv Agron 38: 45–94.

87. Guo XD, Fu BJ, Ma KM, Chen LD (2000) Spatial variability of soil nutrients based on geostatistics combined with GIS – A case study in Zunhua City of Hebei Province. Study on spatial variation of Chinese J Appl Ecol 11:557–563 (in Chinese)

88. Cambardella CA, Moorman TB, Parkin TB, Karlen DL, Novak JM, et al. (1994) Field-scale variability of soil properties in central Iowa soils. Soil Sci Soc Am J 58:1501–1511.

89. Oliver MA, Webster R (1990) Kriging: a method of interpolation for geographical information systems. Int J Geogr Inf Syst 4: 313–332.

90. Cressie N (1985) Fitting variogram models by weighted least squares. Math Geol 17: 563–586.

91. Cressie N (1988) Spatial Prediction and Ordinary Kriging. Math Geol 20: 405–421. Erratum, (1989). Math Geol 21: 493–49.

92. Boccacci P, Torello MD, Botta R, Rovira M (2009) Genetic diversity and relationships among Italian and Spanish hazelnut cultivars. Acta Hort 845: 127–132.

93. Tombesi A, Limongelli F (2002) Varietà e miglioramento genetico del nocciolo. Atti del Convegno Nazionale sul Nocciolo, le frontiere della corilicoltura italiana. Giffoni Valle Piana (SA), 5–6 ottobre; pp. 11–27.

94. Cristofori V, Bignami C, De Salvador R, Rugini E (2011) Il nocciolo in Italia: valorizzazione del prodotto e innovazione colturale per garantire competitività. Frutticol 5: 44–53.

95. Scortichini M, Rossi MP, Marchesi U (2002) Genetic, phenotypic and pathogenic diversity of *Xanthomonas arboricola* pv. *corylina* strains question the representative nature of the type strain. Plant Pathol 51: 374–381.

96. Fabi A, Belli C, Vuono G, Balestra GM, Varvaro L (2005) Innovative strategies in epidemiological studies of Hazelnut dieback by using G.P.S./G.I.S. and A.Sp.I.S. Technology. Acta Hort 686: 427–433.

97. Isaaks EH, Srivastava RM (1989) An Introduction to Applied Geostatistics. Oxford University Press, New York. 561 p.

98. Goovaerts P (1997) Geostatistics for Natural Resources Evaluation. Oxford University Press, New York. 483 p.

99. Stein ML (1999) Interpolation of Spatial Data. Some Theory for Kriging. Springer, New York. 247 p.

100. Franke J, Gebhardt S, Menz G, Helfrich HP (2009) Geostatistical analysis of the spatiotemporal dynamics of powdery mildew and leaf rust in wheat. Phytopathology 99: 974–984.

101. Jaime-Garcia R, Orum TV, Felix-Gastelum R, Trinidad-Correa R, VanEtten HD, et al. (2001) Spatial analysis of *Phytophthora infestans* genotypes and late blight severity on tomato and potato in the Del Fuerte Valley using geostatistics and geographic information systems. Phytopathology 91: 1156–1165.

102. Van Maanen A, Xu XM (2003) Modelling plant disease epidemics. Eur J Plant Pathol 109: 669–682.

103. Ristiano JB, Gumpertz ML (2000) New frontiers in the study of dispersal and spatial analysis of epidemics caused by species in the genus phytopthora. Annu Rev Phytopathol 38: 541–576.

104. Brent KJ, Hollomon DW (1998) Fungicide Resistance: The Assessment of Risk.Brussels, Belgium: Global Crop Prot Fed 48 p.

105. Cooksey DA (1990) Genetics of bactericide resistance in plant pathogenic bacteria. Ann Rev Phytopathol 28: 201–219.

106. Genet JL, Jaworska G, Deparis F (2006) Effect of dose rate and mixtures of fungicides on selection for QoI resistance in populations of *Plasmopara viticola*. Pest Manag Sci 62: 188–194.

107. Gisi U, Waldner M, Kraus N, Dubuis PH, Sierotzki H (2007) Inheritance of resistance to carboxylic acid amide (CAA) fungicides in *Plasmopara viticola*. Plant Pathol 56: 199–208.

108. Martin HL, Hamilton VA, Kopittke RA (2004) Copper tolerance in Australian populations of *Xanthomonas campestris* pv. *vesicatoria* contributes to poor field control of bacterial spot of pepper. Plant Dis 88: 921–924.

109. Sudin GW, Bender CL (1993) Ecological and genetic analysis of copper and streptomycin resistance in *Pseudomonas syringae* pv. *syringae*. Appl Environ Microbiol 59: 1018–1024.

110. Vanneste JL, McLaren GF, Yu J, Cornish DA, Boyd R (2005) Copper and streptomycin resistance in bacterial strains isolated from stone fruit orchards in New Zealand. N Z Plant Prot 58:101–105.

111. Snoeijer SS, Pérez-García A, Joosten MHAJ, De Wit PJGM (2000) The effect of nitrogen on disease development and gene expression in bacterial and fungal plant pathogens. Eur J Plant Pathol 106: 493–506.

112. Bianco M, Danise B (2002) La difesa fitosanitaria del nocciolo. Proceeding of the II National congresso on hazelnut. Giffoni, Valle Piana, Italy. pp. 52–61.

113. Pietrarelli L, Balestra GM, Varvaro L (2006) Effects of simulated rain on *Pseudomonas syringae* pv. *tomato* populations on tomato plants. J Plant Pathol 88: 245–251.

114. Butterworth J, McCartney HA (1991) The dispersal of bacteria from leaf surfaces by water splash. J Appl Microbiol 71: 484–496.

115. Hirano SS, Upper CD (1990) Population biology and epidemiology of *Pseudomonas syringae*. Annu Rev Phytopathology 28: 155–177.

116. Leben C, Daft GC, Schmitthenner AF (1968) Bacterial blight of soybeans: population levels of *Pseudomonas glycinea* in relation to symptom development. Phytopathology 58: 1143–1146

117. Haas JH, Rotem J (1976) *Pseudomonas lachrymans* inoculum on infected cucumber leaves subjected to dew- and rain-typewetting. Phytopathology 66: 1219–1223.

118. MilesWG, Daines RH, Rue JW (1977) Presymptomatic egress of *Xanthomonas prunii* from infected peach leaves. Phytopathology 67: 895–897.

119. Roberts SJ (1985) Bacterial diseases of woody ornamental plants. Ph.D. thesis, The University of Leeds.

120. Aloj B, Bartoletti F, Caporossi U, D'Errico FP, Di Dato F, et al. (1987) Una moria del nocciolo di natura ignota nel Viterbese. Inf Agrario 26: 55–57.

121. Moore LW, Lagerstedt HB, Hartmann N (1974) Stress predisposes young filbert trees to bacterial blight. Phytopathology 64: 1537–1540.

122. Hannan JM (2011) Potassium-magnesium antagonism in high magnesium vineyard soils. Ms Theses and Dissertations. Iowa State University, USA.

123. Jacobsen ST (1993) Interaction between Plant Nutrients: III. Antagonism between Potassium, Magnesium and Calcium. Acta Agric Scandinav, 43: 1–5.

124. Pathak AN, Kalra YP (1971) Antagonism between Potassium, Calcium and Magnesium in Several Varieties of Hybrid Corn. J Plant Nutr Soil Sci 130: 118–124.

125. Voisin A (1963) Mineral balances of soil and mineral balances of grass. In: Thomas CC editor. Grass Tetany. 262 p.

126. Johansson OAH, Hahlin JM (1977) Potassium/Magnesium Balance in Soil for Maximum Yield. Proc. Int. Sem. on Soil Environ. and Fert. Manage. In: Intensive Agric. Soc. Sci. Soil and Manure, pp. 487–495.

127. Marschner H (1995) Mineral Nutrition of Higher Plants. 2nd ed., Academic Press, London. 889p.

128. Pettiet JV (1988) The Influence of Exchangeable Magnesium on Potassium Uptake in Cotton Grown on Mississippi Delta Soils. Proceedings Beltwide Cotton Production Research Conferences. pp. 517–518.

129. Tewari SN, Sinha MK, Mandal SC (1971) Studies on the Interrelationship among Calcium, Magnesium and Potassium in Plant Nutrition. Proc Int Symp Soil Fert Eval New Delhi 1: 317–325.

130. Prince A, Zimmerman A, Bear FE (1947) The Magnesium-supplying Powers of 20 New Jersey Soils. Soil Sci 63: 69–78.

131. Omar MA, El Kobbia T (1966) Some Observations on the Interrelationships of Potassium and Magnesium. Soil Sci 101: 437–440.

132. Barbieri M, Peccerillo A, Poli G, Tolomeo L (1988) Major, trace element and Sr isotopic composition of lavas from Vico volcano central Italy and their evolution in an open system. Contrib Mineral Petrol 99: 485–497.

133. White PJ, Brown PH (2010) Plant nutrition for sustainable development and global health. Ann Bot 105: 1073–1080.

134. Wilkinson SR, Welch RM, Mayland HF, Grunes DL (1990) Magnesium in plants: uptake, distribution, function, and utilization by man and animals. In: Helmut S, editor. Metal Ions in Biological Systems Marcel Dekker, Inc., New York and Basel, Switzerland. pp, 33–56.

135. Scortichini M, Sbaraglia M, Di Prospero P, Angelucci L, Petricca P, et al. (2001) Moria del nocciolo nel Viterbese e terreni acidi. Inf Agrario 21: 85–88.

136. Tombesi A (1991) Il Nocciolo. Frutticol Speciale. Reda Rome, pp. 614–630.

137. Delhaize E, Ryan PR (1995) Aluminium Toxicity and Tolerance in Plants. Plant Physiol 107: 315–321.

138. Foy CD, Chaney RL, White MC (1978) The Physiology of Metal Toxicity in Plants. Ann Rev Plant Physiol 29: 511–566.

139. Kinraide TB, Ryan PR, Kochian LV (1992) Interactive effects of Al^{3+}, H^+, and other cations on root elongation considered in terms of cell-surface electrical potential. Plant Physiol 99: 1461–1468.

140. Cannon R (1998) The implications of predicted climate change for insect pests in the UK, with emphasis on non indigenous species. Global Change Biology 4: 785–796.

141. Pimentel D, McNair S, Janecka J, Wightman J, Simmonds C, et al. (2001) Economic and environmental threats of alien plant, animal and microbe invasions. Agr Ecosyst Environ 84: 1–20.

142. Rosenzweig C, Iglesias A, Yang X, Epstein P, Chivian E (2001) Climate change and extreme weather events; implications for food production, plant diseases and pests. Global change Hum Health 2: 90–104.

143. Anderson P, Cunningham A, Patel N, Morales F, Epstein P, et al. (2004) Emerging infectious diseases of plants: pathogen pollution, climate change and agrotechnology drivers. Trends Ecol Evol 19: 535–544.

144. Berry P, Dawson T, Harrison P, Pearson R (2002) Modelling potential impacts of climate change on the bioclimatic envelop of species in Britain and Ireland. Global Ecol Biogeogr 11: 453–462.

145. Coakley S, Scherm H, Chakraborthy S (1999) Climate change and plant disease management. Ann Rev Phytopathol 37: 399–426.

146. Fuhrer J (2003) Agroecosystem response to combinations of elevated CO_2, ozone and global climate change. Agr, Ecosyst Environ 97: 1–20.

147. Chakraborthy S, Dutta S (2003) How will plant pathogens adapt to host plant resistance at elevated CO_2 under changing climate? New Phytol 159: 733–742.

148. Melloy P, Hollaway G, Luck J, Norton R, Aitken ESC (2010) Production and fitness of *Fusarium pseudograminearum* inoculums at elevated carbon dioxide in FACE. Global Change and Biology 16: 3363–3373.

8

Inorganic Nitrogen Leaching from Organic and Conventional Rice Production on a Newly Claimed Calciustoll in Central Asia

Fanqiao Meng[1]*, Jørgen E. Olesen[2], Xiangping Sun[1], Wenliang Wu[1]

1 College of Resources and Environmental Sciences, China Agricultural University, Beijing, China, 2 Department of Agroecology and Environment, Faculty of Agricultural Sciences, Aarhus University, Tjele, Denmark

Abstract

Characterizing the dynamics of nitrogen (N) leaching from organic and conventional paddy fields is necessary to optimize fertilization and to evaluate the impact of these contrasting farming systems on water bodies. We assessed N leaching in organic versus conventional rice production systems of the Ili River Valley, a representative aquatic ecosystem of Central Asia. The N leaching and overall performance of these systems were measured during 2009, using a randomized block experiment with five treatments. PVC pipes were installed at soil depths of 50 and 180 cm to collect percolation water from flooded organic and conventional paddies, and inorganic N (NH_4-N+NO_3-N) was analyzed. Two high-concentration peaks of NH_4-N were observed in all treatments: one during early tillering and a second during flowering. A third peak at the mid-tillering stage was observed only under conventional fertilization. NO_3-N concentrations were highest at transplant and then declined until harvest. At the 50 cm soil depth, NO_3-N concentration was 21–42% higher than NH_4-N in percolation water from organic paddies, while NH_4-N and NO_3-N concentrations were similar for the conventional and control treatments. At the depth of 180 cm, NH_4-N and NO_3-N were the predominant inorganic N for organic and conventional paddies, respectively. Inorganic N concentrations decreased with soil depth, but this attenuation was more marked in organic than in conventional paddies. Conventional paddies leached a higher percentage of applied N (0.78%) than did organic treatments (0.32–0.60%), but the two farming systems leached a similar amount of inorganic N per unit yield (0.21–0.34 kg N Mg^{-1} rice grains). Conventional production showed higher N utilization efficiency compared to fertilized organic treatments. These results suggest that organic rice production in the Ili River Valley is unlikely to reduce inorganic N leaching, if high crop yields similar to conventional rice production are to be maintained.

Editor: Ben Bond-Lamberty, DOE Pacific Northwest National Laboratory, United States of America

Funding: The study was funded by National Natural Science Foundation of China (No. 30970533) and the National Key Science and Technology Project-Organic Farming Development in the Ili River Valley (No. 2007BAC15B05). The funders had no role in study design, data collection and analysis, decision to publish, or preparation of the manuscript.

Competing Interests: The authors have declared that no competing interests exist.

* E-mail: mengfq@cau.edu.cn

Introduction

Nitrogen (N) leaching is one of the primary pathways of N loss from flooded paddy farmland, representing 30–50% of total N loss from such soils [1]. The amount of leached N may equal up to 15% of the total applied N in rice (*Oryza sativa* L.) production [2,3]. Flooded rice crops typically capture only 20–40% of N applied in the harvested grain [4]. N utilization efficiency can be improved by reducing the various forms of loss, including leaching. Because reactive N is highly soluble, its subsequently high mobility in the environment makes it a major contributor to surface and groundwater contamination. Ju et al. [5] reported that high inputs (550–600 kg N ha^{-1}) associated with current rice production systems in China did not significantly increase crop yields, yet cause a doubling of N losses to the environment, mainly from increased denitrification in waterlogged systems. Accordingly, understanding N loss processes, particularly through leaching and N_2O emission, is necessary to both improve resource use efficiency and to protect the quality of nearby water bodies. In this study, we focused on monitoring N leaching losses at different rice growth

stages and soil depths in organic and conventional rice production systems.

Located in the Xinjiang Uygur Autonomous Region of northwestern China, the Ili River Valley has become an important grain production region, owing to its plentiful surface water, well-developed soils, and extensive meadow grasslands. Of the total land area (5.82 million ha), approximately 1 million ha are under cultivation, and the introduction of new irrigation technologies is facilitating the conversion of meadow grasslands to agricultural uses [6]. However, the Ili River is sensitive to pollution, which is of local and international concern as the river flows into Balkhash Lake in Kazakhstan. The river receives pollutants from both surface run-off and drainage from farming activities throughout the valley. In addition to water quality concerns, the terrestrial ecosystem may also be at risk because it contains shallow soils with a coarse texture. As such, organic farming is being promoted for the conservation of soil and water resources to reduce the overall negative environmental impact of agricultural production, as well as to increase farmer's income. However, it is unclear whether organic production can maintain high rice yields, with or without high fertilizer inputs, and what effect shifting from conventional to organic farming may have on N loading to the Ili River.

Many studies have examined N leaching from organic and conventional production [7,8]. Such research has revealed that organic farming may lower N leaching compared to conventional systems, both by reducing N inputs and by including catch crops (cover crops) in the rotation [8,9], although the magnitude of this effect is subject to high uncertainty due to variation in crop yields and input intensities [9,10]. Although rice dominates grain production in many developing countries including China, little research has been conducted in this Central Asia region about N loss in organic versus conventional rice farming, especially regarding N loss through percolation or leaching [2,4,11].

This study compared the inorganic N (NH$_4$-N and NO$_3$-N) in percolation/leachate water from organic versus conventional production systems with high yields of rice production, by the application of high-loads of animal manure. We hypothesized that organic rice farming leaches less inorganic N compared with conventional rice in the Ili River Valley. In order to calculate the amount of N leached from paddies, we had to address the technical challenge of quantifying the total volume of leachate from a single-cropping rice paddy. To do so, we designed a filtration system using PVC pipes at two depths (50 and 180 cm) to capture the leachate, and quantified the percolation rates throughout the rice growing season using a water balance based on the difference between incoming (i.e., irrigation and precipitation) and outgoing (evapotranspiration and surface runoff) flows. To better understand the potential environmental effects of the treatments, we evaluated N leaching not only in absolute terms, but also in terms of leachate per unit yield of rice and as a percentage of the total N applied.

Materials and Methods

Experiment site

The field experiment was conducted at Chabuchaer Farm (48°65'N, 80°06'E, 634 m ASL) in the Ili River Valley of the Xinjiang Uygur Autonomous Region, China. The region has a temperate continental climate with a mean annual temperature of 9.5°C and annual precipitation of 260 mm, which falls predominantly in June, July, and August. In 2009 when monitoring was implemented, precipitation mainly happened in April to July, November and December (Fig. 1). The farm is located on an alluvial plain off the south bank of the Ili River. The land use was converted to agriculture by the local government in 2005 as part of a national grain supply project. Because this land was newly reclaimed, we could examine the contrasting effects of conventional and organic farming techniques without the usual conversion period from conventional to organic farming, which can confound the results. Our study was initiated in 2008 and the data presented here were collected in the 2009 season.

The soil was a calciustoll with salt content of 1.4%. Rice had been continuously cultivated on this land since 2005, with the aim of reducing soil salinity. The soil is a sandy loam, with 3.2% clay, 44.7% silt, and 52.1% sand. Other important characteristics of the soil (0–20 cm) are as follows: 14.0 g kg^{-1} soil organic matter, 1.15 g kg^{-1} total N, 11 mg kg^{-1} Olsen phosphorous, 264 mg kg^{-1} available potassium, and a pH of 7.9.

Permission to conduct the experiment and sampling at the site was obtained from the Ili Agricultural Technical Extension Station, and the Farmer Liu Ermao. These field studies did not involve endangered or protected species.

Crop management

Rice (Oryza sativa L., cultivar Nonglin-315) was transplanted to the experimental farm on May 29, 2009. The fertilizer used was either a conventional combination of mineral fertilizers with animal manure, as used in the local production system, or only organic fertilizers (composted animal manure [B] with castor (Ricinuscommunis L.) bean meal [A]), as used typically in organic rice production. The level of organic fertilization was manipulated to modify the overall N input. Hence, the five treatments were 1) an unfertilized control treatment (CK), 2) low-level organic fertilization (B1A1): composted animal manure at 15,000 kg ha^{-1} and castor bean meal at 2,250 kg ha^{-1}, 3) mid-level organic fertilization (B2A1): composted animal manure at 45,000 kg ha^{-1} and castor bean meal at 2,250 kg ha^{-1}, 4) high-level organic fertilization (B3A1): composted animal manure at 75,000 kg ha^{-1} and castor bean meal at 2,250 kg ha^{-1}, and 5) conventional production system (LS): composted animal manure at 11,250 kg ha^{-1}, urea at 277.5 kg ha^{-1}, diammonium phosphate at 247.5 kg ha^{-1}, and compounded mineral fertilizer at 75 kg ha^{-1} (Table 1). The treatments were arranged in a randomized block design with three replicates.

Composted animal manure was applied on May 9 as the basal fertilizer for all treatments (except CK), 20 days before transplanting. The total amount of castor bean meal was halved and applied in two stages, first as topdressing at the early tillering stage (June 27) and then at the late tillering stage (July 24). Topdressings for LS were applied on June 27 (120 kg ha^{-1} diammonium phosphate with 75 kg ha^{-1} compounded fertilizer), July 6 (127.5 kg ha^{-1} diammonium phosphate with 127.5 kg ha^{-1} urea), and July 15 (urea, 150 kg ha^{-1}). The nutrient contents of the fertilizers used in the experiments are listed in Table 2.

Each paddy block had an area of 8×10 m $= 80$ m^2. Flood irrigation (about 20 cm of standing water), popular in other Central Asian countries [4], was used continuously from June 1 to September 7, except during the first week after fertilization application. Hand weeding was used in all treatments, along with the addition of herbicides to conventional replicates. The main development stages for the rice were transplant (May 29-Jun 13), tillering and elongation (~Aug 5 for organic paddies / ~Aug 10 for conventional paddies), heading (Aug 5-Sep 1 for organic / Aug 10-Sep 5 for conventional), and ripening (Sep 1-Oct 2 for organic / Sep 5-Oct 13 for conventional).

Measurements of N cycling

Previous studies have shown that inorganic N (NH$_4$-N+NO$_3$-N), rather than particulate N, makes up 50–90% of the total N leached from the soils, and for organic fertilizers or fertile soil, this proportion maybe even higher [12,13,14]. Only inorganic N was analyzed in percolation water collected in our experiment. Percolation water was collected from flooded paddies using polyvinyl chloride (PVC) standpipes [14,15,16]. Two PVC pipes (5 cm in diameter, 150 and 250 cm in length) with sealed bottoms were installed roughly in the center of each field plot to collect drainage water from the saturated soil (Fig. 2). Both pipes were perforated 80 times (6-mm internal diameter) within a 10-cm-wide band, 30 cm from the bottom of the pipe. The porous zone of the pipe was wrapped with nylon textile to prevent sand in-filling. As the average soil layer is 2 m deep in this region, we compared the inorganic N leaching in 50 and 180 cm. At the later depth we considered that inorganic N was not usable by the rice. Thus, the pipes were installed at depths of 50 and 180 cm from the surface to the uppermost pore (Fig. 2). The gap between the collection pipe and soil wall in the porous zone was filled with quartz powder to prevent anything except soil solution from entering the pipe. At a

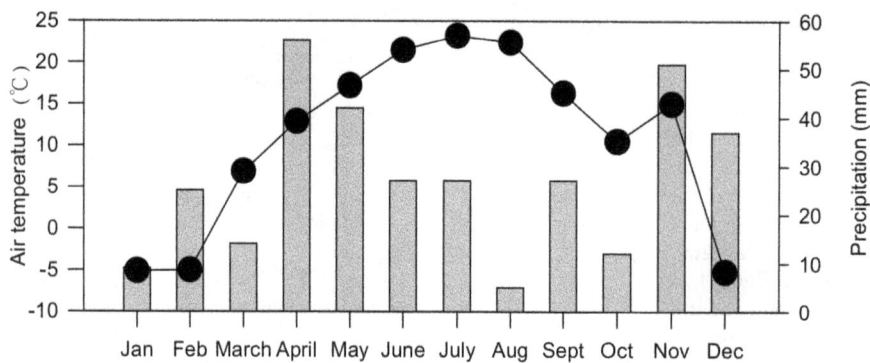

Figure 1. Air temperature and precipitation at the experimental site in Chabuchaer County for 2009.

depth of 20 cm, plastic film was wrapped around the PVC pipe, extending horizontally in a 30-cm radius from the pipe to reduce the preferential flow from the irrigation water.

Sampling was conducted 1 and 3 days after fertilization, and thereafter at intervals of 5 days to 2 weeks, depending on the percolate volume. Water samples were stored in plastic vials and refrigerated until analysis. Percolation and irrigation water were analyzed for inorganic N (NH_4-N+NO_3-N) content using a continuous flow N analyzer (TRAACS 2000). During the irrigation season, average NH_4-N and NO_3-N concentrations in the irrigation water were 0.11 and 0.68 mg L^{-1} respectively; these values were subtracted from the respective inorganic N concentrations in percolation water to calculate net N (i.e., the amount leached).

The volume of drainage water was calculated per unit area of paddy field. We considered the spatial range of percolation into the pipe to be a half-ellipsoid, having a width of 5 cm and a depth of either 50 or 180 cm, with diameters of 27.5 cm (50/2+5/2) or 92.5 cm (180/2+5/2) [15]. The volume of each ellipsoid was calculated as $4/3*\pi*ab^2$, where a and b are the length of the long axis and radius of the ellipsoid, respectively. Each hectare of paddy field included 252,671 or 222,332 half-ellipsoids, at depths of 50 cm and 180 cm, respectively. Thus, total N leached per ha could be calculated by summing these half ellipsoids at each soil depth, as well as across time to calculate total N lost over one growth cycle. In total, percolation water was sampled 15 times from June 1 to September 7, the full growing season of the rice paddies.

The measured volume of percolation water in the paddy field was calibrated using the following equation [17]:

$$P + I = ET + D + R_0 + P_{sd} - V_H \tag{1}$$

Where P is the rainfall (mm) during the rice growth season; I is irrigation (mm); ET is evapotranspiration (mm), calculated using a lysimeter installed in the paddy plot; D is drainage through drain pipe (mm), excluding surface runoff and percolation; R_o is surface runoff (mm); P_{sd} is percolation (mm); and V_H is the change of the water table in the paddy field (mm). Briefly, an iron lysimeter (1 m wide×1 m long×1.2 m high), filled with original paddy soil, was installed in the paddy field. The bottom of the lysimeter was perforated to allow percolated water to drip into a storage tank (0.5 m wide×0.5 m long×0.5 m high) welded to the lysimeter for storage of the percolated water. A meter stick was also fixed inside the lysimeter to measure the water table. Evapotranspiration from the paddy field was calculated according to Equation (1). In our study, there was no artificial drainage or surface runoff, so D and R_o were set to 0. All of the terms are expressed in millimeters and then converted into m^3 ha^{-1}. One growth cycle (June 1 to September 7, 2009) was used to bound all calculations. The difference between P_{sd} calculated from Equation (1) and from the measured percolation water volume was less than 10%; hence, here we presented percolated inorganic N as defined by its concentration (NH_4-N and NO_3-N) multiplied by measured percolate volume.

Relative N leaching loss (% of applied N) for a given plot was determined after first subtracting N leached from the nil treatment

Table 1. Experimental treatments in terms of total nutrient inputs for N, P, and K (kg ha^{-1}).

Treatment	Fertilization	Nutrient input (kg ha^{-1})		
		N	P	K
CK	Control with no fertilization	0	0	0
LS	Local conventional system: composted animal manure (27.6% of water content) at 11,250 kg ha^{-1}, urea at 277.5 kg ha^{-1}, diammonium phosphate at 247.5 kg ha^{-1}, and compounded fertilizer at 75 kg ha^{-1}.	297	83	117
B1A1	Composted animal manure (27.6% of water content) at 15,000 kg ha^{-1} plus castor bean meal (22.5% of water content) at 2,250 kg ha^{-1}	193	52	163
B2A1	Composted animal manure (27.6% of water content) at 45,000 kg ha^{-1} plus castor bean meal (22.5% of water content) at 2,250 kg ha^{-1}	498	125	446
B3A1	Composted animal manure (27.6% of water content) at 75,000 kg ha^{-1} plus castor bean meal (22.5% of water content) at 2,250 kg ha^{-1}	802	199	728

Table 2. Nutrient contents of fertilizers used in the experiments (% dry weight).

	N	P	K
Composted animal manure	1.40	0.34	1.30
Castor bean meal	2.37	0.84	1.10
Urea	46		
Diammonium phosphate	18	20	
Compounded fertilizer	14	6.98	12.45

(CK) and then expressed as a percentage of the total N applied. Similarly, the environmental performance of organic and conventional fertilizer treatments, i.e., N leaching, was also evaluated based on the amount of inorganic N leached during the production of 1 Mg of rice (kg N Mg^{-1} grain). N utilization efficiency (NUE) was estimated using the equation (2):

$$NUE = \left(\begin{array}{l} N\ uptake\ by\ rice\ in\ fertilized\ treatment \\ \left(kg\ N\ ha^{-1}\right) - \\ N\ uptake\ by\ rice\ in\ CK\ treatment \\ \left(kg\ N\ ha^{-1}\right) \end{array} \right) \quad (2)$$

$/N\ input\ from\ the\ fertilizers \left(kg\ N\ ha^{-1}\right).$

Sampling and measurement of fertilizers, soils and plants

Manures were sampled to determine the content of dry matter (DM) and total N before spreading. Plant samples were taken in early October to obtain rice grain with husk and straw, by cutting stems at 2 cm above the ground from two 2 m^2 areas per plot. The plant samples were washed with distilled water, oven-dried at 65°C, and weighed to determine rice biomass. Plant samples were then digested using H_2SO_4-H_2O_2, after which N, P, and K in the digestion solution were determined using the Kjeldahl method, calorimetrically by a Techniconautoanalyzer, and via atomic emission spectrometry (AES), respectively.

Statistical analyses

Cumulative inorganic N leached, net inorganic N leached per Mg of rice produced, and the proportion of net leached inorganic N in total N input were compared among different treatments using the least-significant difference test after a one-way analysis of variance for a randomized block design at the 0.05 significance level. The PROC MIXED procedure in SAS v.9.1 (SAS Institute, Cary, NC, USA) was performed to analyze the effects of fertilization on NH_4-N and NO_3-N concentrations over the sampling period (15 samplings), with the treatment as the fixed effect and sampling time as the random factor. Differences among means were calculated using a Differences of Least Squares Means

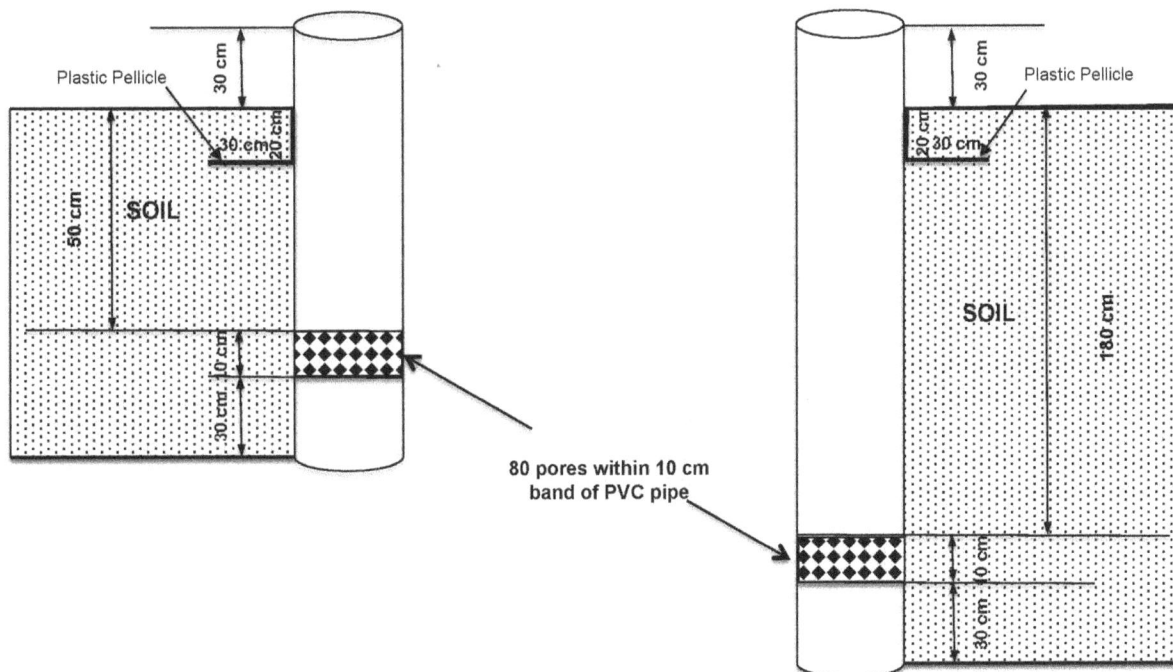

Figure 2. Schematic of standpipes for measuring percolate inorganic N concentrations in the paddy fields. Shaded area indicates the depth below the soil surface. Horizontal lines at 20 cm depth indicate the placement of a plastic film that extended a 30 cm-radium around the pipes.

with the PDIFF option and a Bonferroni adjustment method. The significance level was also at 0.05.

Results

Inorganic N concentrations in percolation water at 50 cm depth

Leaching of NH_4-N and NO_3-N during the rice growing season under different fertilization treatments is shown in Fig. 3. We observed high variance in NH_4-N concentration among three replicates of the same treatment, and this was also the case for NO_3-N (data not shown). Two peak concentrations of NH_4-N were observed in organic treatments, at the beginning of tillering (late June) and at heading (mid-August), and one additional NH_4-N peak at late-tillering (late July) in conventional treatment (LS). By PROC MIXED analysis, we found over the entire growth stage of rice, application of organic fertilizers increased NH_4-N concentration compared to no fertilization (CK, 0.63 mg L^{-1}, DF = 50.9, t = 2.76); this increase was significant for B2A1 (1.37 mg L^{-1}, DF = 50.9, t = 6.04) and B3A1 (1.74 mg L^{-1}, DF = 50.9, t = 7.65) but not for B1A1 (1.14 mg L^{-1}, DF = 50.9, t = 5.04). The concentration of NH_4-N that percolated from LS soils was significantly higher (1.20 mg L^{-1}, DF = 50.9, t = 5.29) than from CK soils, but not different from organic fields (B1A1, B2A1 and B3A1).

Organic and conventional fertilization affected NO_3-N concentrations in percolation water (Fig. 3) differently than they affected NH_4-N. At the 50-cm soil depth, the average NO_3-N concentration was lowest in CK (0.48 mg L^{-1}, DF = 42.2, t = 2.79) and then increased in the order of LS (1.08 mg L^{-1}, DF = 42.2, t = 6.27) < B1A1 (1.96 mg L^{-1}, DF = 42.2, t = 11.33) < B2A1 (2.16 mg L^{-1}, DF = 42.2, t = 12.48) < B3A1 (2.19 mg L^{-1}, DF = 42.2, t = 12.67). Within the percolation waters of both the CK and LS treatments, concentrations of NH_4-N and NO_3-N were similar, whereas in the organic treatments (B1A1, B2A1 and B3A1) average concentrations of NO_3-N were 21–42% higher than NH_4-N. In contrast to the two peak concentrations of NH_4-N, NO_3-N declined from high concentrations at early tillering until the end of the rice harvest

Inorganic N concentrations in percolation water at 180 cm depth

The average NH_4-N concentrations at 180 cm were lowest in the CK treatment (0.41 mg L^{-1}, DF = 48.8, t = 1.75) and then increased as follows: LS (0.88 mg L^{-1}, DF = 48.8, t = 3.75) < B1A1 (1.08 mg L^{-1}, DF = 48.8, t = 4.61) < B2A1 (1.28 mg L^{-1}, DF = 48.8, t = 5.44), and B3A1 (1.57 mg L^{-1}, DF = 48.8, t = 6.67, Fig. 4). These results indicated that NH_4-N leaching increased with the increasing intensity of organic fertilization, although a high temporal variation was also observed over the monitoring period. As was found at 50 cm depth, two NH_4-N peaks were recorded at the beginning of tillering and at the heading stage for the organic treatments, with an additional peak for LS at mid-tilling. Different from the observations at 50 cm, NH_4 concentrations for all treatments increased from early heading, for about 1 month, until the start of paddy filling. NH_4-N concentrations in organic rice leachates at 180 cm were similar to those at 50 cm, except for LS plots, in which values at 180 cm were significantly lower than that at 50 cm depth (0.88 vs. 1.20 mg L^{-1}).

The temporal dynamics of NO_3-N concentrations at 180 cm were similar to those at 50 cm (Fig. 4), *i.e.*, as the quantity of organic fertilizer increased, leached NO_3-N concentrations also increased. However, significantly lower concentrations of NO_3-N were observed for organic treatments than for LS, and this was

opposite to 50 cm depth where organic treatments percolated significantly higher concentrations of NO_3-N than LS. Organic treatments had much lower NO_3-N concentrations at 180 cm than at 50 cm (B1A1: 0.59 vs. 1.96 mg L^{-1}, B2A1: 0.85 vs. 2.16 mg L^{-1} and B3A1: 1.17 vs. 2.19 mg L^{-1}) while NO_3-N concentrations in the LS treatment increased from 50 cm to 180 cm (1.08 vs. 1.70 mg L^{-1}).

Total leached inorganic N

The rice fields were flooded during the growing period from transplant (June 1) to harvest (Sep 7). At the 50 cm depth, leached inorganic N varied from 18.5 to 23.1 kg N ha^{-1} for fertilized organic treatments comparing to 12.1 kg N ha^{-1} for LS (Table 3). Of the total amount of leached inorganic N, the fertilized organic treatments showed a slight higher proportion of N as NO_3-N (51.1~58.9%), compared to 64.2% as NH_4-N in LS. Higher organic fertilization rates led to higher amounts of leached NH_4-N, but this was not the case for NO_3-N, i.e., among the three fertilized organic treatments, the quantities of NO_3-N remained similar. For the net total inorganic N leached, B3A1 was significantly higher than other treatments, especially compared to LS.

At 180 cm, far less total inorganic N was percolated than at 50 cm for all treatments. For fertilized organic treatments (B1A1, B2A1 and B3A1), total inorganic N leached at 180 cm was only 9-13% of the amount leached at 50 cm; for the LS treatment, it was 26%. Relatively higher proportions of inorganic N were leached as NH_4-N (>68%) from fertilized organic production systems, whereas 61% of leached inorganic N from LS was NO_3-N. Among all organic treatments, B3A1 leached the highest quantity of inorganic N, significantly higher than other organic treatments, but interestingly, not higher than LS.

For the purpose of calculating N losses, we considered the inorganic N present at the depth lower than 180 cm (Table 4) to be that which could no longer be utilized by crops, and could potentially reach surface or ground water bodies. These data were used to assess the environmental performance of organic versus conventional rice production using two indicators: the amount of inorganic N leached per unit yield of rice (kg N Mg^{-1} grain) and the percentage of inorganic N leached per unit applied N (% loss) at 180 cm depth. Regarding leachate per unit yield, we found that LS and organic treatments were not significantly different, i.e., a similar quantity of inorganic N was leached per production unit yield of rice (0.21~0.34 kg N Mg^{-1} grain). However, in terms of percent N loss, LS leached significantly more inorganic N in total N input (0.78%) than did organic production (0.32~0.60%).

Discussion

Inorganic N concentration in percolation water

In flooded paddy soils, aerobic and anaerobic metabolisms occur in close proximity [18,19,20]. At the 50 cm depth in the sandy loam soil that we studied, in organic treatments, the presence of the oxidized zone close to the reduced soil zone was conductive for the transformation of NH_4-N mineralized from organic manure to NO_3-N. However, the continuous anaerobic conditions (under flood irrigation, until 1 week before harvest) at the 180 cm depth maintained NH_4-N as the main form of inorganic N, more so than NO_3-N. Particularly, in continuously flooded paddy soils with abundant organic substrate and a limited availability of electron acceptors, the reduction of NO_3-N to NH_4-N would be more efficient than the formation of N_2, so NH_4-N concentrations can be an order of magnitude higher than those of NO_3-N [21,22,23,24]. This situation also occurs in soils with low

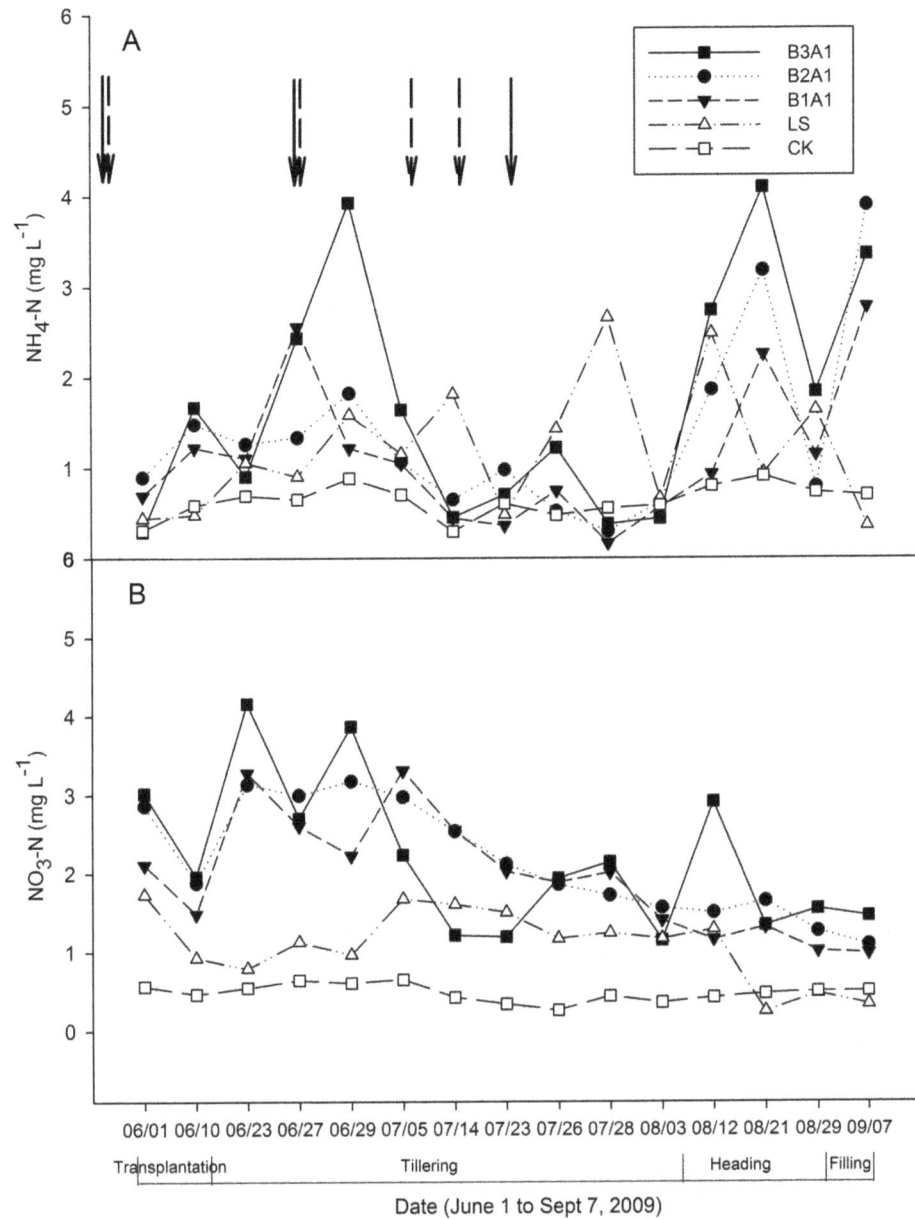

Figure 3. Concentrations of NH₄-N and NO₃-N in leachate at the 50 cm soil depth under different fertilization treatments. A: NH_4-N; B: NO_3-N. The solid and long-dashed arrows indicate the date of fertilizer applications for organic treatments (B1A1, B2A1, and B3A1) and for the conventional treatment (LS), respectively. Bars represent standard deviations ($n = 3$).

CEC and a low content of exchangeable base cations [19,25,26,27]. From the depth of 50 cm to 180 cm, average NH_4-N and NO_3-N concentrations decreased by about 5–10% and 47–70%, respectively, in fertilized organic treatments (B1A1, B2A1 and B3A1). The smaller decrease in NH_4-N concentration, compared to NO_3-N, may resulted from continuous decomposition of organic fertilizer and the limited soil adsorption capacity for NH_4-N [19,25,28]. From 50 cm to 180 cm in LS soils, there was a 27% decrease in NH_4-N, but a 57% increase in NO_3-N, indicating that mineral fertilization may have led to an increased downward movement of NO_3-N, compared to organic fertilizer, and the low dentrification capacity in deeper LS soils would thus have caused NO_3-N concentration to remain high [20]. This substantial loss of inorganic via leaching NH_4-N from continuous flooded rice fields

has also been confirmed by other studies, which reported that NH_4-N might account for up to 92% of the total inorganic N in leachate - this large risk of NH_4-N leaching deserves more attention than the risk of NO_3-N loss [23,27].

During the vegetative phase of rice growth, NO_3-N concentrations were higher than those of NH_4-N, whereas the opposite was observed during the reproductive phase (Figs. 3 and 4). Rice requires more NH_4-N than NO_3-N during the vegetative stage, contributing to a suppression of NH_4-N concentrations in leachate [29,30]. As rice plants shifted into their reproductive stage, NH_4-N was continuously mineralized from the organic fertilizers, but the anaerobic conditions prevented nitrification into NO_3-N, both in the upper and lower soil profiles [23,31]. Unlike NH_4-N, the NO_3-N concentration in percolation water declined from transplant

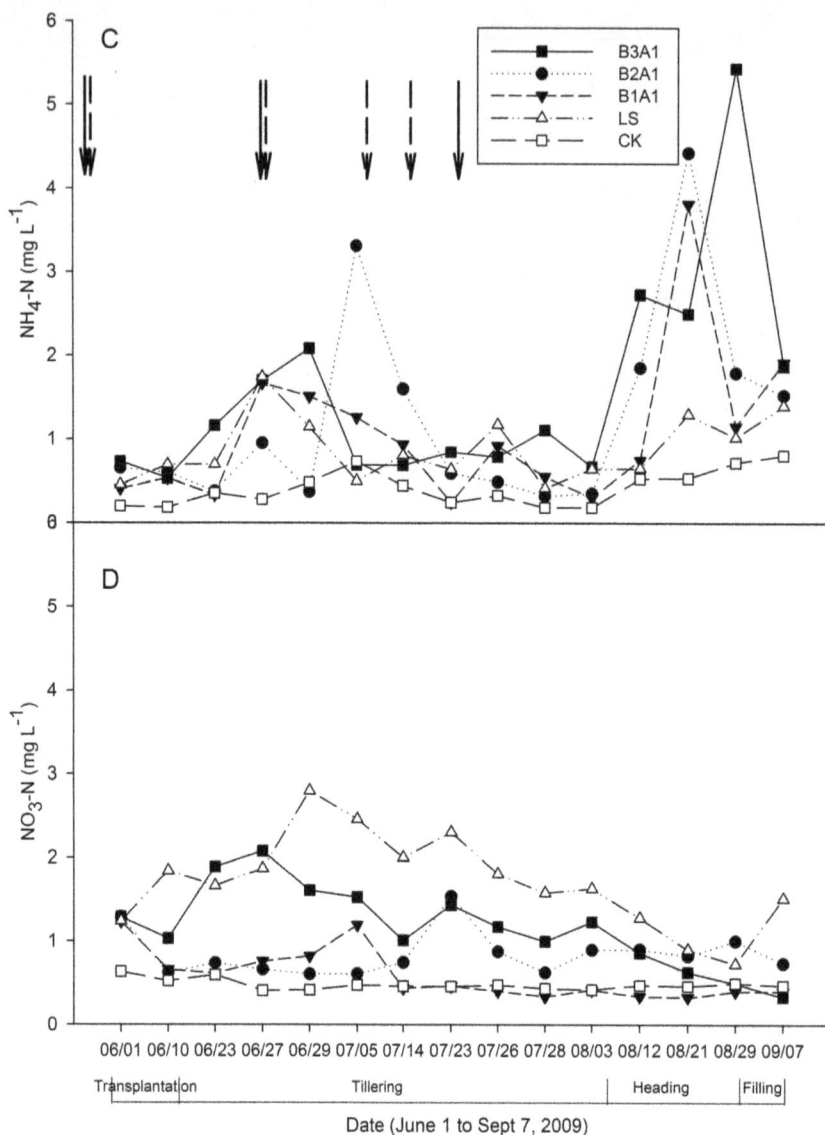

Figure 4. Concentrations of NH₄-N and NO₃-N in leachate at the 180 cm soil depth under different fertilization treatments. C: NH₄-N; D: NO₃-N. The solid and long-dashed arrows indicate the date of fertilizer applications for organic treatments (B1A1, B2A1, and B3A1) and for the conventional treatment (LS), respectively. Bars represent standard deviations (n = 3).

until harvest mainly because of continuous uptake by the plants throughout the growing season [32]. The two peaks observed in NH₄-N concentrations were also driven by fertilization and environmental conditions (higher temperature). At the beginning of tillering (~June 29), air temperature was close to its annual maximum (Fig. 1), which may have promoted the mineralization of organic fertilizer. However, the observed increases in NO₃-N concentrations lagged behind those of NH₄-N, because urea or organic fertilizer must first be transformed to amide N, then to ammonium and further nitrified to NO₃-N [28].

Comparison of N utilization and loss from organic and conventional rice production

Organic and mineral fertilizers influenced rice yield differently, through differential effects on yield components such as the number of panicles per hill, number of grains per panicle, and grain weight. In our experiment, topdressing with castor bean

meal in combination with increasing manure application resulted in higher grain weight, but mineral fertilizer still produced a larger number of panicles and grains per panicle [33]. A previous study revealed that increased mineral N supply to rice increases the amount of dry matter translocated into the grain [34]. However, if organic and mineral fertilizers are used together, higher numbers of panicles and grains per panicle can be achieved because of improved nutrient balance [35]. As the yield increased, more N was removed from the soil, but not in proportion with the increase in N input, such that N utilization efficiency (NUE) declined. Rice grown under conventional production showed the highest NUE (29.6%), which was significantly higher than that of organic treatments (8.7–17.6%, Table 5). This difference in NUE between conventional and organic production has been demonstrated in many other studies [8,32], which is attributed to slow release of N from organic fertilizers that limits N uptake by crops, and also maintains the continuous loss of N via leaching or other channels [22].

Table 3. Cumulative inorganic N leaching (kg N ha^{-1}) at 50 and 180 cm soil depths for control (CK), organic (B1A1, B2A1, and B3A1) and conventional (LS) treatments during a single rice growth cycle.

Depth (cm)	Treatments	NH$_4$-N (kg ha^{-1})	NO$_3$-N (kg ha^{-1})	Total inorganic N leached (kg ha^{-1})	Net inorganic N leached (kg ha^{-1}) [a] [b]
50	CK	3.4 c	0.0 c	3.4 d	n.a.
	LS	7.8 b	4.3 b	12.1 c	8.7 c
	B1A1	7.6 b	10.9 a	18.5 b	15.2 b
	B2A1	9.6 ab	12.6 a	22.2 a	18.8 ab
	B3A1	11.3 a	11.8 a	23.1a	19.8 a
180	CK	0.4 c	0.0 d	0.4 d	n.a
	LS	1.1 bc	1.7 a	2.7 ab	2.3 ab
	B1A1	1.4 ab	0.2 d	1.6 c	1.2 b
	B2A1	1.7 ab	0.5 c	2.2 bc	1.8 ab
	B3A1	2.0 a	1.0 b	3.0 a	2.5 a

[a]After subtraction of the leakage quantity measured in CK.
[b]Different letters in a column denote a significant difference between treatments at the 0.05 level.

Environmental performance per se, in terms of N loss through leaching is a key concern for maintaining water quality and the integrity of organic farming. When comparing organic and conventional paddy rice production, N leaching losses are a tradeoff with rice yield; for example, if high yields are to be achieved, relatively high amounts of inorganic N must be supplied and may be lost. In our study, the highest yield of organic rice (7550 kg ha^{-1}) was achieved using the highest rate of N application (B3A1, 802 kg N ha^{-1}; Table 4); yield in these plots amounted to 80% of that achieved through conventional treatment (LS), which received inputs of only 297 kg of N ha^{-1}. Furthermore, organic production did not always perform well in terms of inorganic N leaching per unit yield of rice. In fact, if organic fertilization was intensified enough to obtain a "conventional" yield, even more N would be leached per unit yield than currently occurs in conventional production. Organic treatments with relatively higher N input (B2A1 and B3A1) tend to have lower overall rates of inorganic nitrogen loss than the LS treatment, but this came at the cost of a much lower rice yield.

In the current study, the N input levels in organic treatments (except for B1A1) were higher than most conventional rice production in China, where for a single growth cycle N application varies from 200 to 300 kg N ha^{-1} [36]. In the Ili River Valley, rice production had an average N application rate of 234 kg ha^{-1}, equal to the national average. This generates additional doubt that organic production can help to reduce N leaching from paddy production. In our experiments, 2.9–7.9% (8.7–19.8 kg ha^{-1}) of total applied N was lost through leaching at 50 cm, and 0.32–0.78% (1.2–2.5 kg ha^{-1}) was lost at 180 cm, whereas other studies involving organic fertilization reported losses of 0.1 to 15% [13,16,31,37,38]. Based on these findings, we believe that organic rice production in the Ili River Valley would not reduce (or may even increase) N leaching compared to conventional production, especially when organic fertilizer input is high enough to achieve conventional yields.

High reliance on external nutrient supply, especially from non-organic/conventional sources, has become a concern for large-scale organic production in recent years [39,40]. Organic

Table 4. Amount of inorganic N leached at 50 cm and 180 cm soil depths, defined as production of 1 Mg rice grains produced and as a % of N input.

Depth (cm)	Treatments	Rice yield (Mg ha^{-1})	N input (kg ha^{-1})	Net leached inorganic N per Mg of rice grains (kg N Mg^{-1}) [b]	Proportion of net leached inorganic N in total N input (%) [a] [b]
	CK	3.6 d	n.a.		
	LS	9.4 a	297	0.93 b	2.9 bc
50	B1A1	5.6bc	193	2.69 a	7.9 a
	B2A1	6.9bc	498	2.72 a	3.8 b
	B3A1	7.6 b	802	2.62 a	2.5 c
	CK	3.6	n.a.	n.a.	n.a.
	LS	9.4	297	0.25 a	0.78 a
180	B1A1	5.6	193	0.21 a	0.60 ab
	B2A1	6.9	498	0.26 a	0.36 b
	B3A1	7.6	802	0.34 a	0.32 b

[a]"Net N leached" was calculated by subtracting the amount of N leached in the control treatment (CK) from total treatment values.
[b]Different letters in a column denote a significant difference between treatments at the 0.05 level.

Table 5. Nitrogen budget for rice production.

Treatments	N input from fertilizers (kg ha^{-1})	N removed by rice (kg ha^{-1})	N removed by straw (kg ha^{-1})	Total N removed by rice crop (kg ha^{-1})	N budget (kg ha^{-1})	N utilization efficiency (%) [a]
CK	0	40	13	53	−53	n.a
LS	297	100	41	141	156	29.6 a
B1A1	193	59	28	87	106	17.6 b
B2A1	498	85	29	114	384	12.2 c
B3A1	802	87	36	123	679	8.7 d

[a]Different letters in a column denote a significant difference between treatments at the 0.05 level.

fertilizers are able to meet crop N requirements, but this is achieved through high application rates, which could be 1) mineralized to release a sufficient amount of N for rice growth, and 2) synchronized between N mineralization and crop uptake [10]. However, higher organic N input can cause higher N losses through leaching, as shown in our experiments. In addition, N in paddy soils can also be denitrified to the powerful greenhouse gas N_2O [14], an additional significant environmental impact that was not measured in our study. Ju et al. [5] summarized that rice production in China using 300 kg N ha^{-1} per season contributed denitrification N losses of approximately 36%, which are markedly higher than losses from volatilization (12%) or leaching (0.5%). Here, we assume that denitrification and volatilization losses for mineral fertilizer in our study would be roughly comparable to those reported by Ju et al. [5] and Guan et al. [41]. Besides inorganic N, organic N in the leachate was not measured in our study, but it could eventually enter water bodies and cause of eutrophication. In addition, given that conventional rice has a higher NUE (29.6%), lower N input, and higher crop yield, N leaching from conventional rice production would not pose as high risk of contamination to local water systems as from organic rice production [10].

Conclusions

This study indicates that inorganic N concentrations in leachate decreased as soil depth increased, but the decrease was significantly larger in organic than in conventional paddies. NO_3-N tended to remain in the upper soil profile of organic paddies, whereas in conventional soils, NO_3-N migrated further downward.

In organic paddy soils, NH_4-N accounted for a substantial portion of inorganic N in the leachate due to the organic manure decomposition and denitrification process under the continuous flooding conditions. In terms of the N leached per unit yield of rice, organic fertilization did not perform better than conventional fertilization in all cases, particularly for the organic production with higher rice yields. Consistent with this, conventional production showed higher N utilization efficiency (29.6%) compared to organic production (8.7–17.6%). We conclude that converting conventional rice production to organic production in the Ili River Valley of Central Asia will not reduce N leaching into local water systems, especially given the high-load application of organic manure to maintain high rice yields. A longer period study with integrated monitoring of N loss through leaching, volatilization, nitrification and denitrification is necessary to compare the overall performance of organic versus conventional rice production.

Acknowledgments

We thank Luo Xinhu and Liu Fuwang of the Ili Agricultural Technical Station, Xinjiang Uygur Autonomous Region for their support during field work. The authors would also like to thank the two anonymous reviewers for their helpful remarks.

Author Contributions

Conceived and designed the experiments: FqM XpS WlW. Performed the experiments: XpS. Analyzed the data: FqM XpS JEO. Contributed reagents/materials/analysis tools: FqM XpS WlW. Wrote the paper: FqM JEO.

References

1. Ghosh BC, Bhat R (1998) Environmental hazards of nitrogen loading in wetland rice fields. Environmental Pollution 102 S1: 123–126.
2. Khind CS, Meelu OP, Singh Y, Singh B (1991) Leaching losses of urea-N applied to permeable soils under lowland rice. Fertilizer Research 28: 179–184.
3. Zhou S, Sugawara S, Riya S, Sagehashi M, Toyota K, et al. (2011) Effect of infiltration rate on nitrogen dynamics in paddy soil after high-load nitrogen application containing ^{15}N tracer. Ecological Engineering 37: 685–692.
4. Devkota KP, Manschadi A, Lamers JPA, Devkota M, Vlek PLG (2013) Mineral nitrogen dynamics in irrigated rice–wheat system under different irrigation and establishment methods and residue levels in arid drylands of Central Asia. European Journal of Agronomy 47: 65–76.
5. Ju XT, Xing GX, Chen XP, Zhang SL, Zhang LJ, et al. (2009) Reducing environmental risk by improving N management in intensive Chinese agricultural systems. Proceedings of the National Academy of Sciences 106: 3041–3046.
6. Yang L, He GH (2008) The impact of Ili irrigation region construction on environment systems. Water Saving Irrigation 3: 34–35. (In Chinese).
7. Jemison JM, Fox RH (1994) Nitrate leaching from nitrogen-fertilized and manured corn measured with zero-tension pan lysimeters. Journal of Environmental Quality 23: 337–343.

8. Askegaard M, Olesen JE, Rasmussen IA, Kristensen K (2011) Nitrate leaching from organic arable crop rotations is mostly determined by autumn field management. Agriculture, Ecosystems and Environment 142: 149–160.
9. Hansen B, Kristensen ES, Grant R, Høgh-Jensen H, Simmelsgaard SE, et al. (2000) Nitrogen leaching from conventional versus organic farming systems-a systems modelling approach. European Journal of Agronomy 13: 65–82.
10. Kirchmann H, Bergström L (2001) Do organic farming practices reduce nitrate leaching? Communications in Soil Science and Plant Analysis 32: 7–8, 997–1028.
11. Chhabra A, Manjunath KR, Panigrahy S (2010) Non-point source pollution in Indian agriculture: Estimation of nitrogen losses from rice crop using remote sensing and GIS. International Journal of Applied Earth Observation and Geoinformation 12: 190–200.
12. Xing G, Cao Y, Shi S, Sun G, Du L, et al. (2001) N pollution sources and denitrification in waterbodies in Taihu Lake region. Science in China Series B: Chemistry 44: 304–314.
13. Tian Y, Yin B, Yang L, Yin S, Zhu Z (2007) Nitrogen runoff and leaching losses during rice-wheat rotations in Taihu Lake Region, China. Pedosphere 17: 445–456.
14. Li F, Pan G, Tang C, Zhang Q, Yu J (2008) Recharge source and hydrogeochemical evolution of shallow groundwater in a complex alluvial fan

system, southwest of North China Plain. Environmental Geology 55: 1109–1122.

15. Li CF, Cao CG, Wang JP, Zhan M, Yuan WL, et al. (2008) Nitrogen losses from integrated rice-duck and rice-fish ecosystems in southern China. Plant Soil 307: 207–217.

16. Zhu JG, Han Y, Liu G, Zhang YL, Shao XH (2000) Nitrogen in percolation water in paddy fields with a rice/wheat rotation. Nutrient Cycling in Agroecosystems 57: 75–82.

17. Luo LG, Wen DZ (1999) Nutrient balance in rice field ecosystem of northern China. Chinese Journal of Applied Ecology 10: 301–304. (In Chinese).

18. Bauder JW, Montgomery BR (1980) N-source and irrigation effects on nitrate leaching. Agronomy Journal 72: 593–596.

19. Xiong Z, Huang T, Ma Y, Xing G, Zhu Z (2010) Nitrate and ammonium leaching in variable-and permanent-charge paddy soils. Pedosphere 20: 209–216.

20. Keeney DR, Sahrawat KL (1986) Nitrogen transformations in flooded rice soils, Fertilizer Research 9: 15–38

21. George T, Ladha JK, Garrity DP, Buresh RJ (1993) Nitrate dynamics during the aerobic soil phase in lowland rice-based cropping systems. Soil Science Society of America Journal 57: 1526–1532.

22. Ouyang H, Xu YC, Shen QR (2009) Effect of combined use of organic and inorganic nitrogen fertilizer on rice yield and nitrogen use efficiency. Jiangsu Journal of Agricultural Science 25: 106–111. (In Chinese).

23. Wang X, Suo Y, Feng Y, Shohag M, Gao J, et al. (2011) Recovery of ^{15}N-labeled urea and soil nitrogen dynamics as affected by irrigation management and nitrogen application rate in a double rice cropping system. Plant Soil 343: 195–208.

24. Islam MM, Lyamuremye F, Dick RP (1998) Effect of organic residue amendment on mineralization of nitrogen in flooded rice soils under laboratory conditions. Communications in Soil Science and Plant Analysis 29: 7–8, 971–981.

25. Qian C, Cai ZZ (2007) Leaching of nitrogen from subtropical soils as affected by nitrification potential and base cations. Plant Soil 300: 197–205.

26. Weier KL, Doran JW, Power JF, Walters DT (1993) Denitrification and the dinitrogen/nitrous oxide ratio as affected by soil water, available carbon, and nitrate. Soil Science Society of America Journal 57: 66–72

27. Stanford G, Smith SJ (1972) Nitrogen mineralization potentials of soils. Soil Science Society of America Journal 36: 465–472

28. Reddy KR, Patrick WH Jr (1986) Denitrification losses in flooded rice fields. Fertilizer Research 9: 99–116.

29. Reddy KR, Patrick WH, Lindau CW (1989) Nitrification-denitrification at the plant root-sediment interface in wetlands. Limnol. Oceanogr 34: 1004–1013.

30. Uhel C, Roumet C, Salsac L (1989) Inducible nitrate reductase of rice plants as a possible indicator for nitrification in water-logged paddy soils. Plant and soil 116: 197–206.

31. Luo L, Itoh S, Zhang Q, Yang S, Zhang Q, et al. (2010) Leaching behavior of nitrogen in a long-term experiment on rice under different N management systems. Environmental monitoring and assessment 177: 141–150.

32. Cassman KG, Peng S, Olk DC, Ladha JK, Reichardt W, et al. (1998) Opportunities for increased nitrogen-use efficiency from improved resource management in irrigated rice systems. Field Crops Research 56: 7–39.

33. Sun XP, Li GX, Meng FQ, Guo YB, Wu WL, et al. (2011) Nutrients balance and nitrogen pollution risk analysis for organic rice production in Ili reclamation area of Xinjiang. Transactions of the CSAE 27: 158–162. (In Chinese).

34. Gao JS, Qin DZ, Liu GL, Xu MG (2002) The impact of long term organic fertilization on growth and yield of rice. Tillage and Cultivation 2: 31–33. (In Chinese).

35. Li X, Liu Q, Rong XM, Xie GX, Zhang YP, et al. (2010) Effect of organic fertilizers on rice yield and its components. Hunan Agricultural Sciences 5: 64–66. (In Chinese).

36. Xi YG, Qin P, Ding GH, Fan WL, Han CM (2007) The application of RCSODS model to fertilization practice of organic rice and its effect analysis. Shanghai Agricultural Sciences 23: 28–33. (In Chinese).

37. Chowdary VM, Rao NH, Sarma PBS (2004) A coupled soil water and nitrogen balance model for flooded rice fields in India. Agriculture, Ecosystems & Environment 103: 425–441.

38. Ma J, Sun W, Liu X, Chen F (2012) Variation in the stable carbon and nitrogen isotope composition of plants and soil along a precipitation gradient in Northern China. PloS one 7: e51894.

39. Kirchmann H, Tterer KT, Bergström L (2008) Nutrient supply in organic agriculture: plant availability, sources and recycling. In: Kirchmann H, Bergström L (eds) Organic crop production–ambitions and limitations. Springer, Netherlands, 89–116.

40. Rodrigues MA, Pereira A, Cabanas JE, Dias L, Pires J, et al. (2006) Crops use-efficiency of nitrogen from manures permitted in organic farming. European journal of agronomy 25: 328–335.

41. Guan JX, Wang BR, Li DC (2009) Effect of chemical fertilizer applied combined with organic manure on yield of rice and nitrogen using efficiency.Chinese Agricultural Science Bulletin 25: 88–92. (In Chinese).

9

The Root Herbivore History of the Soil Affects the Productivity of a Grassland Plant Community and Determines Plant Response to New Root Herbivore Attack

Ilja Sonnemann[1]*, Stefan Hempel[1], Maria Beutel[1], Nicola Hanauer[1], Stefan Reidinger[2], Susanne Wurst[1]

1 Freie Universitaet Berlin, Dahlem Centre of Plant Sciences, Berlin, Germany, **2** University of York, Department of Biology, York, United Kingdom

Abstract

Insect root herbivores can alter plant community structure by affecting the competitive ability of single plants. However, their effects can be modified by the soil environment. Root herbivory itself may induce changes in the soil biota community, and it has recently been shown that these changes can affect plant growth in a subsequent season or plant generation. However, so far it is not known whether these root herbivore history effects (i) are detectable at the plant community level and/or (ii) also determine plant species and plant community responses to new root herbivore attack. The present greenhouse study determined root herbivore history effects of click beetle larvae (Elateridae, Coleoptera, genus *Agriotes*) in a model grassland plant community consisting of six common species (*Achillea millefolium, Plantago lanceolata, Taraxacum officinale, Holcus lanatus, Poa pratensis, Trifolium repens*). Root herbivore history effects were generated in a first phase of the experiment by growing the plant community in soil with or without *Agriotes* larvae, and investigated in a second phase by growing it again in the soils that were either *Agriotes* trained or not. The root herbivore history of the soil affected plant community productivity (but not composition), with communities growing in root herbivore trained soil producing more biomass than those growing in untrained soil. Additionally, it influenced the response of certain plant species to new root herbivore attack. Effects may partly be explained by herbivore-induced shifts in the community of arbuscular mycorrhizal fungi. The root herbivore history of the soil proved to be a stronger driver of plant growth on the community level than an actual root herbivore attack which did not affect plant community parameters. History effects have to be taken into account when predicting the impact of root herbivores on grasslands.

Editor: Alexandra Weigelt, University of Leipzig, Germany

Funding: The work has been funded by the Deutsche Forschungsgemeinschaft (http://www.dfg.de/index.jsp) Priority Program 1374 "Infrastructure-Biodiversity-Exploratories" (DFG-WU 603/3-1). The funders had no role in study design, data collection and analysis, decision to publish, or preparation of the manuscript.

Competing Interests: The authors have declared that no competing interests exist.

* E-mail: i.sonnemann@fu-berlin.de

Introduction

Insect root herbivores can alter plant community structure by affecting the competitive ability of single plants [1]. However, their effects can be modified by the soil environment [2]–[3]. Root herbivory itself may induce changes in the soil biota community [4]–[][6]. It has recently been shown that these changes can affect plant growth in a subsequent season or plant generation [7]. However, it is not known whether these root herbivore history effects (i) are detectable at the plant community level and/or (ii) also determine plant species and plant community responses to new root herbivore attack.

Several studies documented that root symbiotic arbuscular mycorrhizal fungi (AMF) and root herbivores influence each other's performance and effect on the host plant [8]–[][][11]. One study [12] even found consequences of these interactive effects on the plant community level. AMF provide mineral nutrients to the plant in exchange for photosynthates [13]. Their effect on plant growth proved to be plant as well as AMF species specific [14]. As root herbivores can alter AMF community structure [15] their

history effects may potentially be generated through changes in this important soil biota group.

Click beetle larvae (Elateridae, Coleoptera) of the genus *Agriotes* are dominant generalist root herbivores in European grasslands [16]–[][18] and also pests in different economically important crops. While their effects on cropping systems are well studied, their impact on grassland plant communities is almost unknown. Studies that included measurements on background soil biota have so far only been done in a single plant system (*Plantago lanceolata*), and found no effects of *Agriotes* larvae on the microbial carbon source utilization in the rhizosphere [19] and root colonization by arbuscular mycorrhizal fungi [20]–[21].

The presented greenhouse study aimed at determining root herbivore history effects of *Agriotes* spp. larvae on a grassland plant community. The root herbivore history of the soil was generated by growing a model plant community without or with *Agriotes* larvae in soil biota communities from two different grassland sites in a first phase of the experiment. Root herbivore history effects on plant growth and response to a new *Agriotes* attack were determined by growing the model plant community without or with *Agriotes* larvae in the either herbivore untrained (absence of *Agriotes* larvae in phase 1) or trained (presence of *Agriotes* larvae in phase 1) soil, in a second phase of the experiment. We further investigated effects of present or past *Agriotes* herbivory on AMF as one important group within the soil biota community. Measurements included AMF community parameters in soil as well as community parameters and colonization levels in the roots of the model plant *P. lanceolata*. We hypothesized that (i) the root herbivore history of the soil influences plant growth and response to a new root herbivore attack, (ii) this has consequences for plant community structure, and (iii) root herbivore history effects can be explained by herbivore induced shifts in the AMF community.

Materials and Methods

The experiment was conducted within the frame of the Biodiversity Exploratories project [22]. Background soil and inocula were collected from grassland sites in the Schorfheide exploratory, 100 km north of Berlin, Germany. A general field work permit was issued by the Landesumweltamt Brandenburg. The collection did not involve endangered or protected species.

Background soil

The background soil was an alfisol that was collected from the upper 5–30 cm of a mown pasture. Stones and coarse roots were sieved out (1 cm mesh size) and the soil was steamed for 4 h at 90°C prior to usage to kill root herbivores.

Establishment of plant community and *Agriotes* treatment

A total of 120 round 2 L plastic pots (Albert Treppens & Co Samen GmbH, Berlin, Germany) were prepared by sealing the drainage holes with water permeable non-woven material (Plantex®, DuPont, Germany) to prevent escape of *Agriotes* larvae in *Agriotes* treatments. Pots were filled with background soil, and soil biota other than soil macrofauna was reintroduced by means of inocula that were specific for the two phases of the experiment (see detailed experimental set up of phase 1 and 2 below). A plant community consisting of six grassland plant species (*Achillea millefolium* L., *P. lanceolata* L., *Taraxacum officinale* Wiggers, *Holcus lanatus* L., *Poa pratensis* L., *Trifolium repens* L.) was established in each pot, with one individual per species. The plant species chosen were common on the two grassland sites from where the soil biota inocula for phase 1 (see below) were collected. The proportion of plant functional types (two grasses, three herbs and one legume) resembled those at the sites. Pots in each treatment were allocated to three different sowing schemes to account for neighboring effects, and plant species were sown at evenly distanced positions in each pot, with four seeds (Appels Wilde Samen GmbH, Germany) per position, according to the sowing schemes. Each pot was covered with a perforated plastic bag (15×66 cm, EDNA International GmbH, Germany) to prevent invasion of unwanted herbivores. Seedlings were thinned to one per position approximately one week after sowing. Pots were randomized once a week and watered from the bottom as needed during the course of the experiment. To establish the *Agriotes* treatment, *Agriotes* spp. larvae, collected from a fallow grassland app. 10 km south of Berlin, were added to half of the pots in each soil biota/training treatment (see below) four weeks after sowing, with three larvae per pot. Five randomly chosen larvae from that fallow grassland were identified as *A. obscurus* [23]. The composition of instars reflected those at the site but was not assessed in detail. Larvae were randomly allocated to the pots. Larvae were rinsed with tap water prior to addition. The rinsing water was collected and evenly allocated to *Agriotes*-free pots to correct for microorganisms that were introduced with the larvae.

Experimental set up

Phase 1. 40 pots were each filled with 1790+/−1 g of background soil. The upper 790 g of background soil in each pot were mixed with 174+/−1 g of one of two soil biota inocula, with 20 pots per inoculum. Each inoculum consisted of un-steamed soil that had been collected from the upper 10 cm of one of two alfisol grassland sites (SEG 33 and SEG 37). Sites for inocula collection were chosen (i) to be of the same soil type as the background soil to ensure establishment of the soil biota in the pots and (ii) to differ in management (SEG 33: fertilized mown pasture, SEG 37: unfertilized pasture) to maximize differences between the two soil biota communities in general and AMF communities [24] in particular. Macrofauna and coarse material was sieved out (4 mm mesh size) from the inocula, roots were cut to 1 cm pieces and added back, and the inocula were air dried prior to usage to kill remaining root herbivore eggs. The plant community and *Agriotes* treatment were established as described above, resulting in four treatments (two soil biota communities (SEG 33, SEG 37), either without or with *Agriotes* larvae), with 10 replicates each. Plants were grown for two months in a climate chamber at 20/18°C day/night temperature, 69% air humidity and 16 h day length.

Phase 2. 80 pots were each filled with 1140+/−1 g of background soil. The upper 540 g of background soil in each pot were mixed with 825+/−1 g of a soil biota inoculum. Soil biota inocula consisted of root free soil from phase 1 pots. The comparably high amount of inoculum was chosen to transfer not only the soil biota themselves, but also the respective abiotic conditions that potentially resulted from the treatments in phase 1. After the harvest of phase 1, *Agriotes* larvae were removed from the respective treatments and the soil of each phase 1 pot was well mixed and air dried for storage until the set up of phase 2. The soil from one phase 1 pot then served as inoculum for two pots in phase 2. The plant community and *Agriotes* treatment were established as described above, with each pair of pots being split between control and *Agriotes* treatment. This resulted in eight treatments (two soil biota communities (SEG 33, SEG 37), either untrained or *Agriotes* trained during phase 1 (together with respective abiotic soil conditions), each without or with new *Agriotes* larvae in phase 2), with 10 replicates each. Plants were grown for two months in June/July in a greenhouse at 20/19°C minimal day/night temperature with 16 h additional light per day.

Harvests and measurements

Each phase was harvested by cutting the shoots at soil surface level. Shoot biomass for each plant was determined gravimetrically after drying at 56°C for 72 h as gram dry weight (gDW). Shoot biomass data (means (se)) for all plant species grown in different treatments are reported as supporting information (Supporting Tables S1). Shannon's diversity index was calculated per pot as $H' = \sum pi*ln\ pi$, where pi is the share of shoot biomass represented by species i and is calculated as shoot biomass i/shoot biomass total. Roots were sieved (1 mm mesh size) from the soil, and *P. lanceolata* roots were separated as far as possible. As roots were strongly entangled it was not possible to further separate plant species. Thus, total roots per pot were washed, and root biomass (including *P. lanceolata*) per pot was determined gravimetrically after drying at 56°C for 72 as gDW.

Soil carbon (C) and nitrogen (N) content was determined from air dried soil of the original soil biota inocula and of phase 1 pots at harvest by means of complete combustion and chromatographical detection (analyzer EuroEA, HEKATech GmbH, Germany). Soil

CN ratio was calculated as surrogate for nutrient mineralization [25]–[][27].

Mycorrhizal structures in *P. lanceolata* roots from all pots in phase 1 and half of the pots in each treatment in phase 2 were stained with the ink-vinegar method [28]. The percentage of root length colonized [%RLC] by arbuscules and mycorrhizal structures in total (total AMF, including arbuscules, vesicles and intraradical hyphae) was determined at 200 fold magnification using the gridline intersect method [29]. The length of extraradical AMF hyphae in soil (LEH [m/g soil]) was determined for phase 1 and 2 from half of the pots in each treatment applying the methods of [30] and [31]. The AMF community was characterized in phase 1 and 2 from fresh substrate (stored at $-80°C$ until analysis) and dried *P. lanceolata* roots by means of terminal restriction fragment length polymorphism (TRFLP) using the database TRFLP approach as outlined in [32]. Briefly, soil DNA was extracted from root and soil samples of each pot and amplified using AM fungal specific primers for the ribosomal small subunit gene (SSU). PCR amplicons were pooled for root and soil samples separately, cloned and 129 clones were sequenced to obtain a database of AM fungal operational taxonomic units (OTUs) present in the samples. This database was used to calibrate the T-RFLP approach, which was then applied on the PCR products obtained from each sample separately. Obtained TRFLP peaks were compared with the database to identify present OTUs using the TRAMPR package in R [33]. A detailed description of the methods is available as supporting information (Protocol S1). AMF data (means (se)) in different treatments are reported as supporting information (Supporting Tables S1).

Statistical analyses

All statistical analyses were performed using the software program 'R', version 2.12.0 [34]. Data were analyzed separately for phase 1 and 2 of the experiment. Effects of sowing scheme, soil biota- and Agriotes treatments on plant biomass parameters, *P. lanceolata* root length colonized by AMF, length of extraradical AMF hyphae (LEH) and soil CN ratio were analyzed with generalized least square models (GLS) for phase 1 and linear mixed effects models (lme) for phase 2. Identities of phase 1 pots were included as random factor in the models to analyze effects in phase 2. Inhomogeneity of variances among data of different treatments was accounted for by applying the varIdent command [35]. The factor sowing scheme did not affect any of the parameters and was therefore excluded from the models. Effects of soil biota- and Agriotes treatments on AMF community composition were analyzed with permutation tests based on Jaccard distance matrices using the adonis command [36]. AMF species extracted from P. lanceolata roots in phase 1 could not be analyzed due to very low species detection. To visualize AM fungal communities in the pots of phase 2, we calculated a non-metric analysis and plotted mean community composition and standard error for the four treatment combinations of soil biota and *Agriotes* training.

Results

The two phases of the experiment differed in plant biomass and community composition. In phase 1, total plant biomass per pot was on average 3.5 times higher than in phase 2 (Table 1). *T. repens* clearly dominated the plant community at the harvest of phase 1, while in phase 2 the plant community was only slightly dominated by *P. lanceolata*.

Table 1. Plant and AMF parameters and Soil CN ratio (mean (se)) in different soil treatments (soil biota communities SGE 33 and SEG 37, *Agriotes* untrained (−AP1) or trained (+AP1) soil substrate) in phase 1 and 2.

	Phase 1		Phase 2			
	SEG 33	**SEG 37**	**SEG 33**	**SEG 33**	**SEG 37**	**SEG 37**
			(−AP1)	**(+AP1)**	**(−AP1)**	**(+AP1)**
Shannon's H	1.29 (0.04)	1.27 (0.03)	1.53 (0.03)	1.51 (0.03)	1.39 (0.03)	1.40 (0.04)
Plant biomass total (gDW)	20.73 (0.83)	22.18 (0.92)	5.70 (0.33)	6.32 (0.31)	6.13 (0.30)	6.54 (0.18)
Root biomass total (gDW)	7.72 (0.43)	9.23 (0.42)	2.01 (0.17)	2.33 (0.15)	2.38 (0.22)	2.56 (0.16)
Shoot biomass total (gDW)	13.00 (0.59)	12.95 (0.75)	3.70 (0.17)	3.99 (0.17)	3.74 (0.15)	3.99 (0.14)
Shoot biomass (gDW)						
A. millefolium	1.21 (0.16)	0.66 (0.14)	0.42 (0.06)	0.46 (0.05)	0.67 (0.12)	0.47 (0.05)
P. lanceolata	1.82 (0.21)	1.91 (0.27)	0.97 (0.10)	1.10 (0.08)	1.25 (0.14)	1.16 (0.11)
T. officinale	0.25 (0.04)	0.35 (0.07)	0.75 (0.11)	0.81 (0.12)	0.76 (0.14)	1.19 (0.14)
H. lanatus	3.01 (0.24)	3.85 (0.29)	0.91 (0.15)	1.10 (0.18)	0.63 (0.15)	0.93 (0.16)
P. pratensis	0.23 (0.03)	0.25 (0.04)	0.07 (0.02)	0.09 (0.01)	0.02 (0.01)	0.02 (0.01)
T. repens	6.49 (0.74)	5.94 (0.71)	0.57 (0.10)	0.46 (0.08)	0.42 (0.11)	0.22 (0.05)
LEH [m/gDW soil]	0.47 (0.07)	1.24 (0.21)	2.07 (0.47)	1.16 (0.29)	1.40 (0.25)	1.87 (0.42)
No of AMF species in soil	3.75 (0.61)	5.15 (0.23)	2.40 (0.26)	3.33 (0.52)	3.50 (0.49)	2.61 (0.52)
No of AMF species in roots			3.00 (0.48)	2.90 (0.43)	4.25 (0.45)	3.83 (0.46)
Total AMF [% P. lanceolata RLC]	18.95 (2.37)	24.05 (2.50)	71.50 (3.00)	66.80 (3.81)	59.80 (3.53)	46.50 (3.72)
Arbuscules [% P. lanceolata RLC]	8.20 (1.81)	5.70 (0.95)	16.50 (2.66)	17.10 (2.99)	12.50 (1.60)	10.75 (2.12)
Soil CN ratio	12.09 (0.15)	11.98 (0.12)				

gDW: gram dry weight, LEH: length of extraradical AMF hyphae, AMF: arbuscular mycorrhizal fungi, RLC: root length colonized.

Phase 1

Soil biota community effects. Root biomass of the plant community was 20% higher when grown with SEG 37 than with SEG 33, while total shoot and plant biomass as well as plant diversity, measured as Shannon's H' on the basis of shoot biomass was not affected by the soil biota community (Tables 1 and 2). However, the soil biota community affected shoot biomass of three out of six plant species, with *A. millefolium* growing 83% larger with soil biota community SEG 33 than with SEG 37, while *T. officinale* and *H. lanatus* grew 40% and 28% larger with SEG 37 than with SEG 33.

The two soil biota communities differed in their AMF community composition in soil, with 1.4 AMF species more present and extraradical hyphae being 2.6 times longer in soil of SEG 37 than of SEG 33. Total AMF structures of SEG 37 colonized 27% more of *P. lanceolata* roots length than AMF of SEG 33. Soil CN ratio did not differ for the two soil biota communities.

Root herbivory effects. Plant community parameters did not respond to the presence of *Agriotes* larvae. On the level of single plants, *Agriotes* larvae reduced shoot biomass of *P. lanceolata* (Figure 1A) by 31% on average and the effect did not differ significantly for the two soil biota communities. On the contrary, shoot biomass of *A. millefolium* (Figure 1B) was 2.7 times higher in the presence than in the absence of *Agriotes* larvae, but only when grown in soil biota community SEG 37. AMF parameters as well as the soil CN ratio were not affected by the presence of *Agriotes* larvae.

Table 2. Effects of soil biota community (SB, SEG 33 and SEG 37) and *Agriotes* presence (AP1, without and with) on plant and AMF parameters and soil CN in phase 1.

	SB		AP1		SB x AP1	
	df:1		df:1		df:1	
	F	p	F	p	F	p
Shannon's H	0.28	ns	0.75	ns	0.02	ns
Plant biomass total	3.10	ns	0.01	ns	0.00	ns
Root biomass total	13.00	***	1.98	ns	1.46	ns
Shoot biomass total	0.01	ns	0.61	ns	0.55	ns
Shoot biomass						
A. millefolium	30.65	***	9.54	**	7.11	*
P. lanceolata	0.16	ns	9.94	**	2.57	ns
T. officinale	4.21	*	0.16	ns	0.82	ns
H. lanatus	9.66	**	0.57	ns	0.18	ns
P. pratensis	0.35	ns	1.50	ns	2.52	ns
T. repens	0.62	ns	3.10	ns	1.02	ns
LEH	22.65	***	1.68	ns	0.62	ns
AMF community in soil	14.36	**	0.16	ns	2.42	ns
No of AMF species in soil	5.77	*	0.47	ns	0.74	ns
Total AMF (P. lanceolata)	4.15	*	0.01	ns	0.10	ns
Arbuscules (P. lanceolata)	2.84	ns	0.00	ns	0.11	ns
Soil CN ratio	0.62	ns	0.01	ns	0.73	ns

significance level:
*p<0.05,
**p<0.01,
***p<0.001,
ns = not significant, LEH: length of extraradical AMF hyphae, AMF: arbuscular mycorrhizal fungi.

Phase 2

Soil biota community effects. Soil biota community effects on plant community biomasses were similar to phase 1, with root biomass of the plant community being 14% higher when grown with SEG 37 than with SEG 33, while total shoot and plant biomass was not affected (Tables 1 and 3). The plant community was 9% more diverse when grown with SEG 33 than with SEG 37. The effect of the soil biota community on single plant species differed from that in phase 1, with *P. pratensis* and *T. repens* now being affected and growing 400% and 61%, larger with soil biota community SEG 33 than SEG 37, respectively, while responsive plant species in phase 1 (*A. millefolium*, *T. officinale*, *H. lanatus*) were not affected anymore.

The two soil biota communities differed in their AMF community composition in soil and *P. lanceolata* roots, with 1.09 AMF species more present in *P. lanceolata* roots grown in SEG 37 than in SEG 33. The length of extraradical hyphae did not differ for AMF from the two soil biota communities. Contrary to phase 1, AMF from SEG 33 colonized a higher percentage of *P. lanceolata* roots length than AMF from SEG 37. This was true for mycorrhizal structures in total as well as for arbuscules (30% and 45%, higher colonization with SEG 33 than SEG 37, respectively).

Root herbivore history effects. Total plant biomass was higher (9%; Table 1) when grown in *Agriotes* trained soil (presence of *Agriotes* larvae in phase 1) compared to untrained soil. The same tendency was observed for total shoot biomass (7%; lme: p = 0.05). Root biomass and plant diversity were not affected by the herbivore history of the soil. On the level of single plants, *Agriotes* trained soil facilitated shoot growth of *T. officinale* by 32% but reduced shoot growth of *T. repens* by 31% compared to untrained soil.

The effect of the *Agriotes* training on AMF community composition differed for the two soil biota communities (Figure 2). The community of *Agriotes* trained AMF consisted of 0.93 more species and 44% shorter extraradical hyphae for SEG 33 but of 0.42 fewer species and 34% longer hyphae for SEG 37 compared to the respective untrained AMF communities. Total structures of *Agriotes* trained AMF colonized 14% less of *P. lanceolata* roots length than structures of untrained AMF. Other AMF parameters were not affected by the *Agriotes* training.

Root herbivory effects. As in phase 1, plant diversity as well as plant community biomasses did not respond to the presence of *Agriotes* larvae. Also in accordance with phase 1, *Agriotes* larvae reduced shoot biomass of *P. lanceolata* (Figure 1C) by 15% on average and statistically independent of the soil biota community and *Agriotes* training, while shoot biomass of *A. millefolium* (Figure 1D) was not affected by the presence of *Agriotes* larvae when grown in soil biota community SEG 33, but enhanced by 65% when grown in untrained SEG 37. As indicated by a marginally non-significant three way interaction (lme: p = 0.06) of original soil biota community (SB), *Agriotes* training (AP1) and *Agriotes* presence (AP2), the positive effect of the presence of *Agriotes* larvae with soil biota community SEG 37 was lost when the soil was already *Agriotes* trained. Shoot biomass of *T. repens* (Figure 1E) was reduced by 28% by *Agriotes* larvae in phase 2, but only with untrained soil.

Contrarily to phase 1, AMF parameters were affected by the presence of *Agriotes* larvae. For the community composition in soil, the length of extraradical hyphae (LEH) and the percentage of *P. lanceolata* root length colonized by arbuscules the effect depended on the original soil biota community. LEH from SEG 33 was reduced by 27% (from 1.9 (se = 0.3) to 1.4 (se = 0.4) m/g DW soil) by *Agriotes* larvae while LEH from SEG 37 was increased by 49% (from 1.3 (se = 0.3) to 2.0 (se = 0.3) m/g DW soil). For AMF arbuscules an additional interaction of the factors *Agriotes* training

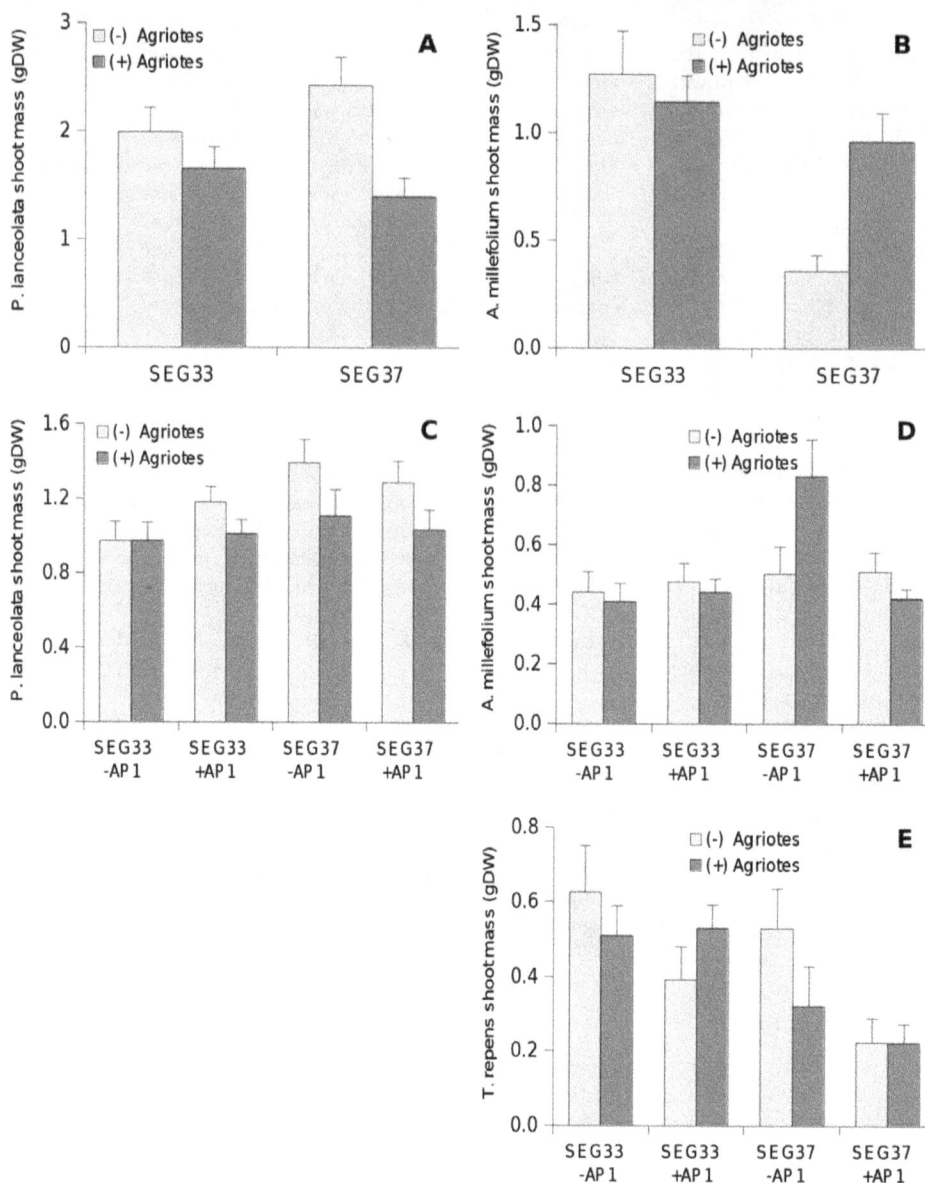

Figure 1. Shoot biomass (mean + se) of plants affected by the presence of *Agriotes* larvae. Phase 1 (A) *P. lanceolata* (B) *A. millefolium*; phase 2 (C) *P. lanceolata* (D) *A. millefolium*, (E) *T repens*; SEG33, SEG37: soil biota from respective grassland sites; −AP1, +AP1: untrained and *Agriotes* trained soil from phase 1, respectively.

and presence of *Agriotes* larvae resulted in colonization by *Agriotes* trained AMF being less negatively affected in SEG 33 and positively affected in SEG 37 by the presence of *Agriotes* larvae compared to arbuscules colonization by untrained AMF (50% reduction (from 22.0 (se = 2.7) to 11.0 (se = 1.0) %RLC) in untrained SEG 33 compared to no effect in trained SEG 33 (mean 17.1 (se = 4.4) %RLC), and 210% increase (from 5.3 (se = 0.5) to 16.3 (se = 1.5) %RLC) in trained SEG 37 compared to no effect in untrained SEG 37 (mean 12.5 (se = 2.4) %RLC)). On the level of total AMF structures, the same two way interaction of the presence of *Agriotes* larvae and *Agriotes* training as well as a three way interaction including original soil biota led to a similar effect for total AMF structures from SEG 37 (38% increase (from 39.0 (se = 3.4) to 54.0 (se = 2.4) %RLC) in trained SEG 37), whereas total AMF structures from SEG 33 were not affected (mean 69.2 (se = 4.9) %RLC).

Discussion

The study aimed at determining root herbivore history effects of *Agriotes* spp. larvae on a grassland plant community. As hypothesized, the root herbivore history of the soil influenced plant growth as well as plant growth response to new root herbivore attack. It affected plant community productivity (but not composition), showing for the first time that root herbivore induced changes in soil conditions can impact plant communities in a subsequent season. The root herbivore history of the soil proved to be a stronger driver of plant growth on the community level than an actual root herbivore attack which did not affect plant community parameters.

The two phases of the experiment differed in plant growth and community composition due to different growth conditions. Pots in phase 2 supported less plant biomass than pots in phase 1,

Table 3. Effects of soil biota community (SB, SEG 33 and SEG 37), *Agriotes* training (AP1, *Agriotes* presence in phase 1, untrained and trained) and *Agriotes* presence (AP2, without and with) on plant and AMF parameters in phase 2.

	SB df:1		AP1 df:1		AP2 df:1		SB X AP1 df:1		SB x AP2 df:1		AP1 x AP2 df:1		SB x AP1 x AP2 df:1	
	F	p	F	p	F	p	F	p	F	p	F	p	F	p
Shannon's H	22.17	***	0.04	ns	2.30	ns	0.44	ns	0.04	ns	0.44	ns	0.02	ns
Plant biomass total	1.90	ns	5.26	*	0.31	ns	0.20	ns	1.03	ns	2.11	ns	0.59	ns
Root biomass total	4.42	*	3.14	ns	0.01	ns	0.34	ns	0.46	ns	2.58	ns	1.06	ns
Shoot biomass total	0.02	ns	3.98	0.05	0.84	ns	0.02	ns	0.45	ns	0.04	ns	0.02	ns
Shoot biomass														
A. millefolium	2.31	ns	2.26	ns	0.02	ns	4.76	*	0.42	ns	3.77	0.06	3.66	0.06
P. lanceolata	3.64	ns	0.05	ns	7.81	**	1.40	ns	2.14	ns	0.35	ns	0.64	ns
T. officinale	3.15	ns	5.32	*	0.92	ns	3.39	ns	0.12	ns	1.01	ns	0.04	ns
H. lanatus	2.90	ns	3.24	ns	3.62	ns	0.25	ns	1.17	ns	0.16	ns	0.54	ns
P. pratensis	28.51	***	0.48	ns	1.74	ns	0.13	ns	0.01	ns	0.00	ns	0.13	ns
T. repens	8.14	**	4.51	*	0.10	ns	0.43	ns	2.21	ns	4.98	*	0.06	ns
LEH	0.00	ns	0.97	ns	0.02	ns	7.12	*	4.89	*	0.67	ns	0.00	ns
AMF community in soil	7.93	**	0.99	ns	−0.39	ns	4.99	**	3.44	*	2.57	ns	0.37	ns
No of AMF species in soil	0.41	ns	0.00	ns	0.04	ns	6.11	*	3.57	ns	0.44	ns	0.03	ns
AMF community in roots	9.30	***	0.06	ns	−0.01	ns	0.26	ns	1.93	ns	0.51	ns	0.11	ns
No of AMF species in roots	9.10	**	0.49	ns	0.74	ns	0.19	ns	0.03	ns	0.47	ns	0.42	ns
Total AMF (P. lanceolata)	15.21	**	4.98	*	0.22	ns	1.20	ns	0.09	ns	6.95	*	4.65	*
Arbuscules (P. lanceolata)	4.91	*	0.05	ns	0.03	ns	0.26	ns	5.72	*	6.40	*	0.17	ns

significance level:
*p<0.05,
**p<0.01,
***p<0.001,
ns = not significant, LEH: length of extraradical AMF hyphae, AMF: arbuscular mycorrhizal fungi.

presumably because they contained a smaller proportion of steamed, nutrient rich background soil.

Soil biota community effects

The original soil biota community affected the root biomass of the plant community and the shoot growth of several plant species in both phases of the experiment. The significant effects of the soil biota treatment confirm that (i) the two soil biota communities differed in their abundance/composition and (ii) the structure of soil biota communities can have pronounced impact on plant community structure [37].

Differences in community composition were also directly confirmed for AMF. Similar CN ratios of soil biota inocula (SEG 33 = 11.0 , SEG 37 = 11.8) suggest that factors other than nutrient provision by the decomposer community, like plant species specific interactions with beneficial or detrimental soil organisms were involved in creating the soil biota effect. The identity of the plant species that were affected by the soil biota community differed between the two phases of the experiment. Additionally, plant diversity was only affected by the soil biota community in phase 2. Differences in soil biota effects between phase 1 and 2 may be ascribed to differences in (i) soil biota/AMF community composition due to plant species specific accumulation of associated organisms (plant-soil feedback effects [38]) during phase 1 and due to the dry dormancy between phases and/or (ii) nutrient content of the growth substrate, with effects of soil organisms being more pronounced under the nutrient poor conditions [39] in phase 2.

Root herbivore history effects

The *Agriotes* training of the soil enhanced plant community productivity (total shoot and plant growth). Effects of the *Agriotes* training indicate changes in plant growth conditions due to the root herbivore that persisted even when the root herbivore was no longer present [7]. Our results show that these root herbivore history effects are even detectable at the plant community level. On the level of single plants *Agriotes* training enhanced shoot growth of *T. officinale* but reduced shoot growth of *T. repens*.

Effects on plant community biomasses and the N indicator species *T. officinale* [40] were positive, pointing to enhanced nutrient supply. Similar CN ratios of untrained and trained soil inocula suggest that nutrient provision by the decomposer community did again not cause the training effect. Instead, AMF communities differed between trained and untrained soils. The difference was not preceded by an actual root herbivore effect on AMF communities in the first phase of the experiment, indicating that it may have been generated through an influence on viability of AMF spores that changed community structure only after the dry dormancy between the two phases of the experiment. Though AMF community shifts due to root herbivore training differed depending on the original soil biota community, they may still have been towards higher abundance of species beneficial for

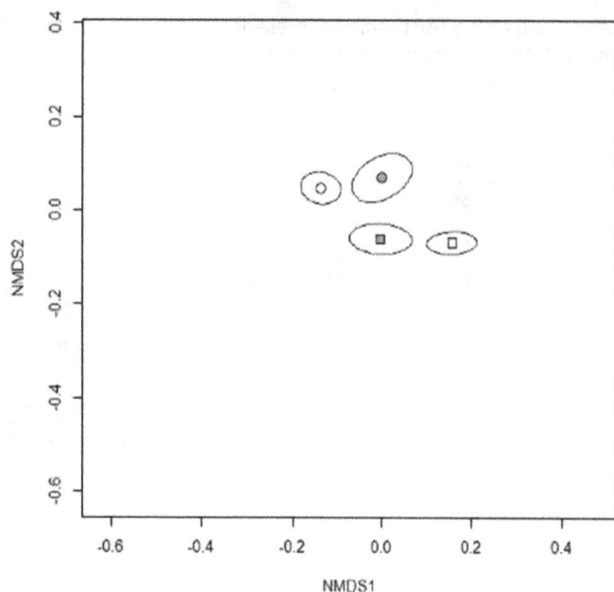

Figure 2. Non-metric multidimensional scaling plot of AMF communities in soils after phase 2. Circles and squares give centroids (means) of AMF communities for SEG33 and SEG37, respectively, open symbols show those of untrained and filled symbols those of trained AMF communities. Ellipsoids give standard error. The two soil biota origins are clearly differentiated along the second axis while the *Agriotes* training leads to opposing community shifts along the first axis.

those plants that contributed to enhanced plant community productivity (e.g. *T. officinale*) in both soil biota communities. The reduction in *T. repens* shoot biomass points to reduced competitive advantage through its symbiotic N fixation at improved nutrient acquisition by AMF for other plants.

Root herbivory effects

Plant community parameters did not respond to the presence of the root herbivore. For single plant species our results show that actual root herbivory effects are influenced by the root herbivore history of the soil, indicating that root herbivore induced changes in plant growth conditions also determine plant responses to new root herbivore attack.

Shoot growth of *A. millefolium* was facilitated by *Agriotes* larvae when grown in untrained soil biota community SEG 37, however, this effect was lost when the soil was already *Agriotes* trained. Shoot biomass of *T. repens* was reduced by *Agriotes* larvae, but only with the two untrained soils in phase 2. AMF communities in phase 2 were affected by the herbivore presence (in interaction with the original soil biota community). However, the actual root herbivore effect on AMF was less strong than the herbivore history effect. When grown individually in SEG 37 soil [21] shoot biomass of *A. millefolium* and *T. repens* did not respond to herbivory by *Agriotes* larvae. Thus, in the present study, responses of *A. millefolium* and *T. repens* to root herbivory may be best explained by (i) herbivore history induced differences in AMF species composition and (ii) shifts in the plant community composition. In untrained SEG 37 soil, roots of *A. millefolium* may have replaced those of other members of the community that were consumed by the root herbivore. Low root losses that did not result in changes in shoot biomass may have been sufficient, as *A. millefolium* responds strongly to release from below ground competition [41]. The shift was not detected in *Agriotes* trained SEG 37 where plants were potentially

better nourished by AMF (see *Root herbivore history effects*) and thus able to compensate the herbivore damage [2], and in SEG 33 where plant community root biomass was lower and release from below ground competition thus less pronounced. *Trifolium repens* did not respond to *Agriotes* presence in phase 1 when it was the strongest competitor in the plant community but in phase 2 where it was less strong. In phase 2 the shoot mass reduction only occurred with untrained, potentially less well nourished plants where *T. repens* presumably benefited more from its N fixing symbiosis, thus, suggesting a root herbivore effect on the symbiosis [42] that became apparent under these conditions. In contrast, shoot biomass of *P. lanceolata* was reduced by *Agriotes* larvae in both phases of the experiment and the effect did not differ significantly for the two original soil biota communities and the training treatments, suggesting independence of soil conditions for the root herbivore effect on this plant species. This was despite the fact that the AMF community in *P. lanceolata* roots differed significantly between the two original soil biota treatments in phase 2 and quantitative root colonization parameters were additionally affected by the herbivore history and herbivore presence. In accordance, this plant species showed similar growth response to several AMF species and no connection of growth response to the degree of root colonization in other studies ([43] and [44], resp.). Substantial root biomass loss in the presence of *Agriotes* larvae in two experiments where *P. lanceolata* was grown individually ([19], [21]) point to a strong direct grazing effect that may have overridden indirect root herbivore effects for this plant species.

In conclusion, our study found evidence for root herbivore history effects on grassland plant communities. The effects may partly be explained by herbivore induced shifts in the AMF community. Interestingly, the root herbivore history of the soil proved to be a stronger driver of plant growth on the community level than an actual herbivore attack. History effects have to be taken into account when predicting the impact of root herbivores in grasslands.

Supporting Information

Tables S1 Includes the tables S1 and S2. Plant biomass data and AMF parameters in different soil treatments in phase1 and phase2.

Acknowledgments

We thank the managers of the three exploratories, Swen Renner, Sonja Gockel, Andreas Hemp and Martin Gorke and Simone Pfeiffer for their work in maintaining the project infrastructure, and Markus Fischer, Elisabeth Kalko, Eduard Linsenmair, Dominik Hessenmöller, Jens Nieschulze, Daniel Prati, Ingo Schöning, François Buscot, Ernst-Detlef Schulze and Wolfgang W. Weisser for their role in setting up the Biodiversity Exploratories project. We thank Uta Schumacher for providing information about the management of Schorfheide sites and Bernhard Warzecha for his work on part of the TRFLP samples. We are very grateful to Uta Schumacher and Andreas Hemp for hands-on help with collecting soil.

Author Contributions

Supervised molecular work: SH. Conceived and designed the experiments: IS SW SR. Performed the experiments: IS MB NH. Analyzed the data: IS SH. Wrote the paper: IS SH SW.

References

1. Brown VK, Gange AC (1990) Insect herbivory below ground. Advances in Ecological Research 20: 1–58.
2. Dosdall LM, Clayton GW, Harker KN, O'Donovan JT, Stevenson FC (2004) The effects of soil fertility and other agronomic factors on infestations of root maggots in canola. Agronomy Journal 96: 1306–1313.
3. Kalb DW, Bergstrom GC, Shields EJ (1994) Prevalence, severity, and association of fungal crown and root rots with injury by the clover root curculio in New York alfalfa. Plant Disease 78: 491–495.
4. Yeates GW, Saggar S, Denton CS, Mercer CF (1998) Impact of clover cyst nematode (*Heterodera trifolii*) infection on soil microbial activity in the rhizosphere of white clover (*Trifolium repens*) - a pulse-labelling experiment. Nematologica 44: 81–90.
5. Bardgett RD, Denton CS, Cook R (1999) Below-ground herbivory promotes soil nutrient transfer and root growth in grassland. Ecology Letters 2: 357–360.
6. Treonis AM, Grayston SJ, Murray PJ, Dawson LA (2005) Effects of root feeding cranefly larvae on soil microorganisms and the composition of rhizosphere solutions collected from grassland plants. Applied Soil Ecology 28: 203–215.
7. Kostenko O, van de Voorde TFJ, Mulder PPJ, van der Putten WH, Bezemer TM (2012) Legacy effects of aboveground–belowground interactions. Ecology Letters 15: 813–821.
8. Gange AC, Brown VK, Sinclair GS (1994) Reduction of black vine weevil larval growth by vesicular-arbuscular mycorrhizal infection. Entomologia Experimentalis et Applicata 70: 115–119.
9. Gange AC (2001) Species-specific responses of a root- and shoot-feeding insect to arbuscular mycorrhizal colonization of its host plant. New Phytologist 150: 611–618.
10. Currie AF, Murray PJ, Gange AC (2006) Root herbivory by *Tipula paludosa* larvae increases colonization of *Agrostis capillaris* by arbuscular mycorrhizal fungi. Soil Biology & Biochemistry 38: 1994–1997.
11. Currie AF, Murray PJ, Gange AC (2011) Is a specialist root-feeding insect affected by arbuscular mycorrhizal fungi? Applied Soil Ecology 47: 77–83.
12. Gange AC, Brown VK (2002) Soil food web components affect plant community structure during early succession. Ecological Research 17: 217–227.
13. Smith SE, Read DJ (2008) Mycorrhizal Symbiosis. London: Elsevier Science Ltd.
14. Klironomos JN (2002) Feedback with soil biota contributes to plant rarity and invasiveness in communities. Nature 417: 67–70.
15. Rodríguez-Echeverría S, de la Peña E, Moens M, Freitas H, van der Putten WH (2009) Can root-feeders alter the composition of AMF communities? Experimental evidence from the dunegrass Ammophila arenaria. Basic and Applied Ecology 10: 131–140.
16. Parker WE, Seeney FM (1997) An investigation into the use of multiple site characteristics to predict the presence and infestation level of wireworms (*Agriotes* spp., Coleoptera: Elateridae) in individual grass fields. Annuals of Applied Biology 130: 409–425.
17. Hemerik L, Gerrit G, Brussaard L (2003) Food preference of wireworms analyzed with multinomial logit models. Journal of Insect Behavior 16: 647–665.
18. Jedlicka P, Frouz J (2007) Population dynamics of wireworms (Coleoptera, Elateridae) in arable land after abandonment. Biologia 62: 103–111.
19. Wurst S, van der Putten WH (2007) Root herbivore identity matters in plant-mediated interactions between root and shoot herbivores. Basic and Applied Ecology 8: 491–499.
20. Wurst S, van Dam NM, Monroy F, Biere A, van der Putten WH (2008) Intraspecific variation in plant defense alters effects of root herbivores on leaf chemistry and aboveground herbivore damage. Journal of Chemical Ecology 34: 1360–1367.
21. Sonnemann I, Baumhaker H, Wurst S (2012) Species specific responses of common grassland plants to a generalist root herbivore (*Agriotes* spp. larva). Basic and Applied Ecology 13: 579–586.
22. Fischer M, Bossdorf O, Gockel S, Hänsel F, Hemp A, et al. (2010) Implementing largescale and longterm functional biodiversity research: The Biodiversity Exploratories. Basic and Applied Ecology 11: 473–485.
23. Klausnitzer B (1994) Die Larven der Kaefer Mitteleuropas. Krefeld: Goecke & Evers Verlag. 325 p.
24. Börstler B, Renker C, Kahmen A, Buscot F (2006) Species composition of arbuscular mycorrhizal fungi in two mountain meadows with differing management types and levels of plant biodiversity. Biol Fertil Soils 42: 286–298.
25. Schoenholtz SH, Van Miegroet H, Burger JA (2000) A review of chemical and physical properties as indicators offorest soil quality: challenges and opportunities. Forest Ecology and Management 138: 335–356
26. Watt MS, Coker G, Clinton PW, Davis MR, Parfitt R, et al. (2005) Defining sustainability of plantation forests through identification of site quality indicators influencing productivity—A national view for New Zealand. Forest Ecology and Management 216: 51–63.
27. Watt MS, Davis MR, Clinton PW, Coker G, Ross C, et al. (2008) Identification of key soil indicators influencing plantation productivity and sustainability across a national trial series in New Zealand. Forest Ecology and Management 256: 180–190.
28. Vierheilig H, Coughlan A, Wyss U, Piche Y (1998) Ink and vinegar, a simple staining technique for arbuscular-mycorrhizal fungi. Applied and Environmental Mircobiology 64: 5004–5007.
29. McGonigle TP, Miller MH, Evans DG, Fairchild GL, Swan JA (1990) A new method which gives an objective measure of colonization of roots by vesicular-arbuscular mycorrhizal fungi. New Phytologist 115: 495–501.
30. Jakobsen I, Abbott LK, Robson AD (1992) External hyphae of vesicular-arbuscular mycorrhizal fungi associated with *Trifolium subterraneum* L. 1. Spread of hyphae and phosphorus inflow into roots. New Phytologist 120: 371–380.
31. Tennant D (1975) A test of a modified line intersect method of estimating root length. Journal of Ecology 63: 995–1001.
32. Dickie IA, FitzJohn RG (2007) Using terminal restriction fragment length polymorphism (T-RFLP) to identify mycorrhizal fungi: a methods review. Mycorrhiza 17: 259–270.
33. FitzJohn RG, Dickie IA (2007) TRAMPR: An R package for analysis and matching of terminal-restriction fragment length polymorphism (TRFLP) profiles. Molecular Ecology Notes 7: 583–587.
34. R Development Core Team (2011) R: A language and environment for statistical computing. R Foundation for Statistical Computing, Vienna, Austria. ISBN 3-900051-07-0, URL http://www.R-project.org/.
35. Zuur AF, Ieno EN, Walker NJ, Saveliew AA, Smith GM (2009) Mixed effects models and extensions in ecology with R. New York: Springer.
36. McArdle BH, Anderson MJ (2001) Fitting multivariate models to community data: A comment on distance-based redundancy analysis. Ecology, 82: 290–297.
37. Wardle DA, Bardgett RD, Klironomos JN, Setälä H, van der Putten WH, et al. (2004) Ecological linkages between aboveground and belowground biota. Science 304: 1629–1633.
38. Bever JD, Westover KM, Antonovics J (1997) Incorporating the soil community into plant population dynamics: the utility of the feedback approach. Journal of Ecology 85: 561–573.
39. Araujo ASF, Leite LFC, Iwata BD, Lira MD, Xavier GR, et al. (2012) Microbiological process in agroforestry systems. A review. Agronomy for Sustainable Development 32: 215–226.
40. Ellenberg H, Leuschner C (2010) Vegetation Mitteleuropas mit den Alpen. Stuttgart: Ulmer Verlag.
41. Kosola KR, Gross KL (1999) Resource competition and suppression of plants colonizing early successional old fields. Oecologia 118: 69–75.
42. Dawson LA, Grayston SJ, Murray PJ, Pratt SM (2002) Root feeding behavior of *Tipula paludosa* (Meig.) (Diptera: Tipulidae) on *Lolium perenne* (L.) and *Trifolium repens* (L.) Soil Biology & Biochemistry 34: 609–615.
43. Zaller JG, Thomas F, Drapela T (2011) Soil sand content can alter effects of different taxa of mycorrhizal fungion plant biomass production of grassland species. European Journal of Soil Biology 47: 175–181.
44. Martin SL, Mooney SJ, Dickinson MJ, West HM (2012) The effects of simultaneous root colonisation by three Glomus species on soil pore characteristics. Soil Biology & Biochemistry 49: 167–173.

10

Selection of Optimal Auxiliary Soil Nutrient Variables for Cokriging Interpolation

Genxin Song, Jing Zhang, Ke Wang*

Institute of Agricultural Remote Sensing and Information Technique, Zhejiang University, Hangzhou, Zhejiang, China; and Ministry of Education Key Laboratory of Environmental Remediation, Ecological and Health, Zhejiang University, Hangzhou, Zhejiang, China

Abstract

In order to explore the selection of the best auxiliary variables (BAVs) when using the Cokriging method for soil attribute interpolation, this paper investigated the selection of BAVs from terrain parameters, soil trace elements, and soil nutrient attributes when applying Cokriging interpolation to soil nutrients (organic matter, total N, available P, and available K). In total, 670 soil samples were collected in Fuyang, and the nutrient and trace element attributes of the soil samples were determined. Based on the spatial autocorrelation of soil attributes, the Digital Elevation Model (DEM) data for Fuyang was combined to explore the coordinate relationship among terrain parameters, trace elements, and soil nutrient attributes. Variables with a high correlation to soil nutrient attributes were selected as BAVs for Cokriging interpolation of soil nutrients, and variables with poor correlation were selected as poor auxiliary variables (PAVs). The results of Cokriging interpolations using BAVs and PAVs were then compared. The results indicated that Cokriging interpolation with BAVs yielded more accurate results than Cokriging interpolation with PAVs (the mean absolute error of BAV interpolation results for organic matter, total N, available P, and available K were 0.020, 0.002, 7.616, and 12.4702, respectively, and the mean absolute error of PAV interpolation results were 0.052, 0.037, 15.619, and 0.037, respectively). The results indicated that Cokriging interpolation with BAVs can significantly improve the accuracy of Cokriging interpolation for soil nutrient attributes. This study provides meaningful guidance and reference for the selection of auxiliary parameters for the application of Cokriging interpolation to soil nutrient attributes.

Editor: Vanesa Magar, Centro de Investigacion Cientifica y Educacion Superior de Ensenada, Mexico

Funding: The study was supported by the National Natural Science Fund of China (Grant No. 31172023). The funders had no role in study design, data collection and analysis, decision to publish, or preparation of the manuscript.

Competing Interests: The authors have declared that no competing interests exist.

* E-mail: kwang@zju.edu.cn

Introduction

The spatial distribution and variation of soil attributes are of considerable interest in soil science. A detailed understanding of the spatial variability of soil is the foundation for precision and variable agriculture management [1]. Soil attribute interpolation is critical for studying the spatial variation and distribution characteristics of soil. Analyzing and forecasting the spatial distribution and dynamics of soil properties are important elements of sustainable land management. In recent years, geostatistics has been widely used to predict the spatial distribution of physical and chemical soil properties [2]. Furthermore, Cokriging interpolation has been increasingly applied to all aspects of soil property prediction because it provides a higher level of prediction accuracy than ordinary kriging interpolation [3]. The Cokriging method, as an extension of the ordinary statistical kriging method, can obtain good results by allowing for more than one variable in a prediction and by considering both self-correlation and cross-correlation between variables [4]. Existing research has demonstrated that calculation of the cross-correlation using the Cokriging method can still achieve accurate prediction results even with a lack of simple correlation between variables [5,6]. Chai Xurong et al. found that the Cokriging method can yield more accurate results than the ordinary kriging method when using auxiliary elevation as a parameter to predict

the spatial distribution of soil exchangeable potassium and pH. Jiang Yong et al. compared the Cokriging method using zinc content (0–10 cm) in the upper soil as an auxiliary variable and the ordinary kriging method for predicting the distribution of zinc content (10–20 cm) in the lower soil, and they found that the Cokriging method yields more accurate results [7]. Liu Bo et al. used the Cokriging method to predict the spatial distribution of soil heavy metals in the city of Kunshan, and they found that use of the Cokriging method yields more accurate results compared to the ordinary kriging method for most heavy metal predictions [8].

The spatial distribution of one soil attribute is often closely related to the spatial distribution of other soil attributes, as all soil attributes are affected by the same regionalization phenomena or space process [9,10]. With the rapid accumulation of data related to soil attributes from other sources, an increasing amount of data is applicable for use as auxiliary parameters for the Cokriging method. These optional auxiliary parameters can be divided into the following three categories: (1) data for various attributes obtained directly from soil samples; (2) soil type, topography, geomorphology, remote-sensing images, and soil spectral data, which are closely related to the collection of soil samples; and (3) influencing factors, which are related to human activities, including land utilization type as well as industrial, mining, and traffic layouts. The Cokriging method has become an effective tool for soil attribute interpolation. However, selection of the best

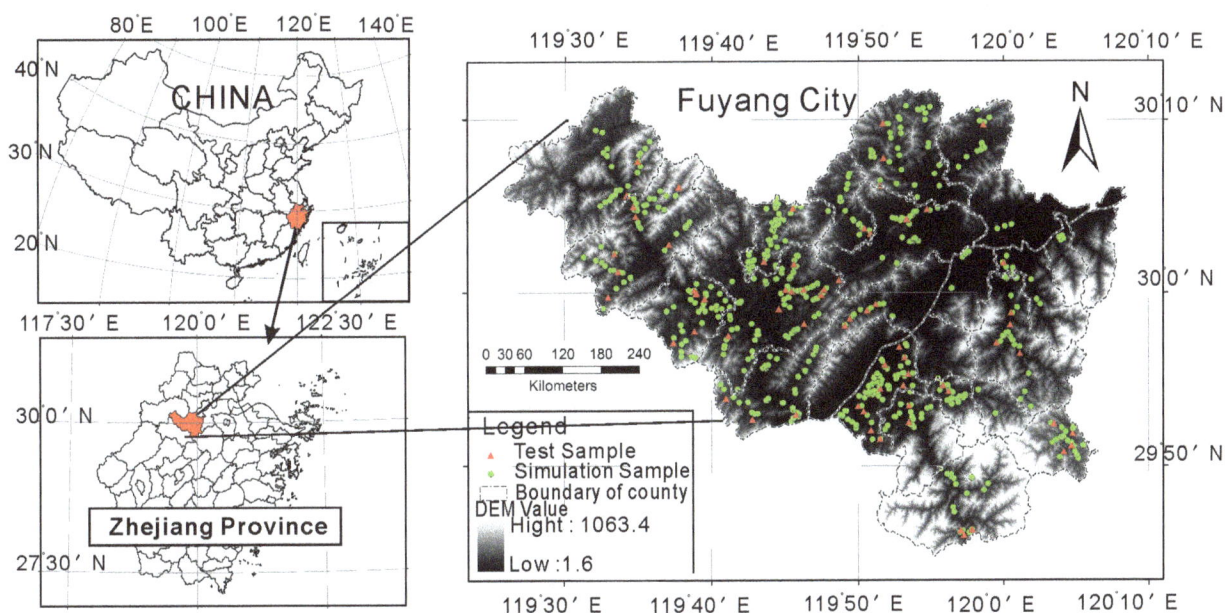

Figure 1. Distribution of sample points.

parameters for Cokriging predictions from the many available auxiliary parameters has become a serious problem due to the lack of understanding of how to select the best auxiliary parameter from the rich related data as well as how to select the second best auxiliary parameter when data for the best auxiliary parameter are unavailable. In this paper, interpolation of soil nutrient elements (i.e., organic matter (OM), total N (TN), available P (AP), and available K (AK)) was used as an example to determine how to select the best auxiliary parameters for Cokriging interpolation. Other relevant data (i.e., soil attribute data and terrain data) were considered as auxiliary parameters. The example also demonstrated how second best data can be used to ensure the accuracy of Cokriging interpolation when the best auxiliary parameters are insufficient.

Materials and Methods

Ethics statement

This study was approved by the City Agricultural Office Department of Fuyang District, which monitors farmland nutrients. All of the data in this study can be published and shared.

Study area

The study was conducted in the Fuyang District of Hangzhou, Zhejiang, China. The Fuyang District is located in northwestern Zhejiang Province, which lies between 45° 44′ 29″ to 30° 11′ 59″ N latitude and 119° 25′ 00″ to 120° 09′ 30″ W longitude. The geographical area of Fuyang is 1,821.10 km^2, and the elevation ranges between 4 and 705 m. The terrain of Fuyang is tilted from southwest to northeast, and the central Fuchun River has an oblique penetration of 52 km. The soil is varied, fertile land with rich agricultural natural resources that are suitable for a variety of crops. The planting industry is also developed.

Soil nutrient sampling data

In this study area, the soil nutrient sampling data for this paper was the Fuyang soil monitoring data from 2006, and this data included 670 sample points. Each sample point included soil OM,

TN, AP, and AK as well as a record of the latitude and longitude from GPS measurements. The sample point distribution is shown in Figure 1.

In this paper, the authors randomly divided 670 sampling points into simulation and test datasets with 600 and 70 points, respectively [11]. This study also simulated the spacing of soil nutrient properties using simulation datasets and evaluated the accuracy of the simulation using test sample datasets [12,13].

Soil trace element data

This article selected soil trace element data that were acquired at the same sample points as the soil nutrient sampling data. These data included laboratory samples tested for effective sulfur (ES), commutative hydrogen (CH), commutative aluminum (CA), commutative magnesium (CM), commutative calcium (CC), effective manganese (EMG), effective copper (EC), effective zinc (EZ), and effective molybdenum (EMO) [14]. In total, nine soil trace elements were considered.

Terrain data

In this paper, the slope and aspect for Fuyang were created based on data from the 25-m resolution DEM for Fuyang and through digital terrain analysis technology to extract soil terrain information for sampling points. Extraction of the terrain index mainly included height (H), slope (β), and aspect (α). However, because the aspect information was from the perspective of the due north direction going clockwise, values ranged from 0 to 360°. Therefore, the aspect was changed to the sine and cosine values [15].

Sampling point processing

Using Fuyang soil sampling point data provided by the agriculture department and according to the GPS coordinates of the sampling points, ArcGIS software was used to identify the sample point location on the Fuyang basic datum and to join the soil nutrient properties to each sample point. Based on the sample point test number, the nine soil trace element data points were

Table 1. Descriptive Statistics for Soil Nutrient Attributes.

	N	Mean (mg/kg)	Std. Deviation	Coefficient of Variation	Skewness	Kurtosis
OM (g/kg)	670	32.54	1.06	32.615	1.39	4.83
SN (g/kg)	670	1.946	0.06	31.579	0.67	0.81
AP (g/kg)	670	36.79	58.25	158.331	3.82	16.87
AK (g/kg)	670	90.18	72.63	80.539	2.94	10.95

then assigned to each sample point and to 670 sampling points. Points were randomly divided into the simulation and test datasets. The simulation dataset was used to interpolate the nutrient elements, and the test datasets were used to evaluate the accuracy of the interpolation results [16].

Extraction and analysis of terrain attributes

The slope and aspect data generated from the 25-m resolution DEM for Fuyang and through digital terrain analysis technology to extract soil terrain information were used to determine the location of the sampling points. Extraction of the terrain index mainly included height (H), slope (β), and aspect (α). However, because the aspect expression was from the northern direction going clockwise, values ranged from 0 to 360°. Therefore, sine and cosine values were used in place of the aspect. Thus, the extracting terrain index included H, β, the sine value of the aspect (sinα), and the cosine value of the aspect (cosα). ArcGIS spatial analysis was used to extract terrain information for each sampling point location. Finally, the correlation analysis for soil OM, TN, AP, and AK as well as terrain parameters was performed using SPSS19 software [17].

Correlation analysis for soil nutrient elements

There may be a certain degree of correlation for OM, TN, AP, and AK with the other elements, which has become a topic of interest for research scholars. In this paper, a correlation analysis was performed with SPSS19 software for OM, TN, AP, and AK to obtain the above four results of the correlation analysis among nutrient elements.

Association correlation analysis between soil trace elements and soil nutrients

During the process of soil formation and development, a correlation exists between soil nutrient elements and soil trace elements. Using SPSS19 software, this study selected soil ES, EH, EA, EMG, ECA, EMA, ECO, EZ, and EMO (a total of nine soil trace elements) to perform a cross-correlation analysis with the four soil nutrient elements (OM, TN, AP, and AK).

Cokriging interpolation method

Cokriging is a variation of the ordinary kriging method that is used when there is a close relationship between the spatial distributions of certain soil properties and other properties at the same position [18]. In particular, Cokriging is useful when it is

Table 2. Correlation Analyses of Soil Nutrient Attributes and Optional Auxiliary Parameters.

	OM	TN	AP	AK
OM	1	0.93**	0.30**	0.32**
TN	0.93**	1	0.18**	0.22**
AP	0.30**	0.18**	1	0.52**
AK	0.32**	0.22**	0.52**	1
H	0.11**	0.09*	0.03	0.05
β	0.03	−0.01	0.00	0.02
sinα	−0.03	0.01	−0.06	−0.09*
cosα	−0.09*	−0.08*	0.03	−0.01
ES	−0.01	−0.01	0.08	0.00
CH	0.11**	−0.03	0.35**	0.33**
CA	−0.07	−0.18**	0.05	0.23**
EMG	0.02	0.11**	0.02	0.03
CC	0.04	0.16**	−0.01	−0.07
EMA	0.16**	0.14**	0.28**	0.13**
EC	−0.06	−0.11**	0.01	0.00
EZ	0.08	0.08*	0.17**	0.16**
EMO	0.20**	0.22**	0.14**	0.11**

**Correlation is significant at the 0.01 level (two-tailed).
*Correlation is significant at the 0.05 level (two-tailed).

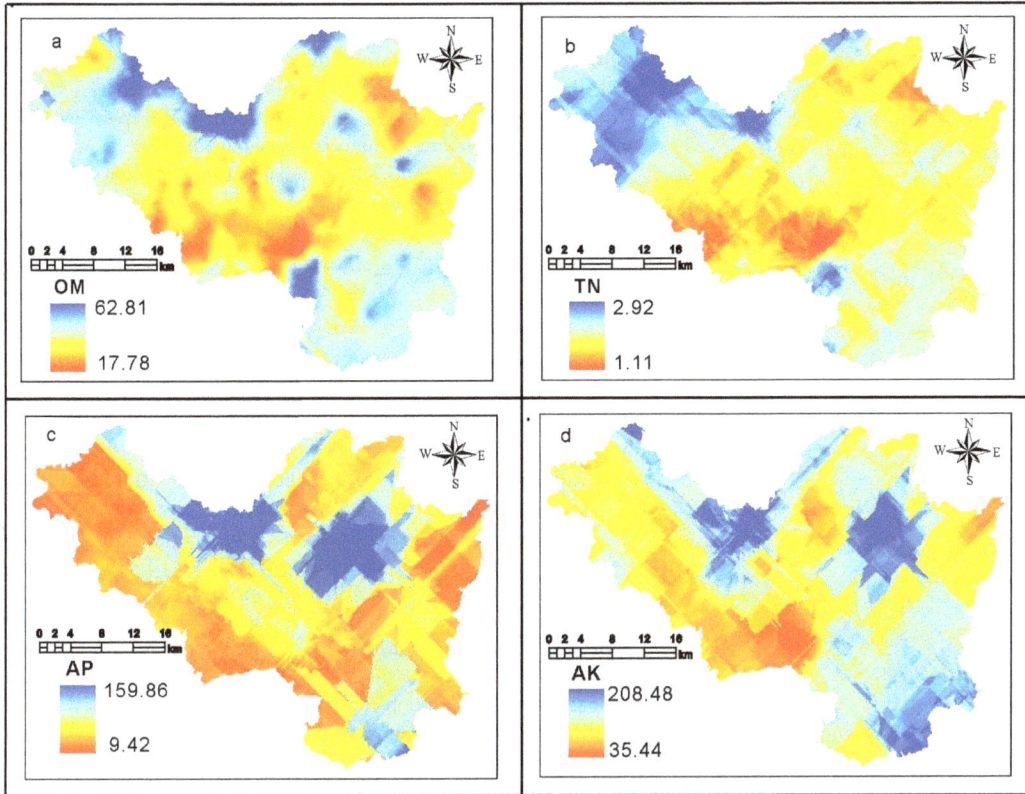

Figure 2. Cokriging prediction results using the BAVs.

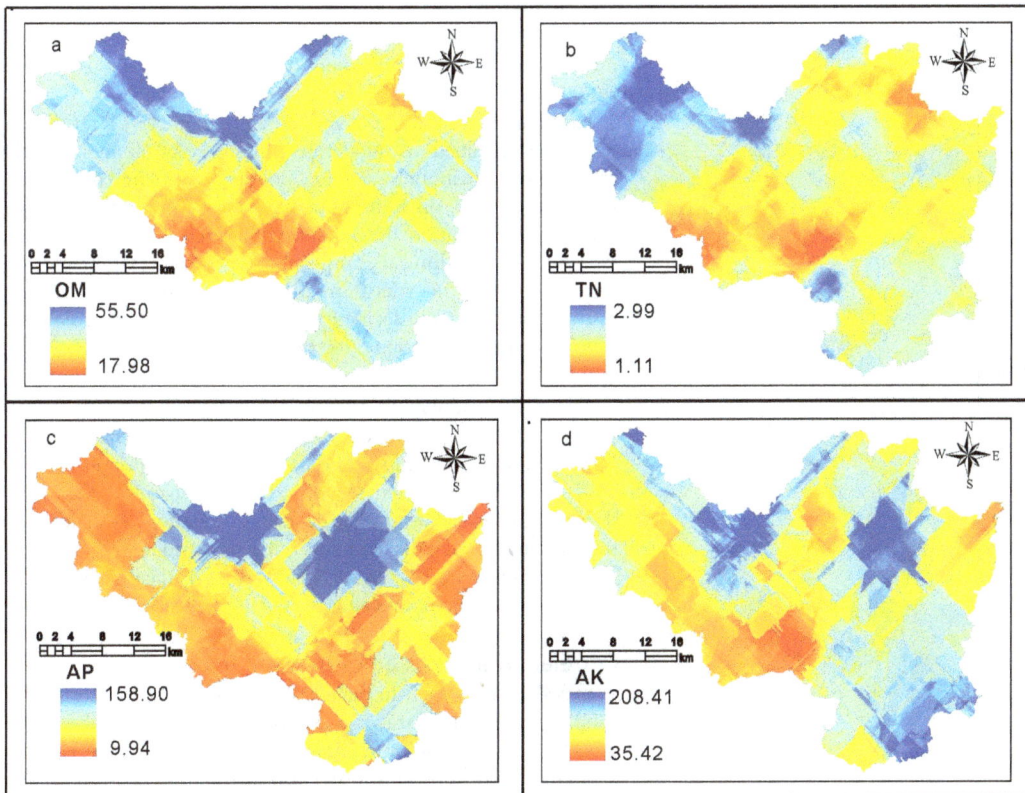

Figure 3. Cokriging prediction results using the PAVs.

Table 3. Results of the Interpolation Accuracy.

	Mean		Std. Deviation		Average Absolute Error		Correlation Coefficient with Measured Values	
	BAV	PAV	BAV	PAV	BAV	PAV	BAV	PAV
OM	3.31	3.26	0.54	0.45	0.02	−0.052	0.58	0.292
TN	0.2	0.2	0.03	0.03	−0.002	0.037	0.504	0.446
AP	29.48	29.56	18.38	18.44	7.616	15.619	0.245	0.244
AK	87.48	87.55	25.9	26	12.47	35.343	0.16	0.16

difficult to obtain certain properties but not others. The Cokriging method is the best valuation method for regionalized variables from a single development to two or more coordinated regional attributes [19], and it uses a space correlation between two or more variables due to the autocorrelation of the main variables and cross-correlation of auxiliary variables. Estimates were made to improve the accuracy and rationality of the estimation.

The Cokriging prediction model can be summarized as follows:

$$Z^*(x_0) = \sum_{i=1}^{n} \omega_{1i} Z_1(x_i) + \sum_{i=1}^{p} \omega_{2j} Z_2(x_j)$$

where $Z^*(x_0)$ is the position of the sample point; ω_{1i} and ω_{2j} are two regionalized variables; and Z_1 (X_i) and Z_2 (X_j) are weight coefficients [20,21,22].

Results obtained from the validation method

This article used test datasets, which contained 70 reserved sample points, to evaluate the accuracy of Cokriging interpolation results that used the best auxiliary variables (BAVs) and poor auxiliary variables (PAVs), and the Cokriging interpolation results were then compared to results from two predictions. Specifically, the BAVs and PAVs from the Cokriging interpolation results were assigned to the 70 test sample points to obtain simulation values for the two interpolation results. A comparative analysis was then performed on the two simulation values and measured values from the 70 test points. The contrast index mainly included correlation coefficients and mean absolute errors for the simulation values and measured values in item 2 [23].

Results and Discussion

Descriptive analysis of soil attributes

SPSS19 was used to perform statistical analyses for the soil nutrient elements for all points (OM, TN, AP, and AK). The results are presented in Table 1.

Based on the coefficient of variation grading scale [24], OM and SN were classified as moderate variable. An analysis of the probability distribution of the original sampling point data indicated that the distribution of the nutrient attribute data from each sampling point exhibited clear deviations. Thus, the original data were logarithmically transformed and BOX-COX transformed so that the transformed data conformed to a normal distribution. The transformed data were used to interpolate the simulated nutrient attribute for each sampling point.

Correlation analysis of terrain factors and soil attributes

ArcGIS was used to generate the slope and aspect map based on Fuyang 25-m resolution DEM data, and the corresponding terrain data were then extracted for each sampling point. SPSS19 was then used to calculate simple Pearson correlation coefficients for the four terrain factors and soil nutrient attributes as shown in Table 2. Soil OM, TN, and AK showed a significant relationship with elevation, slope, and sinα.

Correlation analysis among attributes

SPSS19 was used to calculate simple Pearson correlation coefficients for nine trace elements with the four soil nutrient attributes. A correlation analysis was performed for the four soil nutrient attributes as shown in Table 2. The results indicated that there were significant correlations at the 0.01 level for the four soil nutrient attributes. The correlation between OM and TN was 0.932. Therefore, based on the relevant relationship among the four attributes, a certain attribute can be set as the target attribute to perform the Cokriging interpolation, and the other three soil nutrient attributes can be used as auxiliary parameters. Trace elements had a more significant correlation with nutrient attributes, especially for the correlation of TN, AK, and trace elements with EM, EC, EC, and EH.

Optimizing the choice of auxiliary parameters

The above correlation analyses between soil attributes and terrain factors as well as among soil attributes demonstrated that each soil nutrient attribute had significant correlations with multiple terrain factors, trace elements, and other soil nutrient attributes [25]. Therefore, the terrain factors, trace elements, and other related soil nutrient attributes can be used as auxiliary parameters for a Cokriging interpolation of soil nutrient attributes [26]. The above correlation analyses indicated that there was a high correlation between auxiliary parameters and soil nutrient attributes. Furthermore, the correlations between other parameters and soil nutrients differed. In the process of performing the Cokriging interpolation, we can accord different degrees of correlation to these factors and select the most relevant auxiliary parameters [27].

Interpolation results

Based on the previous results, OM, TN, AP, and AK were set as predicted targets for the Cokriging interpolation, and the three most significant correlation variables were selected as auxiliary variables [28]. TN, AP, and AK were set as auxiliary parameters for the interpolation of OM. OM, AP, and EMO were selected as auxiliary parameters for the interpolation of TN. AK, EH, and OM were selected as auxiliary parameters for the interpolation of AP. AP, EH, and OM were selected as auxiliary parameters for the interpolation of AK. The prediction results are presented in Figure 2.

In order to compare the above interpolation results, OM, TN, AP, and AK were set as predicted targets for Cokriging analysis, and the three poorest significant correlation variables were selected as auxiliary variables [27,28]. ES, EMG, and CC were selected as auxiliary parameters for the interpolation of OM. ES, CH, and β were selected as auxiliary parameters for the interpolation of TN. β, CC, and EC were selected as auxiliary parameters for the interpolation of AP. ES, EC, and cosα were selected as auxiliary parameters for the interpolation of AK. The prediction results are presented in Figure 3 [29].

Accuracy validation

To compare the accuracy of the two methods for the above interpolation, measured test dataset values were compared to the simulation results for the two types of interpolation. The results are presented in Table 3 [30].

According to the above accuracy comparison results, interpolation using the BAVs and PAVs had similar mean and discrete data. However, results using the BAVs had significantly smaller absolute errors than those using the PAVs in addition to higher correlation coefficients with the measured values. Therefore, the BAV method has higher prediction accuracy than the PAV method [31].

Conclusions

This article studied the method to select the best auxiliary parameters when performing a Cokriging interpolation of soil nutrient attributes. The most relevant parameters for Cokriging interpolations of soil nutrient attributes were selected by determining the relationship among soil trace elements, terrain attributes, and soil nutrient attributes by exploring the correlation intensity of multi-source data with soil nutrient data based on the correlations among the optional auxiliary parameters. Finally, this article selected the BAVs and PAVs for the Cokriging interpolation results and verified the optimal parameter selection method for Cokriging interpolation used in this paper.

1) The use of auxiliary variables that are more highly correlated leads to higher prediction accuracy. In the process of choosing soil nutrient attributes for Cokriging prediction based on a correlation analysis between the optional auxiliary parameters and interpolation target attributes, optimizing the auxiliary parameter correlation will yield better prediction results.

2) To select auxiliary parameters, a concrete analysis of the stability of auxiliary parameters and suitable conditions should be conducted. As demonstrated by the results obtained here, the influencing factors of the terrain parameters are relatively stable for optional auxiliary parameters. The relationship of the influencing factors of the terrain parameters to nutrient factors is also relatively constant, so the correlations are similar compared with other auxiliary parameters. Therefore, preference should be given to terrain parameters as auxiliary parameters.

3) Data for multiple soil nutrients are typically obtained at the same time. Therefore, it is more convenient to use nutrient data as the auxiliary parameter. Furthermore, soil trace element data may be limited by the experimental conditions or their availability. Therefore, the selection of auxiliary parameters based on different characteristics must consider the data in light of different research needs for auxiliary parameter selection and optimization.

This article only considered soil trace elements, topography, and soil nutrient elements in the selection of optimal auxiliary parameters. Considering this limited range of auxiliary parameters, future research will consider additional optional auxiliary parameters, such as remote-sensing data and soil spectral data, based on expanding the scope of the research data to find better auxiliary variables in order to improve the precision of soil nutrient kriging interpolation.

Acknowledgments

We would like to thank our colleagues in our laboratory for their valuable comments and other assistance. We also want to thank the anonymous reviewers for their constructive suggestions. Any remaining errors are the authors' own responsibility.

Author Contributions

Conceived and designed the experiments: KW. Performed the experiments: GXS KW JZ. Analyzed the data: GXS JZ. Contributed reagents/materials/analysis tools: GXS KW. Wrote the paper: GXS JZ KW. Designed/drew figures: GXS JZ. Edited manuscript: GXS JZ KW.

References

1. Chunjiang Z, Xuzhang X, Xiu W, Liping C, Yuchun P, et al. (2003) Advance and prospects of precision agriculture technology system. Trans. CSAE 19: 7–12.

2. Ke-lin H, Bao-guo L, De-li C, White RE (2001) Spatial Variability of soil Water and Salt in Field and their Estimations by the CoKriging. Adv Water Resour 12: 460–466.

3. Song G, Zhang L, Wang K, Fang M (2013) Spatial simulation of soil attribute based on principle of soil science. IEEE. pp. 1–4.

4. Hongwei X, Yurong G, Ke W, Bin Z, Qing Z (2007) Investigation of spatial interpolation of available soil nitrogen in paddy field based on rice canopy spectral information. Trans. CSAE 23: 13–17.

5. Atkinson PM, Webster R, Curran PJ (1994) Cokriging with airborne MSS imagery. Remote Sens Environ 50: 335–345.
6. Lei F, Yongzheng C, Guoqing Z, Laigang W, Ting L (2010) Monitoring of winter wheat area by remote sensing based on CoKriging. Trans. CSAE 10: 39.
7. Yong J, Qi L, Xiaoke Z, Wenju L (2006) Kriging prediction of soil zinc in contaminated field by using an auxiliary variable. Chin.J.Appl.Ecol 17: 97–101.
8. Bo L, Dongxiang C (2013) Study on Spatial Variation of Heavy Metals in Soil Based on CoKriging Method-Taking Kunshan City as Example. J Environ Prot Sci 39.
9. Yan L, Zhou S, Renchao W, MIngxiang H (2004) Estimates of Electrical Conductivity for Coastal Saline Soil Profile Using Cokringing under Different Sampling Density. Acta Pedol.Sin 41: 434–443.
10. Behrens T, Schmidt K, Ramirez-Lopez L, Gallant J, Zhu A, et al. (2014) Hyper-scale digital soil mapping and soil formation analysis. Geoderma 213: 578–588.
11. Lin Q, Li H, Luo W, Lin Z, Li B (2013) Optimal Soil-Sampling Design for Rubber Tree Management Based on Fuzzy Clustering. For Ecol Manage 308: 214–222.
12. Gang L, Xu-Dong G, Bo-Jie F, Chen-Xia H (2008) Spatial distribution of Soil Properties in a Small Catchment of the Loess Plateau Based on Environmental Correlation. Acta Geogr Sin 28: 554–558.
13. Mouazen AM, Kuang B, De Baerdemaeker J, Ramon H (2010) Comparison among principal component, partial least squares and back propagation neural network analyses for accuracy of measurement of selected soil properties with visible and near infrared spectroscopy. Geoderma 158: 23–31.
14. Zeiner M, Cindrić IJ, Mikelić IL, Medunić G, Kampić Š, et al. (2013) The determination of the extractability of selected elements from agricultural soil. Environ Monit Assess 185: 223–229.
15. Qiu Y, Fu B, Wang J, Chen L (2002) Variability of the Soil Physical Properties On the Loess Plateau. Acta Geogr Sin 57: 587–594.
16. Willmott CJ (1981) On the validation of models. Prog Phys Geogr 2: 184–194.
17. Zhang W, Dong W (2004) SPSS statistical analysis advanced tutorial: Higher education press Beijing.153p.
18. Laurent L, Boucard P, Soulier B (2013) Generation of a cokriging metamodel using a multiparametric strategy. Comput Mech 51: 151–169.
19. Xiaoqing Y, Miao L, Shuying Z (2013) Spatial Interpolation of the Chlorophyll-a Concentration in Zhalong Wetland Based on Cokringing. Chin.Agric.Sci.Bull. 29: 160–164.
20. Eldeiry AA, Garcia LA (2010) Comparison of ordinary kriging, regression kriging, and cokriging techniques to estimate soil salinity using LANDSAT images. J.Irrig Drain ASCE 136: 355–364.
21. Wang K, Zhang C, Li W (2013) Predictive mapping of soil total nitrogen at a regional scale: A comparison between geographically weighted regression and cokriging. Appl Geogr 42: 73–85.
22. Kunkel ML, Flores AN, Smith TJ, McNamara JP, Benner SG (2011) A simplified approach for estimating soil carbon and nitrogen stocks in semi-arid complex terrain. Geoderma 165: 1–11.
23. Juan P, Mateu J, Jordan MM, Mataix-Solera J, Meléndez-Pastor I, et al. (2011) Geostatistical methods to identify and map spatial variations of soil salinity. J Geochem Explor 108: 62–72.
24. Zhidong L, Shixiu Y, Zhirong X (1985) Preliminary investigation of the Spatial Variability of Soil Properties. J. Hydrul Eng-ASCE 9: 10–21.
25. Yao X, Fu B, Lü Y, Sun F, Wang S, et al. (2013) Comparison of Four Spatial Interpolation Methods for Estimating Soil Moisture in a Complex Terrain Catchment. PloS one 8: e54660.
26. Zhang S, Huang Y, Shen C, Ye H, Du Y (2012) Spatial prediction of soil organic matter using terrain indices and categorical variables as auxiliary information. Geoderma 171: 35–43.
27. Shi W, Liu J, Du Z, Yue T (2012) Development of a surface modeling method for mapping soil properties. J.Geogr.sci 22: 752–760.
28. Yao RJ, Yang JS, Shao HB (2012) Accuracy and uncertainty assessment on geostatistical simulation of soil salinity in a coastal farmland using auxiliary variable. Environ Monit Assess: 1–14.
29. Tittonell P, Corbeels M, Van Wijk MT, Giller KE (2010) FIELD—A summary simulation model of the soil–crop system to analyse long-term resource interactions and use efficiencies at farm scale. Eur J Agron 32: 10–21.
30. Qiao L, Chen L, Duan W, Song R, Wang X (2011) Comparison of three multivariate methods of inferential modeling of soil organic matter using hyper spectra. IEEE. pp. 8124–8127.
31. Brimelow JC, Hanesiak JM, Raddatz R (2010) Validation of soil moisture simulations from the PAMII model, and an assessment of their sensitivity to uncertainties in soil hydraulic parameters. Agric For Meteorol 150: 100–114.

11

Increased Productivity of a Cover Crop Mixture Is Not Associated with Enhanced Agroecosystem Services

Richard G. Smith*, Lesley W. Atwood, Nicholas D. Warren

Department of Natural Resources and the Environment, University of New Hampshire, Durham, New Hampshire, United States of America

Abstract

Cover crops provide a variety of important agroecological services within cropping systems. Typically these crops are grown as monocultures or simple graminoid-legume bicultures; however, ecological theory and empirical evidence suggest that agroecosystem services could be enhanced by growing cover crops in species-rich mixtures. We examined cover crop productivity, weed suppression, stability, and carryover effects to a subsequent cash crop in an experiment involving a five-species annual cover crop mixture and the component species grown as monocultures in SE New Hampshire, USA in 2011 and 2012. The mean land equivalent ratio (LER) for the mixture exceeded 1.0 in both years, indicating that the mixture over-yielded relative to the monocultures. Despite the apparent over-yielding in the mixture, we observed no enhancement in weed suppression, biomass stability, or productivity of a subsequent oat (*Avena sativa* L.) cash crop when compared to the best monoculture component crop. These data are some of the first to include application of the LER to an analysis of a cover crop mixture and contribute to the growing literature on the agroecological effects of cover crop diversity in cropping systems.

Editor: Wen-Xiong Lin, Agroecological Institute, China

Funding: This work was supported by NERA (North East Regional Association of agricultural experiment station directors) project NE-1047 http://nimss.umd.edu/lgu_v2/homepages/home.cfm?trackID=13656. Partial funding was provided by the New Hampshire Agricultural Experiment Station. This is scientific contribution number 2533. The funders had no role in study design, data collection and analysis, decision to publish, or preparation of the manuscript.

Competing Interests: The authors have declared that no competing interests exist.

* E-mail: richard.smith@unh.edu

Introduction

Cover crops are typically sown within annual crop rotations to protect soil from erosion or provide other agroecosystem services such as building soil fertility and organic matter, retaining nutrients, or suppressing weeds during periods when cash crops are not actively growing [1–3]. Typically, these crops are sown as monocultures or simple graminoid-legume bicultures [2]; however, there is increasing interest among growers and researchers in investigating whether there may be additional benefits to growing cover crops in more species-diverse mixtures [4]. While there has been a large number of studies examining the role that crop diversity (including the use of cover crops within diversified rotations) plays with respect to specific agroecosystem services [5–7], few studies have examined the role of cover crop diversity explicitly (but see [4,8]).

There are a number of reasons to expect that a more diverse cover crop mixture might confer enhanced agroecosystem services relative to a monoculture or simple biculture. First, a wide range of plant species can be used as cover crops, including species from the graminoid, legume, brassica, and other broad-leaved families [9]. While each individual species may excel at one or a few services, no species is capable of providing all of the possible services and at the magnitudes likely necessary for substantive benefits to the agroecosystem. Thus, a cover crop mixture that contains a diversity of species, each differing in functional traits (e.g., biological N-fixation, root system, growth rate, tissue C:N, floral display, LAI, etc.) could be expected to provide a greater diversity

of services relative to a monoculture or a two-species cover crop community.

Second, there is often a positive relationship observed between cover crop productivity and its effectiveness for weed suppression [10,11]. Diversity-productivity theory suggests that increased productivity associated with species diversity is due to more efficient resource use [5,12]. Thus, diverse cover crop communities should be expected to produce more biomass than cover crop monocultures. Diverse cover crop communities should also be expected to be more weed suppressive because fewer resources are left available to support weed establishment and growth [13], and compared to monocultures, they may result in a broader spectrum of allelopathic activity toward various weed species or other soil environment modifications that enhance weed suppression [13,14].

Cover crop monocultures are subject to the same risks associated with variable growing conditions as are cash crop monocultures [15]. Therefore, diverse cover crop communities that contain multiple species with differing soil and environmental optima should be expected to be less variable in terms of overall stand productivity and function over space and time than cover crop monocultures or simple bicultures [4,16].

There are also reasons why a diverse cover crop mixture may be less desirable to a farmer than a mono- or biculture. These include increased costs for cover crop seed [17]; difficulty in establishing and managing complex mixtures, particularly if species have very different seed sizes, growth rates, life histories, or termination requirements [8,14]; and the possibility of antagonistic interactions between particular cover crop species or other components of the cash crop rotation [2]. Given the theoretical and practical arguments both for and against diverse cover crop mixtures, there is a clear need for additional research that addresses how diversified cover crop mixtures affect the myriad of agroecosystem functions and services that underpin the sustainability of agriculture.

Recently, Wortman et al. [4] reported land equivalent ratio (LER) and stability indices for multi-species mixtures of legume and brassica cover crops. Their study provided the first evidence available that cover crop mixtures are capable of over-yielding (i.e., LER>1) relative to the component species grown as monocultures. While these data help to confirm some of the suspected benefits of multi-species cover crop plantings, the study was limited to only two plant functional groups, legumes and brassicas. They also did not report on other agroecosystem services beyond productivity and stability, such as weed suppression. Also unknown are the effects diverse cover crop mixtures have on the growth of subsequent cash crops which would be planted after the cover crop mixture is terminated. Thus, the generality of the findings reported by Wortman et al. [4], and the potential for diverse cover crop mixtures to provide agroecosystem services relative to weed suppression and cash crop productivity remain unclear.

The objective of this study was to determine whether a mixture containing a functionally diverse group of spring-sown cover crops representing four plant families (Polygonaceae, Brassicaceae, Fabaceae, and Poaceae) could provide enhanced agroecosystem services relative to the component cover crops grown in monoculture. Specifically, we were interested in testing the following hypotheses:

1. A diverse cover crop mixture will be more productive than the most productive component crop grown in monoculture.
2. A diverse cover crop mixture will suppress weeds better than the most suppressive component crop grown in monoculture.
3. A diverse cover crop mixture will be more stable, in terms of biomass production and weed suppression, than the component crops grown in monoculture.
4. The biomass production of a subsequent crop will be higher following a cover crop mixture compared to monocultures of the component crops.

Materials and Methods

Site Description

The experiment was conducted at the University of New Hampshire Kingman Research Farm in Madbury, NH (43°11′N 70°56′W). Dominant soil type at this site is a Hollis-Charlton fine sandy loam (Hollis = loamy, mixed, mesic Entic Lithic Haplorthods; Charlton = coarse-loamy, mixed, mesic Entic Haplorthods) [18]. Mean monthly precipitation during the growing season (May–September) ranges from 89.9 to 107.4 mm and high and low temperatures range from 21 to 28°C and 7 to15°C, respectively. For several years prior to the experiment the site had been under a conventionally managed vegetable-winter rye (*Secale cereal* L.) cover crop rotation as part of a squash and pumpkin (Cucurbitaceae) breeding program.

Experimental Design

The experiment was conducted in 2011 and again in 2012 at an adjacent site. The experimental design both years was a randomized complete block with eight cover crop treatments, each replicated four times. The cover crop treatments were monocultures of buckwheat (*Fagopyrum esculentum*), mustard (*Brassica juncea*), sorghum-sudangrass (*Sorghum bicolor* var. sudanense), cereal rye (*Secale cereale*), hairy vetch (*Vicia villosa*) (2011 only) or field pea (*Pisum sativum*) (2012 only), and a mixture of all five species in which individual seeding rates were 20% of the monoculture rate (Table 1). A mixture of all five species in which individual seeding rates were 100% of the monoculture rate and a weedy fallow treatment in which no cover crops were planted were also included in the experimental design; however, these treatments were not germane to the present study and were therefore excluded from this analysis. Cover crop species were chosen to represent a diversity of plant families (i.e., Polygonaceae, Brassicaceae, Poaceae, and Fabaceae) corresponding to different plant functional groups (broadleaf forbs, C4 and C3 grasses; nitrogen-fixers; [19]). Field pea was substituted for hairy vetch in 2012 due to poor hairy vetch performance in 2011. In 2011, the experimental units were 2.5 m by 4 m, and in 2012 they were 4 m by 4.9 m. Prior to establishing each run of the experiment, the site was moldboard plowed and the seedbed was prepared using a Perfecta II field cultivator (Unverferth Equipment, Kalida, OH). Cover crops were broadcast seeded by hand in late spring (14 June 2011 and 19 June 2012) and seeds were incorporated into the soil with a rake.

Monoculture treatments were seeded at recommended rates for each species (Table 1). To evaluate cover crop mixture effects relative to the component species, we used a substitutive approach (i.e., proportional replacement design) such that seeding rates for each species in the mixture were proportional to their monoculture rate [20]. Therefore, the seeding rates for individual species in the mixture were determined by dividing each recommended seeding rate by the total number of species in the mixture (i.e., five). This approach minimizes potentially confounding effects of a higher overall seeding rate in the mixture and preserves the ability to use well established intercropping indices such as the LER [4,20]. No fertilizers or pesticides were applied to the experimental plots during the duration of the experiment.

Cover Crop Productivity

We quantified cover crop productivity for all treatments using two metrics: total aboveground biomass per unit area (dry weight, kg ha^{-1}) and the LER, which is traditionally used to evaluate the productivity of crop mixtures relative to component monocultures [4,21]. Both metrics were assessed by harvesting cover crop biomass at the soil surface from six 0.5 by 0.5 m quadrats located semi-randomly within each treatment replicate (the replicate was divided into six zones and one quadrat was positioned randomly in each zone). The six quadrat locations from which cover crop biomass was measured corresponded to the location of four "surrogate weed" subplots (described below) and two additional locations, all of which were located at least 0.5 m from the edge of the plot. The cover crop biomass harvest occurred at 43 (2011) and 72 (2012) days after the cover crop treatments were planted. The harvest in the 2012 study was delayed relative to the 2011 study in an effort to generate higher overall cover crop biomass. Harvested biomass was separated to species, dried at 65°C to constant biomass, and weighed to the nearest 0.01 g. Plot-level cover crop biomass (shoot dry weight) was then calculated for each treatment replicate by averaging the six subsamples. It is important to note that difference in harvest time (and substitution of legume species) between years does not impact our ability to quantify effects of the mixture relative to its component species, but does restrict our ability to make statements regarding the importance of climate factors as drivers of between-year differences in cover crop performance.

Table 1. Seeding rates of the cover crops used to create the mixture and component monoculture treatments in 2011 and 2012.

Cover crop treatment	Species	Family	Seeding rate (kg ha^{-1})
Buckwheat	Fagopyrum esculentum	Polygonaceae	67.2
Hairy vetch (2011 only)	Vicia villosa	Fabaceae	44.8
Field pea (2012 only)	Pisum sativum 'maxum'	Fabaceae	224
Mustard	Brassica juncia 'Pacific gold'	Brassicaceae	6.72
Sorghum-sudangrass	Sorghum bicolor x S. Bicolor var. sudanese	Poaceae	33.6
Cereal rye	Secale cereale	Poaceae	112
Mixture	All	All	All at 20% of full-rate

Plot-level cover crop biomass data were used to calculate the LER for the mixture, which represents the amount of land area that would be required to grow the individual component species as monocultures so as to achieve the same level of productivity as was attained in the mixture [21]. The LER is calculated as:

$$LER = pLER_i + pLER_j \ldots . + pLER_n$$

where $pLER_i$ is the partial LER of cover crop species i, $pLER_j$ is the partial LER of cover crop species j, and so forth for all n number of cover crop species present in both the mixture and monoculture. The partial LER of a given cover crop species ($i\ldots n$) is calculated as:

$$pLERi$$

$$= \text{biomass of species i in mixture}$$

$$\div \text{biomass of species i in monoculture}$$

A total LER for a mixture>1 indicates that more land area would be required to grow the cover crops as monocultures than growing an equivalent biomass using a mixture (i.e., the mixture "over-yielded" relative to the component monocultures). Conversely, a LER <1 indicates that less land would be necessary for monocultures than for a mixture to achieve an equivalent biomass yield (i.e., the mixture "under-yielded" relat'ive to the component monocultures). The partial LER for individual species can be used to compare their relative contribution to the total LER. In our case, because the mixture contained five species, the pLER for each species would be 0.20 in the absence of any interspecific interactions. Thus, the pLER indicates whether each species was positively (i.e., facilitation, when pLER>0.20) or negatively (i.e., antagonism, when pLER <0.20) affected by the other mixture components relative to its performance in monoculture [4]. Partial and total LERs were calculated at the block-level to enable statistical analysis (see below).

Weed Suppression

We used two approaches to quantify how the cover crop mixture and component monocultures affected the agroecosystem function of weed suppression. First, weeds that emerged from the soil seed bank (i.e., "ambient weeds") were harvested at the same time and from the same six quadrats used to collect cover crop biomass samples. Weeds were also sorted to species, dried at 65°C

to constant biomass, and weighed to the nearest 0.01 g. Second, we also established a "surrogate weed community" within four of the six 0.5 m by 0.5 m subplots located within each experimental unit. The surrogate weed community consisted of a total of 50 seeds made up of five crop plant species (Table 2). The purpose of the surrogate weeds was to create a uniform weed density and composition to more thoroughly assess competitive effects of the cover crop treatments. The rationale for planting crop species, as opposed to "weed" species, was to differentiate between seeds we added from those that emerged from the ambient seed bank and to ensure rapid germination and growth. With the exception of *Helianthus annuus*, which is in the Asteraceae family, the surrogate weed species were chosen to represent the same plant families included in the cover crop treatments. Seeds of the surrogate weed community were sown by hand into each subplot at two times ("early" to simulate weeds emerging at the same time as the cover crop, and "late" to simulate weeds emerging several weeks later). Within each plot, two subplots were designated as "early" and two were designated as "late". Surrogate weed communities were planted on 17 June 2011 and 20 June 2012 for "early" subplots and 30 June 2011, and 12 July 2012 for "late" subplots. Subplot locations were marked with stakes to facilitate relocation at the time of sampling. Surrogate weed biomass was collected at the same time and in the same manner as the cover crop and ambient weed biomass.

Analysis of cover crop mixture and component monoculture effects on the abundance of "ambient" weeds were based on samples collected from the two quadrats that did not contain surrogate weeds and the two quadrats that contained the "late" surrogate weed subplots. The "late" subplots were included in the analysis of the ambient weeds because emergence of surrogate weeds from those quadrats was effectively zero (data not shown). Cover crop treatment effects on surrogate weed abundance were thus restricted to the two "early" subplots within each replicate.

Stability

Cover crops that exhibit variable performance (i.e., are not stable) across space or time are not likely to be adopted by farmers, who often tend to be risk averse. We assessed spatial (plot to plot) and temporal (year to year) stability of the mixture and component treatments using the approach described by Tilman [22] and Wortman et al. [4]. Stability was assessed for both cover crop biomass production and ambient weed suppression by calculating the coefficient of variation (CV) for each cover crop treatment pooled across replications (n = 4) and years (n = 2). The CV was calculated as the standard deviation of cover crop biomass (or weed biomass) divided by the average cover crop biomass (or weed

Table 2. Crop species used to create "surrogate weed community" subplots in 2011 and 2012.

Surrogate weed	Species	Family	Density
			(No. m^{-2})
Sorrel	*Rumex sanguineus*	Polygonaceae	40
Field pea (2011)	*Pisum sativum*	Fabaceae	40
Red clover (2012 only)	*Trifolium pratense* 'mammoth'	Fabaceae	40
Canola	*Brassica napus*	Brassicaceae	60
Wheat (2011 only)	*Triticum aestivum*	Poaceae	40
Oats (2012 only)	*Avena sativa*	Poaceae	40
Sunflower	*Helianthus annuus* 'Zebulon'	Asteraceae	20

biomass). A lower CV indicates lower variability and hence greater stability in biomass production or weed suppression [22].

Carryover Effects on the Productivity of a Subsequent Crop

We examined carryover effects of the 2012 treatments by quantifying growth of a subsequent oat (*Avena sativa*) crop. The 2012 study site was cut with a sickle bar mower on 16 November 2012 to a height of 6 cm, and residues were allowed to remain within the plot over winter. In spring (20 May 2013), the plots were georeferenced and the entire site was chisel plowed. Following tillage, the field was prepared using a Perfecta II field cultivator (Unverferth Equipment, Kalida, OH), and oats were broadcast at a rate of 168 kg ha^{-1} on 31 May 2013. No fertilizer or herbicides were applied. On 30 July 2013 the location of the boundaries corresponding to the previous year's cover crop treatment plots were geolocated and oat biomass was harvested from two 0.5 by 0.5 m quadrats placed within 1.25 m of the center of each plot. Weed species were removed from the harvested oats and the biomass was dried at 65°C to constant biomass, and weighed to the nearest 0.01 g. The oat response was not initially an objective of the study, and thus was not implemented following the 2011 treatments.

Statistical Analyses

Cover crop, ambient weed, and surrogate weed dry biomass data were analyzed with the MIXED procedure in SAS (SAS Institute, Cary, NC, USA). The factors cover crop treatment, year, and the treatment x year interaction were all considered fixed effects. The block effect was considered random. The oat biomass data were analyzed as above, but without including the year and treatment x year interaction factors in the model. If significant treatment differences were detected, pairwise comparisons were made using least squares means. Distributions of the raw data did not deviate significantly from normal but were heteroscedastic. Transformations did not result in homogeneity of variance and tended to result in departures from a normal distribution; therefore, data were analyzed untransformed and presented as box-plots to enable visualization of the distribution of responses. One sample t-tests were used to determine whether the total LER of the mixture and partial LERs of the mixture components differed from 1 and 0.20, respectively. The weedy fallow treatment and mixture treatment in which all species were seeded at 100% of the full rate were not included in any of the analyses.

Results

Productivity

Analysis of the cover crop biomass dry weight data indicated a significant treatment by year interaction (treatment x year: $F_{5,33} = 11.76$, $P<0.0001$); therefore, the data from each year were analyzed separately. The subsequent analyses indicated that cover crop treatment effects were significant in both years (2011: $F_{5,15} = 12.65$, $P<0.0001$; 2012: $F_{5,15} = 16.28$, $P<0.0001$). In 2011, the treatment effects were driven primarily by buckwheat ($2{,}478\pm363$ kg ha^{-1}, mean ±1 SE) and hairy vetch (15 ± 4 kg ha^{-1}), which produced significantly higher and lower biomass than the other five treatments, respectively ($P<0.05$). The biomass of the mixture ($1{,}062\pm174$ kg ha^{-1}) did not differ from the mustard, sorghum-sudangrass, or cereal rye monocultures (Figure 1). In 2012, there was more differentiation in biomass between the treatments. Biomass of the sorghum-sudangrass monoculture ($7{,}200\pm926$ kg ha^{-1}) was significantly higher than all the other treatments ($P<0.05$), except the buckwheat monoculture. Biomass of the mixture ($4{,}476\pm720$ kg ha^{-1}) was not significantly different from the buckwheat or field pea monocultures, but was higher than the mustard and cereal rye monocultures (Figure 1).

In 2011 the mean LER of the mixture was 1.26, while in 2012 it was 1.12. Pooled across both years, the LER for the mixture was significantly greater than 1 (LER = 1.19 ± 0.09; t-test, $P=0.035$), indicating the mixture over-yielded relative to the component monocultures (Figure 2). This result means the mixture resulted in more efficient use of the land than the alternative of growing the component species as monocultures [4]. Investigation of the partial LERs indicated that only buckwheat had a pLER greater than 0.2, suggesting this species contributed most to the over-yielding response (pLER = 0.39 ± 0.07; t-test, $P=0.017$), and that its growth may have been facilitated by interspecific interactions. Conversely, only one species, cereal rye, had a pLER that was less than 0.2 (pLER = 0.05 ± 0.02; t-test, $P<0.001$), suggesting that its growth may have been limited by the other species in the mixture (Figure 2). The pLER for the other three species did not differ from 0.2 ($P>0.05$).

Weed Suppression

The ambient weed biomass present at harvest differed by year ($F_{1,13} = 21.27$, $P<0.0001$) and cover crop treatment ($F_{5,33} = 4.75$, $P=0.0022$), but there was no interaction between year and treatment. Across the two years, ambient weed biomass was lower in buckwheat monoculture plots compared to legume, mustard, and sorghum-sudangrass monocultures ($P<0.05$). Weed abun-

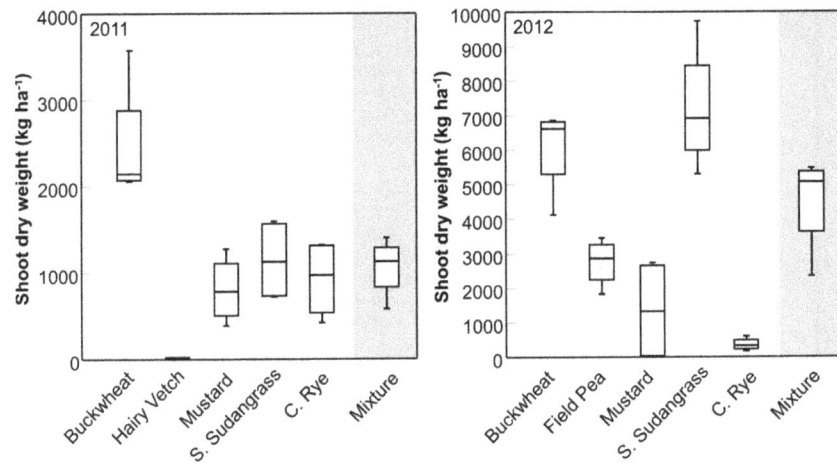

Figure 1. Productivity of a cover crop mixture and component monocultures. Box plots showing variation around the median for shoot dry weight of five cover crops grown in monoculture and a mixture containing all five species in 2011 and 2012. The line within the box represents the median; the box represents 50% of the data; whiskers represent the 10th and 90th percentiles; n = 4.

dance in the mixture was not significantly different from buckwheat or other component monocultures (Figure 3).

Surrogate weed response indicated an interaction between year and cover crop treatment ($F_{5,33} = 3.22$, $P = 0.0178$); therefore, the data were analyzed separately for each year. In 2011, surrogate weed biomass tended to be lower in the buckwheat monoculture, but differences among treatments were not statistically significant ($F_{5,15} = 2.48$, $P = 0.079$). In 2012, treatment differences were

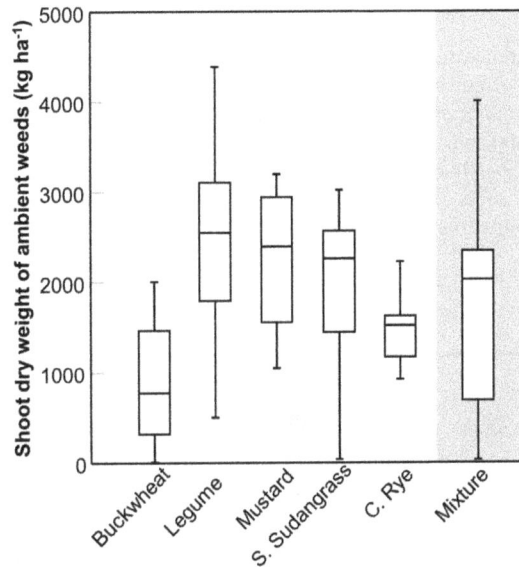

Figure 3. Biomass of weeds that emerged from the ambient weed seed bank. Box plots showing variation around the median for ambient weed biomass in five cover crops grown in monoculture and a mixture containing all five species across the two study years. The line within the box represents the median; the box represents 50% of the data; whiskers represent the 10th and 90th percentiles; n = 8.

significant ($F_{5,15} = 6.05$, $P = 0.003$) and were driven primarily by the mustard monoculture. Surrogate weed biomass was higher in the mustard monoculture than any of the other monocultures or the mixture ($P < 0.05$). Surrogate weed biomass in the mixture was not significantly different from the other four monocultures, despite a trend toward lower biomass in the buckwheat monoculture (Figure 4).

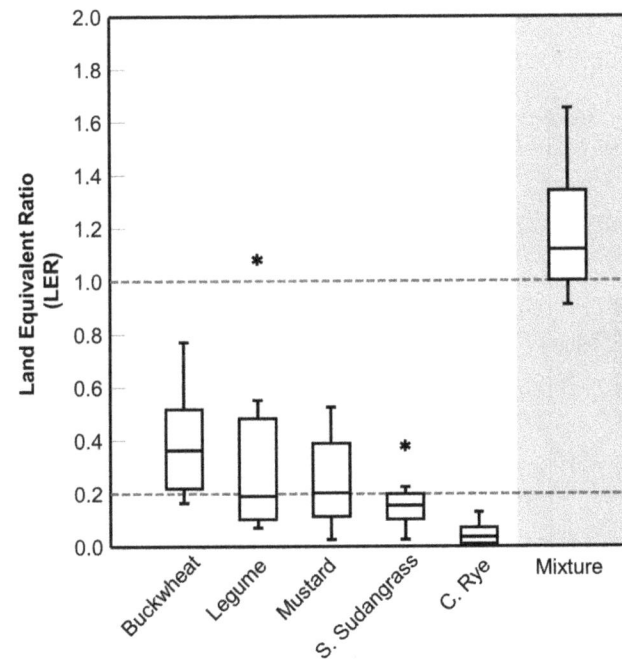

Figure 2. Land equivalent ratio (LER) of the mixture. Box plot showing variation around the median for the partial (individual species contributions) and total LER of the cover crop mixture across the two study years. The grey dotted lines at 0.2 and 1.0 indicate "break even" points above which partial and total LER indicate over-yielding, respectively. The line within the box represents the median; the box represents 50% of the data; whiskers represent the 10th and 90th percentiles; asterisks indicate outliers; n = 8.

Stability of Productivity and Weed Suppression

The CV was used as a measure of the relative stability of the different cover crop treatments in terms of their productivity and weed suppression. Buckwheat (CV = 50%) and cereal rye (CV = 70%) monocultures had the least variable biomass produc-

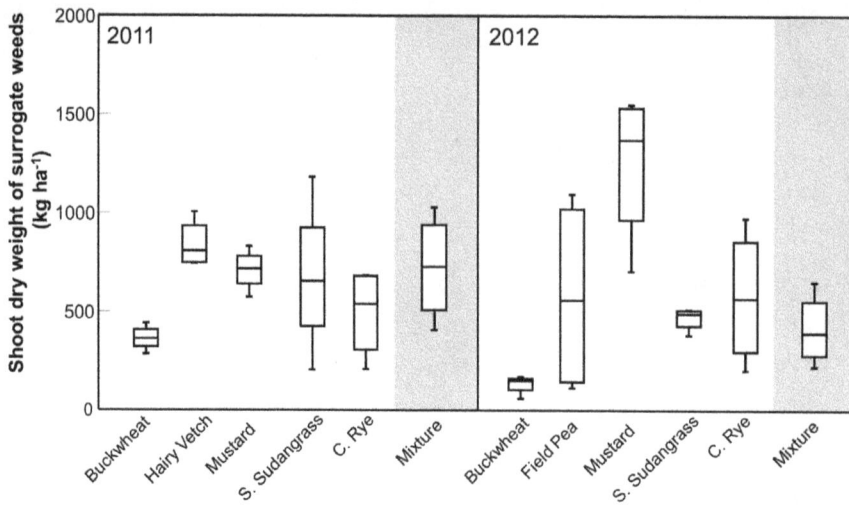

Figure 4. Biomass of surrogate weeds. Box plots showing variation around the median for surrogate weed biomass in five cover crops grown in monoculture and a mixture containing all five species in 2011 and 2012. The line within the box represents the median; the box represents 50% of the data; whiskers represent the 10th and 90th percentiles; n = 4.

tion across replicates and years. The mixture (CV = 75%) was also less variable in space and time than legume, mustard, and sorghum-sudangrass monocultures (Figure 5). With respect to weed suppression, the cereal rye monoculture had the least variable weed abundance (CV = 27%). In contrast, weed abundance was most variable in the buckwheat monoculture and mixture treatments (buckwheat CV = 82%; mixture CV = 72%; Figure 5).

Carryover Effects on Subsequent Crop Productivity

An oat crop was planted uniformly across the 2012 study site to quantify the potential carryover effects of the previous cover crop treatments on oat productivity. Oats following field pea monoculture tended to have higher biomass than following the other treatments, including the mixture (Figure 6), but the effect was not statistically significant ($P = 0.168$).

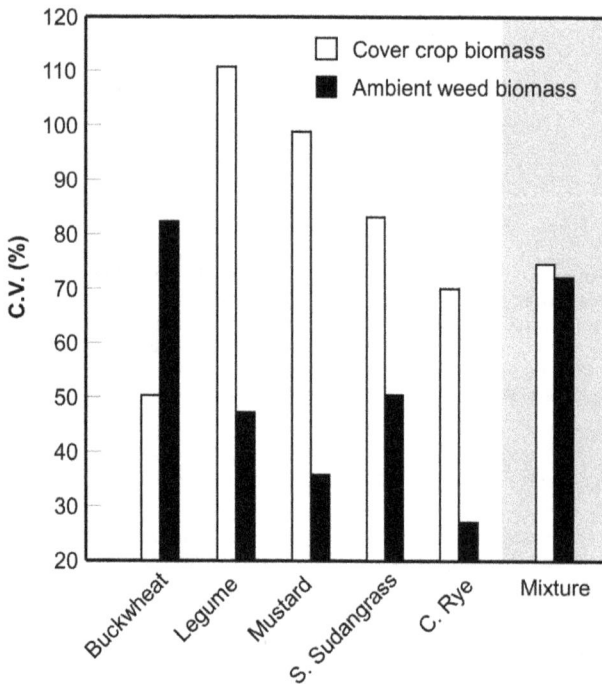

Figure 5. Variability in weed suppression in space and time. Coefficient of variation (CV) calculated across replicates (n = 4) and years (n = 2) for cover crop and ambient weed biomass in each cover crop monoculture and a mixture containing all five cover crop species.

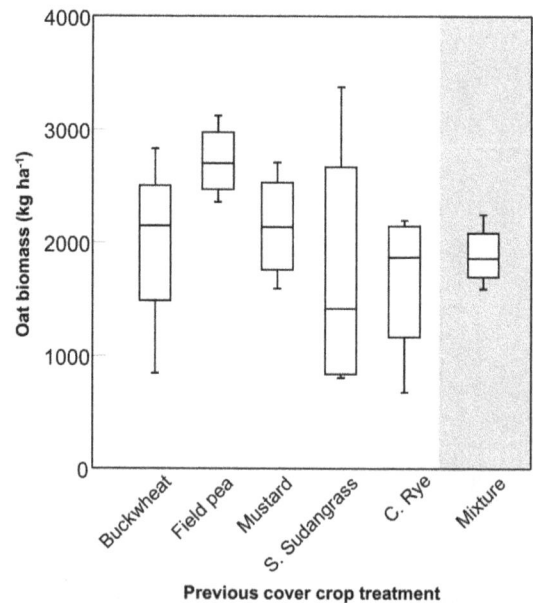

Figure 6. Carryover effects on oat growth. Box plots showing variation around the median for biomass of an oat phytometer sown in 2013 on plots that were previously sown with five cover crops grown in monoculture and a mixture containing all five species. The line within the box represents the median; the box represents 50% of the data; whiskers represent the 10th and 90th percentiles; n = 4.

Discussion

The objective of this study was to assess whether a cover crop mixture containing five species from four different plant families would provide enhanced agroecosystem services relative to the same species grown as monocultures. The services we measured included cover crop productivity, weed suppression, biomass stability, and carryover effects on the productivity of a subsequent crop. Although our study included five different cover crop species, our intent was not to examine all possible levels of diversity. Instead, we were interested in the extreme ends of the diversity gradient (i.e., monocultures vs. a mixture of all species), assuming these extremes would correspond to the greatest differences in agroecosystem functional response [23,24]. The time period for our study was designed to simulate a summer fallow period that might precede a late summer/fall or subsequent spring cash crop. Thus, our results may not apply to all possible cover crop niches (e.g., fall-sown cover crops), cover crop species, or cover crop combinations.

While our data do not support the hypothesis that a diverse cover crop mixture would produce more biomass on a per unit area basis than the most productive component crop grown in monoculture, we did observe an increased biomass yield (i.e., over-yielding) with the mixture relative to the component monocultures. This seemingly contradictory result deserves further clarification. With regard to productivity responses in plant biodiversity studies, Schmid et al. [25] distinguish between 'over-yielding' and 'transgressive over-yielding'. Over-yielding occurs when the biomass production of the mixture is greater than the average monoculture yield [25]. Similar to Wortman et al. [4], we observed this type of response in our mixture treatments, as indicated by LER values greater than 1.0 in both 2011 and 2012 (Figure 2). In contrast, transgressive over-yielding occurs when biomass of the mixture is greater than that produced by the most productive monoculture [25]. Like Wortman et al. [4], we did not observe biomass production in the mixture to be higher than the most productive monoculture treatment (i.e., buckwheat in 2011 and sorghum-sudangrass in 2012; Figure 1). This response suggests higher biomass yields may not be a realistic outcome of cover crop mixtures constructed using a substitutive approach.

The over-yielding observed in our study was attributed mainly to buckwheat, which produced proportionately more biomass per unit area when grown in mixture than in monoculture. This was confirmed by a partial LER that exceeded 0.20. The only species that appeared to be negatively affected by being grown in mixture was cereal rye. Since that crop is typically sown in the fall, it was not unexpected for it to be less competitive when sown in spring [26]. To our knowledge, ours is the first study to assess yield response of cover crop mixtures containing buckwheat, sorghum-sudangrass, and cereal rye using the LER. Therefore, we are not certain that the partial LER values reported here are typical for these species. Wortman et al. [4], who used cover crop species from the Fabaceae and Brassicaceae families, observed apparent antagonism between the mustard, hairy vetch, and field pea when grown in their mixtures; however, we observed no evidence of antagonism between the mustard and legumes. These differences could be due to site-specific differences in climate, soil type, or other factors that varied between the two studies.

The over-yielding we observed with the mixture did not appear to enhance any of the other agroecosystem services typically associated with cover crops. For example, the mixture did not suppress ambient weed abundance compared to the most suppressive component crop (i.e., buckwheat; Figure 3). Similarly, biomass of the surrogate weeds was not lower in the mixture compared to the most suppressive monoculture (Figure 4). These results are in agreement with those of Teasdale and Abdul-Baki [27], who found that cover crop mixtures containing two legumes (hairy vetch and crimson clover, *Trifolium incarnatum* L.) and cereal rye reduced weed biomass compared to the legume monocultures but not the cereal rye monoculture. The trend for lower ambient and surrogate weed biomass in the buckwheat monoculture, suggests that this species is particularly effective in suppression of late spring and summer-emerging weeds; a result that has been observed in previous studies [28].

Despite strong theoretical and empirical support for a link between plant species diversity and agroecosystem stability [16,29], the over-yielding we observed with the mixture did not enhance stability, either in terms of cover crop biomass or weed suppression, relative to the most stable component monoculture (Figure 5). These results are also in agreement with those reported by Wortman et al. [4] who found that the relative stability of cover crop biomass production was not associated with the number of different legume and brassica cover crop species grown in mixtures. Considered along with that study, our results suggest that greater cover crop functional group richness (i.e., mixtures with four plant families) does not necessarily improve stability. This conclusion is in accordance with recent work by Cardinale et al. [30] suggesting that effects of species diversity on biomass production can be independent of diversity effects on stability.

Our results suggest that of the five cover crop species examined, buckwheat grown in monoculture should be preferred over this specific mixture if a producer's goal is to produce consistent (spatially and temporally stable) summer cover crop biomass and to maximize summer weed suppression. Cereal rye, which did not produce excessive biomass, did provide a fairly consistent (stable) level of weed suppression across replicates and years, and would likely be preferable to the mixture examined here, if the primary goal was weed suppression.

One of the primary motivations for growing cover crops is to improve growing conditions for a subsequent cash crop [2]. Relative to a monoculture, a cover crop mixture should be expected to contribute residues that vary in quality and biochemical composition, which in turn could affect soil processes (and their microbial drivers) that influence crop growth [31]. We used common oats as a "phytometer" to assess whether the cover crop mixture resulted in carryover effects that would improve cash crop productivity compared to the component species grown as monocultures. We found no evidence to suggest the mixture enhanced oat growth more than the component species, although due to land and labor constraints the carryover study was not conducted following the 2011 study. Therefore, these results should be interpreted with some degree of caution. Despite this caveat, our results are congruent with a recent study showing no difference in crop yields associated with cover crop mixtures differing in the number of legume and *Brassica* species [8,17]; however, that study did not include cover crop monocultures.

How do we explain the fact that over-yielding with the cover crop mixture resulted in no apparent enhancement of other agroecosystem services relative to the component monocultures? One possible explanation is that the increased yields appeared to be driven primarily by a single species, buckwheat. Therefore, the potential for concomitant effects on other functions was relatively limited. Another possible explanation is the metric used to assess yield response. The LER has primarily been used to measure the yields of cash crops grown in polyculture [21]. When applied to cash crops, the proportional yield is a relevant metric, providing information about the amount of land area that would be required to produce an equivalent yield of each crop in monoculture as can be obtained by growing those crops in mixture, and has important implications for improving the efficiency of agricultural land use [32]. However, when applied to cover crops, a proportional metric such as LER does not align with the purpose of cover crops, as the goal is often to maximize total cover crop biomass per unit area rather than minimize the total land area required to grow a certain number of different cover crop species. Thus, if a particular agroecosystem function, such as weed suppression, is strongly linked to cover crop biomass [10,11], a cover crop mixture can over-yield but still not "out-perform" the most productive monoculture. Finally, we utilized a limited number of potential cover crop species (five) and quantified only a relatively small subset of the possible agroecosystem services associated with cover crop use in a single season. For example, we did not assess beneficial insect populations, soil-borne disease, or soil organic matter quality and diversity, all of which could be affected by our cover crop treatments and which may manifest over longer time periods of cover crop use [31,33,34]. Additional research will be

necessary to determine the full range of mixture combinations and cover crop planting niches, and their effects on a wider range of possible agroecosystem functions, including food-web dynamics, biological control, and weed-crop competitive interactions.

Acknowledgments

The authors thank John McLean, Evan Ford, Dan Tauriello, Josh Cain, Kelsey Juntwait, Lucie Worthen, Luke Douglas, Tessa Wheeler, and Jennifer Wilhelm for help with site preparation, field work, and sample processing.

Author Contributions

Conceived and designed the experiments: RGS. Performed the experiments: RGS LWA NDW. Analyzed the data: RGS LWA NDW. Contributed reagents/materials/analysis tools: RGS. Wrote the paper: RGS LWA NDW.

References

1. Teasdale JR (1996) Contribution of cover crops to weed management in sustainable agricultural systems. J Prod Agric 9: 475–479.
2. Snapp SS, Swinton SM, Labarta R, Mutch D, Black JR, et al. (2005) Evaluating cover crops for benefits, costs and performance within cropping system niches. Agron J 97: 322–332.
3. Smith RG, Gareau TP, Mortensen DA, Curran WS, Barbercheck ME (2011) Assessing and visualizing agricultural management practices: A multivariable hands-on approach for education and extension. Weed Technol 25: 680–687.
4. Wortman SE, Francis CA, Lindquist JL (2012) Cover crop mixtures for the Western Corn Belt: opportunities for increased productivity and stability. Agron J 104: 699–705.
5. Trenbath BR (1974) Biomass productivity of mixtures. Adv Agron 26: 177–210.
6. Smith RG, Gross KL, Robertson GP (2008) Effects of crop diversity on agroecosystem function: crop yield response. Ecosystems 11: 355–366.
7. Tooker JF, Frank SD (2012) Genotypically diverse cultivar mixtures for insect pest management and increased crop yields. J Appl Ecol 49: 974–985.
8. Wortman SE, Francis CA, Bernards MA, Blankenship EE, Lindquist JL (2013) Mechanical termination of diverse cover crop mixtures for improved weed suppression in organic cropping systems. Weed Sci 61: 162–170.
9. Clark A (2007) Managing Cover Crops Profitably, 3rd Ed. Sustainable Agriculture Research and Education (SARE) program handbook series. Available: http://www.sare.org/Learning-Center/Books/Managing-Cover-Crops-Profitably-3rd-Edition.
10. Mohler CL, Teasdale JR (1993) Response of weed emergence to rate of *Vicia villosa* Roth and *Secale cereale* L residue. Weed Res 33: 487–499.
11. Mirsky SB, Ryan MR, Teasdale JR, Curran WS, Reberg-Horton CS, et al. (2013) Overcoming weed management challenges in cover crop-based organic rotational no-till soybean production in the Eastern United States. Weed Technol 27: 193–203.
12. Tilman D (1999) The ecological consequences of changes in biodiversity: a search for general principles. Ecology 80: 1455–1474.
13. Liebman M, Davis AS (2000) Integration of soil, crop and weed management in low-external-input farming systems. Weed Res 40: 27–47.
14. Creamer NG, Bennett MA, Stinner BR (1997) Evaluation of cover crop mixtures for use in vegetable production systems. HortScience 32: 866–870.
15. Lin BB (2011) Resilience in agriculture through crop diversification: adaptive management for environmental change. BioScience 61: 183–193.
16. McCann KS (2000) The diversity-stability debate. Nature 405: 228–233.
17. Wortman SE, Francis CA, Bernards M, Drijber R, Lindquist JL (2012) Optimizing cover crop benefits with diverse mixtures and an alternative termination method. Agron J 104: 1425–1435.

18. Freyre R, Loy JB (2000) Evaluation and yield trials of tomatillo in New Hampshire. HortTechnol 10: 374–377.
19. Lavorel S, McIntyre S, Landsberg J, Forbes TDA (1997) Plant functional classifications: from general groups to specific groups based on response to disturbance. Trends Ecol Evol 12: 474–478.
20. Jolliffe PA (2000) The replacement series. J Ecol 88: 371–385.
21. Vandermeer J (1989) The Ecology of Intercropping Systems. Cambridge University Press. 237 pgs.
22. Tilman D (1996) Biodiversity: population versus ecosystem stability. Ecology 77: 350–363.
23. Lehman CL, Tilman D (2000) Biodiversity, stability, and productivity in competitive communities. Am Nat 156: 534–552.
24. Cardinale BJ, Matulich JK, Hooper DU, Byrnes JE, Duffy E, et al. (2011) The functional role of producer diversity in ecosystems. Am J Bot 98: 572–592.
25. Schmid B, Hector A, Saha P, Loreau M (2008) Biodiversity effects and transgressive overyielding. J Plant Ecol 1: 95–102.
26. Ateh CM, Doll JD (1996) Spring-planted winter rye (*Secale cereal*) as a living mulch to control weeds in soybean (*Glycine max*). Weed Technol 10: 347–353.
27. Teasdale JR, Abdul-Baki AA (1998) Comparison of mixtures vs. monocultures of cover crops for fresh-market tomato production with and without herbicide. HortScience 33: 1163–1166.
28. Kumar V, Brainard DC, Bellinder RR (2009) Effects of spring-sown cover crops on establishment and growth of hairy galinsoga (*Galinsoga ciliata*) and four vegetable crops. HortScience 44: 730–736.
29. Trenbath BR (1999) Multispecies cropping systems in India-predictions of their productivity, stability, resilience and ecological sustainability. Agroforest Syst 45: 81–107.
30. Cardinale BJ, Gross K, Fritschie K, Flombaum P, Fox J, et al. (2013) Biodiversity simultaneously enhances the production and stability of community biomass, but the effects are independent. Ecology 94: 1697–1707.
31. Smith RG, Mortensen DA, Ryan MR (2010) A new hypothesis for the functional role of diversity in mediating resource pools and weed-crop competition in agroecosystems. Weed Res 50: 37–48.
32. Tilman D, Balzer C, Hill J, Befort BL (2011) Global food demand and the sustainable intensification of agriculture. P Natl Acad Sci USA 108: 20260–20264.
33. Trenbath BR (1993) Intercropping for the management of pests and diseases. Field Crops Res 34: 381–405.
34. Mitchell CE, Tilman D, Groth JV (2002) Effects of grassland plant species diversity, abundance, and composition on foliar fungal disease. Ecology 83: 1713–1726.

12

Long-Term Effect of Manure and Fertilizer on Soil Organic Carbon Pools in Dryland Farming in Northwest China

Enke Liu[1,2], Changrong Yan[1,2]*, Xurong Mei[1,2], Yanqing Zhang[1,2]*, Tinglu Fan[3]

1 Institute of Environment and Sustainable Development in Agriculture, Chinese Academy of Agricultural Sciences, Beijing, China, 2 Key Laboratory of Dryland Farming g Agriculture, Ministry of Agriculture of the People's Republic of China (MOA), Beijing, China, 3 Dryland Agricultural Institute, Gansu Academy of Agricultural Sciences, Lanzhou, Gansu, China

Abstract

An understanding of the dynamics of soil organic carbon (SOC) as affected by farming practices is imperative for maintaining soil productivity and mitigating global warming. The objectives of this study were to investigate the effects of long-term fertilization on SOC and SOC fractions for the whole soil profile (0–100 cm) in northwest China. The study was initiated in 1979 in Gansu, China and included six treatments: unfertilized control (CK), nitrogen fertilizer (N), nitrogen and phosphorus (P) fertilizers (NP), straw plus N and P fertilizers (NP+S), farmyard manure (FYM), and farmyard manure plus N and P fertilizers (NP+FYM). Results showed that SOC concentration in the 0–20 cm soil layer increased with time except in the CK and N treatments. Long-term fertilization significantly influenced SOC concentrations and storage to 60 cm depth. Below 60 cm, SOC concentrations and storages were statistically not significant between all treatments. The concentration of SOC at different depths in 0–60 cm soil profile was higher under NP+FYM follow by under NP+S, compared to under CK. The SOC storage in 0–60 cm in NP+FYM, NP+S, FYM and NP treatments were increased by 41.3%, 32.9%, 28.1% and 17.9%, respectively, as compared to the CK treatment. Organic manure plus inorganic fertilizer application also increased labile soil organic carbon pools in 0–60 cm depth. The average concentration of particulate organic carbon (POC), dissolved organic carbon (DOC) and microbial biomass carbon (MBC) in organic manure plus inorganic fertilizer treatments (NP+S and NP+FYM) in 0–60 cm depth were increased by 64.9–91.9%, 42.5–56.9%, and 74.7–99.4%, respectively, over the CK treatment. The POC, MBC and DOC concentrations increased linearly with increasing SOC content. These results indicate that long-term additions of organic manure have the most beneficial effects in building carbon pools among the investigated types of fertilization.

Editor: Vishal Shah, Dowling College, United States of America

Funding: This study was supported by the National Basic Research Program of China (973 Program) (No.2012CB955904), the Chinese National Scientific Foundation (No. 31000253, 31170490), the 12th five-year plan of the National Key Technologies R&D Program (No. 2012BAD09B01). The authors declare that no additional external funding was received for this study.The funders had no role in study design, data collection and analysis, decision to publish, or preparation of the manuscript.

Competing Interests: The authors have declared that no competing interests exist.

* E-mail: yancr@ieda.org.cn (CY); zhangyq@ieda.org.cn(YZ)

Introduction

Soil organic matter (SOM) plays a key role in the improvement of soil physical, chemical and biological properties [1]. Conservation of the quantity and quality of soil organic matter (SOM) is considered a central component of sustainable soil management and maintenance of soil quality [2]. Organic manure and inorganic fertilizer are the most common materials applied in agricultural management to improve soil quality and crop productivity [3]. Many studies have shown that balanced application of inorganic fertilizers or organic manure plus inorganic fertilizers can increase SOC and maintain soil productivity [4–7].

However, SOM is not sensitive to short-term changes of soil quality with different soil or crop management practices due to high background levels and natural soil variability [8]. Labile soil organic carbon pools like dissolved organic C (DOC), microbial biomass C (MBC), and particulate organic matter C (POC) are the fine indicators of soil quality which influence soil function in specific ways (e.g., immobilization–mineralization) and are much more sensitive to change in soil management practices [9,10]. Because these components can respond rapidly to changes in C supply, they have been suggested as early indicators of the effects of land use on SOM quality [11]. Recently, many studies have reported responses of labile SOC pools to management practices [5,12,13], though limited to tillage practices or cropping intensity and rotations management [14]. Few studies have focused on the effect of labile organic C after long-term fertilizer application in northwest China.

In most cases, studies for SOC and SOC fractions have mostly focused on shallow surface soil [15]. The limited information on soil profile SOC and its fractions distribution is a hindrance to conclusive identification of beneficial effects after long-term fertilizer application. Thus, further research is needed to clarify fertilizer application impacts on SOC and SOC fractions for the entire soil profile. Documenting increased SOC levels at deeper depths in the soil profile, however, has been difficult due to a lack of studies where sampling occurred below 30 cm. Nayak et al. [13] found that applications of combined inorganic fertilizers with or without manure can sequester carbon in the 0–60 cm soil layer at the Indian sub-Himalayas. In hot humid subtropical eastern India, Majumder et al. [16] found that after 19 y in a puddle rice-wheat (Triticum aestivum L.) system, NPK+FYM treated plots had 14% larger labile C pools compared with the control plots in the 0–60 cm soil layer.

Northwest China is a vast semi-arid area with average annual precipitation ranging from 300 to 600 mm and more than 90% of the cropland depends on rain fall. Dryland farming has prevailed for several decades in this region. The dry climate and sparse vegetation are mainly responsible for the low SOM. A long period of cultivation and severe erosion in northwest China are likely other potential causes of low SOM [17]. SOC content in the 0–20 cm soil layer of this region is about 11.4 t C ha^{-1} [18]. In recent years, there has been a large increase in the use of inorganic fertilizer with a concomitant decrease in the use of manure. Challenges for dryland farming in Northwest China are low SOC and nutrient retention [19]. However, little is known about the long-term application of inorganic fertilizers either alone or with organic manure on SOC and the distribution of labile organic C fractions at different profile depths. Thus, it is crucial to collect SOC data from long-term experiments in order to understand and estimate the contribution of manure and fertilizer to soil C dynamics. This study provided a unique opportunity to examine the long-term effects of manure and fertilizer on soil organic carbon pools for dryland farming in Northwest China. We hypothesized that long-term fertilizer and manure application would influence the SOC and labile carbon. Moreover, we considered labile organic C fractions would be responsive indicators to SOC change with long histories of fertilizer managements. Our objective was to study the changes of the depth distribution (0–100 cm) in SOC and SOC fractions under a 30-year field experiment in the north of China, and to explain the relationship between different SOC fractions and SOC concentrations. Improved understanding of labile organic matter fractions will provide valuable information for establishing sustainable fertilizer management systems to maintain and enhance soil quality.

Materials and Methods

Experimental Site

The research was based on a long-term fertilizer experiment started in 1979 at the Gaoping Agronomy Farm (35°16′N, 107°30′E, 1254 m altitude), Pingliang, Gansu, China (Fig. 1). Under average climatic conditions, the area has an aridity index (P/PET: precipitation/potential evapotranspiration) of 0.39 and receives 540 mm precipitation, about 60% of which occurs in the summer from July through September. May through June is the driest period for crop growth and little precipitation occurs during the winter months of December and January. The mean annual temperature is 9.8°C. The mean annual sunshine period is 2834 h. The soil is a dark loessial soil classified as calcarid regosols [20]. Analysis of soil samples taken from the experimental area in October 1978 indicated that the surface 15 cm of soil had a pH of 8.2, SOC content of 6.2 g kg^{-1}, total N of 0.95 g kg^{-1}, total P content of 0.57 g kg^{-1}, available P of 7.2 mg kg^{-1} and available K of 165 mg kg^{-1}.

Experimental Design and Treatments

The experiment began in 1979 with a maize crop on land that had been cropped to maize during the previous year, one crop per year. Six fertilization treatments were arranged in a randomized complete block design with three replications. Maize was grown in 1979 and 1980, wheat from 1981 to 1984, maize in 1985 and 1986, wheat from 1987 to 1990, maize in 1991 and 1992, wheat from 1993 to 1998, soybean (Glycine max (L.) Merr.) in 1999, sorghum (Sorghum bicolor (L.) Moench) in 2000, wheat from 2001 to 2004, maize in 2005 and 2006, and wheat in 2006 to 2008.

Winter wheat (Qingxuan 8271, Longyuan 935, and Ping 93-2) was seeded in rows 14.7 cm apart at rates of 165 kg ha^{-1} on about 20 September each year when wheat followed wheat, and in early October when wheat followed maize. Maize was seeded about 20 April each year that maize was grown and Zhongdan 2 was seeded by hand in clumps every 33 cm in rows 66.5 cm apart. About 3 weeks after seeding, maize plants were thinned to one plant per clump. Later, if tillers developed, they were removed to avoid competition. Hand weeding was done to control weeds and plant protection measures were applied when needed. Crops were harvested manually close to the ground and all harvested biomass was removed from the plots. Grain yields were determined by harvesting 20 m^2 for wheat and 40 m^2 for maize at centers of the plots. Grain samples were air-dried on concrete, threshed, and oven-dried at 70°C for 48 hrs, and then weighed.

The experimental area was 0.44 ha. Each plot was 16.7 m×13.3 m with a buffer zone of 1.0 m between each plot. The six treatments were (1) CK, unfertilized control, (2) N, nitrogen fertilizer annually, (3) NP, nitrogen and phosphorus (P) fertilizers annually, (4) NP+S, straw (S) plus N added annually and P fertilizer added every second year, (5) FYM, farmyard manure added annually, and (6) NP+FYM, farmyard manure plus N and P fertilizers added annually. Urea was the N source and was applied to supply 90 kg N ha^{-1} yr^{-1}. Superphosphate was the P source and applied to supply 30 kg P ha^{-1} yr^{-1}. Farmyard manure was added at rate of 75 t ha^{-1} (wet weight). Deep plowing of approximately 23 cm was performed in July after wheat harvest or in October after maize harvest except for the years in which wheat followed maize. In those years, shallow disk tillage was done after maize harvest and wheat was seeded immediately.

Generally, the farmyard manure was a mixture of about 1:5 ratio of wet cattle manure to loess soils and so its nutrient content was quite variable from year to year. The SOC, N, P, and K contents of manure were 11.37, 1.07, 0.69, and 12.3 g kg^{-1} in dry weight, indicating that manure is very low in N, and high in P and K. Although the specific amounts of nutrients added with manure each year were not determined, an application of approximately 75 t ha^{-1} (wet weight) supplied roughly 425 kg C ha^{-1}, 40 kg N ha^{-1}, 26 kg P ha^{-1}, and 460 kg K ha^{-1} in manure annually to crops. For NP+S treatment, mean 5.61 t ha^{-1} y^{-1} of straw (winter wheat and maize) approximately 10 cm in length was returned to the soil prior to plowing, and P fertilizer was added every second year. The straw contained 2.45 t C ha^{-1}, 29.8 kg N ha^{-1}, 7.4 kg P ha^{-1}, and 34.0 kg K ha^{-1}.

Soil Sampling

For SOC trend (0–20 cm layer)study from 1979 to 2008, the composite soil sample (0–20 cm) for each plot was prepared by mixing ten soil cores (4-cm inner diameter) collected randomly after the harvest during 1979 through 1991 and 1996 through 2008 at about 15 d after harvest. The fresh soil was mixed thoroughly, air dried for 7 d, sieved through a 2.0 mm sieve at field moisture content, mixed, and stored in sealed plastic jars for

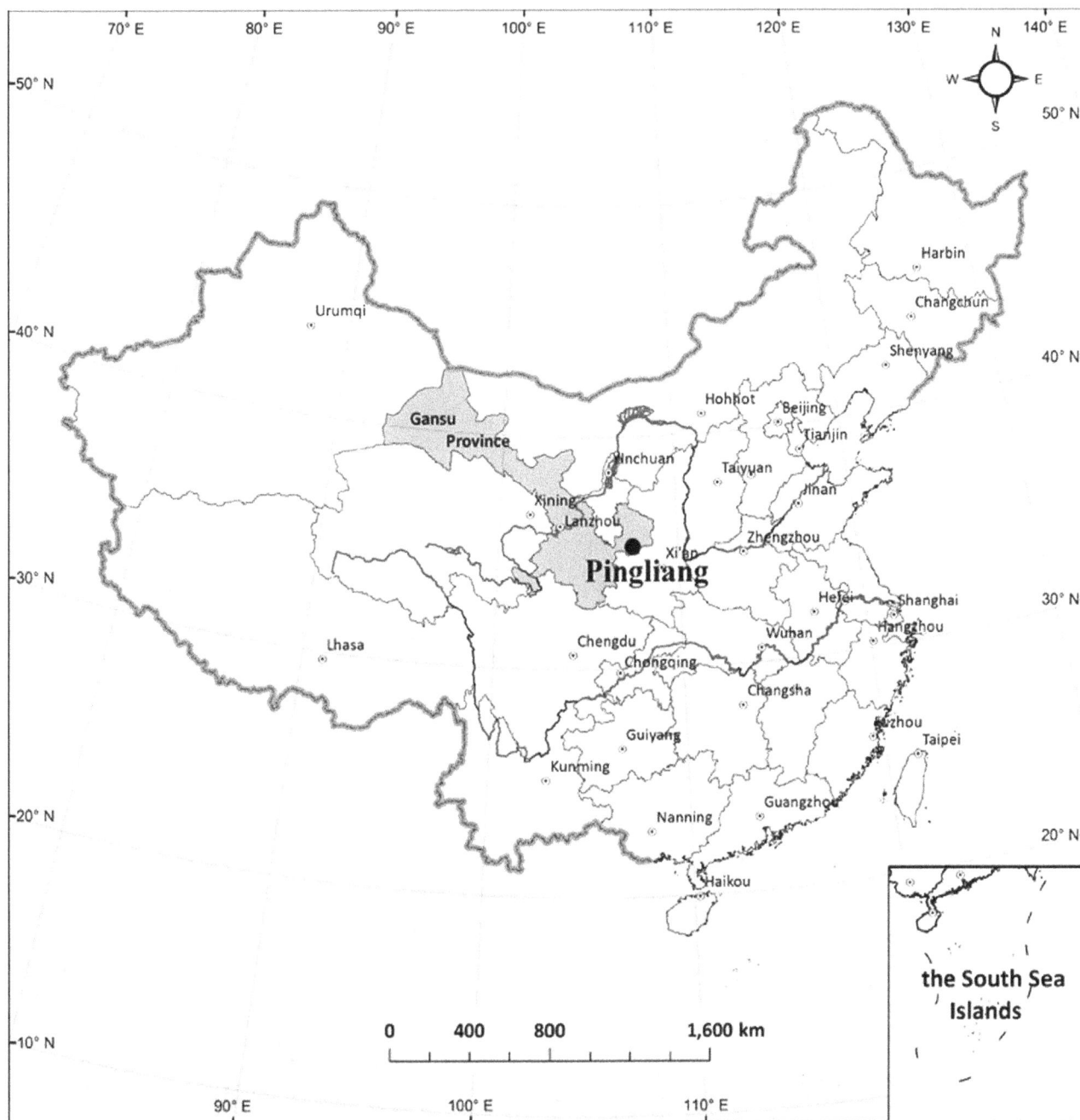

Figure 1. Location of the study site.

analysis. Sub-samples were drawn to determine SOC in the 0–20 cm soil layer. Soil organic C was not determined during 1992 through 1995.

For the distribution of SOC and labile organic C fractions at different profile depths study, soil profile samples (0–20, 20–40, 40–60, 60–80, and 80–100 cm) were collected from six treatments before wheat sowing in September 2008. In each plot the soil was collected from ten points randomly, and mixed into one sample. After carefully removing the surface organic materials and fine roots, each mixed soil sample was divided into two parts. One part of the soil sample was air-dried for the estimation of soil chemical properties and the other part was sieved through a 2 mm wide screen and immediately transferred to the laboratory for bio-

chemical analysis. Soil fresh samples were kept at 4°C in plastic bags for a few days to stabilize the microbiological activity and analyzed within 2 weeks.

Soil Analyses

Soil organic C and bulk densities measured using the method of Blake [21]. DOC was measured using the method of Jiang et al. [22]. POM-C was determined by the method of Cambardella and Elliott [23]. MBC was estimated by fumigation-extraction [24].

Table 1. Estimated mean annual crop biomass and carbon input (Mg C ha^{-1} yr^{-1}) to soil under different fertilizer treatments.

Treatments	Mean annual crop biomass (Mg C ha^{-1} yr^{-1})					Mean annual carbon input (Mg C ha^{-1} yr^{-1})					
	grain yield	straw biomass	root biomass	stubble biomass	Rhizodeposition	Straw-C	Roots-C	Stubble-C	Rhizodeposition-C	FYM-C	Total C
CK	2.11	2.92	0.71	0.29	0.73	0.00	0.29	0.13	0.27	0.00	0.69
N	2.76	3.71	1.09	0.37	0.84	0.00	0.44	0.16	0.35	0.00	0.95
NP	4.59	5.58	1.46	0.56	1.22	0.00	0.59	0.24	0.50	0.00	1.34
FYM	4.33	5.15	1.48	0.52	1.19	0.00	0.60	0.22	0.47	0.43	1.72
NP+S	4.88	5.61	1.68	0.56	1.27	2.45	0.68	0.24	0.52	0.00	3.89
NP+FYM	5.49	6.63	1.96	0.66	1.45	0.00	0.80	0.29	0.60	0.43	2.11

Carbon Inputs

To compute the fraction of added C stabilized cumulative C input was estimated from exogenous supply of C to the soil through straw and FYM and plant input of C through root biomass, stubble and rhizodeposition (Table 1).

The straw and stubble were collected from three 6 m^2 areas for each plot immediately after the harvest of the grains in 1997 and 2005. The straw, stubble and samples were then oven-dried at 60°C for 72 h and weighed. The straw contributed 56.1, 55.6, 54.0, 54.2, 53.0 and 54.6 per cent of total harvestable above ground biomass in winter wheat, and 59.9, 59.5, 56.1, 54.5, 54,2 and 54.8 per cent of total harvestable above ground biomass in maize for CK, N, NP, FYM, NP+S and NP+FYM, respectively. The stubble on an average constituted 10 per cent of the straw.

Root biomasses in winter wheat and maize were calculated using the root: shoot ratio. After harvest, four soil cores (8 cm diameter by 100 cm depth) per plot (two from rows and the other two from between rows) were collected from the 0 to 100 cm soil depths to measure root biomass. The root biomass represented 20.6, 23.6, 19.4, 20.6, 21.6, and 21.8 per cent of the harvestable above ground biomass in winter wheat, and 7.1, 8.0, 7.0, 8.9, 7.7, and 8.8 per cent of the harvestable above ground biomass in maize, respectively in the treatments listed above.

In 1997 and 2005, portions of air-dried straw, stubble and straw were passed through a 0.25 mm sieve for the determinations of the C concentrations. The root biomass, stubble and straw contained 40.4, 42.9 and 42.9 per cent C for winter wheat and 41.5, 44.6 and 44.6 per cent C for maize, respectively. While calculating total rhizodeposition derived from different crops in this study, we used the values mentioned by Bronson et al. [25]. Root exudates therefore represented 15% of aboveground biomass at maturity with a C concentration of 36% in CK and N treatments and 33% in NP, NP+S, FYM and NP+FYM treatments.

Estimation of Soil Organic Carbon Stock

Total SOC stock of profile for each of the five depths (0–20, 20–40, 40–60, 60–80, 80–100 cm) was computed by multiplying the SOC concentration by the bulk density, depth, and factor by 10 (Equation 1).

$$\text{Profile SOC stock} = \text{SOC concentration} (\text{g kg}^{-1})$$
$$\times \text{bulk density} (\text{Mg m}^{-3}) \quad (1)$$
$$\times \text{depth (m)} \times 10$$

Statistical Analysis

The effects of fertilizer treatments on SOC and labile SOC fractions (MBC, DOC, POC) within each depth were analyzed using one-way ANOVA. Differences were considered significant at P<0.05. Pearsons linear correlation were used to evaluate the relationships between SOC and POC, MBC and DOC. Linear-regression analyses were performed to determine trends using composite soil data from three replicate plots to assess trends of SOC (0–20 cm layer) over the years. For statistical analysis of data, Microsoft Excel (Microsoft Corporation, USA) and SPSS window version 11.0 (SPSS Inc., Chicago, USA) packages were used. Unless otherwise stated, the level of significance referred to in the results is P<0.05.

Results

Soil Organic Carbon (SOC) Trends

The SOC concentrations in the 0–20 cm soil layer for CK, N, NP, FYM, NP+S and NP+FYM treatments at the beginning of the study in 1979 were 5.97, 6.15, 5.92, 6.38, 6.09, 6.03 g kg^{-1}. Above data was the source of the growth rates. Although there were large fluctuations of SOC content with time, the SOC content in CK and N treatments generally constant with time and in NP, slightly increased (Fig. 2). The SOC concentration significantly increased with the lapse of year in the C input treatments (FYM, NP+S and NP+FYM). Across the 30 cropping and fertilization periods, annual SOC concentration rates (slopes of the linear regression vs. time) in Fig. 2 indicated that 0.15, 0.16 and 0.19 g kg^{-1} yr^{-1} were increased each year in NP+S, FYM and NP+FYM treatments, respectively.

Bulk Density

Long-term application of manure and fertilizer significantly affected soil bulk density (BD) to a depth of 40 cm (Fig. 3A). The addition of FYM or straw (FYM, NP+FYM and NP+S) treatments decreased soil bulk density significantly in comparison to that in control plots in all the layers. However, the decrease was more in upper soil layers (0–20 and 20–40 cm) than in the lower layers (40–60, 60–80 and 80–100 cm). Similar was the case with NP treatment, where BD was lower than that in CT treatment at 0–20 and 20–40 cm depths. There were no statistically significant differences in BD among treatments below 40 cm depth.

Depth Distribution of Soil Organic Carbon

The distribution of SOC with depth was dependent on the use of various fertilizers (Fig. 3B). The highest SOC concentration was obtained for 0–20 cm depth and decreased with depth for all

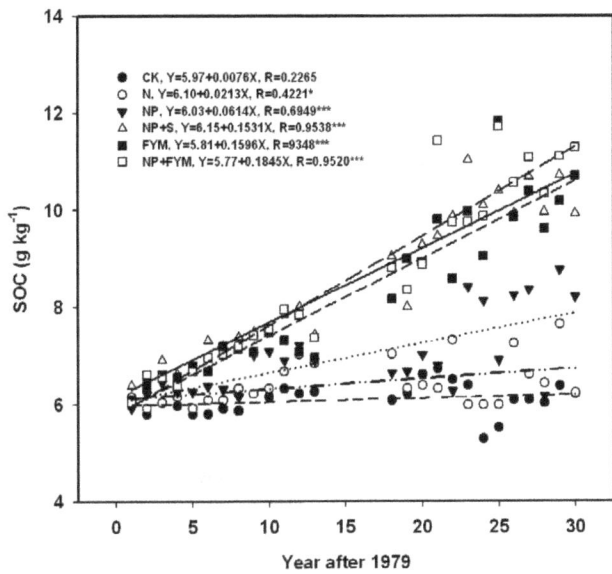

Figure 2. Trend changes of soil organic carbon (SOC) at 0–20 cm top soil layer in a long-term (1979–2008) fertilization experiment in Pingliang, Gansu, China.

treatments. The SOC concentration in 0–20, 20–40 and 40–60 cm depths increased significantly by farmyard manure or straw application. At the 0–20 and 20–40 cm soil depths, SOC was highest in NP+FKM followed by NP+S and FYM treatments and the least in CK treatment. However, the SOC concentration below 60 cm depth was statistically similar among different treatments.

Soil Organic Carbon Storage

The effects of fertilization on SOC storage showed a similar trend to SOC concentration (Fig. 3C). The topsoil (0–20 cm) had the maximum levels of cumulative SOC storage in the 1 m soil depth for the CK, N, NP, FYM, NP+S and NP+FYM treatments, accounting for 24%, 23%, 27%, 30%, 31% and 31%, respectively. At the 20–40 cm and 40–60 cm soil layers, the SOC stocks of the NP, FYM, NP+S and NP+FYM treatments were significantly higher by 17%, 21%, 25% and 37% and 5.3%, 8.1%, 7.3% and

11%, respectively, than that of the CK. The differences of SOC storage between different treatments were not significant in the 60–80 cm and 80–100 cm soil layers. SOC storages were significantly different between fertilization treatments in the 0–100 cm profile. Compared with the CK treatment, SOC storages of the NP+FYM, NP+S, FYM and NP treatments within the 0–100 cm soil depth were increased by nearly 30, 24, 20 and 12%, respectively.

Particulate Organic Carbon

Particulate organic C was found stratified along the soil depth. A higher POC was found in surface soil decreasing with depth (Fig. 4A). At the 0–20 cm, POC content under NP+FYM, NP+S and FYM were 103, 89 and 90% greater than under CK, respectively. In 20–40 cm and 40–60 cm soil layers, NP+FYM had maximum POC which was significantly higher than NP+S and FYM treatments. Even though POC below 60 cm depth was statistically similar among fertilization treatments, the general trend was for increased POC with farmyard manure or straw application down to 100 cm soil depth.

Dissolved Organic Carbon

Irrespective of soil depths, NP+FYM invariably showed higher content of DOC over all other treatments. The CK and N treatments showed lower content of DOC. The DOC concentrations in 0–20 cm, 20–40 cm and 40–60 cm depths were observed highest for NP+FYM followed by NP+S and FYM, and both of them were significant higher than NP (Fig. 4B). However, in the deeper layers (60–80 cm and 80–100 cm), the difference in DOC among the treatments was not significant.

Microbial Biomass Carbon

The MBC differences among treatments not only presented in the surface soil layers, but also presented at deeper depths in the profile. In our study, MBC showed a significant effect at different fertilizer treatments (Fig. 4C). The SOC concentration in 0–80 cm depth increased significantly by farmyard manure or straw application. The mean MBC content in 0–80 cm profile was 82% higher in NP+FYM treatment than in CK treatment.

Comparison of Labile Organic Carbon Pools

POC, MBC and DOC concentrations increased linearly with increasing soil SOC content (Fig. 5), suggesting that total organic

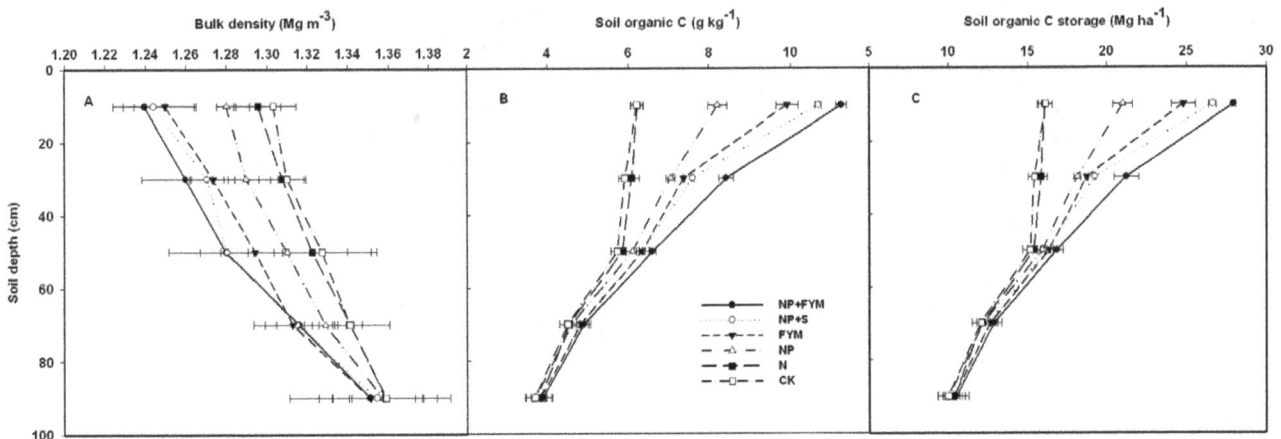

Figure 3. Effect of long-term fertilizer applications on depth distribution of bulk density (A), soil organic C (B) and soil organic C storage (C).

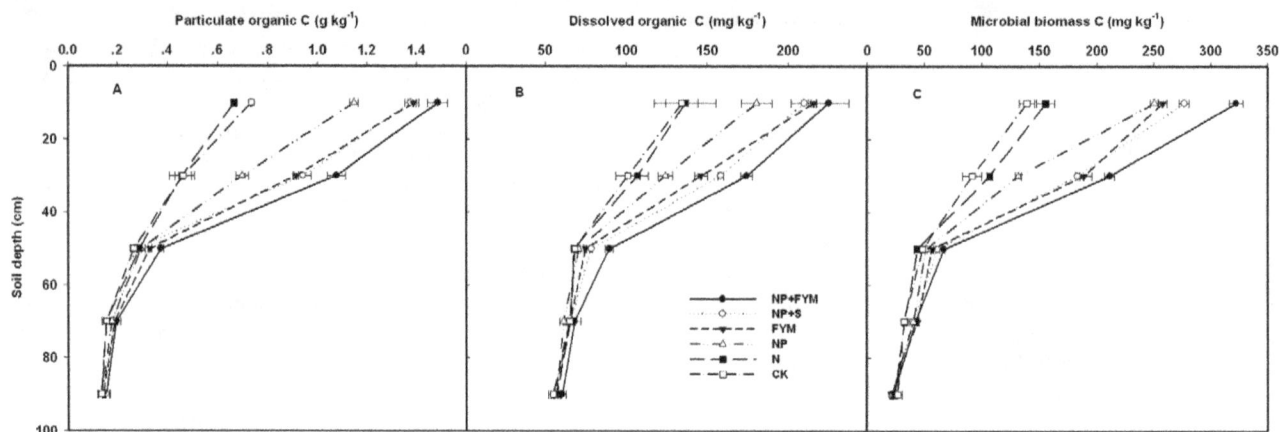

Figure 4. Effect of long-term fertilizer applications on depth distribution of particulate organic C (A), dissolved organic C (B) and microbial biomass C (C).

matter content was a major determinant of the amount of POC, MBC and DOC present. Of the reported C pools, POC was most highly correlated with SOC, followed by MBC then DOC.

Discussion

Soil Total C

From 1979 to 2008, SOC concentrations (0–20 cm layer) increased significantly for all treatments except the CK and N treatments, and the greatest increases occurred for the three plots of FYM, NP+FYM and NP+FYM treatments that received organic materials (Fig. 2). Apparently, application of only N fertilizer did not increase SOC content over long-term cropping. This observation was consistent with that of Goyal et al. [26], who reported that no significant increase in SOC by the addition of only N fertilizer. In the present study, SOC in the CK and N soil were at par, presumably because of lower crop productivity that results in significantly lower accumulation of root biomass. The NP treatment increased SOC concentrations but at a much lower rate of 0.061 g kg^{-1} yr^{-1}. This finding indicated that long-term chemical NP fertilizer alone can increase soil C sequestration, which has been confirmed by other long-term fertilizer experiments in China [27]. A before linear relationship was found between SOC and organic material (i.e., manure and straw) input after 30 y of continuous cropping in northwest China. Most of the soil organic matter (SOM) models assume a linear increase in SOC levels with increasing C input [28].

The SOC concentration differences among treatments not only presented in the surface soil layers, but also presented at deeper depths in the profile. SOC concentration was highest in NP+FYM plots and the least in unfertilized control (CK) at all the sampling depths. Though the average SOC concentration decreased with soil depth, the NP+FYM, NP+S and FYM treatments resulted in a significant increase in organic C even in 40–60 cm soil layer. Manjaiah and Singh [29] and also found that inorganic fertilizers plus organic material increased the SOC content of the soil. The reasons for the higher SOC in manure soils at deeper depths include the following. First, the crop rooting depth between organic manure and inorganic fertilizer soils differ. The organic manure soils can be favorable for the growth of roots into deeper layers due to the relatively loose soil and high soil water content. Second, SOC in organic manure soils can also move to lower depths through earthworm burrows and leaching [30]. Applying straw with N and P fertilizer (NP+S) had the highest total C input

(Table 1), yet decreased SOC concentration over the FYM with NP fertilizer treatment. As NP+FYM treatment significantly increased SOC concentration, this suggests that animal manure is more effective in building soil C than straw, possibly due to the presence of more humified and recalcitrant C forms in animal manure as compared to the straw. For the inorganic fertilizer treatment, the optimum application of inorganic fertilizer NP treatment showed a higher SOC concentration over the application of inorganic fertilizer N treatment at all the sampling depths. The optimum fertilization results in better plant growth including the root biomass (Table 1), which could have added to the SOC particularly as indicated in the lower layers [31].

Similar to the concentration, SOC stock of the profile was also significantly ($P<0.05$) higher in the organic manure treatments (FYM, NP+FYM and NP+S) compared with the only inorganic fertilizer (N, NP) and control treatments. Gami et al. [32] also reported a significant increase in SOC stocks to 60 cm depth under three 23–25-year-old long-term fertility experiments in the Nepal, with application of manure and inorganic fertilizer. Within 1 m soil depth, the cumulative distribution of SOC in the CK, N, NP, FYM, NP+S and NP+FYM treatments were by 50%, 46%, 51%, 53%, 54% and 55% in the 0–40 cm layer, and 68%, 68%, 71%, 72%, 73% and 74% in the 0–60 cm layer, respectively. The SOC storage in the 60–100 cm layer was statistically similar among different treatments. On average the estimate of soil C accumulation to 60 cm depth were 267% and 41% greater than that for soil C accumulated to 20 cm depth and to 40 cm depth, respectively. These findings suggest that the estimate of soil C accumulation to 60 cm depth was more effective than that for soil C accumulated to 40 cm. In this study, C input was increased under the N treatment compared to the CK. However, neither SOC concentration nor C storage was significantly changed under the N treatment. The reason for this is that the N treatment may stimulate soil microbial activity, therefore increasing the C output. The increase in C mineralization might offset the increase in C input. Similar results were also found by Halvorson et al. [33], Su et al. [34], and Lou et al. [35].

Soil C Fractions

The POC fraction has been defined as a labile SOC pool mainly consisting of plant residues partially decomposed and not associated with soil minerals [23,36]. In the present study, the soil amended by FYM or straw contained significantly higher POC in

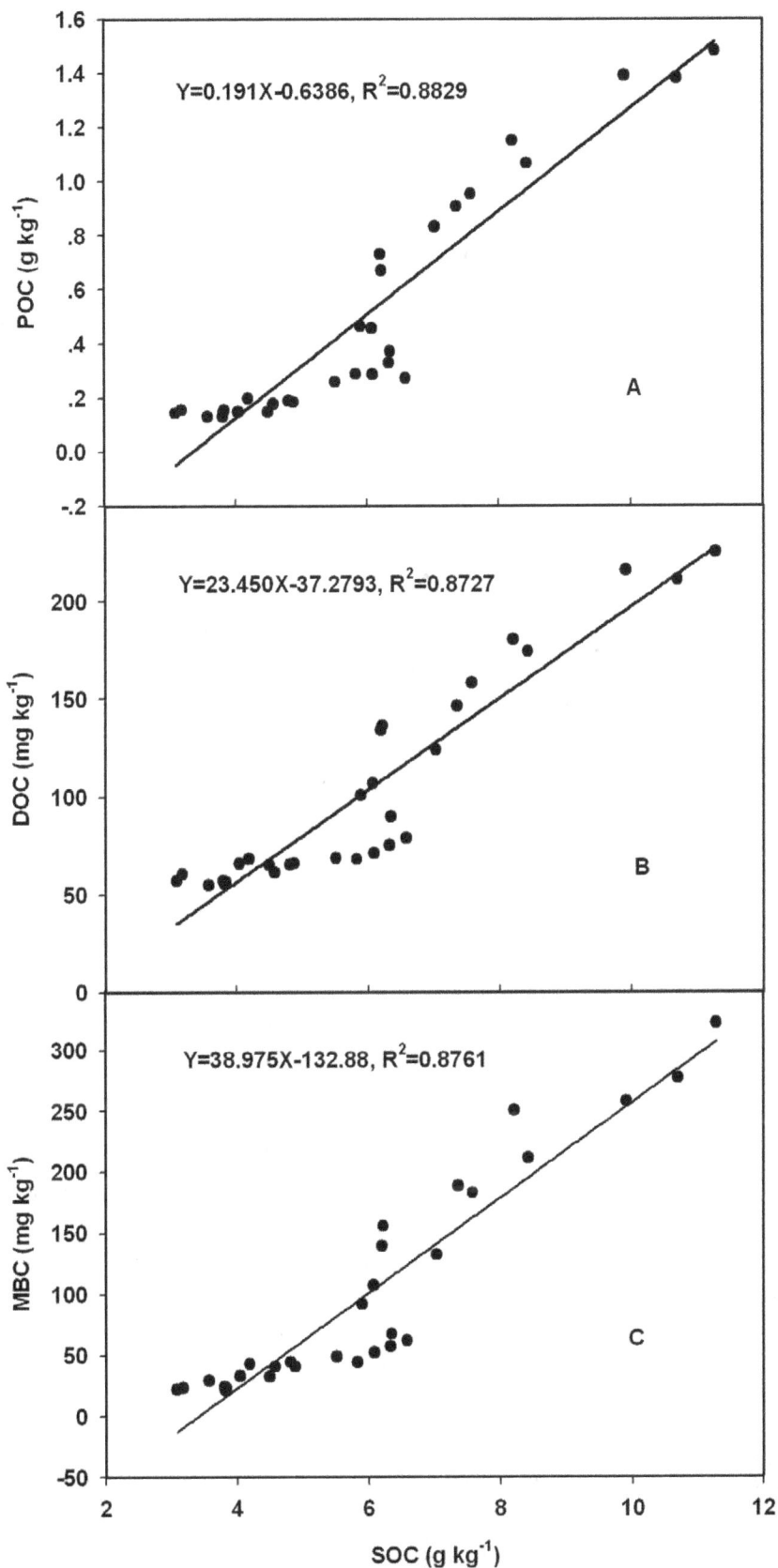

Figure 5. Relationship between soil organic carbon and different labile SOC pools: (A) dissolve organic C (DOC), (B) particulate organic C (POC), and (C) microbial biomass C (MBC).

the 0–60 cm than that in the inorganic fertilizer treatments. Rudrappa et al. [31] reported that the additional organic carbon input could enhance the POC accumulation. Purakayastha et al. [5] concluded that FYM can increase the root biomass and microbial biomass debris which is the main source of POC. It is suggested that the greater biochemical recalcitrance of root litter [37] might have also increased the POC contents in soil depending upon the root biomass produced. The continuous replacement of organic manure on the soil creates a favourable environment for the cycling of C and formation of macroaggregates. Furthermore, POC acts as a cementing agent to stabilise macroaggregates and protect intra-aggregate C in the form of POC [36,38]. Below 60 cm soil layer, the POC declined with increase in soil depth. Chan [39] also found that straw application increased POC in surface soil but not at lower depths.

DOC is believed to be derived from plant roots, litter and soil humus and is a labile substrate for microbial activity [40,41]. The concentration of DOC varied widely among all the treatments and a significant increase was observed in surface soils under different fertilizer treatments compared with CK. In the long-term, the quantity of organic residues are the main factors influencing the amount and composition of DOC. Likewise, in our study, the upper 60 cm soil layer had more DOC concentration than that of lower layer. Below 60 cm, the DOC concentration sharply decreased with soil depth. DOC in subsurface soils may be a result of decomposition of crop residues or translocation from surface soil [42]. Several field studies have sown that concentration and fluxes of DOM in soil solution decrease significantly with soil depth [41].

In our study, MBC was highest in the farmyard manure plus inorganic fertilizer treatment in top soil, an increased MBC content after farmyard application was also reported by Chakraborty et al. [43] and Marschner et al. [44]. This indicated the activation of microorganisms through carbon source inputs consisting of organic residues. Increases in soil organic matter are usually associated with similar increases in microbial biomass because the SOM provides principal substrates for the microorganisms [45]. Among the investigated fertilizer treatments, straw plus inorganic fertilizer had impact on the microbial biomass. This effect is mainly due to the input of straw manure as an organic carbon source. Lynch and Panting [46] reported that eight months after application of straw manure to a loamy arable soil the microbial biomass was almost twice as high as compared to a control. Also, Ocio et al. [47] have demonstrated rapid and significant increases in microbial biomass following straw inputs in field conditions. The MBC was not only correlated with SOC concentration near the surface but also at deeper depths. Though the average MBC decreased with soil depth, the NP+FYM, NP+S and FYM treatments resulted in significant increase in microbial biomass C even in 60–80 cm soil layer. The main source of MBC in deep soil was mainly the left over root biomass and increased microbial biomass debris. It is suggested that the greater biochemical recalcitrance of root litter [37] might have also increased the MBC contents in soil depending upon the root biomass produced. However, the MBC content was lower in the 80–100 cm profile. The reason is that roots may be difficult extend to lower depths. The imbalanced use of fertilizers (CK and N) decreased MBC due to limitation imposed by major nutrients like P and K, which are essential for higher crop production as well as for microbial cell synthesis.

SOC was highly correlated with labile carbon (POC, MBC and DOC). POC, MBC and DOC were significantly and positively correlated with SOC. The correlation coefficient was highest between POC and SOC ($R^2 = 0.883$), followed between MBC and SOC ($R^2 = 0.876$) and DOC and SOC ($R^2 = 0.873$). Such high correlations have also been reported by Rudrappa et al. [31] and

Liang et al. [12], and it is not surprising that the three measures of labile organic matter were closely correlated since they are closely interrelated properties. This result confirms the value of these fractions as sensitive indicators for detecting changes in SOM in the short term, before they are readily measurable in total C. Likewise, these correlations also indicated that SOC was a major determinant of the labile C fractions present.

Conclusions

Fertilizer application has played an important role in improving the total SOC and labile C pools content in the soil after 30 years. Because there was low SOC content in the Northwest of China, the long-term application of organic manure and inorganic fertilizer increased the content of SOC. SOC concentrations and storage were highest in surface soil and depth interval down to 60 cm under NP+FYM and NP+S, below which concentrations did not change with depth. At the same time, on average the estimate of soil C storage to 60 cm depth was higher than that for soil C accumulated to 20 cm depth and to 40 cm depth, respectively. These findings suggest that the estimate of soil C accumulation to 60 cm depth was more effective than that for soil C accumulated to 20 cm depth and to 40 cm depth. NP+FYM was the most efficient management system for sequestering SOC. A large amount of C was also sequestered in soil under NP+S treatment. Soil microbial biomass C, POC and DOC were all significantly greater under organic manure (farmyard manure or straw) plus inorganic fertilizers, especially in the surface. The labile fraction organic C contents decreased significantly with increasing soil depth. These labile pools were highly correlated with each other and SOC, indicating that they were sensitive to changes in SOC.

In Northwest China, the effects of manure and fertilizer application practices on soil C sequestration were studied so that dryland farming soil could contribute to both sustainable food production and mitigation of greenhouse gas emissions through soil C sequestration. Our results have very significant implications for soil C sequestration potential in semiarid agro-ecosystems of northwest China. SOC concentration in surface soil (0–20 cm) and SOC storage of the profile (0–100 cm) were not significantly or slightly increased by the 30 yr of fertilizer treatments (N and NP), but they were sharply increased by the manure and straw amendment (FYM, NP+S and NP+FYM). Thus, returning crop residue to the soil or adding farmyard manure on the soil surface is crucial to improving the SOC level. The large scale implementation of the straw or manure plus inorganic fertilizer amendments will help to enhance the capacity of carbon sequestration and promote food security in the region. Therefore, local government should encourage farmers to manage the nutrients and soil fertility based on integrated nutrient management by combining organic matter with inorganic fertilizer to improve soil carbon pools and increase crop productivity for long-term.

Acknowledgments

We wish to thank Prof. Bing So for his revision of the manuscript. We thank the staff of Pingliang Prefecture Institute of Agricultural Sciences. We are grateful for the support and technical assistance from colleagues at Institute of Environment and Sustainable Development in Agriculture, CAAS.

Author Contributions

Conceived and designed the experiments: EKL CRY YQZ. Performed the experiments: EKL TLF. Analyzed the data: EKL CRY XRM. Contributed reagents/materials/analysis tools: EKL TLF YQZ. Wrote the paper: EKL CRY XRM.

References

1. Ouédraogo E, Mando A, Brussaard L, Stroosnijder L (2007) Tillage and fertility management effects on soil organic matter and sorghum yield in semi-arid West Africa. Soil Till Res 94: 64–74.
2. Doran JW, Sarrantonio M, Liebig MA (1996) Soil health and sustainability. Adv Agron 56: 1–54.
3. Verma S, Sharma PK (2007) Effect of long-term manuring and fertilizers on carbon pools, soil structure, and sustainability under different cropping systems in wet-temperate zone of northwest Himalayas. Biol Fertility Soils 44: 235–240.
4. Blair N, Faulkner RD, Till AR, Korschens M, Schulz E (2006) Long-term management impacts on soil C, N and physical fertility: Part II: Bad Lauchstadt static and extreme FYM experiments. Soil Till Res 91: 39–47.
5. Purakayastha TJ, Rudrappa L, Singh D, Swarup A, Bhadraray S (2008) Long-term impact of fertilizers on soil organic carbon pools and sequestration rates in maize–wheat–cowpea cropping system. Geoderma 144: 370–378.
6. Gong W, Yan X, Wang J, Hu T, Gong Y (2009) Long-term manure and fertilizer effects on soil organic matter fractions and microbes under a wheat–maize cropping system in northern China. Geoderma 149: 318–324.
7. Powlson DS, Bhogal A, Chambers BJ, Coleman K, Macdonald AJ, et al. (2012) The potential to increase soil carbon stocks through reduced tillage or organic material additions in England and Wales: A case study. Agric Ecosyst Environ 146: 23–33.
8. Haynes RJ (2005) Labile organic matter fractions as central components of the quality of agricultural soils: An overview. Adv Agron 85: 221–268.
9. Saviozzi A, Levi-Minzi R, Cardelli R, Riffaldi R (2001) A comparison of soil quality in adjacent cultivated, forest and native grassland soils. Plant Soil 233: 251–259.
10. Xu M, Lou Y, Sun X, Wang W, Baniyamuddin M, et al. (2011) Soil organic carbon active fractions as early indicators for total carbon change under straw incorporation. Biol Fertility Soils 47: 745–752.
11. Gregorich EG, Monreal CM, Carter MR, Angers DA, Ellert BH (1994) Towards a minimum data set to assess soil organic matter quality in agricultural soils. Can J Soil Sci 74: 367–385.
12. Liang Q, Chen H, Gong Y, Fan M, Yang H, et al. (2011) Effects of 15 years of manure and inorganic fertilizers on soil organic carbon fractions in a wheat–maize system in the North China Plain. Nutr Cycl Agroecosyst 92: 1–13.
13. Nayak AK, Gangwar B, Shukla AK, Mazumdar SP, Kumar A, et al. (2012) Long-term effect of different integrated nutrient management on soil organic carbon and its fractions and sustainability of rice–wheat system in Indo Gangetic Plains of India. Field Crop Res 127: 129–139.
14. Dou F, Wright AL, Hons FM (2008) Sensitivity of labile soil organic carbon to tillage in wheat-based cropping systems. Soil Sci Soc Am J 72: 1445–1453.
15. West TO, Post WM (2002) Soil organic carbon sequestration rates by tillage and crop rotation. Soil Sci Soc Am J 66: 1930–1946.
16. Majumder B, Mandal B, Bandyopadhyay PK, Gangopadhyay A, Mani PK, et al. (2008) Organic amendments influence soil organic carbon pools and rice–wheat productivity. Soil Sci Soc Am J 72: 775–785.
17. Janzen HH, Campbell CA, Ellert BH, Bremer E (1997) Soil organic matter dynamics and their relationship to soil quality. In: Gregorich, EG, Carter MR, editors. Soil Quality for Crop Production and Ecosystem Health. Amsterdam: Developments in Soil Science 25, Elsevier, 277–292.
18. Xing NQ, Zhang YQ, Wang LX (2001) The study on dryland agriculture in North China (in Chinese). Beijing: Chinese Agriculture Press.
19. Zhang X, Quine TA, Walling DE (1998) Soil erosion rates on sloping cultivated land on the Loess Plateau near Ansai, Shaanxi Province, China: an investigation using 137Cs and rill measurements. Hydrol Processes 12: 171–189.
20. FAO/UNESCO (1988) Soil map of the world. Paris: UNESCO.
21. Black CA (1965) Methods of soil analysis, Part I and II. Madison, Wis: American Society of Agronomy.
22. Jiang PK, Xu QF, Xu ZH, Cao ZH (2006) Seasonal changes in soil labile organic carbon pools within a Phyllostachys praecox stand under high rate fertilization and winter mulch in subtropical China. Forest Ecol Manag 236: 30–36.
23. Cambardella CA, Elliott ET (1992) Particulate soil organic-matter changes across a grassland cultivation sequence. Soil Sci Soc Am J 56: 777–783.
24. Vance ED, Brookes PC, Jenkinson DS (1987) An extraction method for measuring soil microbial biomass C. Soil Biol Biochem 19: 703–707.
25. Bronson KF, Cassman KG, Wassmann R, Olk DC, Noordwijk MV, et al. (1998) Management of carbon sequestration in soil. In: Lal R, Kimble J, Follet RF, Stewart BA, editors, Soil Carbon Dynamics in Different Cropping Systems in Principal Ecoregions of Asia. Boca Raton: CRC/Press. 35–57.
26. Goyal S, Mishra MM, Hooda IS, Singh R (1992) Organic matter-microbial biomass relationships in field experiments under tropical conditions: Effects of inorganic fertilization and organic amendments. Soil Biol Biochem 24: 1081–1084.
27. Wu T, Schoenau JJ, Li F, Qian P, Malhi SS, et al. (2005) Influence of fertilization and organic amendments on organic-carbon fractions in Heilu soil on the loess plateau of China. J Plant Nutr Soil Sci 168: 100–107.
28. Zhang W, Xu M, Wang B, Wang X (2009) Soil organic carbon, total nitrogen and grain yields under long-term fertilizations in the upland red soil of southern China. Nutr Cycl Agroecosyst 84: 59–69.
29. Kanchikerimath M, Singh D (2001) Soil organic matter and biological properties after 26 years of maize–wheat–cowpea cropping as affected by manure and fertilization in a Cambisol in semiarid region of India. Agric Ecosyst Environ 86: 155–162.
30. Lorenz K, Lal R (2005) The depth distribution of soil organic carbon in relation to land use and management and the potential of carbon sequestration in subsoil horizons. Adv Agron 88: 35–66.
31. Rudrappa L, Purakayastha TJ, Singh D, Bhadraray S (2006) Long-term manuring and fertilization effects on soil organic carbon pools in a Typic Haplustept of semi-arid sub-tropical India. Soil Till Res 88: 180–192.
32. Gami SK, Lauren JG, Duxbury JM (2009) Soil organic carbon and nitrogen stocks in Nepal long-term soil fertility experiments. Soil Till Res 106: 95–103.
33. Halvorson AD, Wienhold BJ, Black AL (2002) Tillage, nitrogen, and cropping system effects on soil carbon sequestration. Soil Sci Soc Am J 66: 906–912.
34. Su YZ, Wang F, Suo DR, Zhang ZH, Du MW (2006) Long-term effect of fertilizer and manure application on soil-carbon sequestration and soil fertility under the wheat–wheat–maize cropping system in northwest China. Nutr Cycl Agroecosyst 75: 285–295.
35. Lou Y, Wang J, Liang W (2011) Impacts of 22-year organic and inorganic N managements on soil organic C fractions in a maize field, northeast China. Catena 87: 386–390.
36. Six J, Conant RT, Paul EA, Paustian K (2002) Stabilization mechanisms of soil organic matter: Implications for C-saturation of soils. Plant Soil 241: 155–176.
37. Puget P, Drinkwater LE (2001) Short-term dynamics of root-and shoot-derived carbon from a leguminous green manure. Soil Sci Soc Am J 65: 771–779.
38. Sá JCM, Lal R (2009) Stratification ratio of soil organic matter pools as an indicator of carbon sequestration in a tillage chronosequence on a Brazilian Oxisol. Soil Till Res 103: 46–56.
39. Chan KY (1997) Consequences of changes in particulate organic carbon in vertisols under pasture and cropping. Soil Sci Soc Am J 61: 1376–1382.
40. Liang BC, Mackenzie AF, Schnitzer M, Monreal CM, Voroney PR, et al. (1997) Management-induced change in labile soil organic matter under continuous corn in eastern Canadian soils. Biol Fertility Soils 26: 88–94.
41. Kalbitz K, Solinger S, Park J.H, Michalzik B, Matzner E (2000) Controls on the dynamics of dissolved organic matter in soils: a review. Soil Sci 165: 277–304.
42. Dou F, Wright AL, Hons FM (2007) Depth distribution of soil organic C and N after long-term soybean cropping in Texas. Soil Till Res 94: 530–536.
43. Chakraborty A, Chakrabarti K, Chakraborty A, Ghosh S (2011) Effect of long-term fertilizers and manure application on microbial biomass and microbial activity of a tropical agricultural soil. Biology and Fertility of Soils 47: 227–233.
44. Marschner P, Kandeler E, Marschner B (2003) Structure and function of the soil microbial community in a long-term fertilizer experiment. Biol Fertility Soils 35: 453–461.
45. Melero S, López-Garrido R, Murillo J.M, Moreno F (2009) Conservation tillage: Short-and long-term effects on soil carbon fractions and enzymatic activities under Mediterranean conditions. Soil Till Res 104: 292–298.
46. Lynch JM, Panting LM (1980) Variations in the size of the soil biomass. Soil Biol Biochem 12: 547–550.
47. Ocio JA, Martinez J, Brookes PC (1991) Contribution of straw-derived N to total microbial biomass N following incorporation of cereal straw to soil. Soil Biol Biochem 23: 655–659.

13

Effects of Crop Canopies on Rain Splash Detachment

Bo Ma[1,2], Xiaoling Yu[2], Fan Ma[3], Zhanbin Li[1], Faqi Wu[2]*

1 State Key Laboratory of Soil Erosion and Dryland Farming on Loess Plateau, Institute of Soil and Water Conservation, Northwest A&F University, Yangling, Shaanxi Province, China, **2** College of Resources and Environment, Northwest A&F University, Yangling, Shaanxi Province, China, **3** Institute of Desertification Control, Ningxia Academy of Agriculture and Forestry Science, Yinchuan, Ningxia Hui Autonomous Region, China

Abstract

Crops are one of the main factors affecting soil erosion in sloping fields. To determine the characteristics of splash erosion under crop canopies, corn, soybean, millet, and winter wheat were collected, and the relationship among splash erosion, rainfall intensity, and throughfall intensity under different crop canopies was analyzed through artificial rainfall experiments. The results showed that, the mean splash detachment rate on the ground surface was 390.12 $g/m^2 \cdot h$, which was lower by 67.81% than that on bare land. The inhibiting effects of crops on splash erosion increased as the crops grew, and the ability of the four crops to inhibit splash erosion was in the order of winter wheat>corn>soybeans>millet. An increase in rainfall intensity could significantly enhance the occurrence of splash erosion, but the ability of crops to inhibit splash erosion was 13% greater in cases of higher rainfall intensity. The throughfall intensity under crop canopies was positively related to the splash detachment rate, and this relationship was more significant when the rainfall intensity was 40 mm/h. Splash erosion tended to occur intensively in the central row of croplands as the crop grew, and the non-uniformity of splash erosion was substantial, with splash erosion occurring mainly between the rows and in the region directly under the leaf margin. This study has provided a theoretical basis for describing the erosion mechanisms of cropland and for assisting soil erosion prediction as well as irrigation and fertilizer management in cultivated fields.

Editor: Vanesa Magar, Centro de Investigacion Cientifica y Educacion Superior de Ensenada, Mexico

Funding: This research for this thesis has been supported by the National Basic Research Program of China (973 Program, 2007CB407201-5) and the National Natural Science Foundation of China under Grant No. 41330858. The funders had no role in study design, data collection and analysis, decision to publish, or preparation of the manuscript.

Competing Interests: The authors have declared that no competing interests exist.

* Email: wufaqi@263.net

Introduction

Soil-particle splashing caused by raindrop impacts on the ground during rainfall is usually evenly distributed if farmlands are not covered with crops. However, crop growth and coverage disturb this uniformity. The course of rainfall through crop canopies can be divided into three parts: throughfall, stemflow, and canopy interception [1]. Among these, throughfall has the strongest influence on soil splash erosion. Studies have confirmed that, due to the wide row spacing of cultivated crops, coverage in the center of the between-row area is usually very low during a large part of the growing season, and that therefore throughfall in this position was significantly greater than in regions closer to the plants. The uniformity of throughfall distribution under a densely planted crop canopy was higher than under intertilled crops [2]. Therefore, emphasis should be placed on intertilled crops when studying the effects of crops on splash erosion.

Rainfall splash erosion is the initial stage of water erosion and occupies a prominent position in the formation and evolution of erosion [3,4,5,6]. The power of splash erosion is related to the size, shape, terminal velocity, and kinetic energy of raindrops. In addition, it is closely related to slope gradient, slope aspect, soil properties, and vegetation cover [7]. On cultivated land, if other conditions were relatively uniform, the biological characteristics of crops would become the main factor affecting splash erosion. Armstrong and Mitchell [8] indicated that in some positions under the crop canopy, throughfall intensity increased considerably compared with rainfall intensity higher in the canopy. The median

diameter (D_{50}) of rainfall under corn and soybean canopies was larger than that of natural rainfall, and many large-diameter raindrops (≥ 50 mm) dropped from a height of more than one meter, creating substantial erosion. Although crop canopies can reduce rainfall energy, if vegetation cover at a distance of 0.3 meter from the surface is not yet fully developed or completely canopied, soil detachment caused by large raindrops will still occur [9,10]. Generally, splash erosion amounts decreased with increasing coverage, and the closer the cover is to the ground, the lower are the splash erosion amounts [11]. Morgan [12,13] found that splash erosion yields under wheat canopies over one hundred days decreased as the kinetic energy of rainfall increased and that soil protection by crops under high-intensity rainfall was stronger than under low-intensity rainfall. It was also noted that the effects of crops on the number of throughfalling raindrops, the size distribution of raindrops, and raindrop energy characteristics were the major factors affecting splash erosion under canopies. This viewpoint was confirmed by Finney (1984), who observed and analyzed the mean diameter of raindrops, rainfall kinetic energy, and splash erosion yields under the canopies of several vegetable crops [14].

The Loess Plateau in China is one of the areas in the world which is seriously plagued by soil erosion. Sloping land in this region covers 875.97 ha, accounting for 55.69% of total land area. Erosion is the main source of soil and water loss, and erosion yield is approximately 50%–60% of total erosion yield in this region [15,16]. Therefore, studies of the mechanisms of soil and water loss in sloping fields as well as possible protective measures have become very important. However, in the current research situation, studies of the mechanism of crop protection from erosion are still relatively rare. This study has focused mainly on analyzing the variation and spatial distribution of splash erosion rates under corn, soybean, millet, and winter wheat canopies at different growth stages to reveal the effects of soil splash erosion and to provide a basis for soil erosion prediction based on previous studies.

Materials and Methods

Study Site

This study was carried out from 2007 to 2009 at the Soil and Water Conservation Engineering Laboratory, Northwest A&F University (Shaanxi province, P.R. China), situated in the southern fringe of the Chinese Loess Plateau. The exact geographical position is 113.08° East longitude and 34.58° North latitude, with an elevation 468 m above the mean sea level. Soil in the study area, according to the Chinese Soil Taxonomy, is Eum-Orthic Anthrosols, which is a kind of Cinnamon soil [17]. The climate of the study area is semi-humid monsoon. Most precipitation (nearly 60% of total rainfall) typically occurs between July and October, and the annual rainfall ranges from 635 to 646 mm. The mean monthly maximum temperature is 26.1°C in July, and the mean minimum temperature is -1.2°C in January. The main crops in the study area are corn (*Zea mays L.*), soybeans (*Glycine max merr.*), cotton (*Gossypium hirsutum Linn.*), winter wheat (*Triticum aestivum Linn.*), and millet (*Setaria italica Beauv.*).

General Information

Since 2006, corn, soybeans, millet, and winter wheat have been planted according to their sowing seasons. The corn used in this study was Zhengdan-958, and seeding started on June 20, 2009. According to local conditions, the line and row spacings of corn land are 60 cm and 25 cm. The soybean used in this research was Zhonghuang-13, and seeding began on June 30, 2007, with a planting density of 20 cm×40 cm. The millet used in this study was Jingu-29, and seeding began in 2008 with a planting density of 10 cm (plant spacing) and 20 cm (row spacing). The wheat used in this research was Xiaoyan-22 and was sowed using drill seeding with a seeding quantity of 130 kilograms per hectare. Planting management was conducted according to local customs. The soil used in the study was Eum-Orthic Anthrosols, and rainfall intensities were 40 mm/h and 80 mm/h, with 30 minutes of rain at one time according to the characteristics of local storms which are concentrated in summer and autumn. The crop growth, vegetative growth stages and average leaf area for each sample time are shown in table 1.

The rainfall simulator in this study was designed and constructed by the Institute of Soil and Water Conservation, Yangling, China. For indoor rainfall simulation, the downward-facing sprinkling rainfall simulation system was similar to that used by Jin *et al.* [18]. Four nozzles were positioned at a drop fall height of 4 m. The rainfall simulator consisted of two 3 m-long sprinkler booms, positioned at a distance of 30 cm from each other. On each sprinkler boom, two nozzles were fixed at a distance of 1.5 m from each other. A range of 20–140 mm/h rainfall intensity can be achieved by changing the hydrostatic pressure by moving the valve system horizontally. The mean drop size of the rainfall simulator was 1.8 mm, and the kinetic energy of the rainfall simulator was approximately 75% that of natural rainfall [19]. The effective rainfall area of the simulator was 3 m×3 m, and rainfall uniformity was >80%. The side-sprinkling rainfall simulation system was used for outdoor rainfall simulation.

Rainfall devices included the rainfall system and the water supply system. The rainfall system consisted of two single rainfall vertical brackets. A rainfall vertical bracket includes the side sprinkler nozzle, nozzle stents, and pressure-control section. The side-sprinkler nozzle was made up of a nozzle body, steam breaker, and outflow orifice. The nozzle was installed on the rainfall vertical bracket and fixed by a tripod. Each nozzle was 6 m above the ground, and the raindrop spray height was 1.5 m as it sprayed out of the outflow orifice. Therefore, the height from which raindrops reached the ground was 7.5 m, and the effective rainfall area was 5×7 m^2. The simulated rainfall pattern was created by opposing sprays from these two single rainfall vertical brackets, forming a superimposed rainfall area. The kinetic energy of the side-sprinkling rainfall simulator was similar to that of natural rainfall, and rainfall uniformity was >80%. Supply pressure was controlled by a pressure gauge, and rainfall intensity was controlled mainly by adjusting the supply pressure and bore diameter of the outflow orifice. Rainfall intensity could be controlled over a range of 30–140 mm/h.

Determination of throughfall

On each sampling date, experimental crops were cut off at the ground, quickly moved indoors, and fixed under the rainfall simulator. Each crop was arranged according to its row spacing (corn, 25×60 cm; soybeans, 20×40 cm; millet, 10×20 cm) and fixed upright in the steel frame. Then the rain gauges were located under the crop canopies in a matrix pattern (Figure 1). Each rain gauge had a diameter of 5.5 cm and a height of 7 cm. After 30 minutes of simulated rainfall, rainwater in each rain gauge was collected and calculated, and the throughfall amount and throughfall intensity of each point were calculated. Finally, the crops were removed, glasses were situated in their place, and rainfall was continued for another 30 minutes under the same rainfall intensities. The design rainfall intensities were 40 mm/h and 80 mm/h according to the characteristics of local storms which are concentrated in summer and autumn.

Due to tilling, throughfall for winter wheat could not be simulated indoors, and therefore these experiments were carried out in the field. Rain gauges were placed in an "S" pattern under the crop canopies, which were formed by two rows of crops, at a rate of 10 gauges per row (Figure 1d). Meanwhile, 20 rain gauges were also located in a bare field close to the row of wheat plants to determine rainfall amount. Each rain gauge was placed using the method of inner and outer sleeves. The outer cylinders were embedded in soil (diameter of 6.5 cm, height of 7.5 cm), and the top cylinder was embedded flush with the ground. When measuring, rain gauges were placed in the outer cylinders as the inner sleeve collected rain water (Figure 2).

Determination of splash detachment rate

The splash detachment rate was determined using the method of splash cups [14]. Splash cups were used to test splash erosion under the canopy at the same place after every throughfall test. Each splash cup was 5 cm high with a diameter of 7 cm and permeable holes evenly located in the bottom of the cup. Each soil sample was sifted using 5-mm sieves and then oven-dried at 105°C until the weight became constant. The splash cup base was paved with filter papers, then filled with sieved soil, and then weighed. Similarly to the throughfall determination, splash cups were placed under the crop canopies. After 30 minutes of rainfall at rainfall intensity of 40 mm/h or 80 mm/h, the splash cups were taken out and oven-dried at 105°C to a constant weight. The difference in the weight of soil in the splash cup before and after artificial rainfall was defined as the splash amount per cup. The splash

Table 1. Crop growth and vegetative stage at each sampling date.

Crops	Observing date	Growth stage	Symbol	Average plant height (cm)	Leaf area(cm² plant⁻¹)	LAI
Corn	2009/7/10	Seedling stage	V4	35	470	0.31
	2009/7/25	Early jointing stage	V6	92	2220	1.48
	2009/8/3	Middle jointing stage	V9	128	4250	2.83
	2009/8/10	Late jointing stage	V12	161	4830	3.22
	2009/8/17	Tasseling stage	VT	215	6470	4.31
Soybean	2007/7/30	Initial blossoming stage	R1	38	1730	2.16
	2007/8/10	Full flowering stage	R2	46	3020	3.77
	2007/8/20	Initial pod-filling stage	R4	76	4170	5.21
	2007/8/28	Pod-bearing stage	R6	79	5210	6.51
Millet	2008/5/25	Fifth leaf stage	GS2	43	170	0.86
	2008/6/6	Flag leaf visible stage	GS4	68	310	1.54
	2008/6/19	50% stigma emergence	GS6	85	440	2.18
	2008/6/29	Milk stage	GS7	112	620	3.11
Winter wheat	2008/3/15	Stem elongation stage	Feekes 6.0	16	50	2.32
	2008/4/1	Jointing stage	Feekes 9.0	38	70	3.61
	2008/4/15	Early heading stage	Feekes 10.1	52	100	4.82
	2008/5/2	Anthesis flowering stage	Feekes 10.52	86	120	6.12

If the one digit was 5 or greater, rounded to the greater ten in column of leaf area.

A. Corn

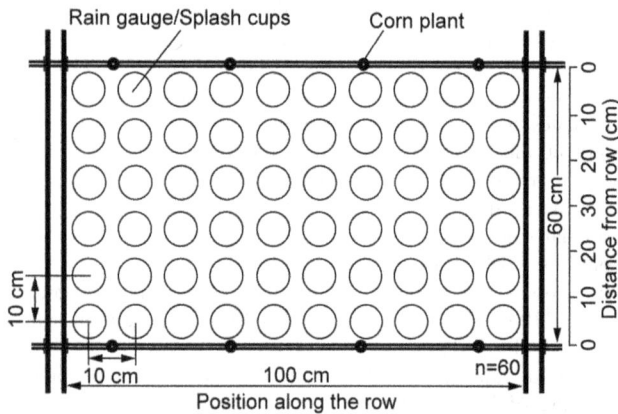

B. Soybean

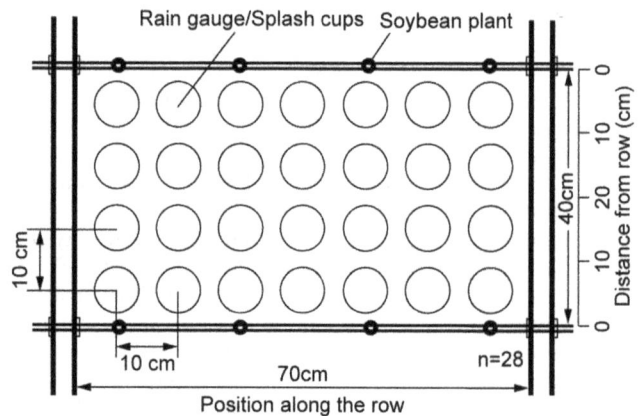

C. Millet

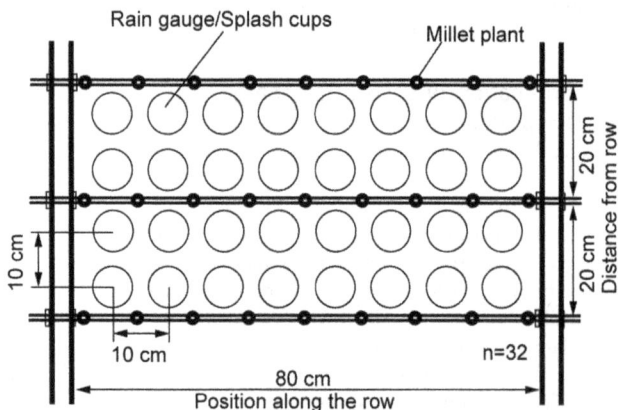

D. Winter wheat

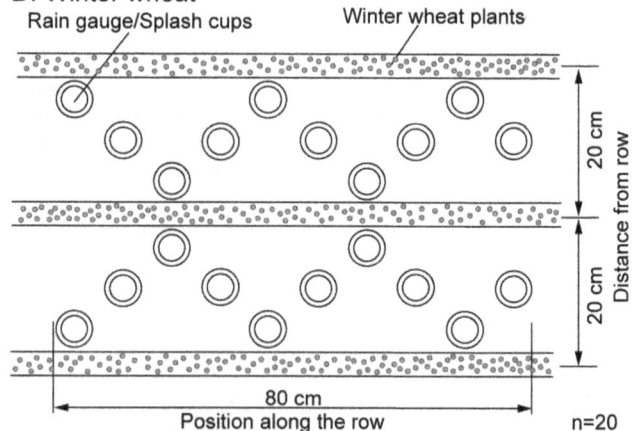

Figure 1. Schematic diagram of measurement for throughfall and splash detachment under crop canopy.

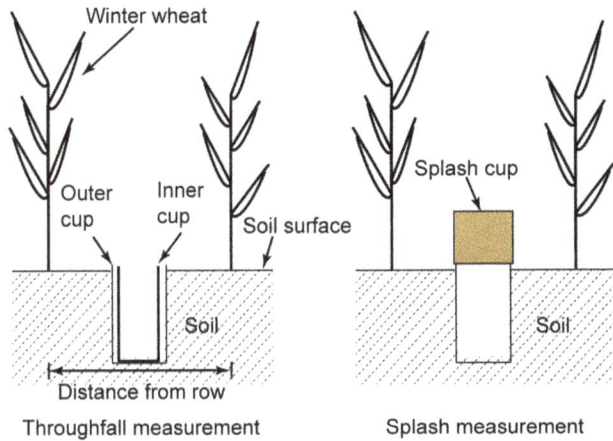

Figure 2. Measurement for throughfall and splash detachment under wheat canopy.

erosion amount per unit area and per unit time (the splash detachment rate, SDR) was calculated according to the diameter of the splash cup and the rainfall duration. The splash detachment rate on bare soil was determined using the same method. The rainfall amount and intensity were also determined using a gauge located as shown in Figure 1. All these steps were followed to calculate the intensities of throughfall, rainfall, and splash under design rainfall rates of 40 mm h^{-1} and 80 mm h^{-1}.

Measurement of Leaf Area Index

Splash erosion yields in the bare field were determined using the same method. Furthermore, at the end of each stage, the leaf areas of corn, millet, and winter wheat were determined by the method of length-to-width ratio, and total leaf area was determined according to the following formula:

$$A_L = \sum_{i=1}^{n} (K \times L_i \times W_i), \qquad (1)$$

where A_L is the total area of each plant (cm^2), K is a modification coefficient (corn 0.75, millet and winter wheat 0.85), L_i is the length of the i-th leaf (cm), W_i is the width at the widest point of the i-th leaf (cm), and n is the number of leaves on a plant.

The leaf area of soybeans was determined by the method of special leaf weight, and total leaf area was determined using the following formula:

$$LSI = T_{Ai}/D_{Wi}, \qquad (2)$$

$$A_L = W \cdot LSI, \qquad (3)$$

where LSI is the special leaf area (cm^2/g), T_{Ai} is a unit of leaf area in some growth period (cm^2), D_{Wi} is the dry weight of a unit of leaf area in some growth period (g), A_L is leaf area (cm^2), and W is the dry weight of leaves of the test crops (g).

The leaf area index (LAI) was calculated as the green leaf area per unit ground area in broadleaf canopies [20].

Results and Analysis

Influence of the crops on splash detachment rate

The characteristics of splash detachment rate and throughfall intensity under the four crop canopies during the whole growth period are given in Table 2. Under experimental conditions, the mean splash detachment rate under the four crop canopies during the whole growth period was 390.12 g/m^2·h, which was lower by 67.81% than on bare land, suggesting that crops can intensively inhibit splash erosion. This ability indicated that the initial phase (raindrop erosion process) was very weak when water erosion occurred on cultivated land with crop cover. The mean splash detachment rate under the corn canopy during its growth period was 380.43 g/m^2·h (68.32% less than bare land), under soybeans, 489.56 g/m^2·h (60.76% less than bare land), under millet, 627.84 g/m^2·h (17.94% less than bare land), and under winter wheat, 62.66 g/m^2·h, the lowest value and 94.75% less than bare land. These values show that splash erosion by confluence on the soil surface differs by crop variety and depends on rainfall intensity, throughfall intensity, and crop growth status.

Table 2 also shows that the splash detachment rate under high rainfall intensity (80 mm/h) was two to five times higher than under low rainfall intensity (40 mm/h). The effect of rainfall intensity on splash erosion was greatest under the soybean canopy, where splash erosion was 4.67 times higher under 80 mm/h rainfall intensity than under 40 mm/h rainfall intensity. The change in splash detachment rate under winter wheat at the two rainfall intensities was the least, 2.03 times. The ability of crops to inhibit splash erosion increased by 13% on average as rainfall intensity increased. The ability of corn, millet, and winter wheat to inhibit splash erosion rose by 19.02%, 30.04%, and 5.34% respectively as rainfall intensity increased, while the ability of soybeans to inhibit splash erosion decreased by 2% on average as rainfall intensity increased.

Figure 3 shows the relationship between throughfall intensity and splash detachment rate under different crop canopies at the testing stage. Clearly, there was a significant relationship between throughfall intensity and splash detachment rate under the crop canopies. Under low rainfall intensity (40 mm/h), except for the R4 stage of soybeans (LAI = 5.21), splash detachment rate increased with rainfall intensity. However, this relationship became complex when rainfall intensity was high (80 mm/h), depending on plant height during different growth periods and the ability of branches and leaves to bear raindrop impact. At low rainfall intensity, because of fewer raindrop impacts and longer time to form large raindrops from rain convergence on leaves, large raindrops contributed less to splash erosion, and the regularity became obvious. However, this phenomenon was reversed at high rainfall intensity, and therefore the regularity was not obvious.

Figure 3 also shows that the leaf area index increased and the ability of crops to inhibit splash erosion also increased with crop growth. The effects of corn on splash erosion changed significantly with crop growth, but the variation was random (Figure 3a). Under 40 mm/h rainfall intensity, splash detachment rates fluctuated more widely under corn canopies in the whole growth season, with the splash detachment rate being low in the V6 and V9 stages (LAI = 1.48–2.83), then increasing in the V12 stage (LAI = 3.22); splash erosion in the VT stage (LAI = 4.31) was very close to that in the V4 stage (LAI = 0.31), although the leaf area index was larger. Under 80 mm/h rainfall intensity, the splash detachment rate also fluctuated irregularly. However, the splash detachment rate in the VT stage decreased by 30% compared to the V4 stage, which indicated that the corn canopy could

Table 2. Splash detachment and throughfall under crop canopy in different growth stage.

Rainfall intensity (mm/h)	Crop type	Observation date	Average rainfall intensity up the canopy (mm/h)	Average throughfall intensity under the canopy (mm/h)	Ratio of throughfall amount to the total rainfall (%)	Splash detachment rate on bare soil (g/m²·h)	Average splash detachment rate under crop canopy (g/m²·h)	Coefficient of dispersion
40	Corn	2009/7/10~8/17	40.37	29.19	72.40	427.97	202.48	1.42
	Soybean	2007/7/30~8/28	40.49	32.02	79.07	458.09	172.82	1.35
	Millet	2008/5/25~6/26	40.36	34.35	85.87	423.80	325.57	1.15
	Winter wheat	2008/3/15~5/2	39.89	31.52	78.80	430.20	41.43	1.26
80	Corn	2009/7/10~8/17	80.24	57.66	71.82	1973.57	558.39	1.69
	Soybean	2007/7/30~8/28	80.58	66.12	82.05	2037.08	806.31	1.27
	Millet	2008/5/25~6/26	80.59	59.94	74.93	1988.20	930.11	0.91
	Winter wheat	2008/3/15~5/2	80.46	66.13	82.66	1957.86	83.90	0.97

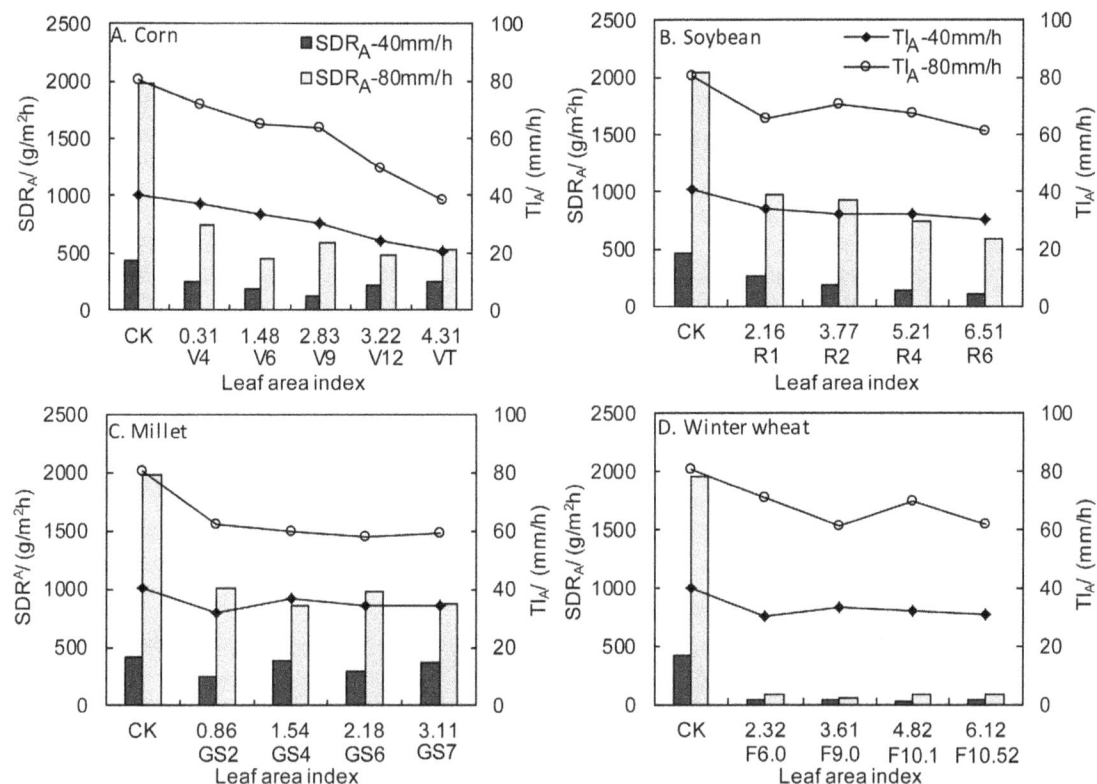

Figure 3. Relationship between throughfall and average splash detachment rate under crop canopy.

significantly reduce the splash detachment rate, but there was no regularity in its inhibition of splash erosion.

As for soybeans (Figure 3b), the mean splash detachment rate under the soybean canopy tended to decrease as the soybeans grew under both rainfall intensities, and the splash detachment rate decreased by 59.68% at 40 mm/h rainfall intensity and 40.18% at 80 mm/h rainfall intensity from the R1 stage (LAI = 2.16) to the R6 stage (LAI = 6.51), which indicated that the inhibition effects on splash erosion increased more regularly as the soybeans grew. Rainfall intensity had a marked effect on splash erosion. Splash erosion in the R1 stage at 80 mm/h rainfall intensity was 3.69 times greater than at 40 mm/h rainfall intensity, but this differences increased to 5.48 times in the R6 stage, when the splash detachment rate at high rainfall intensity was far higher than at low rainfall intensity, and this differences increased further as the soybeans grew. Maybe this phenomenon occurred because rainfall kinetic energy at high rainfall intensity could increase the bare areas between rows and reduce energy dissipation by the canopy. Compared to corn, splash erosion was lower under the soybean canopy at rainfall intensity of 40 mm/h, suggesting that the inhibition ability of the soybean canopy on splash erosion was greater than that of the corn canopy under low rainfall intensity, but less under high rainfall intensity. Under 80 mm/h rainfall intensity, rainfall kinetic energy caused more throughfall between soybean rows and increased the bare area between rows compared to corn. This increased bare-field splash erosion, which caused a higher splash detachment rate under the soybean canopy compared to corn under high rainfall intensity. However, under 40 mm/h rainfall intensity, lower rainfall kinetic energy affected the soybean canopy less, which together with the short and dense form of the soybean canopy and its better surface

coverage, which greatly reduced rainfall kinetic energy, reduced the effects of soybeans on splash erosion compared to corn.

Splash erosion changed under the millet canopy as the plants grew, with significant fluctuations. Under 40 mm/h rainfall intensity, splash erosion fluctuated greatly and increased as the plants grew, but decreased when rainfall intensity was 80 mm/h (Figure 3c), which indicates that the inhibition effect of the millet canopy on splash erosion varied significantly at different rainfall intensities. The inhibition effects on splash erosion were increased by 53.22% compared with bare land under high rainfall intensity, but were reduced by 30% under 40 mm/h rainfall intensity, which indicates that the inhibition effect of millet on splash erosion increased at higher rainfall intensity.

Splash erosion changed under the winter wheat canopy as the plants grew, with significant fluctuations (Figure 3d). Under 40 mm/h rainfall intensity, splash detachment rate under the winter wheat canopy varied from 35.58 $g/m^2{\cdot}h$ to 50.39 $g/m^2{\cdot}h$. Under 80 mm/h rainfall intensity, detachment rate varied from 65.71 $g/m^2{\cdot}h$ to 93.25 $g/m^2{\cdot}h$. This suggests that the inhibition effect of winter wheat on splash erosion varies with rainfall intensity. Under 40 mm/h rainfall intensity, splash erosion yield decreased by 90% compared to bare land, but by 96% under 80 mm/h rainfall intensity compared to bare land, suggesting that the inhibition effect of winter wheat on splash erosion increased with rainfall intensity.

The inhibition effects of corn, millet, and winter wheat on splash erosion were greater under 80 mm/h rainfall intensity than under 40 mm/h rainfall intensity. The inhibition effect on splash erosion was increased by 30% under 80 mm/h rainfall intensity over that at 40 mm/h rainfall intensity, the biggest jump among the four crops, while erosion was least under the winter wheat canopy and increased by only 5%. Inhibition of splash erosion by soybeans differed little between the two rainfall intensities. The inhibition ability of the four crops on splash erosion was in the order of winter wheat > corn > soybeans > millet. Because of its short plants and dense canopy, winter wheat could inhibit splash erosion markedly and thus protect the soil effectively.

A regression analysis of the effects of leaf area index and throughfall intensity on splash detachment rate during the whole growth stage was carried out, with the results shown in Table 3. The relationships among average splash detachment rate, LAI, and average throughfall intensity were highly significant ($P<0.01$). Therefore, splash detachment rates of different crops at different growth stages can be estimated using this regression equation.

Spatial distribution of splash erosion

The coefficient of dispersion is an index which describes the degree of dispersion of a sample. According to statistical data, the degree of dispersion of point splash erosion yield increased with crop growth (Figure 4). Clearly, the spatial distribution of splash erosion was relatively uniform over cultivated land at early crop growth stages. However, the distribution was not uniform in middle and later growth stages.

The CV of splash detachment rate under corn canopies increased from 0.37 at the V4 stage to 2.65 at the VT stage, suggesting that the spatial distribution of splash detachment rate became less uniform as the corn grew. The CV of splash detachment rate under soybean canopies increased from approximately 1.0 at the R1 stage to approximately 1.7 at the R4 stage; this non-uniformity increased with soybean growth, but decreased at the R6 stage. The CV of splash detachment rate under millet canopies ranged from 0.67 to 1.28 during the whole growth period, and this variation was not regular with crop growth, suggesting that the variability of splash detachment rate under millet canopies was low and that the spatial distribution of splash detachment rate differed only slightly under both rainfall intensities during the whole growth period compared to corn and soybeans. The CV of splash detachment rate under winter wheat canopies ranged from 0.83 to 1.47 during the whole growth period, suggesting that the spatial distribution uniformity of splash detachment rate under winter wheat canopies was lower. The CV increased from 1.09 at the Feekes 6.0 stage to 1.47 at the Feekes 10.52 stage under 40 mm/h rainfall intensity, whereas the variation was irregular under 80 mm/h rainfall intensity and the differences were small. This indicated that the non-uniformity of splash detachment rate increased with winter wheat growth under low rainfall intensity, but that non-uniformity was stable and differences were small under high rainfall intensity.

Let us define the 0–20 cm band nearest the corn plants as the region directly under the canopy (the 0–20 cm and 20–0 cm bands in Figure 1a) and the 20–30 cm band between the plants as the region in the central row position (the 20–30–20 cm band in Figure 1a). The area between two rows of soybean plants is divided into a 0–10 cm band nearest the soybean plants (the region directly under the canopy, the 0–10 cm and 10–0 cm bands in Figure 1b) and a 10–20 cm band (the region in the central row position, the 10–20–10 cm band in Figure 1b). If the SDR_{CR}/SDR_{DUC} ratio were reflected in the concentration and spatial distribution of splash erosion, it could be suspected that splash erosion occurs mainly between crop rows and this phenomenon becomes more obvious as crops grow, as shown in Table 4.

Splash detachment rate in the central row position (20–30 cm band) increased by 28.11% compared with the regions directly under the canopy (0–20 cm band) during the whole period of corn growth. The SDR_{CR}/SDR_{DUC} ratio increased from 0.9 to more than 2, suggesting that splash erosion under canopies tends to occur in central row positions as corn grows, with the largest observed values at the V12 stage. The splash detachment rate slightly decreased at the VT stage, and the SDR_{CR}/SDR_{DUC} ratio decreased by approximately 0.4–0.7 compared to the V12 stage. Splash erosion was focused mainly in the 0–20 cm band at the V4 stage of corn. Under 40 mm/h and 80 mm/h rainfall intensities, average splash detachment rates were 1.06 and 1.09 times higher than in the 20–30 cm band, but the difference was small. This indicated that the distribution of splash erosion under the corn canopy was uniform when the corn plants were small. As the corn grew, the corn canopies changed, and therefore the splash concentration moved from the 0–20 cm band to the 20–30 cm band. It reached its maximum at the V12 stage, when average splash detachment rates were 2.05 and 1.67 times higher than in the 0–20 cm band under 40 mm/h and 80 mm/h rainfall intensity respectively. The above discussion shows that splash erosion under corn canopies is highly concentrated in the central row position, with splash erosion yield in the central row position accounting for more than half the total splash erosion yield under canopies. When corn was in the VT stage, the corn leaves grew to their greatest extent, and canopy breadth also reached a maximum. Under these conditions, large water drops formed and fell from the leaf edge and apex, and the concentration moved from the 20–30 cm band to the 0–20 cm band, leading to greater splash erosion yield in some positions (10–20 cm band) near the between-row space. Under 40 mm/h and 80 mm/h rainfall intensities, average splash detachment rates in the 20–30 cm band were 1.34 and 1.25 times higher than in the 0–20 cm band in the VT stage. This indicated that splash erosion yield was concentrated mainly in the 20–30 cm band and in some positions in the 10–20 cm band, while the splash erosion yield in the 0–10 cm band nearest the corn plants gradually decreased with crop growth.

Unlike corn, the splash detachment rate in the 10–20 cm band was 194.49% higher than in the 0–10 cm band during soybean growth. Under 40 mm/h rainfall intensity, the SDR_{CR}/SDR_{DUC} ratio expanded its range from 1.4 to 2.2, and from 2.1 to 6.3 under 80 mm/h rainfall intensity. This suggests that splash erosion yield

Table 3. Regression about the average splash detachment rate under crop canopies.

Crop types	Regression formula	R^2	F value
Corn	$SDR_A = 47.618LAI + 11.659TI_A - 241.592$	0.812	15.131[**]
Soybean	$SDR_A = -49.046LAI + 18.496 TI_A - 201.658$	0.982	133.732[**]
Millet	$SDR_A = 13.274LAI + 23.593 TI_A - 509.926$	0.976	103.160[**]
Winter wheat	$SDR_A = 2.470LAI + 1.227 TI_A - 7.660$	0.884	19.037[**]

where, SDR_A was average splash detachment rate under crop canopy, g/m^2·h; LAI was leaf area index; TI_A was average throughfall intensity under crop canopy, mm/h.

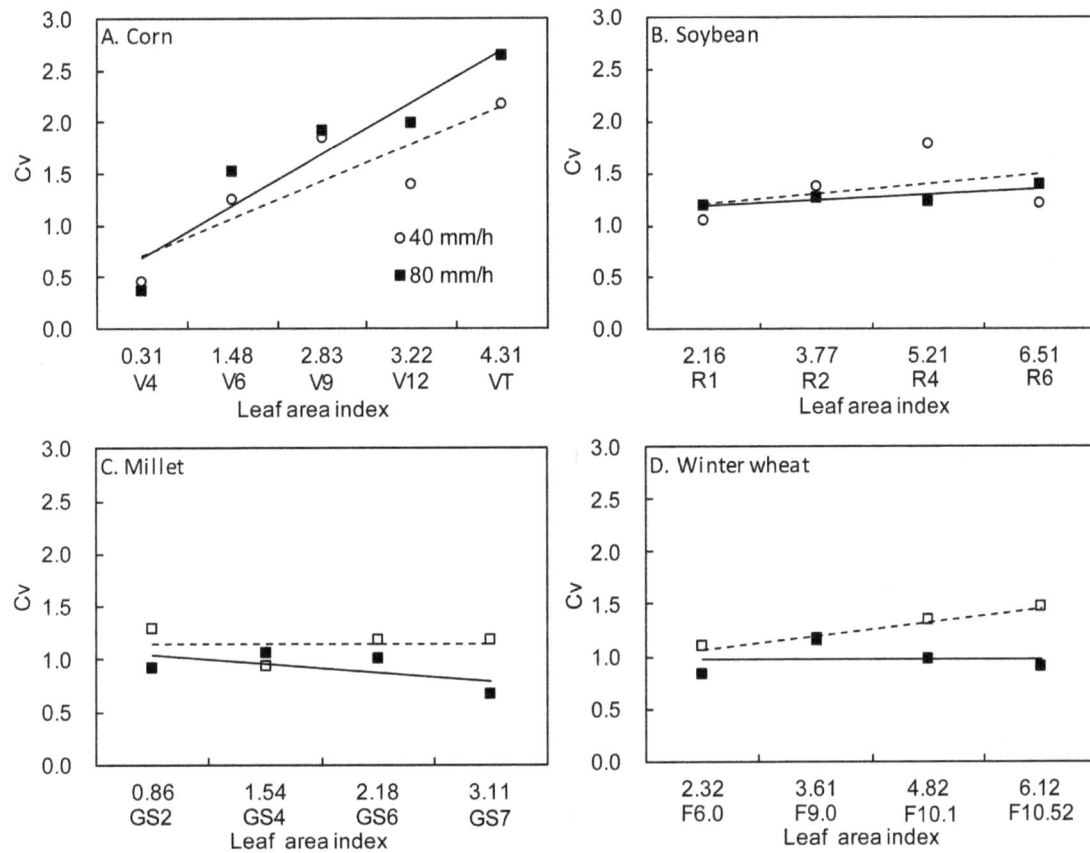

Figure 4. Relationship between Cv of splash detachment and crop growth.

was concentrated mainly in the central row position and that this concentration increased greatly at high rainfall intensity. The average splash detachment rate in the 10–20 cm band was 1.54 times higher than in the 0–10 cm band in the R1 stage. From the R1 to the R2 stage, splash erosion was mostly concentrated in the central row area and appeared in the form of a zonal distribution. Afterwards, in the R4 stage, although the maximum splash detachment rate data points still occurred in the 10–20 cm band, the probability of occurrence of extreme splash detachment rates was decreased, in particular that of the higher splash detachment rates shown as a dotted distribution in the R4 and R6 stages. The splash erosion yield at these high-concentration points accounted for much of the erosion, 48.36% at the R4 stage and 37.96% at the R6 stage of total splash erosion yield. This indicated that the splash erosion yield gradually decreased with soybean growth, but that the splash erosion yield remained mostly concentrated at a few points in the central row position with a large proportion of total splash erosion.

Considering the above analysis, the splash detachment and its spatial distribution under crop canopy were related closely to the big raindrops which make up the throughfall under crop canopy. The big raindrops form from the leaf edge and apex are a source of kinetic energy which can bring about erosion, and leading to uneven distribution of splash detachment. The leaf shapes are different with different crops, and leading to the different big raindrops form ability. Corn, millet and winter wheat are all gramineous plants with long and narrow leaves, but the leaves of corn are relatively large with waving leaf edge. The leaves could deflexed with the corn growth, and caused rain water on the leaves

flowed to apex and the hollow of leaf edge. Thus forming considerable big raindrops under the canopy, and facilitating the splash erosion. Soybean is a kind of leguminous plant with wide and soft leaves. The leaves were apt to bend down when they undertake the rainfall, thus form considerable big raindrops from the leaf apex. Therefore, throughfall under corn and soybean canopies had higher numbers of big raindrops, which can increased the splash detachment. Besides that, there are great differences between different crops height. Corn could reach a maximum height of 2.2 m in the observation period, while 0.9~1.2 m of max plant height in soybean, millet and winter wheat growth period. It has strong kinetic energy when big drops falling from higher corn leaves, and increasing splash erosion sharply. It indicated that, high stalk crops such as corn have serious splash erosion under the canopy compare other crops which have lower canopy height.

Conclusions and Discussion

Crop canopies can effectively reduce rainfall kinetic energy and protect soil surfaces from raindrop impact, thus inhibiting splash erosion. However, for different crop types and growth status, the effects on splash erosion were different. Results indicated that the average splash detachment rate under crop canopies was 390.12 g/m²·h, which represented a decrease of 67.81% compared with bare land. Crop coverage had some effect on reducing raindrop impact and preventing splash erosion, and different types of crops had different effects on reducing splash erosion. In this study, winter wheat had the strongest inhibition effects on splash erosion, followed by corn and soybeans, with millet the lowest. As

Table 4. Distribution of splash erosion under crop canopy in whole growth stage.

Rainfall intensity (mm/h)	Corn Growth stage	LAI	Average splash detachment rate under canopy (SDR_{AV}) g m^{-2} h^{-1}	Splash detachment rate in the region direct under the canopy (SDR_{DUC}) g m^{-2} h^{-1}	Splash detachment rate in the central row position (SDR_{CR}) g m^{-2} h^{-1}	SDR_{CR}/SDR_{DUC}	Soybean Growth stage	LAI	Average splash detachment rate under canopy (SDR_{AV}) g m^{-2} h^{-1}	Splash detachment rate in the region direct under the canopy (SDR_{DUC}) g m^{-2} h^{-1}	Splash detachment rate in the central row position (SDR_{CR}) g m^{-2} h^{-1}	SDR_{CR}/SDR_{DUC}
40	V4	0.31	247.77	252.70	237.92	0.94	R1	2.16	264.56	208.53	320.59	1.54
	V6	1.48	183.41	187.47	171.33	0.91	R2	3.77	187.38	69.39	305.38	4.40
	V9	2.83	125.29	114.97	145.91	1.27	R4	5.21	132.65	84.23	181.08	2.15
	V12	3.22	211.87	156.99	321.62	2.05	R6	6.51	106.68	93.14	120.22	1.29
	VT	4.31	244.05	219.06	294.04	1.34						
	Average	2.43	202.48	186.24	234.16	1.26	Average	4.41	172.82	113.82	231.82	2.34
80	V4	0.31	748.33	768.75	707.48	0.92	R1	2.16	976.99	623.75	1330.24	2.13
	V6	1.48	452.05	425.13	505.88	1.19	R2	3.77	919.29	392.21	1446.38	3.69
	V9	2.83	592.06	543.93	688.31	1.27	R4	5.21	744.53	474.95	1014.10	2.14
	V12	3.22	478.74	391.52	653.17	1.67	R6	6.51	584.42	161.78	1007.05	6.22
	VT	4.31	520.76	480.52	601.22	1.25						
	Average	2.43	558.39	521.97	631.21	1.21	Average	4.41	806.31	413.17	1199.44	3.55

the crops grew, the splash erosion yield under their canopies decreased, which was in accordance with the results reported by Miao [21], but different from those of Morgan [13]. Morgan conducted experiments on splash erosion under corn and soybean canopies and concluded that splash erosion increased with the height of the corn canopy, but the reverse occurred for soybeans. On the contrary, in this study, the average splash erosion yield under corn canopies decreased as the plants grew, while splash erosion at some points between corn rows was far greater than on bare land. In addition, rainfall intensity had a significant effect on inhibiting splash erosion; the splash erosion inhibition ability of crops was 13% greater under 80 mm/h rainfall intensity than under 40 mm/h rainfall intensity.

Rainfall was intercepted by crop canopies, was divided into three parts (throughfall, stemflow, and canopy interception), and then fell into the soil surface or was dissipated [8]. The effects of crops on splash erosion were influenced mainly by throughfall intensity, raindrop diameter distribution, and energy variations [12]. Throughfall intensity under crop canopies was closely related to splash detachment rate, which indicated that splash detachment rate increased with throughfall intensity. Under 40 mm/h rainfall intensity, the relationship between splash detachment rate and throughfall intensity was more significant, while the relationship became complex under 80 mm/h rainfall intensity because of other factors.

The spatial distribution of splash erosion under crop canopies became less uniform between rows, with splash erosion evidently tending to occur intensively in central row positions. Although throughfall under crop canopies decreased with crop growth, it tended to converge in central rows, which caused marked splash erosion and created a concentration of splash erosion in the central row position. This was consistent with reports by Armstrong and Mitchell [8] and Quinn and Laflen [9]. The latter studied the effects of throughfall on splash erosion under crop canopies, made a comparison to data from the USLE model, and concluded that

raindrops from the leaf margin and apex were an important factor in soil erosion. Armstrong and Mitchell [8] believed that large raindrops formed by rainfall convergence from crop canopies and focusing on a small impact plot would cause higher soil loss under canopies. Therefore, large throughfall raindrops formed from raindrops on the leaf margin and apex were an important kinetic energy source for splash erosion occurrence and distribution. However, this paper did not include a quantitative analysis of the relationship between large throughfalling raindrops and splash erosion occurrence and distribution. The results showed significant linear relationships among splash erosion, leaf area index, and throughfall intensity, suggesting that crop growth made the relationship between soil surface splash erosion and rainfall more random and complex.

Based on the above analysis, under suitable local conditions, rational close planting and make full use of the population dominance will be a good way to reduce splash erosion while boost crop yields. The compact crop types of corn and sorghum would be selected to increase the LAI, thus reduce throughfall by crisscrossed leaves. Furthermore, corn intercropping with other crops such as soybean and potato could be considered to increase the coverage between the rows of high stalk crops thereby reduces splash erosion.

Acknowledgments

The author would like to thank Jian Wang for useful discussions related to this study. The authors thank the anonymous reviewers and editors for their constructive comments and suggestions.

Author Contributions

Conceived and designed the experiments: BM FM FW. Performed the experiments: BM FM. Analyzed the data: XY. Contributed reagents/materials/analysis tools: ZL FW. Wrote the paper: BM FM. Provided help on language: XY.

References

1. Lamm RF, Manges LH (2000) Portioning of sprinler irrigation water by a corn canopy. Transactions of the ASABE 43 (3): 909–918.
2. Haynes LJ (1940) Ground Rainfall Under Vegetative Canopy of Crops. Journal of the American Society of Agronomy 32 (3): 176–184.
3. Ellison WD (1944) Studies of raindrop erosion. Aric. Eng. 25: 131–136.
4. Ellison WD (1947) Soil erosion study-part II: soil detachment hazard by raindrop splash. Aric. Eng. 28: 197–201.
5. Ellison WD (1947) Soil erosion Study-Part V: Soil transport in splash process. Aric. Eng. 28: 349–351, 353.
6. Ellison WD, Ellison OT (1947) Soil erosion Study-Part VI: Soil detachment by surface flow. Aric. Eng. 28: 402–405, 408.
7. Wu FQ, Zhao XG, Liu BZ (2001) Dynamic mechanism and erosion environment in sloping fields. Xian: Shaanxi Science Technology Press.
8. Armstrong LC, Mitchell KJ (1987) Transformations of Rainfall by Plant Canopy. Transactions of the ASABE 30 (3): 688–696.
9. Quinn WN, Laflen MJ (1983) Characteristics of raindrop throughfall under corn canopy. Transactions of the ASABE 26 (5): 1445–1450.
10. Moss AJ, Green TW (1987) Erosive effects of the large water drops (gravity drops) that fall from plants. Australian Journal of Soil Research 25: 9–20.
11. Sreenivas L, Johnston JR, Hill HO (1947) Some relationships of vegetation and detachment in the erosion process. Soil Science Society Proceedings 12: 471–474.
12. Morgan RPC (1982) Splash detachment under plant covers: results and implications of a field study. Transactions of the ASABE 25 (4): 987–991.
13. Morgan RPC (1985) Effect of corn and soybean canopy on soil detachment by rainfall. Transactions of the ASABE 28 (4): 1135–1140.
14. Finney HJ (1984) The effect of crop covers on rainfall characteristics and splash detachment. Journal of Agricultural Engineering Research 29: 337–343.
15. Tang KL (2004) Soil and Water Conservation in China. Bijing: Science Press.
16. Xie JQ (2005) The slope Land in China. Beijing: Chinese Dadi press.
17. Gong ZT (1999) Theory, Methodology and Application of Chinese Soil Taxonomy. Beijing: Science Press.
18. Jin K, Cornelis WM, Gabriels D, Schiettecatte W, Neve SD, et al. (2008) Soil management effects on runoff and soil loss from field rainfall simulation. Catena 75 (2): 191–199.
19. Ma F (2009) Effects of crop cover on soil erosion on slope land [Doctor]. Xi'an: Northwest A&F University.
20. Myneni R, Hoffman S, Knyazikhin Y, Privette J, Glassy J, et al. (2002) Global products of vegetation leaf area and fraction absorbed PAR from year one of MODIS data. Remote Sensing of Environment 83: 214–231.
21. Miao QA, Cao CG, Wang JP, Gao C, Li CF (2011) Effects of different crop systems on soil splash erosion from sloping land in Danjiangkou Reservoir region. Science of Soil and Water Conservation 9 (5): 11–14.

14

Automatic Detection of Regions in Spinach Canopies Responding to Soil Moisture Deficit Using Combined Visible and Thermal Imagery

Shan-e-Ahmed Raza[1]*, Hazel K. Smith[2], Graham J. J. Clarkson[3], Gail Taylor[2], Andrew J. Thompson[4], John Clarkson[5], Nasir M. Rajpoot[1,6]*

1 Department of Computer Science, University of Warwick, Coventry, United Kingdom, 2 Centre for Biological Sciences, Life Sciences, University of Southampton, Southampton, United Kingdom, 3 Vitacress Salads Ltd., Lower Link Farm, St Mary Bourne, Andover, United Kingdom, 4 Soil and Agri-Food Institute, School of Applied Sciences, Cranfield University, Bedford, United Kingdom, 5 School of Life Sciences, University of Warwick, Wellsbourne, United Kingdom, 6 Department of Computer Science and Engineering, Qatar University, Doha, Qatar

Abstract

Thermal imaging has been used in the past for remote detection of regions of canopy showing symptoms of stress, including water deficit stress. Stress indices derived from thermal images have been used as an indicator of canopy water status, but these depend on the choice of reference surfaces and environmental conditions and can be confounded by variations in complex canopy structure. Therefore, in this work, instead of using stress indices, information from thermal and visible light imagery was combined along with machine learning techniques to identify regions of canopy showing a response to soil water deficit. Thermal and visible light images of a spinach canopy with different levels of soil moisture were captured. Statistical measurements from these images were extracted and used to classify between canopies growing in well-watered soil or under soil moisture deficit using Support Vector Machines (SVM) and Gaussian Processes Classifier (GPC) and a combination of both the classifiers. The classification results show a high correlation with soil moisture. We demonstrate that regions of a spinach crop responding to soil water deficit can be identified by using machine learning techniques with a high accuracy of 97%. This method could, in principle, be applied to any crop at a range of scales.

Editor: Roeland M. H. Merks, Centrum Wiskunde & Informatica (CWI) & Netherlands Institute for Systems Biology, Netherlands

Funding: S.E.A.R was funded by the Horticultural Development Company (HDC) and by the Department of Computer Science, University of Warwick. Work on sustainable water use in salad crops in the laboratory of GT is funded by Vitacress Salads Ltd. with the award of a PhD to H.K.S. Research on baby leaf salads in the lab of Gail Taylor is funded by Biotechnology and Biological Sciences Research Council (BBSRC). Field experiments were designed with the help of Vitacress Salads Ltd UK. All the other funders had no role in study design, data collection and analysis, decision to publish, or preparation of the manuscript.

Competing Interests: S.E.A.R was funded by the Horticultural Development Company (HDC) and by the Department of Computer Science, University of Warwick. Work on sustainable water use in salad crops in the laboratory of GT is funded by Vitacress Salads Ltd. with the award of a PhD to H.K.S. Graham Clarkson is an employee of Vitacress Salads Ltd.

* E-mail: s.e.a.raza@warwick.ac.uk (SEAR); n.m.rajpoot@warwick.ac.uk (NMR)

Introduction

Infrared thermometers have been used in the past by researchers to determine temperature differences in both individual plants and their canopies for irrigation scheduling purposes. The development of thermal imagers has extended the opportunities for analysis of thermal properties of plants and canopies [1]. The non-contact, non-destructive nature and repeatability of measurements makes thermal imaging useful in agriculture, the food industry and forestry [2,3]. Imaging has been used as a tool in plants for predicting crop water stress, early disease detection, predicting fruit yield, bruise detection and detection of foreign bodies in food material. Under soil water deficits beyond a critical threshold, plants tend to close their stomata, and the rate of transpiration is reduced. This reduction in transpiration leads to an associated increase in leaf temperature. It also widens the range of temperature variation within the canopy which can be detected using infrared thermometry or by the use of thermal imagers [4]. There has been a lot of work focused on water stress analysis of plants using thermal imaging; however few researchers have

exploited the information from the visible light images for analysis. Most of the work conducted uses stress indices [5,6] and researchers have conducted various experiments to investigate the relationship between different stress indices and temperature values determined by thermal imaging [7,8]. The use of thermal imaging as an indicator of plant stress has also been tested in a number of environmental conditions and the conditions best suited to its successful application have been explored. Leaf energy balance equation was formulated to estimate stomatal conductance [9], but the proposed energy balance equation was dependent on a range of environmental factors and plant variables such as emissivity of the leaf surface, air density and specific heat capacity. The complexity, and associated difficulty of measuring these variables accurately, made it difficult to obtain accurate estimates of stomatal conductance from leaf temperature. Consequently, leaf energy balance equation was rearranged to derive thermal indices based on 'wet' and 'dry' reference surfaces [10,11], using the 'Crop Water Stress Index' (CWSI) [5,6], thus making stomatal conductance more straightforward to calculate from leaf temperatures. There is a debate within the scientific community as

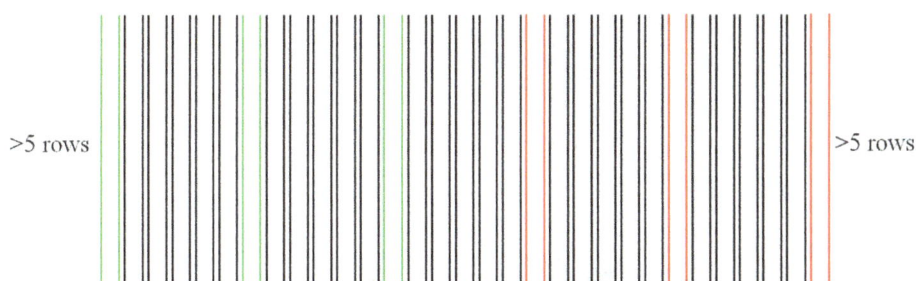

Figure 1. Sampling layout for the collection of thermal images and soil moisture measurements, 2010. The rows represent beds of Spinach (cv. Racoon), with those marked in green showing irrigated sample rows and the red indicating non-irrigated sample rows. Point measurements were made every 20 m for the full length of each bed (n = 54 for each treatment).

to the ideal choice of reference surfaces and much work has been undertaken to find the best choice for reference surfaces and in what conditions they must be used [12].

The robustness, sensitivity and limitations of thermal imaging for detecting changes in stomatal conductance and leaf water status in plants has been analysed by researchers in various conditions [13]. The temperature of surfaces within the canopy is highly dependent on whether they are shaded or in direct sunlight; this variation has been investigated and various options have been suggested to minimise the effect. It was suggested that the average temperature of the canopy was more useful to reduce the effect of leaf angles and other environmental factors when compared to individual leaf temperatures [14]. Researchers have also compared various techniques for image acquisition and have performed experiments to investigate the potential of infrared thermography for irrigation scheduling and to evaluate the consistency and repeatability of measurements under a range of environmental conditions [15]. It was suggested to exclude pixels which are outside the wet-dry threshold range to allow for semi-automated analysis of a large area of canopy. In addition, the authors proposed using thermal data from shaded leaves for improved data consistency, since there is less variability in temperature within an image, and smaller errors resulting from differences in radiation absorbed by reference and transpiring shaded leaves. Variation coefficients of stress indices were found to be of considerable importance and discriminatory powers of the techniques for estimates of stomatal conductance were found to be limited. In a later study, it was proposed that sunlit leaves show a wider range of temperatures because, although natural leaf orientation has little effect on the energy balance of shaded leaves, there is a large effect

on exposed leaves [16]. Based on these observations, the information from temperature distribution can be combined with the leaf orientation for thermal analysis in high resolution images.

Combining information from thermal and visible light images has the potential to provide a better estimate of stress indices and to identify regions in the canopy responding to soil water deficit. The use of thermal and visible imaging has been studied to maintain mild to moderate water stress levels in grapevine [17]. To estimate the canopy temperature, different sections of the canopy were used, including: the whole canopy, all of the sunlit canopy, the centre of the canopy and only sunlit leaves from the centre of the canopy. The best correlation between CWSI and stomatal conductance was calculated from the centre of the canopy measurements (or its sunlit fraction). The authors observed that CWSI computed with wet and dry references was the most robust index and suggested that the fusion of thermal and visible imaging can not only improve the accuracy of remote CWSI determination but also provide precise data on water status and stomatal conductance of grapevine.

Partly automated methods have also been used in the past to study plant stress indices [18]. The authors exploited colour information from visible light images to identify leaf area, as well as sunlit and shaded parts of the canopy. As a pre-processing step, images of constant temperature background were subtracted from the actual image to correct for relative errors in calibration of the camera caused by internal warming. Ground Control Points (GCPs) were manually selected to overlay the thermal image on the visible light image. Different regions in the visible light images were classified, using a supervised classification method, into pixels which represent leaves, other parts of the plant and background.

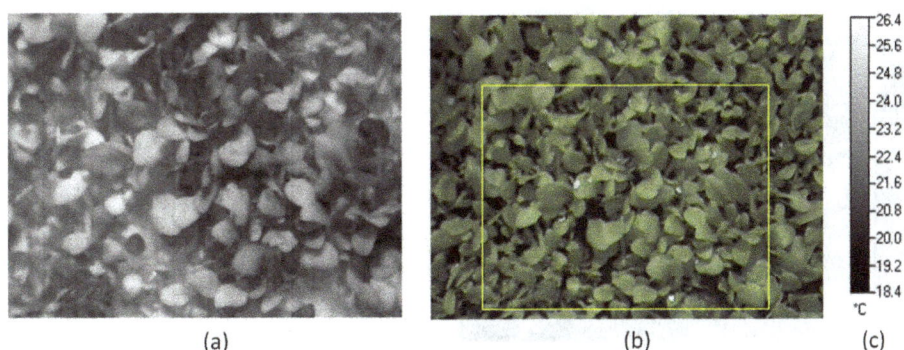

(a)   (b)   (c)

Figure 2. Image(s) obtained using a thermal imaging camera (NEC Thermo TracerTH9100 Pro). (a) thermal image with pixel values ranging from 0–255. (b) Region (rectangle) corresponding to the thermal image in the visible light image. (c) corresponding temperature range.

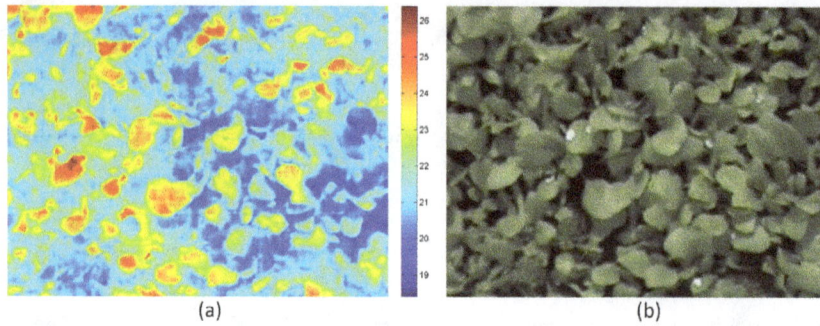

Figure 3. Visible light thermal images of **Figure 2** obtained after pre-processing; (a) the thermal image in **Figure 2**(a) has been replaced by temperature values. (b) visible light image in **Figure 2**(b) has been transformed to match thermal image in a way that same pixel locations correspond to same point located on the plant.

Statistical parameters and stress indices were calculated based on temperature values from the corresponding classified regions of the plant. The results showed that temperature distribution can be used as an indicator of stomatal conductance and plant stress. More recently, researchers have used automated methods to estimate water status using aerial thermal images of palm tree canopies [19]. The authors used watershed segmentation of thermal images to detect the palm trees, and found the detected temperature to be a good indicator of the tree's water status.

Here, we aim to use combined information from thermal and visible light images of a spinach canopy to classify well-watered and water deficient plants. We present a new technique to enhance the 'discriminatory power' of thermal imaging to identify parts of the canopy which have reduced their transpiration rates in response to soil moisture deficit. Instead of using stress indices to identify stress regions, we combine information from visible light and thermal images and use machine learning techniques to

classify between canopies growing in well-watered soil or under soil moisture deficit. Furthermore, we have acquired information about the light intensity and green-ness of the plant from the visible light images. These data are subsequently used, along with statistical information from thermal images, to classify between crop irrigation treatments using 1. Support Vector Machines (SVM), 2. Gaussian Processes Classifier (GPC) and 3. a combination of both classifiers. All three classifiers show promising results with the set of features extracted using combined information from thermal and visible light images.

Materials and Methods

Image Acquisition

Spinach (cv. Racoon) was drilled on 11 March 2010 at Mullens Farm, Wiltshire and was maintained with commercial practice. Permission for this study was given by the farm manager (Graham

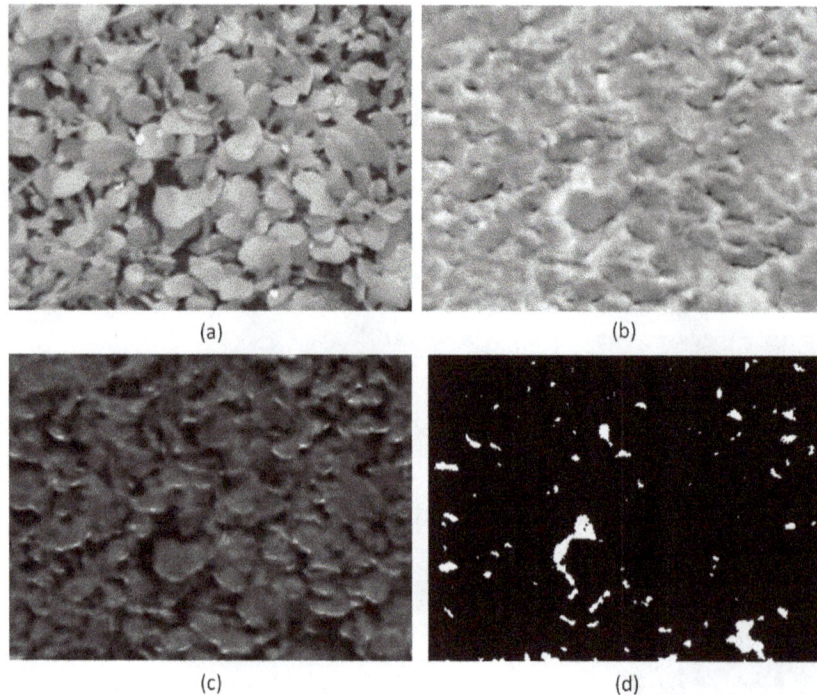

Figure 4. (a), (b) and (c) 'L', 'a' and 'b' channels of the visible light image. (d) thresholded a-channel.

Table 1. Features selected for our experiments.

	Symbol	Description	Type	p-value
1.	μ_{LT}	Luminance has been found to be a major factor which affects the thermal profile of an image [16]. In this work the temperature values were linearly scaled (multiply) with the corresponding L-channel of the colour image so that the effect of light intensity was incorporated into the model. After scaling temperature data with the L-channel, mean temperature values of an image was used as a feature.	C/T	0.154
2.	μ_a	The colour information indicates the amount of area covered by the plants or by other types of region. In Figure 4(b), lower intensities corresponded to green parts of the plant whereas the background shows a higher intensity value. For this reason the mean of the a-channel in our set of features was used.	C	1.92×10^{-07}
3.	μ_b	Similar to Feature 2, in Figure 4 (c) darker regions corresponded to background and hence the mean of b-channel was included in the set of features.	C	1.67×10^{-04}
4.	σ_{nT}	The amount of variation present in an image is also important [30]. Each row of the temperature data was therefore normalised by its median and then the standard deviation of the temperature values employed as a feature, to determine the amount of variation in the canopy region covered by the image.	T	2.89×10^{-19}
5.	μ_{aT}	In Lab colour space, lower values in a-channel corresponded to green regions. The a-channel was thresholded using Otsu's method [31] to find the background regions as represented by white pixels in Figure 4 (d). Temperature values corresponding to the background were discarded and the mean of the temperature values corresponding to the rest of pixels calculated, as a measure of the mean temperature of green parts of the plant.	C/T	1.88×10^{-21}
6.	σ_{aT}	Similar to Feature 5, the temperature values corresponding to background were discarded and the standard deviation of temperature values corresponding to the rest of the pixels calculated as a measure of variation in thermal intensities of green parts of the plant.	C/T	1.024×10^{-04}
7.	μ_T	Mean of temperature values	T	1.46×10^{-21}
8.	σ_T	Standard deviation of temperature values	T	1.12×10^{-04}

Feature type shows that the corresponding feature contains information about colour (C) or thermal (T) data or both (C/T). The rightmost column shows p-values of the features calculated using analysis of variance (ANOVA).

Clarkson) who is also a contributing author to this manuscript. Measurements were taken on 27 April of two treatment areas in bright and clear conditions; well-watered and water-deficient. The former treatment had been irrigated during the preceding week, while the latter had not, and were both harvested the following week for market. Both treatment areas were crops of spinach of the same age and variety and both had reached full canopy cover. Sampling consisted of taking a single image and soil moisture measurement at 20 m intervals for the length of each row. Three rows were sampled per treatment, with five rows separating the sampled rows (Figure 1). Soil moisture measurements were made using a Delta-T ML2x Thetaprobe connected to a HH2 moisture meter (Delta-T Devices, Cambridge, UK), with the probe position being in the centre of the bed at a depth of approximately 7 cm. The infra-red thermal images were taken using a TH9100WR thermal camera (NEC, Metrum) from a fixed distance of approximately 1 m above the crop. The camera operated in the region of 8–14 μm with 0.1°C thermal resolution and a spatial resolution of 320 (V) and 240 (H) pixels. Emissivity was set at 1.0 because it has been reported to induce errors of less than 1°C [20,21]. All measurements were taken between 11:00 and 13:00 hrs on a single day.

Pre-processing

Information from both thermal and visible light images (Figure 2) was used for classification. Thermal images were obtained as images with pixel intensity values ranging from 0 to 255. Initially, the image values were transformed to temperature values. A character recognition algorithm based on cross correlation was used, which automatically recognised the charac-

ters in the temperature bar (Figure 2c) and identified the temperature range for the thermal image [22]. This made it possible to replace the image values, which ranged from 0 to 255, with temperature values. In order to extract useful information from thermal and visible light images, both must be aligned so that the pixel location in both images corresponds to the same physical location with respect to the plant. Since both thermal and visible light images are acquired using a single device, there is a fixed transformation between thermal and visible light images. In order to compute this transformation, the transformation between a single pair of thermal and visible light images was calculated by manually selecting control points. To reduce the amount of noise present in the visible light image, anisotropic diffusion filtering was applied [23]. These pre-processing steps resulted in the images shown in Figure 3 and further calculations were conducted on these images.

Feature Computation

In order to get good classification results, we extracted information from the data in the form of features which carry discriminating information from different treatments and similar information from the same treatment type. Features were selected on the basis of observations made by various researchers [13–18]. Average values and variation in the thermal profile of the canopy were selected and combined with information from the visible light image. As a first step, the colour space of the visible light image from RGB to Lab colour space was transformed (Figure 4). In Lab colour space, instead of Red, Green and Blue channels, an L-channel exists for luminance, as well as 'a' and 'b' channels for the

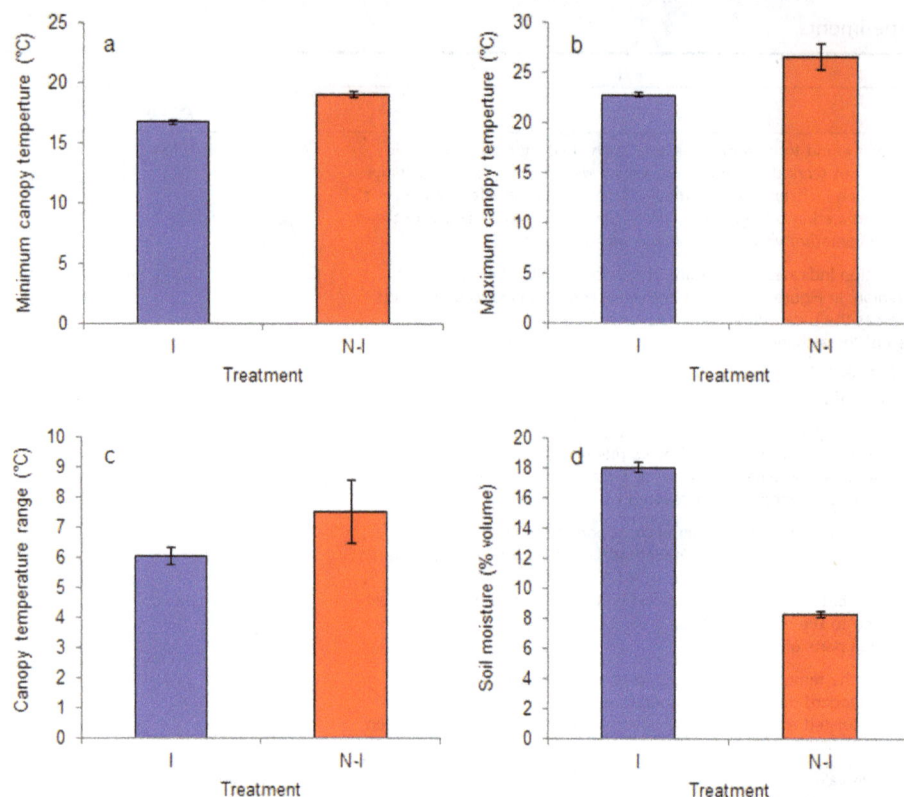

Figure 5. Crop canopy thermal properties (a–c) and soil moisture (d) of irrigated (I) and non-irrigated (N-I) beds of spinach. Crops were grown commercially at Mullens Farm, UK in April 2010. Each bar represents the mean value $\pm$ SE n = 3.

colour components. Features selected for experiments are given in Table 1.

Support Vector Machines (SVM)

SVM is a supervised learning method used for classification and regression analysis [24]. SVM constructs a hyperplane in high dimensional space and tries to find the hyperplane which maximises the separation between two classes of training data points. In this work, we used linear SVM which uses the model,

$$y = \mathbf{w}^T \mathbf{x} + b \qquad (1)$$

where $\mathbf{x} = [\mu_{LT}, \mu_a, \mu_b, \sigma_{nT}, \mu_{aT}, \sigma_{aT}, \mu_T, \sigma_T]$ denotes the input feature vector and y denotes the classification output (+1 for plants undergoing water stress, and −1 for well-watered plants). SVM models the parameters b and $\mathbf{w}$ to find the maximum margin hyperplane between data points from two classes.

Gaussian Processes for Classification (GPC)

Gaussian Processes (GP) can be defined as a class of probabilistic models comprised of distributions over functions instead of vectors [25–27]. A Gaussian distribution can be expressed by a mean vector and a covariance matrix. A GP is fully characterised by its mean and covariance functions. In machine learning, GPs have been used for regression analysis and classification. Similar to SVM, GPCs also belong to the class of supervised classification methods. However, instead of giving discriminant function values it produces output with probabilistic interpretation, i.e., a prediction for $p(y = +1|\mathbf{x})$ which denotes the probability of assigning a label (y) value +1 to the input feature

vector $\mathbf{x}$ [28]. GPCs do not calculate this probability directly on the input variables and assume that the probability of belonging to a class is linked to an underlying GP in the form of a latent function. Given a training set $D = \{(\mathbf{x}_i, y_i) | i = 1, 2, \dots n\}$ consisting of training images of both classes (water deficit and well-watered), with manually assigned labels y_i to the corresponding feature vectors $\mathbf{x}_i$ extracted from those images, GPC makes prediction about the label of the feature vector computed from an unseen image $\mathbf{x}_*$, using posterior probability,

$$p(y_* = +1|D, \mathbf{x}_*) = \int p(y_* = +1|f_*) p(f_*|D, \mathbf{x}_*) df_* \qquad (2)$$

The probability of belonging to a class $y_i = +1$ for an input $\mathbf{x}_i$ (known data point) is related to the value f_i of a latent function f [29]. This relationship is defined with the help of a squashing function. In this case, a Gaussian cumulative distribution function was used as the squashing function.

$$p(y = +1|f_i) = \frac{1}{2} \left[1 + \frac{erf(y_i f_i)}{\sqrt{2}} \right] \qquad (3)$$

where $erf(z)$ is the error function defined as $erf(z) = \frac{2}{\sqrt{\pi}} \int_0^z e^{-t^2} dt$.

The second term in the integral in equation (2) is given by,

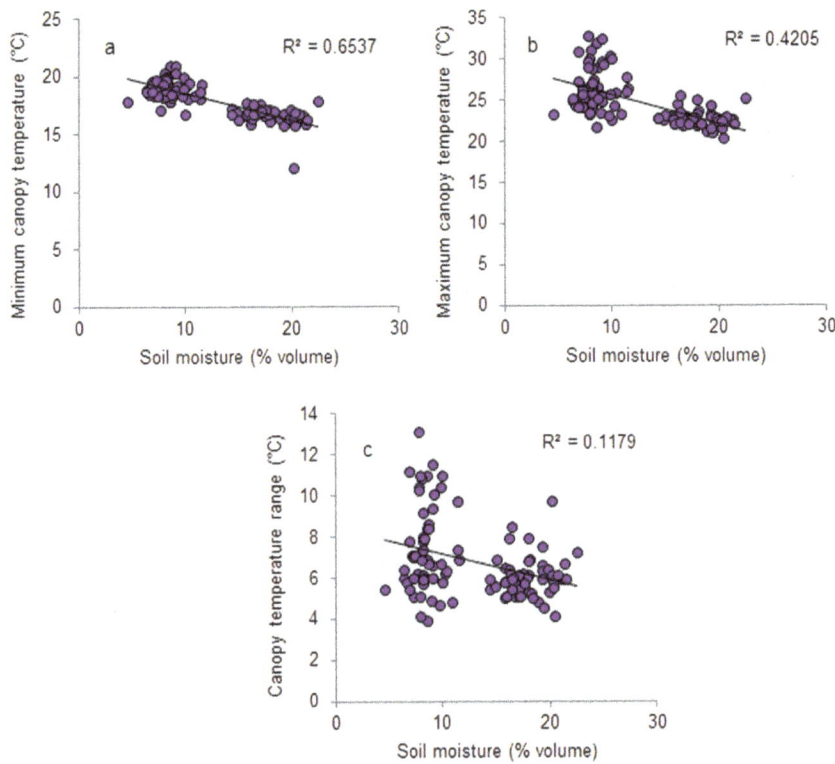

Figure 6. Regressions of crop canopy thermal properties (temperature minimum (a), maximum (b) and range (c)) and soil moisture measurements of irrigated and non-irrigated spinach beds. Crops were grown commercially in April 2010. Trend lines are shown when p< 0.005 and the R² value is given.

$$p(f_*|D,\mathbf{x}_*) = \int p(f_*|\mathbf{X},\mathbf{x}_*,f)p(f|D)df \qquad (4)$$

where $\mathbf{X} = [\mathbf{x}_1,\mathbf{x}_2,...,\mathbf{x}_n]$ and $f = [f_1,f_2,....f_n]$, n is the number of samples. $p(f|D)$ can be formulated by the Bayes' rule as follows,

$$p(f|D) = \frac{p(f|\mathbf{X})}{p(\mathbf{y}|\mathbf{X})} \prod_{i=1}^{n} p(y_i|f_i) \qquad (5)$$

and $p(y_i|f_i)$ can be calculated by equation (3) and $p(f|\mathbf{X})$ is the GP prior over latent function. Since a GP is characterised by a mean function and a covariance function, a zero mean was used for symmetry reasons, and a linear covariance function selected which has been found to be effective in classification problems [26]. The

normalisation term in the denominator is the marginal likelihood given by,

$$p(\mathbf{y}|\mathbf{X}) = \int p(f|\mathbf{X}) \prod_{i=1}^{n} p(y_i|f_i) \qquad (6)$$

where $\mathbf{y} = \{y_1,y_2,...y_n\}$. The second term in the above equation is not Gaussian and this makes the posterior in equation (5) analytically intractable. However, analytical approximations or Monte Carlo methods can be used. Two commonly used approximation methods are Laplace approximation and Expectation Propagation (EP). EP minimises the local Kullback-Leibler (KL) divergence between the posterior and its approximation and has been found to be more accurate in predicting than Laplace

Table 2. Total variance explained by Principle Component Analysis when both well-watered and droughted spinach crops were measured for their thermal properties (maximum, minimum and range of temperatures) and soil moisture.

Component	Initial Eigenvalues			Extraction Sums of Squared Loadings		
	Total	% of Variance	Cumulative %	Total	% of Variance	Cumulative %
1	2.863	71.567	71.567	2.863	71.567	71.567
2	.830	20.742	92.309			
3	.308	7.691	100.000			
4	6.967E-16	1.742E-14	100.000			

Extraction Method: Principal Component Analysis.

Table 3. Component Matrix[a] from Principle Component Analysis when both irrigated and non-irrigated spinach crops were measured for their thermal properties (maximum, minimum and range of temperatures) and soil moisture.

	Component
	1
Soil moisture	−.750
Minimum temperature	.869
Maximum temperature	.969
Temperature range	.779

approximation and hence EP was used for approximation in these experiments [25,26].

Experiments and Results

Classification using Machine Learning Methods

A total of 108 images of spinach canopies and corresponding soil moisture point measurements were acquired, with 54 images of well-watered beds and 54 images of droughted beds. The thermal images demonstrated significant variation between the

two treatments when judged by soil moisture and thermal canopy properties as taken from the primary thermal images (Figure 5).

Well-watered canopies exhibited lower minimum ($F_{1,5} = 59.74$, p = 0.002) and maximum ($F_{1,5} = 8.71$, p<0.05) temperatures than droughted beds. However, the range of temperatures did not differ between treatments when the droughted beds were compared to irrigated spinach plots ($F_{1,5} = 1.80$, p>0.05). Additionally, it was confirmed that soil moisture differed significantly between treatments ($F_{1,5} = 556.19$, p<0.0001). All analyses were conducted using 1-way ANOVA.

Regressions demonstrated a number of relationships linking crop canopy thermal properties, taken from the primary thermal images, to direct soil moisture measurements (Figure 6). Moreover, there was a clear segregation into two clusters, accounting for the two treatments. To establish how these relationships interacted, PCA was performed upon the four traits of: soil moisture, minimum temperature, maximum temperature and range of temperature. Components were extracted when their Eigenvalue exceeded a threshold value of 1. One component was extracted which explained 71.6% of total variance (Table 2). This component measured all four traits thus showing their tight coupling and the need for more complex analysis if they are to be used for the detection of soil water deficits. All thermal properties were strongly, positively related to each other while soil moisture was negatively related to all thermal traits. These results implied that the thermal properties of spinach canopies can be used as an

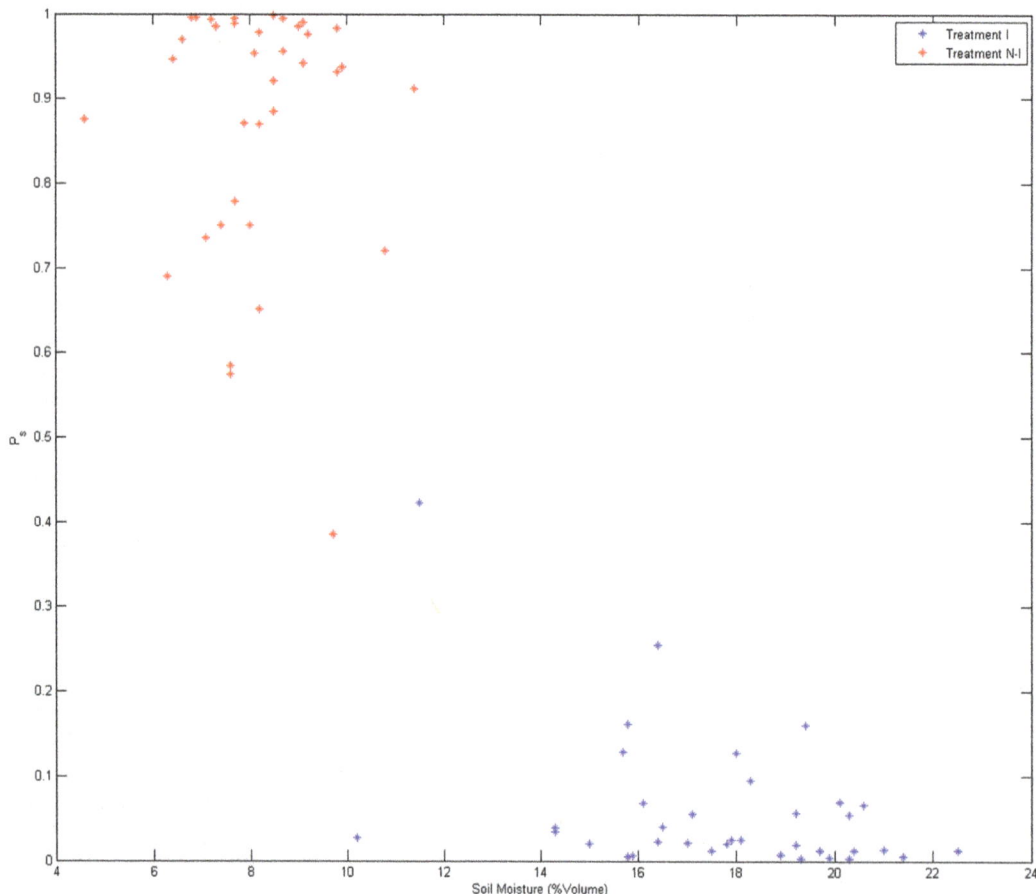

Figure 7. Probability of belonging to treatment N-I (P_s) versus Soil moisture values (correlation value = −0.89, High moisture means less probability of stress). Soil moisture is given as percentage soil water content v/v. Classification accuracy for this particular set of training and testing data was 98.6% as given by GPC.

Table 4. Comparison of average classification results of different classifiers using the proposed set of features.

Feature(s) selected	Classifier	Sensitivity (%)	Specificity (%)	PPV (%)	Accuracy (%)	$\sigma_{accuracy}$
Color only (μ_a, μ_b)	SVM	67.28	70.29	70.98	67.74	3.36
	GPC	80.68	52.96	21.68	56.87	3.55
	Both Classifiers	67.32	70.42	71.11	67.80	3.40
Thermal only (μ_T, σ_T)	SVM	93.35	91.28	90.89	92.14	1.92
	GPC	93.06	80.30	76.67	85.42	2.29
	Both Classifiers	93.35	91.28	90.88	92.14	1.92
Features (1–8) Table 1.	SVM	95.52	96.39	96.30	95.85	1.97
	GPC	96.38	97.39	97.30	96.79	1.56
	Both Classifiers	96.62	96.93	96.84	96.70	1.60
Features (1–6) Table 1.	SVM	95.86	96.86	96.80	96.27	1.58
	GPC	96.53	96.99	96.90	96.68	2.00
	Both Classifiers	**96.97**	**97.38**	**97.31**	**97.12**	**1.52**

indicator of soil water content (Table 3) yet that this approach is not able to accurately detect soil moisture status using primary thermal images. A more complex analysis method is required which is able to utilise both visual and thermal image data to improve soil moisture detection.

The same 108 images of spinach canopies were used for the image processing approach, with the 54 images of well-watered beds being designated treatment I, while the 54 images of the water deficient canopy were designated treatment N-I. The identity of the two treatments was not known during the development of image analysis. After pre-processing, six different features (1–6, Table 1) were obtained from each image. SVM and GPCs were used to classify the test images into water deficient and well-watered. For SVM linear kernel was used and for GPC a zero mean and a linear covariance function were chosen. As discussed before, SVM gives discrete classification results and classifies each image as treatment I or treatment N-I, whereas GPC gives the probability (likelihood) of each image belonging to a particular treatment. Figure 7 shows the probability of an image belonging to treatment N-I (P_s) versus the values of soil moisture for one set of training and testing data. It was clear that the probability (P_s) was highly related to manually calculated soil moisture values (correlation value = −0.89 for Figure 7). Based on the probabilities given by GPC, each image was classified as an image from either treatment I or treatment N-I.

Since two different types of classifier were used, disagreement between the results of both the different classifiers could be assessed, which occurred in some cases. This disparity was utilised to further refine the classification results; although this refinement is not very significant, it produces better results. Information from both classification methods was combined to reduce the error from classification. If an image was classified by SVM as treatment I and its probability of belonging to treatment N-I according to GPC was higher than 80% then this image was classified as treatment N-I. On the other hand, if an image was classified as treatment N-I and its probability according to GPC was less than 20%, the image was classified as treatment I. It was found experimentally that the 80–20% threshold gave the best results.

200 iterations were employed to test the accuracy of the classifiers for different pairs of training and testing sets. In each iteration, 36 images were chosen at random (18 from each treatment) for training purposes and the proposed algorithm was tested on the other 72 images. Results showed that GPC demonstrated a higher level of accuracy than the SVM classifier (Table 4); however if information from the results of both of the classifiers was combined, results were improved in terms of sensitivity, specificity, positive predictive value (PPV) and accuracy. An average accuracy of 96.3% was obtained for SVM, 96.7% by using GPC and a slightly higher 97.1% when information from both classifiers was combined. When the results of colour-only and temperature-only features were compared, it was found that combining information from both temperature and colour data increased the accuracy of classification. Furthermore, including mean and standard deviation of temperature values without

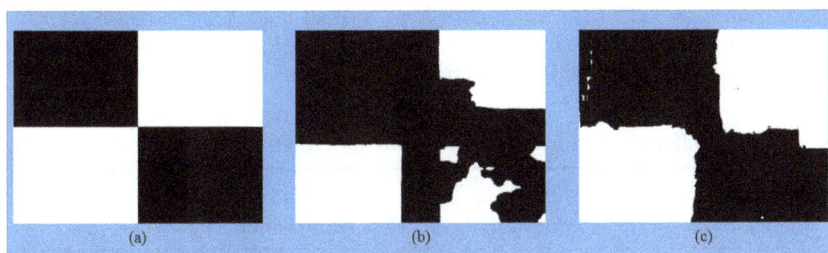

Figure 8. (a) The ground truth pattern for mixed condition mosaicked image. Black colour represents image region corresponding to treatment I and white colour represents the image region which corresponds to treatment N-I. (b) & (c) show classification results obtained using combined classifier with thermal only and proposed feature set respectively.

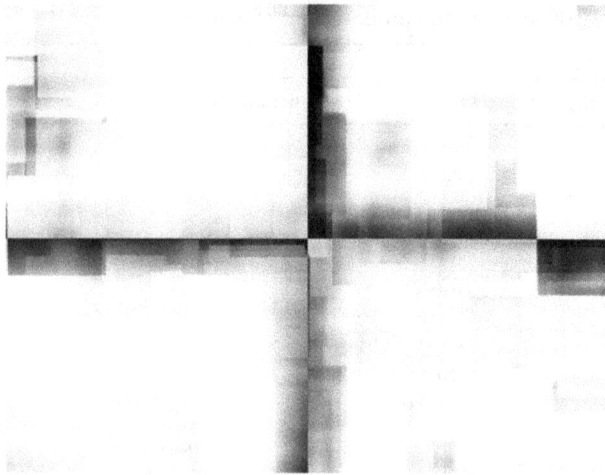

Figure 9. GPC classification result in terms of confidence score (C_s). Bright shade represents high confidence in classification results and dark shade represents low confidence in the classification. The classifier has higher confidence in the region with image from treatment I or treatment N-I, however the confidence value is low, as depicted by darker shade, around the boundary of two merging images from different treatments.

combining them with colour information diminished the accuracy of results; thus the mean (μ_T) and standard deviation (σ_T) were removed from the set of features.

To further investigate the strength of classifier with the proposed set of features, we created an artificial image with mixed conditions by combining randomly picked thermal and visible light images from Treatment I and Treatment N-I to form a mosaic. The ground truth pattern for the mosaicked image is shown in **Figure 8** (a). Black colour represents image region corresponding to treatment I and white colour represents the image region which corresponds to treatment N-I. A block of size 50×50 pixels was defined at each pixel location in the mosaicked image and the classifier was tested using the features extracted from each of these small blocks (307,200 blocks in total). The classifier for this experiment was trained in a similar way as for the real data (i.e., on randomly selected 36 original images). By using 50×50 blocks to simulate mixed conditions, we reduced the amount of information available, so the accuracy of classification is expected to deteriorate. However, the results show robustness of our proposed feature set when compared to thermal only features. The classification results using the combined classifier with thermal only and the proposed feature set are shown in **Figure 8** (b) & (c) respectively. The classification accuracy using SVM, GPC and the combined classifier was calculated to be **89.1%**, **94.1%** and **92.5%** using the proposed feature set compared to **78.3%**, **54.1%** and **76.3%** when using thermal only features. The classification accuracy for the combined

classifier is less than GPC in the proposed feature set and less than SVM in the thermal only feature set in mixed conditions, however, we still consider this classifier to be important as it gives the best results on real data. **Figure 9** shows GPC classification results using the proposed set of features in terms of the confidence score (C_s). For treatment I, $C_s = 1 - P_s$ and for treatment N-I, $C_s = P_s$, where P_s is the probability of belonging to treatment N-I as given by GPC. The bright shade represents high confidence in classification results and dark shade represents low confidence in the classification. It can be observed that the classifier has higher confidence in the region where the image is from treatment I or treatment N-I, however the confidence value is low, as depicted by low grey values around the boundary of two merging images from different treatments. The mean and standard deviation of C_s was calculated to be **90.5%** and **17.8%** using proposed feature set and **51.1%** and **32.3%** using thermal only features respectively.

Discussion and Conclusions

Our results show that by combining information from thermal and visible light images and using machine learning techniques, canopies which are experiencing water deficits can be identified with high accuracy – more than 97%. Thus we have considerably improved the use of remote images in the detection on canopy stress using this combined approach. The purpose of this study was to test a new dimension of automated classification methods for the detection of regions of a crop canopy that are responding to soil water deficit and to go beyond the restrictions of commonly used statistical approaches. We showed that extraction of a good set of image features can be useful for classifications of this type. In this study, we were able to detect regions of the canopy which were experiencing soil moisture deficit by using a machine learning approach instead of stress indices. Initially, the effect of reflected light and background information was reduced in order to extract features. In the second step these features were classified using SVM, GPC and a combination of both classifiers. The colour information in visible light images provides information about the amount of reflected light intensity from the plant. Using this information, temperature values were scaled on the basis of reflected light. Plant regions can also be identified in the registered thermal image using colour information. This helped to discard temperature values belonging to the background and extract useful information from plant regions in [15]. Based on information from visible light and thermal images, a worthy set of features can be extracted. In these experiments, it was found that scaling with luminance intensity (μ_{LT}) plays an important role in classification. When the luminance intensity scaling feature was removed from our set of features, we found that the accuracy of the classifiers decreased (Table 5). In the case of GPC classification, accuracy fell by up to 7%. This showed that the selection of suitable features is critical when data from thermal images are classified for stress analysis. We have also tested the proposed classifier on an artificially generated mixed condition image. The classification

Table 5. Comparison of average classification results of different classifiers without using light intensity scaling feature (μ_{LT}).

	Sensitivity (%)	Specificity (%)	PPV (%)	Accuracy (%)	$\sigma_{accuracy}$
SVM	94.98	95.01	94.83	94.84	2.01
GPC	88.21	91.84	92.05	89.70	2.61
Both Classifiers	95.28	95.27	95.08	95.12	1.89

results in this image show a significant improvement using the proposed feature set when compared to the thermal only feature set. We found the proposed set of features robust to amount of input information and to mixed-condition images.

In the future, we plan to extend this work to identify canopies under multiple levels of stress. Furthermore, information about leaf angles and distance of the plant from the camera will be used to estimate a more accurate model of the thermal profile, which in this case was linear scaling with light intensity values. For information about depth and leaf angles, a stereo image setup is needed in order to model the effect of leaf angles and distance of leaves from the camera. This model can be combined with more sophisticated machine learning techniques for early water stress detection in crops, and, if automated, could be used to improve irrigation efficiency by optimising the timing and spatial distribution of irrigation events. Other plant stresses such as disease could also potentially be detected rapidly and pre-symptomatically using these methods.

Acknowledgments

The authors would like to thank the staff at Mullens Farm where the experimental field work was conducted.

Author Contributions

Conceived and designed the experiments: SEAR HKS GJJC. Performed the experiments: SEAR HKS. Analyzed the data: SEAR HKS. Contributed reagents/materials/analysis tools: GJJC. Wrote the paper: SEAR HKS NMR GT GJJC JC AJT. Performed computational analysis under supervision of NMR, JC and AJT: SEAR. Performed experiments in the field under supervision of GJJC and GT: HKS.

References

1. Hackl H, Baresel JP, Mistele B, Hu Y, Schmidhalter U (2012) A Comparison of plant temperatures as measured by thermal imaging and infrared thermometry. J Agron Crop Sci 198: 415–429.
2. Eberius M (2011) Automated image based plant phenotyping - Challenges and chances. 2nd Int Plant Phenotyping Symp.
3. Pierce L, Running S, Riggs G (1990) Remote detection of canopy water stress in coniferous forests using the NS 001 Thematic Mapper Simulator and the thermal infrared multispectral scanner. Photogramm Eng Remote Sensing: 579–586.
4. Fuchs M (1990) Infrared Measurement of Canopy Temperature and Detection of Plant Water Stress. Theor Appl Climatol 261: 253–261.
5. Idso SB, Jackson RD, Pinter Jr PJ, Reginato RJ, Hatfield JL (1981) Normalizing the stress-degree-day parameter for environmental variability. Agric Meteorol 24: 45–55. doi:10.1016/0002-1571(81)90032-7.
6. Jackson RD, Idso SB, Reginato RJ, Pinter PJ (1981) Canopy temperature as a crop water-stress indicator. Water Resour Res 17: 1133–1138.
7. Alchanatis V, Cohen Y, Cohen S, Moller M, Sprinstin M, et al. (2009) Evaluation of different approaches for estimating and mapping crop water status in cotton with thermal imaging. Precis Agric 11: 27–41.
8. Reinert S, Bögelein R, Thomas FM (2012) Use of thermal imaging to determine leaf conductance along a canopy gradient in European beech (Fagus sylvatica). Tree Physiol 32: 294–302.
9. Jones HG (1992) Plants and Microclimate: a quantitative approach to environmental plant physiology. 2nd ed. Cambridge: Cambridge University Press.
10. Jones HG (1999) Use of infrared thermometry for estimation of stomatal conductance as a possible aid to irrigation scheduling. Agric For Meteorol 95: 139–149. doi:10.1016/S0168-1923(99)00030-1.
11. Jones HG (1999) Use of thermography for quantitative studies of spatial and temporal variation of stomatal conductance over leaf surfaces. Plant, Cell Environ 22: 1043–1055. doi:10.1046/j.1365-3040.1999.00468.x.
12. Leinonen I, Grant OM, Tagliavia CPP, Chaves MM, Jones HG (2006) Estimating stomatal conductance with thermal imagery. Plant, Cell Environ 29: 1508–1518.
13. Grant OM, Chaves MM, Jones HG (2006) Optimizing thermal imaging as a technique for detecting stomatal closure induced by drought stress under greenhouse conditions. Physiol Plant 127: 507–518. doi:10.1111/j.1399-3054.2006.00686.x.
14. Grant OM, Tronina L, Jones HG, Chaves MM (2007) Exploring thermal imaging variables for the detection of stress responses in grapevine under different irrigation regimes. J Exp Bot 58: 815–825. doi:10.1093/jxb/erl153.
15. Jones HG (2002) Use of infrared thermography for monitoring stomatal closure in the field: application to grapevine. J Exp Bot 53: 2249–2260. doi:10.1093/jxb/erf083.
16. Stoll M, Jones H (2007) Thermal imaging as a viable tool for monitoring plant stress. Int J Vine Wine Sci 41: 77–84.
17. Möller M, Alchanatis V, Cohen Y, Meron M, Tsipris J, et al. (2007) Use of thermal and visible imagery for estimating crop water status of irrigated grapevine. J Exp Bot 58: 827–838. doi:10.1093/jxb/erl115.
18. Leinonen I, Jones HG (2004) Combining thermal and visible imagery for estimating canopy temperature and identifying plant stress. J Exp Bot 55: 1423–1431. doi:10.1093/jxb/erh146.
19. Cohen Y, Alchanatis V, Prigojin A, Levi A, Soroker V (2011) Use of aerial thermal imaging to estimate water status of palm trees. Precis Agric 13: 123–140.
20. Jackson RD (1982) Canopy temperature and crop water stress. Adv Irrig 1: 43–85.
21. López a., Molina-Aiz FD, Valera DL, Peña A (2012) Determining the emissivity of the leaves of nine horticultural crops by means of infrared thermography. Sci Hortic (Amsterdam) 137: 49–58. doi:10.1016/j.scienta.2012.01.022.
22. Eikvil L (1993) Optical Character Recognition. Oslo.
23. Perona P, Malik J (1990) Scale-space and edge detection using anisotropic diffusion. IEEE Trans Pattern Anal Mach Intell 12: 629–639.
24. Cortes C, Vapnik V (1995) Support-vector networks. Mach Learn 20: 273–297.
25. Rasmussen C (2004) Gaussian processes in machine learning. Adv Lect Mach Learn: 63–71.
26. Rasmussen CE, Williams CK (2006) Gaussian processes for machine learning. Cambridge: MIT Press.
27. Haranadh G, Sekhar CC (2008) Hyperparameters of Gaussian process as features for trajectory classification. Neural Networks (IEEE World Congr Comput Intell IEEE Int Jt Conf: 2195–2199.
28. Bazi Y, Melgani F (2010) Gaussian process approach to remote sensing image classification. Geosci Remote Sensing, IEEE Trans 48: 186–197.
29. Ebden M (2008) Gaussian Processes for Regression: A Quick Introduction.
30. Jones HG, Stoll M, Santos T, Sousa C de, Chaves MM, et al. (2002) Use of infrared thermography for monitoring stomatal closure in the field: application to grapevine. J Exp Bot 53: 2249–2260.
31. González RC, Woods RE (2008) Digital image processing. Pearson/Prentice Hall.

15

Methane, Carbon Dioxide and Nitrous Oxide Fluxes in Soil Profile under a Winter Wheat-Summer Maize Rotation in the North China Plain

Yuying Wang[1], Chunsheng Hu[1]*, Hua Ming[1], Oene Oenema[2], Douglas A. Schaefer[3], Wenxu Dong[1], Yuming Zhang[1], Xiaoxin Li[1]

1 Key Laboratory of Agricultural Water Resources, Center for Agricultural Resources Research, Institute of Genetics and Developmental Biology, Chinese Academy of Sciences, Shijiazhuang, Hebei, China, 2 Department of Soil Quality, Wageningen University, Alterra, Wageningen, The Netherlands, 3 Key Lab of Tropical Forest Ecology, Xishuangbanna Tropical Botanical Garden, Chinese Academy of Sciences, Menglun, Yunnan, China

Abstract

The production and consumption of the greenhouse gases (GHGs) methane (CH_4), carbon dioxide (CO_2) and nitrous oxide (N_2O) in soil profile are poorly understood. This work sought to quantify the GHG production and consumption at seven depths (0–30, 30–60, 60–90, 90–150, 150–200, 200–250 and 250–300 cm) in a long-term field experiment with a winter wheat-summer maize rotation system, and four N application rates (0; 200; 400 and 600 kg N ha^{-1} $year^{-1}$) in the North China Plain. The gas samples were taken twice a week and analyzed by gas chromatography. GHG production and consumption in soil layers were inferred using Fick's law. Results showed nitrogen application significantly increased N_2O fluxes in soil down to 90 cm but did not affect CH_4 and CO_2 fluxes. Soil moisture played an important role in soil profile GHG fluxes; both CH_4 consumption and CO_2 fluxes in and from soil tended to decrease with increasing soil water filled pore space (WFPS). The top 0–60 cm of soil was a sink of atmospheric CH_4, and a source of both CO_2 and N_2O, more than 90% of the annual cumulative GHG fluxes originated at depths shallower than 90 cm; the subsoil (>90 cm) was not a major source or sink of GHG, rather it acted as a 'reservoir'. This study provides quantitative evidence for the production and consumption of CH_4, CO_2 and N_2O in the soil profile.

Editor: Dafeng Hui, Tennessee State University, United States of America

Funding: This research is supported by the "Strategic Priority Research Program of Chinese Academy of Sciences" (Grant No. XDA0505050202 and XDA05050601) and the "National Science & Technology Pillar Program" (Grant No. 2012BAD14B07-5). It is also supported by the "National Basic Research Program of China" (Grant No. 2010CB833501); the "National Natural Science Foundation of China" (Grant No. 30970534) and the "Main Direction Program of Knowledge Innovation of Chinese Academy of Sciences" (Grant No. KSCX2-EW-J-5). The funders had no role in study design, data collection and analysis, decision to publish, or preparation of the manuscript.

Competing Interests: The authors have declared that no competing interests exist.

* E-mail: cshu@sjziam.ac.cn

Introduction

Atmospheric concentrations of carbon dioxide (CO_2), methane (CH_4) and nitrous oxide (N_2O) have increased considerably since the industrial revolution, and are still increasing annually by about 0.5%, 1.1% and 0.3%, respectively [1]. Worldwide concerns about the increased greenhouse gases (GHGs) concentrations in the atmosphere and its effects on our future environment require a better understanding of the cause of these emissions [2]. Agricultural lands occupy 37% of the earth's land surface; about 13.5% of global anthropogenic GHG was emitted from agricultural production [1]. It was estimated that 84% of N_2O and 52% of CH_4 emitted from agriculture activities [3]. In China, agriculture tends to produce more emissions than the global average over the last 30 years due to increased chemical and manure N inputs. Gaining a better understanding of GHG production and emission processes, and developing methods for mitigating emissions from agroecosystems are essential steps in order to mitigate climate change [4].

Agricultural soils are main sources and sinks of GHG emissions, depending on their characteristics and management. Many studies have been conducted to quantify the net fluxes of CO_2, CH_4, and N_2O across the soil/atmosphere interface [3,5–9]. These studies provide an integrative estimate of the net production and consumption of CH_4, CO_2 and N_2O in the soil, but do not provide information on the depth-distribution of CH_4, CO_2 and N_2O production-consumption patterns within soil profiles. It has been suggested that subsurface processes exert a significant control on carbon (C) and nitrogen (N) dynamics and hence on CO_2, CH_4, and N_2O emissions from soil [10], but few studies have elucidated the role of the subsoil so far. Understanding these processes might also provide a better insight into the possibilities and effectiveness of measures to reduce GHG emissions. For example, a temporary accumulation of GHG in the soil profile influences GHG flux patterns at the soil surface over time, and thereby may confuse empirical relationships between agricultural activities and measured GHG emissions [11]. Thus, measurements of CH_4, CO_2 and N_2O concentration profiles may be helpful for increasing the understanding of the net exchanges of these gases between soil and atmosphere.

Though few studies have examined the production and consumption of CO_2 [12,13], CH_4 [14] and N_2O [15] within individual soil horizons and their transports between soil horizons so far, very few studies have made combined measurements of the dynamics of CO_2, CH_4 and N_2O production and emission processes in soil profiles in agro-ecosystems, especially in China [16]. It has been well-established that N fertilizer applications increase crop growth and N_2O emissions, and tend to decrease CH_4 emissions into the atmosphere, but there is little information about the combined effects of N fertilizer application and irrigation on subsoil N_2O, CO_2 and CH_4 production, consumption and transport.

Recently, Wang et al [16] presented bi-weekly measured CH_4, CO_2 and N_2O concentration profiles down to a depth of 300 cm in a winter wheat-summer maize rotation in the North China Plain, with four N application rates (0; 200; 400 and 600 kg N ha^{-1}). Here, we build on the results of that study, and present calculated subsurface fluxes of CH_4, CO_2 and N_2O over a whole-year period. The purpose of this study is to evaluate the effects of seasonal cropping, N applications, irrigation, soil temperature, and soil moisture on net subsurface transport of CO_2, N_2O and CH_4.

Materials and Methods

Site description

The study was conducted at Luancheng Agroecosystem Experimental Station (37°53′N, 114°41′E, elevation 50 m), Chinese Academy of Sciences. This area is at the piedmont of the Taihang Mountains, in the North China Plain. Mean annual precipitation is about 480 mm, 70% of which is in the period from July to September. Annual average air temperature is 12.5 °C. The dominant cropping system in the region is a winter wheat-summer maize double-cropping system (two crops harvested in a single year) without fallow between the crops.

Field experimental design

The field experiment with a randomized complete block design was laid down in a winter wheat (*Triticum aestivum L.* Wheat variety Kenong 199)-summer maize (*Zea mays L.* Maize variety Xianyu 335) double-cropping system in 1998. It had four N fertilizer (urea) application treatments in triplicate: 0 (N0), 200 (N200), 400 (N400) and 600 (N600) kg N ha^{-1} year^{-1}. Plot size was 7 m×10 m. Results of the present study refer to the period March 2007 to January 2008. Details on fertilizer application and crop management activities are presented in Table 1. Crops are flood-irrigated with pumped groundwater about five times per year, depending on rainfall distribution.

Soil sampling, analysis and climate data collection

The soil has a silt-loam texture in the upper 90 cm and clay-loam to clay texture at depth of 90–300 cm (Table 2). All soil samples were collected from different depths of the soil profile before the GHG measurements (on 5 December 2006); soil samples were mixed to make a specific representative soil sample for each depth; and all analyses of soil chemical properties in Table 2 were based on the standard methods for soil analyses described by Sparks [17]. Soil bulk density was determined using the cutting ring method. Soil particle size analysis was done by the Bouyoucos Hydrometer Method [18]. Soil pH was measured in a suspension of 5 g soil with 25 ml distilled water after 1 h after shaking.

Soil core samples were collected from different depths (0–30, 30–60, 60–90, 90–150, 150–200, 200–250, 250–300 cm) of the soil profile in the farmland described above on 5 December 2006 (before the GHG measurements), 16 June 2007 (after the winter wheat harvest) and 11 October 2007 (after the summer maize

harvest), respectively. Three different sub-samples, taken from a cross-section around the soil auger (3 meter in length), were mixed to make a specific representative soil sample for each depth from each point. The soil profile samples were sealed in dark plastic bags immediately after sampling and stored at 4°C until NO_3 extraction. Samples of soil NO_3-N were extracted with 1 M KCl solution (1:5 w/v) by shaking for 1 h. The extracts were then filtered and the concentrations of NO_3-N in the soil extracts were measured colorimetrically using a UV spectrophotometer (UV-2450, Shimadzu, Japan). Each measurement was replicated three times.

Soil temperature was measured using seven CS107b soil temperature probes (Cambell Scientific Inc., Logan, UT) installed at depths of 30, 60, 90, 150, 200, 250 and 300 cm. Three-meter neutron access tubes were installed at each plot. Soil moisture at seven depths (30, 60, 90, 150, 200, 250 and 300 cm.) was measured using a neutron moisture meter when gas samples were collected. Soil temperature and water content were used to explore the relationships between calculated CO_2, N_2O and CH_4 fluxes and soil water-filled pore space (WFPS) and soil temperature at various depths. Daily rainfall was recorded at a weather station on the experimental site.

Soil gas sampling and measurements

Measurements of CO_2, N_2O and CH_4 concentrations in soil started in March 2007, i.e., 9 years after the start of the field experiment, assuming that by then the CO_2, N_2O and CH_4 production-consumption dynamics in the subsoil had been adjusted to the experimental treatments. Seven subsurface soil air equilibration tubes were installed at each site with sampling ports at 30, 60, 90, 150, 200, 250 and 300 cm in December 2006 (for more details, see reference 16). Soil-air samples were taken twice a week between 9:00 AM and 11:00 AM, using 100 ml plastic syringes connected to the tubes via the three-way stopcocks at the surface. The surface air was concurrently sampled at a height of 5 cm above the soil surface. The gas samples were analyzed by gas chromatography (Agilent GC-6820, Agilent Technologies Inc. Santa Clara, California, US) with separate electron capture detector (ECD at 330°C) for N_2O determination and flame ionization detector (FID at 200°C) for CH4 and CO2 determinations.

Calculations of gas fluxes

The basic method of our study followed that of Campbell [19]. It was assumed that the soil conditions are uniform in horizontal direction, and that the gas diffusion in soil is in one-dimensional vertical flow, that fundamentally follows Fick's law [20,21]:

$$q = -D_p \frac{\delta c}{\delta x} \tag{1}$$

Where q is the gas flux density (g gas m^{-2} soil s^{-1}), D_p is the soil-gas diffusivity (m^3 soil air m^{-1} soil s^{-1}), $\frac{\delta c}{\delta x}$ is the concentration gradient between two soil layers (g gas m^{-3} soil air m^{-1} soil).

D_p was derived from the following equation [22,23]:

$$D_p = (\varepsilon^{10/3}/E^2) \cdot D_0 \tag{2}$$

Where, D_0 is the gas diffusivity (m^2 air s^{-1}). We estimated the diffusion coefficient D_0 of CH_4, CO_2 and N_2O at 298 K and 1 kPa

Table 1. Fertilization treatments (A); and timing of crop management activities (B).

(A) Treatments	Basal fertilization, applied at wheat sowing (kg·ha^{-1})			Supplementary N fertilization (kg·ha^{-1})	
	N	P$_2$O$_5$	K$_2$O	Wheat (in April)	Maize (in July)
N0	0	65	0	0	0
N200	50	65	0	50	100
N400	100	65	0	100	200
N600	150	65	0	150	300
(B) Timing	**Crop management activities**				
	Winter-wheat season				**Summer-maize season**
October 3, 2006	Basal N fertilization and irrigation (60 mm)				
October 10, 2006	Seeding				
April 7, 2007				Supplementary N fertilization and irrigation (94 mm)	
May 19, 2007					Irrigation (60 mm)
June 1, 2007					Seeding
June 14, 2007				Harvest	
June 19, 2007					Irrigation (60 mm)
July 27, 2007					Supplementary N fertilization
July 29, 2007					Irrigation (60 mm)
October 1, 2007					Harvest
October 2, 2007	Basal N fertilization and irrigation				
October 7, 2007	Seeding				

at 1.79×10^{-5}, 1.32×10^{-5}, 1.29×10^{-5} m^2 s^{-1}, respectively, by using a semiempirical equation by Gilliland et al [24]. Parameter ε is the soil air filled porosity (m^3 air m^{-3} soil), and E is the soil porosity (m^3 voids m^{-3} soil).

The Millington-Quirk model was used to compute ε and E [25]:

$$E = 1 - \frac{\rho_b}{\rho_s} \tag{3}$$

$$\varepsilon = E - \theta \tag{4}$$

Where ρ_b is the dry bulk density (g m^{-3}) at each soil depth (Table 2), ρ_s is the average bulk density of surface soil (2.65 g m^{-3}); θ is the volumetric soil water content which was measured using a neutron moisture meter at each depth.

Calculations of annual cumulative gas fluxes

The annual cumulative emissions were obtained by multiplying the average daily flux from two consecutive measurements within a week by the number of days between the measurements, and then summing the fluxes of these periods to an accumulative flux for the whole year [26]:

$$T = \sum_{i=1}^{n} (X_i \times 24) \quad (n = 1,2,3...) \tag{5}$$

Where: T (kg ha^{-1}), X_i (kg ha^{-1} h^{-1}) and i are the accumulative GHG emission, the average daily GHG emission rate, and the number of days, respectively.

Data analyses

All data were subjected to statistical analysis (SPSS 13.0). Differences between treatments were analyzed using ANOVA, followed by LSD at the 0.05 probability level. Regression analysis was used to identify relationships between CH$_4$, CO$_2$ and N$_2$O fluxes and the climatic variables.

Results

Concentrations of CH$_4$, CO$_2$ and N$_2$O

Mean concentrations and its standard deviations of CH$_4$, CO$_2$ and N$_2$O at each depth are shown in Figure 1 A, as function of N application rates. Mean CH$_4$ concentration decreased with soil depth. Ambient air CH$_4$ concentration in the area was about 2.2 ppmv. At a depth of 30 cm, CH$_4$ concentration ranged between 1.4 and 1.6 ppmv and at depth of 60 to 300 cm between 0.3 and 0.6 ppmv. There were no clear effects of N fertilizer application on the mean CH$_4$ concentration (Figure 1 A). Mean CH$_4$ concentrations decreased significantly at soil depths of 0, 30, 60 and 90 cm (P<0.05); changes in mean concentration below a depth of 90 cm were not significant (Figure 1 C).

Mean CO$_2$ concentration increased with soil depth. At a depth of 30 cm, CO$_2$ concentration ranged between 8400 and 9900 ppmv and at depth of 60 to 300 cm between 16000 and 21000 ppmv. Mean CO$_2$ concentrations increased significantly at soil depths of 0, 30 and 60 cm (P<0.05); changes in mean concentration below a depth of 60 cm were not significant

Table 2. Soil characteristics at the experimental site in 2007.

Depth (cm)	pH (H$_2$O)	Sand (%)	Silt (%)	Clay (%)	Textural class[a]	Dry bulk Density (g cm^{-3})	Total organic matter (g kg^{-1})	Total nitrogen (g N kg^{-1})	Available nitrogen (mg N kg^{-1})	Available phosphorus (mg P kg^{-1})	Available potassium (mg K kg^{-1})
0–30	8.5	25	58	17	SSL	1.47	16.7	1.40	148	4.1	79.3
30–60	7.74	22	60	18	L	1.40	10.9	0.85	72.0	2.1	54.4
60–90	7.78	31	55	14	L	1.45	7.8	0.64	70.1	0.44	36.9
90–150	7.76	15	59	26	SCL	1.57	6.5	0.38	49.8	0.17	27.8
150–200	7.74	18	47	35	GCL	1.43	5.4	0.28	38.9	0.13	21.9
200–250	7.77	15	35	50	C	1.51	4.2	0.15	30.1	0.09	13.9
250–300	7.75	12	35	53	C	1.50	3.0	0.86	25.9	0.04	7.1

[a]SSL: Silty sandy loam; L: Loam; SCL: Silty clay loam; GCL: Gravely clay loam; C = Clay.

(Figure 1 C). There were no clear effects of N fertilizer application on the mean CO$_2$ concentrations.

Concentrations of N$_2$O were strongly influenced by agricultural management activities such as N application and irrigation. Fertilizer N applications increased the mean N$_2$O concentrations. Mean N$_2$O concentrations at depth of 30 to 300 cm ranged from 600 to 1500, 1100 to 1700, 1600 to 2100 and 2500 to 3000 ppbv for the N0, N200, N400 and N600 treatments, respectively (Figure 1 A). Mean N$_2$O concentrations increased significantly from soil surface to a depth of 30 cm (P<0.05), but changes in mean concentration below a depth of 30 cm were not significant (Figure 1 C). Fertilizer N application increased soil NO$_3$-N content; differences in mean N$_2$O concentrations were correlated with differences in mean NO$_3$-N contents in the four fertilizer N treatments (Figure 1 B).

Fluxes of CH$_4$ in soil

Diffusive fluxes between soil layers and between soil and atmosphere were calculated from the concentration gradients, using equation 1. There was a net influx of atmospheric CH$_4$ into the top 0–60 cm (Figure 2), suggesting consumption of CH$_4$ by methanotropic bacteria. Interestingly, the calculated fluxes into the soil were rather similar for the 0–30 and 30–60 cm soil layers, suggesting similar CH$_4$ uptake rates. Uptake of CH$_4$ apparently also occurred in the layers 60–90, 90–150 and 150–200 cm during the first one or two months of the measurement period (Figure 2). However, we cannot exclude the possibility that this apparent uptake of CH$_4$ in the subsoil during the first two months is an artifact related to the installation of the samplers when atmospheric CH$_4$ may have diffused into the subsoil. Fluxes between soil layers were negligible small during most of the maize growing season (Figure 2).

Annual cumulative fluxes of CH$_4$ for all soil layers and N fertilizer treatments are shown in Table 3 A. Evidently, the influx of atmospheric CH$_4$ decreased with soil depth. During the study period, mean calculated uptake was about 176 g CH$_4$ per ha by the top 30 cm, 252 g CH$_4$ per ha by the soil layer 30–60 cm, 98 g CH$_4$ per ha by the layer 60–90 cm and 22 g CH$_4$ per ha below a depth of 90 cm; mean calculated uptakes in the layers 0–30 and 30–60 cm were both significantly higher than that in the layer 90–300 cm (P<0.05) (Figure 1 C). Annual cumulative CH$_4$ uptake in the layer 0–90 cm (526 g CH$_4$ per ha per year) contributed about 96% to that in the layer 0–300 cm (547 g CH$_4$ per ha per year). Annual cumulative uptake in 0–30 cm layer is relatively low compared to literature data [4–6,27].

Fluxes of CO$_2$ in soil

There was a large efflux of CO$_2$ from the top 30 cm of soil to the atmosphere from March till June, i.e., during the second half of the wheat growing season, and from August till October, i.e., during the second half of the maize growing season (Figure 3). The same holds for the upward flux from the layer 30–60 to the layer 0–30 cm. These patterns were related to the crop growing seasons of wheat and maize, and to the changes in water filled pore space (WFPS) and soil temperature. There were no clear relationships between N treatments and CO$_2$ fluxes. Treatment N200 had the smallest flux from the layer 0–30 cm to the atmosphere, but the largest from 30–60 cm to 0–30 cm from April to May. Upward fluxes from the layers 60–90 cm and especially below this layer were much smaller. There were small but significant changes in fluxes in the subsoil at the transition of the winter-wheat growing season to the summer-maize growing season (Figure 3).

Surprisingly, annual cumulative fluxes of CO$_2$ tended to decrease with increasing N fertilizer application rates (Table 3

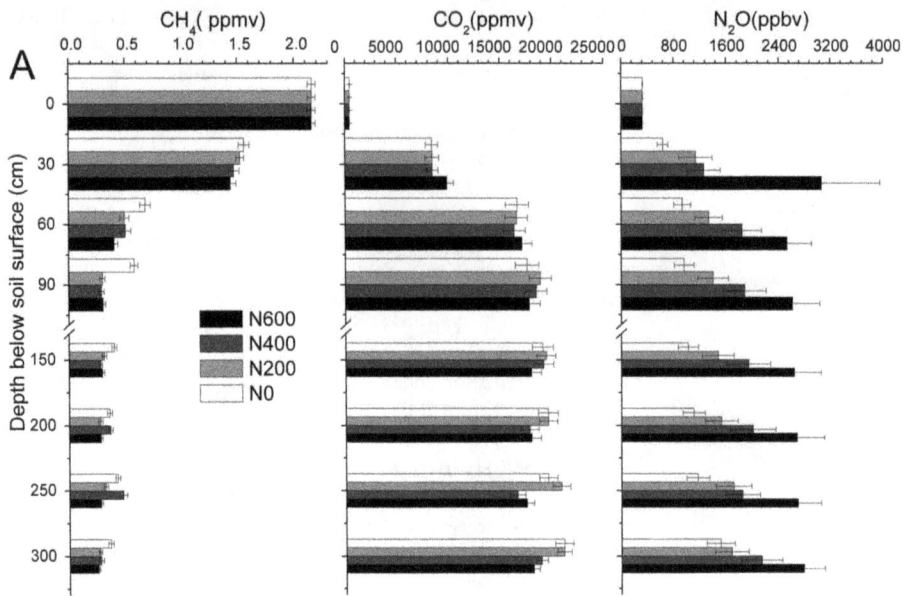

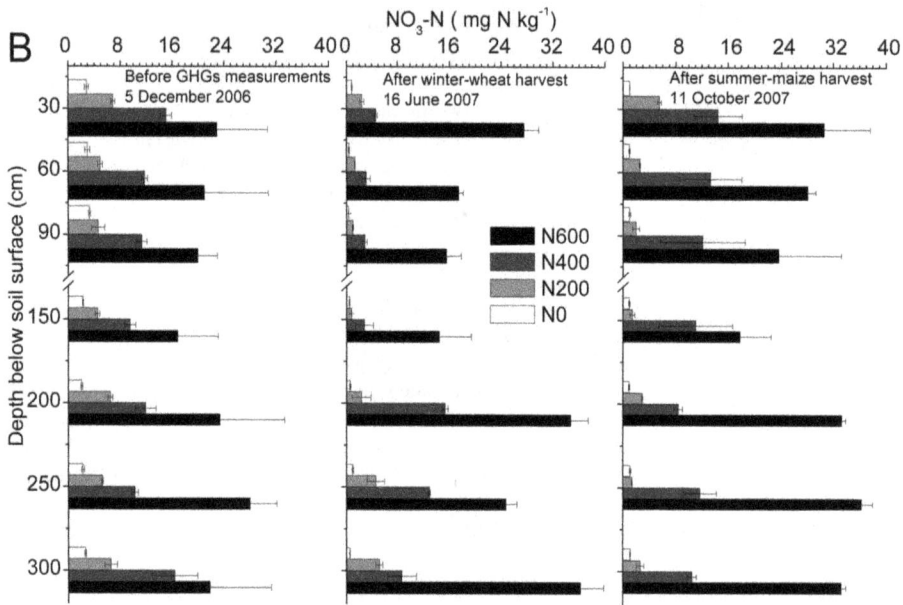

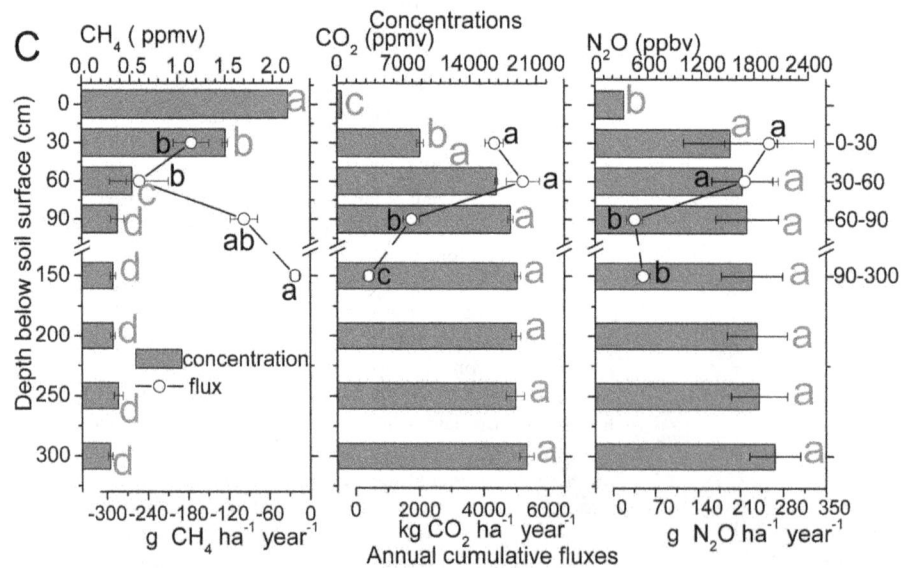

Figure 1. CH_4, CO_2 and N_2O concentrations (mean ± standard deviations, n = 3) in soil air at various soil depths in a winter wheat-summer maize double cropping rotation receiving 0, 200, 400 and 600 kg of N ha^{-1} year^{-1}, in 2007–2008 (A); NO_3-N contents (mean ± standard deviations, n = 3) at various soil depths as function of N fertilizer application rate, in 2007–2008 (B); Profiles of concentration and annual cumulative flux of CH_4, CO_2 and N_2O, in 2007–2008 (mean ± standard deviations, n = 4). Same letters next to the bars indicated no significant differences between slope positions (P<0.05). (C). Note the differences in X-axes.

B). Moreover, cumulative upward fluxes were somewhat larger from the layer 30–60 cm (mean 5,227; range 3,800–6,000 kg CO_2 per ha) than from the layer 0–30 cm to the atmosphere (mean 4,331; range 3,800–5,000 kg CO_2 per ha) (Figure 1 C; Table 3 B).

This suggests that a relatively large portion of total respiration in soil took place in the layer 30–60 cm. However, we can not exclude the possibility that the calculated CO_2 efflux from the top layer is underestimated, because the concentration gradient in the

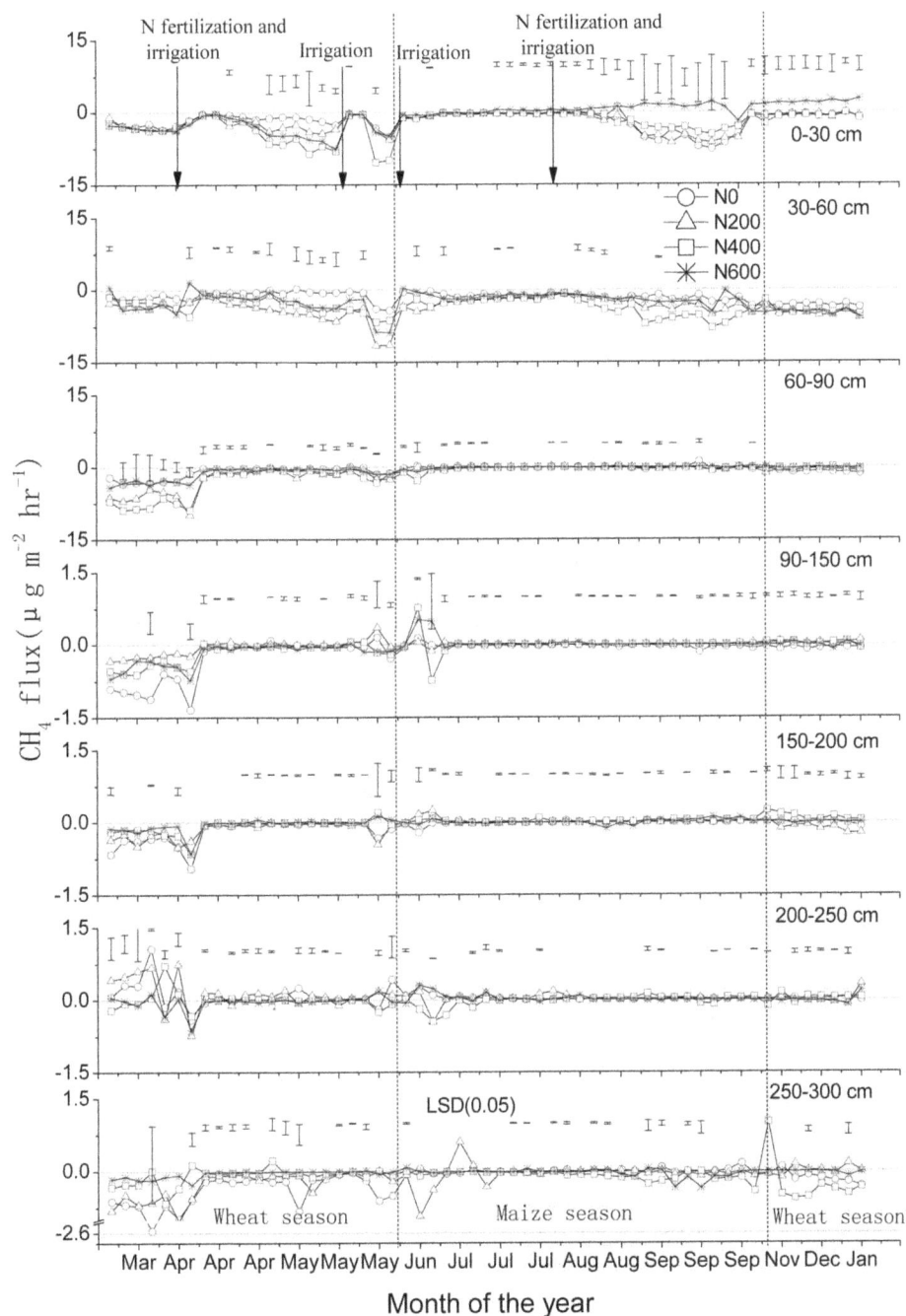

Figure 2. CH_4 flux rates (means ± standard deviations, n = 3) at various soil depths in a winter wheat-summer maize double cropping rotation receiving 0, 200, 400 and 600 kg of N ha^{-1} year^{-1}, in 2007–2008. Vertical dashed lines indicate a change in crop. Bars in figures indicate 1 standard deviation (n = 3). Note the differences in Y-axes.

Table 3. Annual cumulative emissions of CH_4 and N_2O (in g ha^{-1} yr^{-1}) and of CO_2 (in kg ha^{-1} yr^{-1}) between soil layers.

(A) CH₄

Treatments	0–30 cm	30–60 cm	60–90 cm	90–150 cm	150–200 cm	200–250cm	250–300 cm	0–300cm
N0	−167 (12) b	−138 (13) a	−63 (2) a	−15 (0.5) c	−9 (0.1) d	5 (0.04) a	−21 (0.3) d	−408
N200	−201 (9) bc	−322 (10) c	−123 (3) b	−2 (0.07) a	−7 (0.5) c	5 (0.08) a	−12 (0.6) c	−662
N400	−228 (3) c	−318 (8) c	−140 (10) b	−6 (0.5) b	−1 (0.1) a	−1 (0.06) b	−10 (0.4) b	−704
N600	−106 (15) a	−231 (5) b	−64 (7) a	−6 (0.5) b	−2 (0.1) b	−1 (0.05) b	−4 (0.1) a	−414

(B) CO₂

Treatments	0–30 cm	30–60 cm	60–90 cm	90–150 cm	150–200 cm	200–250cm	250–300 cm	0–300cm
N0	4,986 (401) a	6,060 (394) a	1,730 (121) b	163 (9) a	92 (1) a	12 (0.1) b	294 (15) a	13,337
N200	3,819 (652) a	6,019 (424) b	2,194 (177) a	194 (19) a	19 (0.5) c	263 (8) a	198 (4) c	12,706
N400	4,640 (225) a	4,955 (174) c	2,100 (112) ab	152 (15) a	−48 (1) d	−1 (0.01) c	251 (12) b	12,049
N600	3,880 (383) a	3,874 (147) c	1,049 (120) c	11 (2) b	33 (0.6) b	−79 (0.5) d	116 (4) d	8,884

(C) N₂O

Treatments	0–30 cm	30–60 cm	60–90 cm	90–150 cm	150–200 cm	200–250cm	250–300 cm	0–300cm
N0	93 (7) c	90 (2) d	23 (3) b	6 (0.3) a	9 (0.3) b	4 (0.4) d	62 (1) a	287
N200	226 (23) b	199 (7) c	24 (1) b	−1 (0.02) b	12 (0.6) a	51 (2) a	−20 (2) d	491
N400	263 (27) b	358 (5) a	23 (2) b	−6 (0.2) d	−2 (0.2) c	20 (1) b	35 (0.5) b	691
N600	447 (1) a	222 (2) b	74 (4) a	−3 (0.1) c	−2 (0.1) c	11 (0.7) c	22 (2) c	771

Values (means with SE in the brackets) followed by the same letter are not significantly different within columns (one-way ANOVA with LSD; $P < 0.05$)

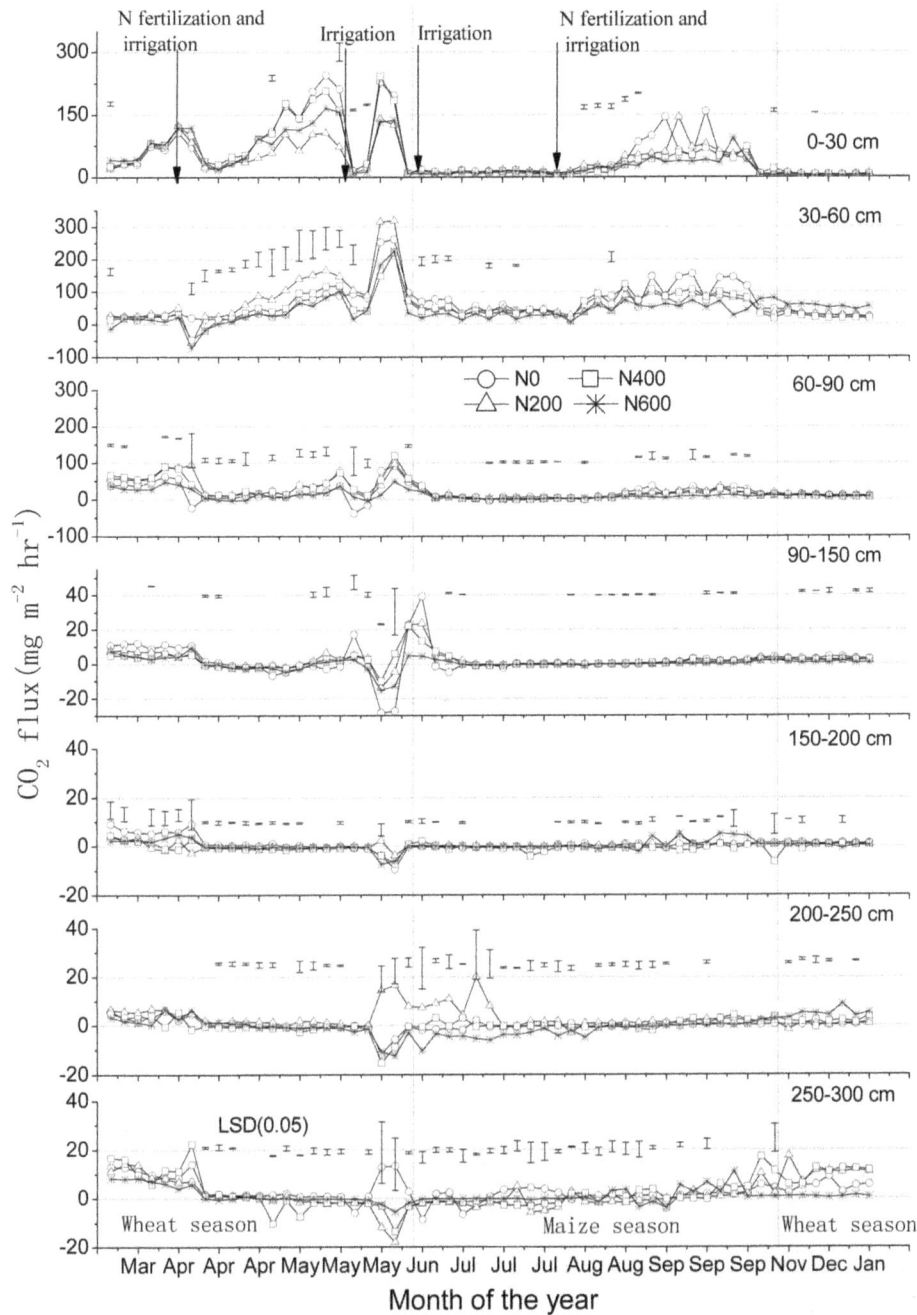

Figure 3. CO₂ flux rates (means ± standard deviations, n = 3) at various soil depths in a winter wheat-summer maize double cropping rotation receiving 0, 200, 400 and 600 kg of N ha⁻¹ year⁻¹, in 2007–2008. Vertical dashed lines indicate a change in crop. Bars in figures indicate 1 standard deviation (n = 3). Note the differences in Y-axes.

upper 0–30 cm soil layer was averaged, and soil diffusivity may be higher in the top few cm than the bulk of the top 30 cm of soil [28]. Annual cumulative CO_2 flux in the layer 0–90 cm (11,327 kg CO_2 per ha per year) contributed about 97% to that in the layer 0–300 cm (11,744 kg CO_2 per ha per year). Mean calculated fluxes in the layers 0–30, 30–60 and 60–90 cm were all significantly higher than that in the layer 90–300 cm (P<0.05); mean annual cumulative fluxes from the soil below 90 cm were very small and in upwards direction (Figure1 C; Table 3 B).

Fluxes of N₂O in soil

Fertilizer application, irrigation and precipitation events triggered an efflux of N_2O from the topsoil to the atmosphere (Figure 4). The peak efflux, associated with the supplemental N fertilizer application and flooding in early April (wheat growing season), was accompanied with significant downward directed fluxes below the topsoil layer (0–30 cm). There was another relatively large efflux of N_2O into the atmosphere during the relatively moist and warm August summer month (maize growing season) (Figures 4 and 5), but this peak was not accompanied with significant downward directed fluxes below the topsoil layer. In the

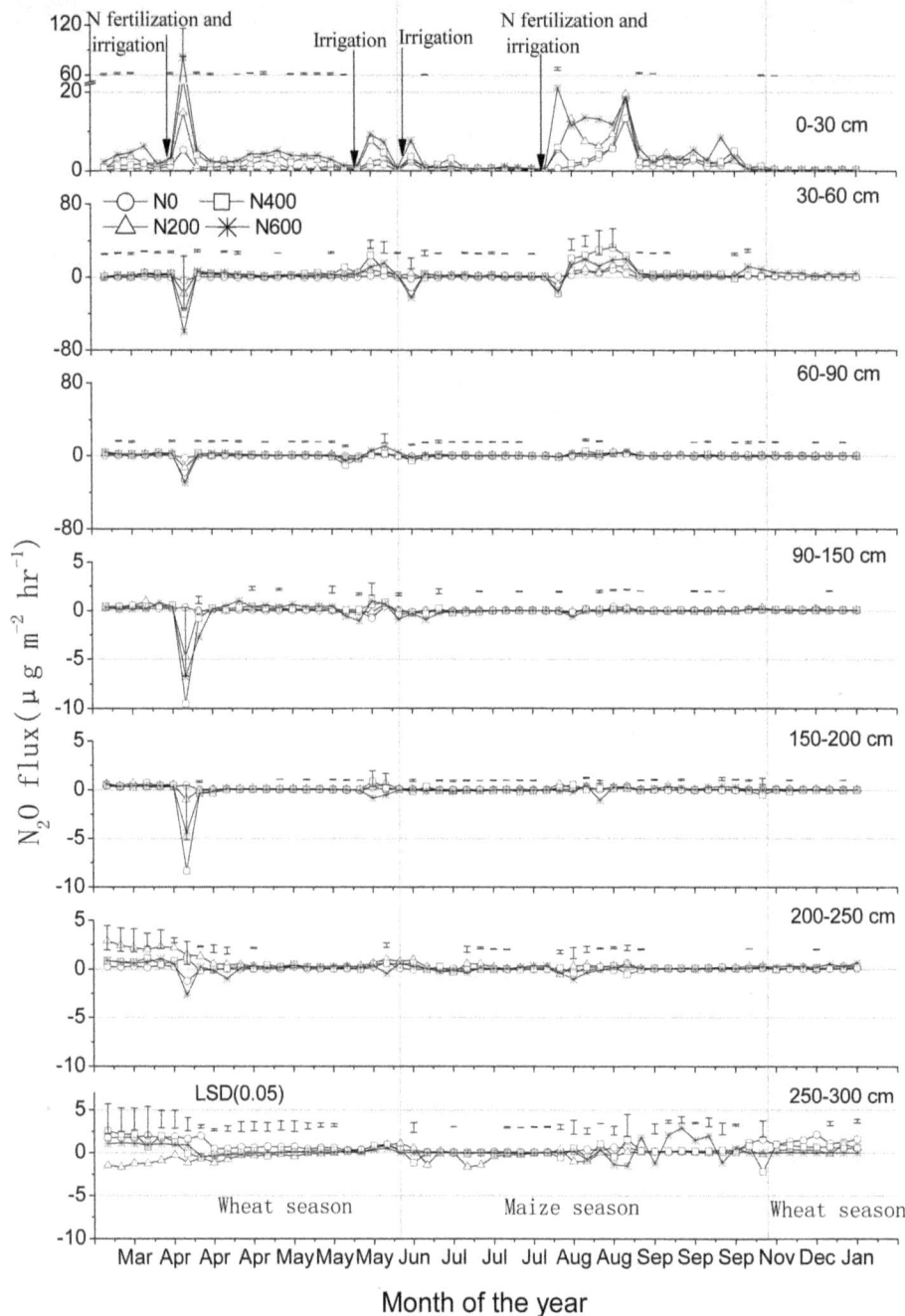

Figure 4. N₂O flux rates (means ± standard deviations, n = 3) at various soil depths in a winter wheat-summer maize double cropping rotation receiving 0, 200, 400 and 600 kg of N ha⁻¹ year⁻¹, in 2007–2008. Vertical dashed lines indicate a change in crop. Bars in figures indicate 1 standard deviation (n = 3). Note the differences in Y-axes.

subsoil, fluxes were relatively small and directions variable (Figure 4). Essentially all seasonal fluctuations of N_2O flux rates in the subsoil (60–200 cm) seem to be related to fertilizer application, irrigation and rainfall events and changes in WFPS; therefore, there was no clear evidence of N_2O production in the subsoil after excluding these influence of interfering factors [16].

Annual cumulative fluxes of N_2O increased with increasing N fertilizer application rates; calculated total emissions at the soil surface were 93, 226, 263 and 447 g N_2O per ha for the N0, N200, N400 and N600 treatments, respectively; net upward fluxes

from the 30–60 cm layer were almost as large (90, 199, 358 and 222 g N_2O per ha for the N0, N200, N400 and N600 treatments, respectively) as the fluxes from the 0–30 cm layer to the atmosphere (Table 3 C). Mean calculated fluxes in the layers 0–30 and 30–60 cm were both significantly higher than those in the layers 60–90 and 90–300 cm (P<0.05) (Figure 1 C); mean annual cumulative fluxes from the soil below 90 cm were small but mostly in upwards direction, suggesting that the subsoil was a small source of N_2O, and/or that accumulated N_2O from the previous season contributed to the net upward directed fluxes.

Relations between WFPS and temperature and CH_4, CO_2 and N_2O fluxes

Linear regression relationships between WFPS and CH_4 fluxes (positive) and between WFPS and CO_2 fluxes (negative) were statistically significant (p<0.05) for almost all soil layers (Table 4). Uptake of CH_4 by the soil was relatively high when WFPS was relatively low, probably because the diffusion rate of CH_4 into the soil was high when soil was dry, and vice versa [6]. Similarly, the upward transport of CO_2 was low when WFPS was high, and vice versa. This indicates that soil moisture exerted a dominant control on CH_4 and CO_2 fluxes. The linear relationship between WFPS and N_2O flux was also significant for the soil layers 200–250 and 250–300 cm (p<0.05), but not for the other layers. Significant downward directed fluxes below the topsoil were only observed down to a deep of 200 cm (Figure 4); note that N_2O fluxes were very low in 200–300 cm soil layer. Apparently, in 200–300 cm deep soil profile, soil moisture exerted a dominant control on nitrification and denitrification processes. But in 0–200 cm soil layer, WFPS was not the dominant controlling factor for the diffusive N_2O flux, likely the combination of WFPS, ammonia, nitrate and metabolizable carbon, because these factors commonly control nitrification and denitrification processes.

Relationships between soil temperature and CH_4, CO_2 and N_2O fluxes showed relatively large scatter (Table 4). Evidently, high temperatures are associated with the summer season, which is relatively moist (Figure 5). The significant relationships (p<0.05) between soil temperature and CH_4 fluxes at depth of 60–300 cm may be the result in part of the covariance between WFPS and soil temperature. Fluxes of N_2O in soil were not significantly related to soil temperature (Table 4).

Discussion

Fluxes of CH_4, CO_2 and N_2O at the interface of soil and atmosphere are the net result of production, consumption and transport in the soil [11]. In this study, we inferred fluxes in the soil profile from changes in concentrations with depth and over time,

so as to identify soil horizons of CH_4, CO_2 and N_2O production and consumption, and thereby to increase the understanding of the dynamics of the net fluxes at the interface of soil and atmosphere. The study is unique in the sense that the inference of subsurface fluxes of CH_4, CO_2 and N_2O in 300 cm deep soil profiles at high temporal resolution over a full year has not been reported before in such comprehensive manner.

Though the C and N cycles are intimately linked in the biosphere, there were significant differences in the dynamics of CH_4, CO_2 and N_2O production, consumption and transport in the studied soil. The concentration profiles have distinct characteristics (Figure 1 A); the seasonal dynamics were much larger in the topsoil than subsoil. Moreover, the seasonal dynamics in inferred fluxes occurred during distinct periods (Figures 2, 3 and 4), and these were related to changes in WFPS (Figure 5 A), following rainfall and irrigation events. Fertilizer N application affected N_2O fluxes greatly, but not those of CH_4 and CO_2. The soil under the winter wheat-summer maize double cropping system was a net sink of atmospheric CH_4 and a net source of N_2O. It was also a large source of CO_2 but it is unknown whether the efflux compensated the influx of C into the soil via plant growth, as the latter influx was not measured.

The inferred fluxes at the soil-atmosphere interface (Figure 1 C; Table 3) were relatively small compared to those observed in other studies [5,6,29]. Our estimated soil surface fluxes are very likely underestimates because the depth resolution of the gas samplers in the top soil was too low to capture the curvature of the concentration profile properly. Hence, our study may have underestimated the dynamics of the CH_4, CO_2 and N_2O fluxes in the top soil. The depth resolution of the sampling below 90 cm appeared to be adequate. Below, we discuss the dynamics of the CH_4, CO_2 and N_2O fluxes in the soil profile in more detail.

CH_4 flux

Application of fertilizer N has been shown to inhibit CH_4 oxidation in soil [30,31], and several studies noted that non amended soils act as sink of CH_4 [32–34]. In our study, seasonal

Table 4. Linear regressions for the relationship between climatic variables and GHG fluxes.

Climatic variable	Soil depth (cm)	CH_4 ($\mu g\ m^{-2}\ hr^{-1}$)	CO_2 ($mg\ m^{-2}\ hr^{-1}$)	N_2O ($\mu g\ m^{-2}\ hr^{-1}$)
Soil water filled pore space	0–30	0.853**	−0.775**	−0.149
	30–60	0.645**	−0.372**	0.084
	60–90	0.787**	−0.852**	0.067
	90–150	0.637**	−0.447**	0.093
	150–200	0.771**	−0.268*	0.084
	200–250	−0.146	−0.289*	−0.692**
	250–300	0.763**	−0.532**	−0.579**
Soil temperature	0–30	−0.001	0.042	0.104
	30–60	0.270*	0.356**	0.211
	60–90	0.635**	−0.348**	0.093
	90–150	0.620**	−0.225	0.062
	150–200	0.575**	−0.266*	−0.075
	200–250	−0.122	−0.299*	−0.493**
	250–300	0.473**	−0.323*	−0.263*

Pearson's correlation coefficient, 2-tailed tests of significance.
**Significant correlation at a <0.01.
*Significant correlation at a <0.05.

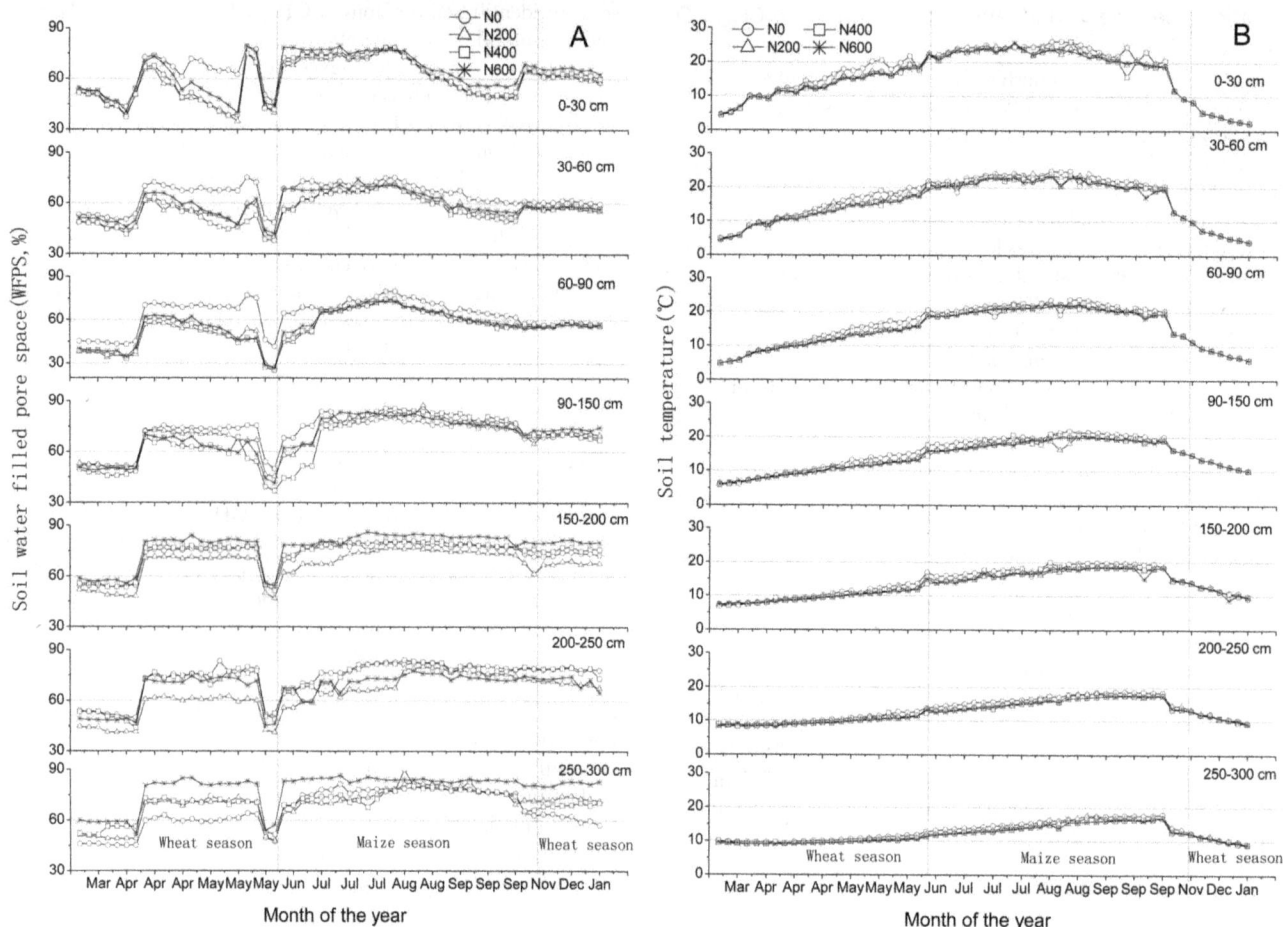

Figure 5. Water-filled pore space (WFPS) at various soil depths in a winter wheat-summer maize double cropping rotation receiving 0, 200, 400 and 600 kg of N ha^{-1} year^{-1}, in 2007–2008. Bars in figures indicate 1 standard deviation (n = 3). (A); Soil temperatures at various soil depths in winter wheat-summer maize double cropping rotation receiving 0, 200, 400 and 600 kg of N ha^{-1} year^{-1}, in 2007–2008.(B)

mean emission rates and annual cumulative fluxes of CH_4 for all soil layers and N fertilizer treatments were consistently directed downward (Figures 1 C and 2; Table 3 A). Though statistical significant differences in cumulative CH_4 fluxes between fertilizer N treatments were observed (Table 3 A), there was no clear trend that an increase in total N application decreased CH_4 uptake by soil. Inferred uptake was higher in the N200 and N400 treatments than in the N0 and N600 treatments at depth of 30 to 90 cm.

The magnitude of methane uptake by soils is largely controlled by diffusion of atmospheric methane into the soil [35], which in turn is strongly influenced by soil moisture [28]. The rate of diffusion of CH_4 in soil was high when WFPS was low. Our results showed a significant negative linear correlation between CH_4 uptake rate and WFPS for almost all layers; and the highest CH_4 uptake rates took place when WFPS was under 70% (Figures 2 and 5 A; Table 4). This is in agreement with the studies by Guo et al, Wu et al and Wang et al [4,6,16]. Inferred downward CH_4 fluxes decreased with depth (Figures 1 C and 2; Table 3 A). It has been reported that methanotrophic activity is most pronounced in the top soil [4,27], but our study suggests that significant uptake took place up to depths of 60 to 90 cm; below 90 cm, inferred fluxes of CH_4 were negligibly small (Figure 1 C).

CO_2 flux

Application of 200 kg fertilizer N per ha per year and more roughly doubled grain yields relative to the control treatment [16], but did not have statistical significant effects on the CO_2 efflux from the soil and the diffusive flux in the soil profile (Figure 3; Table 3 B). Apparently, fertilizer N application affected predominantly aboveground biomass production, and not so much underground biomass production and respiration. Yet, we may have missed some of the topsoil dynamics, also because the incorporation of the stubbles by ploughing was in the top 15 cm of soil only. A relatively large portion of total respiration in soil took place in the layer 30–60 cm; and the total respiration in the layer was significantly higher than those in the layers 60–90 and 90–300 cm (P<0.05) (Figure 1 C).

When soil WFPS ranged between 40 and 70% and soil temperature was >10 °C (Figure 5), highest CO_2 fluxes took place at depth of 0–60 cm during the second half of the growing seasons of wheat and maize, i.e., from mid-April to mid-May and from mid-August to mid-September (Figures 3 and 5 A). These elevated CO_2 emissions are attributed to root respiration and to enhanced mineralization of soil organic matter by increased microbial activity [36], but also to changes in the stability and formation of soil aggregates and in the microbial community structure [37]. Sufficient soil moisture is needed to allow and support substrate diffusion to the sites of microbial activity. However, if soil moisture values exceed certain thresholds (which do depend on soil properties such as porosity, bulk density and SOC content) microbial soil respiration can get O_2 limited due to diffusion constrains [6]. In a saturated soil, air is pushed out of soil pore spaces and root respiration further depletes O_2 in the soil air [38,39]. In our study, a significant negative linear correlation was found between CO_2 flux rate and WFPS (40–70%) in all layers (Table 4); and low CO_2 fluxes took place when WFPS exceeded 70%, especially after irrigation or heavy rainfall events, i.e., from June to August (Figures 3 and 5 A). This assertion is also supported by results presented by Davidson et al, Jassal et al and Fang et al [38–40].

Generally, the CO_2 evolution from soil is directly correlated with soil temperature, though within a certain temperature range [41,42], and depending on the presence of active roots [43]. In our study, the relationship of soil temperature and CO_2 fluxes was variable, likely because of the dominant effect of WFPS (Table 4).

N_2O flux

Nitrogen application and irrigation/rainfall are main triggers for increased N_2O concentrations in a soil profile and for increased emissions [15,16]. The top soil was the source of N_2O production. The combined urea applications and irrigations in early April and by the end of July 2007 strongly increased NO_3^- (Figure 1 B), NH_4^+ (data not shown) contents and WFPS (Figure 5 A) in soil, and induced large upward directed fluxes in the upper 0–30 cm soil layer; interestingly, relatively large downward directed fluxes in the subsoil only took place in April (up to the depths of 60 and 90 cm) (Figure 4), we cannot exclude the possibility that the apparent downward directed peaks during the first two months probably related to the soil structure disturbance that resulted from the installation of the samplers in December 2006.

In soil, N_2O is mainly produced by nitrification and denitrification processes. The most important factors controlling these processes are NH_4^+ and NO_3^- contents, O_2 partial pressure, and available carbon to fuel heterotrophic denitrification [44,45]. The rapid increases of WFPS (Figure 5 A), NO_3-N content (Figure 1 B) and N_2O productions in the subsoil (Figures 1 A and 4) would suggest that convective transport contributed to the downward transport of water and solutes (especially in the maize growing season), which is in line with other observations [2,46]. For instance, significant upward directed fluxes of N_2O took place during the relatively moist and warm August (maize growing season) than in the preceding wheat season in 30–60 cm soil layer (Figure 4). Several studies have demonstrated that higher values of soil moisture and temperature result in higher N_2O fluxes [44–46] Also, Li et al found while carrying out a three-year field experiment at the same study sites that significant NO_3- leaching events occur predominantly during August to October (maize growing season) [47]. Zhu et al found while carrying out a four-year field experiment in a hillslope cropland that soil NO_3-concentrations in the subsurface soil (15–30 cm) were higher than in the topsoil (0–15 cm) during most of the maize season, indicating a rapid and effective transport of NO_3- to the subsurface soil following over irrigation or rainfall events [48]. Our results indicate that NO_3-N contents in soil layers after the maize harvest were higher than after the wheat harvest (Figure 1 B); but due to missing measurements, we can not fully rule out that high NH_4^+ content [47] may have contributed to relatively high N_2O emissions in the warm and wet maize season. Although N_2O concentration increased with soil depth, changes in inferred N_2O flux below a depth of 60 cm were relatively small (Figures 1 A and 4). It may be related to the variation of the vertical N_2O

concentration gradient; changes in mean N_2O concentration below a depth of 30 cm were not significant (Figure 1 C).

It has been frequently observed that high rates of N_2O emissions take place when WFPS ranges between 30 and 70% [49]. According to Zou et al the N_2O production in dry land soil of Northern China is mostly driven by nitrification [50]. Wang et al suggests that nitrification is likely a main source of the N_2O production in the soil profile when WFPS varied between 45 and 70% at the study site [16]. N_2 starts being emitted through denitrification at a WFPS of 70%, and is the main N gas emitted when WFPS exceeds 75% [51]. This may explain the relatively high inferred N_2O flux from late April to mid-May (WFPS, 40–70%) and the very low flux from late June to late July (before applying nitrogen) (WFPS, >70%) in the layer 0–30 cm (Figures 4 and 5 A).

The accumulated N_2O fluxes were significantly related to N application rate. This was most apparent in the top 30 to 90 cm of soil (Table 3 C). Annual cumulative N_2O flux in the layer 0–90 cm (511 g N_2O per ha per year) contributed about 90% to that in the layer 0–300 cm (560 g N $_2O$ per ha per year). The 90 cm thick cinnamon top soil overlays the so-called Shajiang layer (90–140 cm) with silty clay loam texture [52]. The Shajiang layer has no crop roots, contains many iron-manganese nodules and has high bulk density (Table 2). This compacted subsoil may explain that fertilizer application and irrigation mainly affected N_2O fluxes down to 90 cm (Figure 1 C; Table 3 C). In this study, calculated total emissions in the layer 0–90 cm were 206, 449, 644 and 743 g N_2O per ha for the N0, N200, N400 and N600 treatments, respectively; these fluxes translate into fertilizer-derived emissions of 0.14, 0.10 and 0.07% for the N200, N400 and N600 treatments, respectively. The fertilizer induced emission factors (0.07–0.14%) were lower than the 0.30–0.39% measured by Ding et al [53] over the maize-wheat rotation year in a long-term mineral nitrogen addition field experiments (150–300 kg N ha^{-1} year^{-1}, over 20-years) in the North China Plain. A reason for the lower fertilizer induced emission factor is probably related to the likely underestimates of soil surface N_2O fluxes, because the concentration gradient in the upper 0–30 cm soil layer was averaged, and soil diffusivity may be higher in the top few cm than the bulk of the top 30 cm of soil [28]. Furthermore, due to missing measurements we can not fully rule out the indirect N_2O emissions from leaching and atmospheric deposition [47,53].

Conclusions

Our study is one of few that inferred CH_4, CO_2 and N_2O transport between soil layers from changes in CH_4, CO_2 and N_2O concentrations in the upper 300 cm of soil, measured at (bi)-weekly time intervals for one year in a winter wheat-summer maize double crop rotation. The top 30 to 60 cm of soil was a sink of atmospheric CH_4, and a source of both CO_2 and N_2O. There was little or no evidence that the subsoil (>90 cm) acted as a sink or source of GHG; rather it acted as "reservoir".

Nitrogen fertilizer application increased N_2O fluxes but did not affect CH_4 and CO_2 fluxes. The fertilizer-derived N_2O flux was small, likely because our sampling design may have missed N_2O production in the top 15 cm of soil. This holds as well for the CH_4 consumption by soil and the CO_2 emissions from soil; both are likely underestimated. Soil moisture (WFPS) was found to play an important regulating role for CH_4, CO_2 and N_2O fluxes in soil and between soil and atmosphere. Both CH_4 consumption and CO_2 fluxes in and from soil all tended to decrease with increasing WFPS.

More than 90% of the annual cumulative GHG fluxes originated at depths shallower than 90 cm. Mostly because the productive soil of our study site in the North China Plain had two distinct layers (0–90 and >90 cm), with different texture and bulk density. These differences showed up in characteristic differences in GHG concentration profiles and fluxes.

Acknowledgments

The authors would like to thank Luancheng Agroecosystem Experimental Station, Chinese Academy of Sciences for tireless efforts with maintaining the long-term fertilizer experiments. The authors would also like to thank two anonymous reviewers, whose suggestions and comments greatly improved the manuscript.

Author Contributions

Conceived and designed the experiments: CSH YYW HM YMZ XXL WXD OO DAS. Performed the experiments: YYW HM CSH. Analyzed the data: YYW. Wrote the paper: YYW.

References

1. IPCC (2007) Agriculture. In: Metz, B., D.O.R., Bosch P.R. (Eds.), Climate Change 2007: Mitigation, Contribution of Working Group III to the Fourth Assessment Report of the Intergovernmental Panel on Climate Change. Cambridge University Press, Cambridge. United Kingdom and New York, NY, USA.

2. Jassal RS, Black TA, Trofymow AJ, Roy R, Nesic Z (2010) Forest-floor CO_2 and N_2O flux dynamics in a nitrogen-fertilized Pacific Northwest Douglas-fir stand. Geoderma 157(3–4): 118–125.

3. Smith P, Martino D, Cai ZC, Gwary D, Janzen H, et al. (2008) Greenhouse gas mitigation in agriculture. Philosophical Transactions of the Royal Society B: Biological Sciences 363: 789–813.

4. Guo JP, Zhou CD (2007) Greenhouse gas emissions and mitigation measures in Chinese agroecosystems. Agricultural and Forest Meteorology 142: 270–277.

5. Kim Y, Ueyama M, Nakagawa F, Tsunogai U, Harazono Y, et al. (2007) Assessment of winter fluxes of CO_2 and CH_4 in boreal forest soils of central Alaska estimated by the profile method and the chamber method: a diagnosis of methane emission and implications for the regional carbon budget. Tellus 59B: 223–233.

6. Wu X, Yao Z, Brüggemann N, Shen ZY, Wolf B, et al. (2010) Effects of soil moisture and temperature on CO_2 and CH_4 soil-atmosphere exchange of various land use/cover types in a semi-arid grassland in Inner Mongolia, China. Soil Biology & Biochemistry 42: 773–787.

7. Banger K, Tian HQ, Lu CQ (2012) Do nitrogen fertilizers stimulate or inhibit methane emissions from rice fields? Global Change Biology 18: 3259–3267.

8. Sanz-Cobena A, Sánchez-Martín L, García-Torres L, Vallejo A (2012) Gaseous emissions of N_2O and NO and NO_3– leaching from urea applied with urease and nitrification inhibitors to a maize (Zea mays) crop. Agriculture, Ecosystems and Environment 149: 64–73.

9. Sanz-Cobena A, García-Marco S, Quemada M, Gabriel JL, Almendros P, et al. (2014) Do cover crops enhance N_2O, CO_2 or CH_4 emissions from soil in Mediterranean arable systems? Science of the Total Environment 466–467: 164–174.

10. Valentini R, Matteucci G, Dolman AJ, Schulze ED, Rebmann C, et al. (2000) Respiration as the main determinant of carbon balance in European forests. Nature 404: 861–865.

11. Bowden WB, Bormann FH (1986) Transport and loss of nitrous oxide in soil water after forest cutting. Science 233: 867–869.

12. Tang JW, Baldocchi DD, Qi Y, Xu LK (2003) Assessing soil CO_2 efflux using continuous measurements of CO_2 profiles in soils with small solid-state sensors. Agricultural and Forest Meteorology 118: 207–220.

13. Fierer N, Chadwick OA, Trumbore SE (2005) Production of CO_2 in Soil Profiles of a California Annual Grassland. Ecosystems 8: 412–429.

14. Gebert J, Röer IU, Scharff H, Roncato CDL, Cabral AR (2011) Can soil gas profiles be used to assess microbial CH_4 oxidation in landfill covers? Waste Management 31: 987–994.

15. Reth S, Graf W, Gefke O, Schilling R, Seidlitz HK, et al. (2008) Whole-year-round Observation of N_2O Profiles in Soil: A Lysimeter Study. Water Air Soil Pollut: Focus 8:129–137.

16. Wang YY, Hu CS, Ming H, Zhang YM, Li XX, et al. (2013) Concentration profiles of CH_4, CO_2 and N_2O in soils of a wheat–maize rotation cosystem in North China Plain, measured weekly over a whole year. Agriculture, Ecosystems & Environment 164: 260–272.

17. Sparks DL (1996) Methods of soil analysis, part 3–chemical methods. In: Sparks, D.L. (Eds.), SSSA book series, No. 5. SSSA, Inc and American Society of Agronomy, Inc. Madison, WI. pp. 475–1185.

18. Gee GW, Bauder JW (1986) Particle-size analysis. In: Klute, A. (Ed.), Methods of Soil Analysis Part 1-Physical and Mineralogical Methods. American Society of gronomy, Madison, WI. pp. 383–409.

19. Campbell GS (1985) Soil Physics with BASIC. Elsevier Science Publishers BV, Amsterdam. Soil Science Society of America Book Series: 5 Methods of Soil Analysis Part I-Physical and Mineralogical Methods.

20. Marshall TJ (1959) The diffusion of gas through porous media. J Soil Sci 10: 79–82.

21. Rolston DE (1986) 47 Gas Flux, 1103–1109. American Society of Agronomy, Inc. Soil Science Society of America, Inc. Publisher. Madison, Wisconsin USA.

22. Sallam A, Jury WA, Letey J (1984) Measurement of gas-diffusion coefficient under relatively low air-filled porosity. Soil Sci Soc Am J 48: 3–6.

23. Jury WA, Gardner WR, Gardner WH (1991) Soil Physics, fifth ed. Wiley, New York.

24. Gilliland E, Baddour R, Perkinson G, Sladek KJ (1974) Diffusion on surfaces. I. Effect of concentration on the diffusivity of physically adsorbed gases. Ind Eng Chem Fundam 13(2): 95–100.

25. Millington R, Quirk JP (1961) Permeability of porous solids. Trans Faraday Soc 57: 1200–1207.

26. Wang YY, Hu CS, Zhu B, Xiang HY, He XH (2010) Effects of wheat straw application on methane and nitrous oxide emissions from purplish paddy fields. Plant, Soil and Environ 56 (1): 16–22.

27. Stiehl-Braun PA, Powlson DS, Poulton PR, Niklaus PA (2011) Effects of N fertilizers and liming on the micro-scale distribution of soil methane assimilation in the long-term Park Grass experiment at Rothamsted. Soil Biology and Biochemistry 43(5): 1034–1041.

28. Shrestha BM, Sitaula BK, Singh BR, Bajracharya RM (2004) Fluxes of CO_2 and CH_4 in soil profiles of a mountainous watershed of Nepal as influenced by land use, temperature, moisture and substrate Addition. Nutrient Cycling in Agroecosystems 68: 155–164.

29. Chu H, Hosen Y, Yagi K (2004) Nitrogen oxide emissions and soil microbial properties as affected by N-fertilizer management in a Japanese Andisol. Soil Science and Plant Nutrition 50: 287–292.

30. Steudler PA, Bowden RD, Melillo JM, Aber JD (1989) Influence of nitrogen fertilisation on methane uptake in temperate forest soils. Nature 341: 314–316.

31. Kravchenko I, Boeckx P, Galchenko V, Van Cleemput O (2002) Shortand medium-term effects of $NH_4{}^+$ on CH_4 and N_2O fluxes in arable soils with a different texture. Soil Biology & Biochemistry 34: 669–678.

32. Flessa H, Beese F (2000) Laboratory estimates of trace gas emissions following surface application and injection of cattle slurry. J Environ Qual 29: 262–268.

33. Sherlock RR, Sommer SG, Khan RZ, Wood CW, Guertal EA, et al. (2002) Emissions of ammonia, methane and nitrous oxide from pig slurry applied to a pasture in New Zealand. J Environ Qual 31: 1491–1501.

34. Rodhe L, Pell M, Yamulki S (2006) Nitrous oxide, methane and ammonia emissions following slurry spreading on grassland. Soil Use Manage 22: 229–237.

35. Koschorreck M, Conrad R (1993) Oxidation of atmospheric methane in soil: measurements in the field, in soil cores and in soil samples. Global Biogeochemical Cycles 7: 109–121.

36. Borken W, Matzner E (2009) Reappraisal of drying and wetting effects on C and N mineralization and fluxes in soils. Global Change Biology 15: 808–824.

37. Denef K, Six J, Bossuyt H, Frey SD, Elliott ET, et al. (2001) Influence of dry-wet cycles on the interrelationship between aggregate, particulate organic matter, and microbial community dynamics. Soil Biology and Biochemistry 33: 1599–1611.

38. Davidson EA, Belk E, Boone RD (1998) Soil water content and temperature as independent or confounded factors controlling soil respiration in a temperate mixed hardwood forest. Global Change Biol 4: 217–227.

39. Jassal RS, Black TA, Drewitt GB, Novak MD, Gaumont-Guay D, et al. (2004) A model of the production and transport of CO_2 in soil: predicting soil CO_2 concentrations and CO_2 efflux from a forest floor. Agric For Meteorol 124: 219–236.

40. Fang YT, Gundersen P, Zhang W, Zhou GY, Christiansen JR, et al. (2009) Soil-atmosphere exchange of N_2O, CO_2 and CH_4 along a slope of an evergreen broad-leaved forest in southern China. Plant and Soil 319: 37–48.

41. Bajracharya RM, Lal R, Kimble JM (2000) Erosion effect on carbon dioxide concentration and carbon flux from an Ohio Alfisol. Soil Sci. Soc. Am. J 64: 694–700.

42. Fang C, Moncrieff JB (2001) The dependence of soil CO_2 efflux on temperature. Soil Biology & Biochemistry 33: 155–165.

43. Kelting DL, Burger JA, Edwards GS (1998) Estimating root respiration, microbial respiration in the rhizosphere, and root-free soil respiration in forest soils. Soil Biology and Biochemistry 30: 961–968.

44. Clough TJ, Sherlock RR, Kelliher FM (2003) Can liming mitigate N_2O fluxes from a urine-amended soil? Aust J Soil Res 41: 439–457.

45. Clough TJ, Kelliher FM, Sherlock RR, Ford CD (2004) Lime and soil moisture effects on nitrous oxide emissions from a Urine Patch. Soil Sci Soc A J 68: 1600–1609.

46. Grandy SA, Robertson PG (2006) Initial cultivation of a temperate-region soil immediately accelerates aggregate turnover and CO_2 and N_2O fluxes. Global Change Biology 12: 1507–1520.

47. Li XX, Hu CS, Delgado JA, Zhang YM, Ouyang ZY (2007) Increased nitrogen use efficiencies as a key mitigation alternative to reduce nitrate leaching in north China plain. Agricultural Water Management 89: 137–147.

48. Zhu B, Wang T, Kuang F, Luo Z, Tang J, et al. (2009) Measurements of nitrate leaching from a hillslope cropland in the Central Sichuan Basin, China. Soil Sci Soc Am J 73: 1419–1426.

49. Dobbie KE, Smith KA (2003) Nitrous oxide emission factors for agricultural soils in Great Britain: the impact of soil water-filled pore space and other controlling variables. Global Change Biology 9: 204–218.

50. Zou GY, Zhang FS, Chen XP, Li XH (2001) Nitrification–denitrification and N_2O emission from arable soil. Soil Environ Sci 10 (4): 273–276 (in Chinese).

51. Davidson EA (1992) Sources of nitric oxide and nitrous oxide following wetting of dry soil. Soil Sci Soc Am J 56: 95–102.

52. Zhu HJ, He YG eds (1992) Soil Geography. Higher Education Press, Beijing, China (In Chinese).

53. Ding WX, Luo JF, Li J, Yu HY, Fan JL, et al. (2013) Effect of long-term compost and inorganic fertilizer application on background N_2O and fertilizer-induced N_2O emissions from an intensively cultivated soil. Science of the Total Environment 465: 115–124.

16

Stratification of Carbon Fractions and Carbon Management Index in Deep Soil Affected by the Grain-to-Green Program in China

Fazhu Zhao, Gaihe Yang*, Xinhui Han*, Yongzhong Feng, Guangxin Ren

College of Agronomy, Northwest A&F University, Yangling, Shaanxi, China; and The Research Center of Recycle Agricultural Engineering and Technology of Shaanxi Province, Yangling, Shaanxi, China

Abstract

Conversion of slope cropland to perennial vegetation has a significant impact on soil organic carbon (SOC) stock in A horizon. However, the impact on SOC and its fraction stratification is still poorly understood in deep soil in Loess Hilly Region (LHR) of China. Samples were collected from three typical conversion lands, *Robinia psendoacacia* (RP), *Caragana Korshinskii Kom* (CK), and abandoned land (AB), which have been converted from slope croplands (SC) for 30 years in LHR. Contents of SOC, total nitrogen (TN), particulate organic carbon (POC), and labile organic carbon (LOC), and their stratification ratios (SR) and carbon management indexes (CMI) were determined on soil profiles from 0 to 200 cm. Results showed that the SOC, TN, POC and LOC stocks of RP were significantly higher than that of SC in soil layers of 0–10, 10–40, 40–100 and 100–200 cm ($P<0.05$). Soil layer of 100–200 cm accounted for 27.38–36.62%, 25.10–32.91%, 21.59–31.69% and 21.08–26.83% to SOC, TN, POC and LOC stocks in lands of RP, CK and AB. SR values were >2.0 in most cases of RP, CK and AB. Moreover, CMI values of RP, CK, and AB increased by 11.61–61.53% in soil layer of 100–200 cm compared with SC. Significant positive correlations between SOC stocks and CMI or SR values of both surface soil and deep soil layers indicated that they were suitable indicators for soil quality and carbon changes evaluation. The Grain-to-Green Program (GTGP) had strong influence on improving quantity and activity of SOC pool through all soil layers of converted lands, and deep soil organic carbon should be considered in C cycle induced by GTGP. It was concluded that converting slope croplands to RP forestlands was the most efficient way for sequestering C in LHR soils.

Editor: Raffaella Balestrini, Institute for Plant Protection (IPP), CNR, Italy

Funding: This work was supported by Special Fund for forest-scientific Research in the Public Interest (201304312). The funders had no role in study design, data collection and analysis, decision to publish, or preparation of the manuscript.

Competing Interests: The authors have declared that no competing interests exist.

* E-mail: ygh@nwsuaf.edu.cn (GY); hanxinhui@nwsuaf.edu.cn (XH)

Introduction

Soil organic carbon (SOC) is a dynamic component of the terrestrial system, with internal changes in both vertical and horizontal directions and external exchanges between the atmosphere and the biosphere [1]. SOC storage is estimated at approximately 1500 Pg globally, which is about two and three times the size of carbon pools in the atmosphere and vegetation, respectively [2]. Since carbon uptake and storage is tightly linked to the nitrogen (N) cycle, it is equally important to understand how N pools and fluxes are affected by land use change [3]. Moreover, more than 50% of the total SOC is stored in the subsoil [4]. The proportion of soil organic matter (SOM) stored in the first meter of the world soils below 30 cm depth ranges 46%~63%, except for Podzoluvisols, where 30% of SOC is stored below the depth of 30 cm [4]. A recent study also suggests that in the northern circumpolar permafrost region, at least 61% of the total soil C is stored below 30 cm [5]. Therefore, subsoil C may be even more important in terms of source or sink for CO_2 than topsoil C [6]. Considering the potential role of SOC in atmospheric CO_2 sink, it is important to understand what leads to sequestration of large amounts of SOC in the subsoil or even in deep soil. However, the

SOC contents in deep soil layers are not fully understood in LHR of China to date.

As an indicator of soil quality, SOM stratification, which is related to the rate and amount of SOC sequestration [7], is common in many natural ecosystems [8] and managed grasslands and forests [9–10]. Stratification ratio (SR) is defined as the ratio of a soil property at the surface layer to that at a deeper layer. In general, high SR values indicate good soil quality and are usually used to assess agricultural practices [7]. For instance, SR values for SOC at depths of 0–5 cm and 20–40 cm range from 1.1 to 1.5 under traditional tillage (TT) while from 1.6 to 2.6 under conservation tillage (CT) [11]. Little information is available on natural ecosystems and managed shrubs or forests land. Additionally, under semiarid climate, SOC in active fractions, such as particulate organic carbon (POC) and labile organic carbon (LOC) was more sensitive to soil management practices than total SOC [12]. Previous researches have indicated that changing rate of POC and LOC was faster than SOC in whole soil [11], and they could be an early indicator for SOC change in soil [13]. Meanwhile, the carbon management index (CMI), which is derived from the total soil organic C pool and C lability, had been extensively used as a sensitive indicator of SOC variation rate in response to soil management changes [14–15]. Therefore, under semi-arid climate, using SR of total SOC and of different SOC fractions may be useful to reveal how soil management affects soil quality and helpful to understand the mechanism of SOC transformation and cycling in subsoil as well as in deep soil. In LHR of China, soil erosion and desertification are causing a loss of net primary productivity that was estimated as high as 12 kg C $ha^{-1}y^{-1}$ [16]. To counteract soil erosion and other environmental problems, an environmental protection policy was implemented by Chinese central government, which was known as the Grain to Green Program (GTGP). The purpose of GTGP was to convert up to 26.87 million ha low-yield sloped croplands (>25°) into forests, shrubs or grasslands by the end of 2008 [17]. It is the first and the most ambitious "payment–for–ecosystem–services" program in China to date [18]. Although the initial goal of GTGP was to control soil erosion in China, it also plays a significant role in circulation of SOC and total nitrogen (TN). In recent years, a few studies estimated the effects of GTGP on vegetation structure, economic benefits, soil physiochemical properties, and niche characteristics [19–21]. However, SR values of SOC and/or TN and CMI value among different land use types are rarely reported. Especially, information on dynamics of C in deep soil is largely ignored in this region.

This study aimed to: 1) analyze the contents of SOC, TN, POC and LOC and their vertical distributions at the depths of 0–200 cm; 2) assess the stocks of SOC and TN at different soil depths of three land use types; and 3) evaluate the soil quality of different land use types using SR and CMI values as the main assessment parameters.

Materials and Methods

All sites in the watershed we were selected for study was determined through interviews with local farmers (Mr. Yibin Zhang, Soil and Water Conservation Experiment Station, Northwest A&F University, Ansai County, Shaanxi, NW China).We state clearly that no specific permissions were required for the location. We confirm that the location is not privately-owned or protected in any way. We confirm that the field studies do not involve endangered or protected species.

Research area

The study was conducted in the Zhifanggou catchment (36°46′42″–36°46′28″N, 109°13′46″–109°16′03″E), which is located in Ansai county, central LHR (see Fig. 1). Ansai is a typical county characterized by semi-arid climate and hilly loess landscape in the Loess Plateau with an annual average temperature of 8.8°C, and an average annual precipitation of 505 mm. 60% of precipitation occurs between July and September (~300 mm in dry years while >700 mm in wet years). Accumulated temperatures above 0°C and 10°C are 3733°C and 3283°C, respectively. On average, there are about 157 frost-free days and 2415 h sunny time each year. Arable farming mostly occurs on

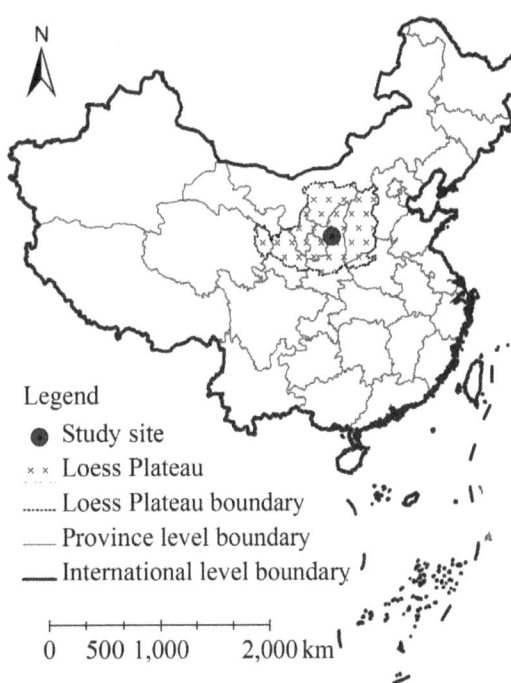

Figure 1. Location of the Loess Plateau and the study site.

sloping lands without irrigation. The loess parent material at the site has an average thickness of approximately 50–80 m and the soil in this region is classified as Calciustepts soil [22]. Sand (2–0.05 mm) and silt (0.05–0.002 mm) account for approximately 29.22% and 63.56% in soil depth of 0–20 cm, respectively. The soil is highly erodible, with an erosion modulus of 10,000–12,000 $Mg{\cdot}km^{-2}{\cdot}yr^{-1}$ before the start of restoration efforts [23]. After 30 years vegetation restoration, the area of forest lands significantly increased from 5% to 40% [24].

The Zhifanggou catchment has been an experimental site of the Institute of Soil and Water Conservation, Chinese Academy of Science (CAS) since 1973 [25]. The major agricultural land use type in LHR is slope cropland. Agricultural management in this region, including the major crop types grown, has not been changed significantly since the 1970s. The main crops grown in these sites were millet (*Setaria italica*) and soybean (*Glycine max*) rotation, and no irrigation was provided in grown season (depend on rainfall). One crop was grown each, and fertilizer was applied (mainly manure). After more than 30 years of comprehensive management, the ecological environment of the catchment has been significantly improved [26]. Since late 1970s, slope cropland is replanted with shrubs and woods, mainly *Robinia pseudoacacia* L. (RP) and *Caragana Korshinskii Kom (CK)*, to control soil erosion (see Table 1). Abandoned cropland was also generated during this period due to its extremely low productivity and long

Table 1. Characteristics of different vegetation types.

Vegetation types	Age	Canopy closure (%)	Litter accumulation (t.ha^{-2})	Undergrowth Vegetation[a]	Species diversity indices
Robinia pseudoacacia L.	30	58	20.5	*Lespedeza dahurica - Stipa bungeana*	6.5
Caragana Korshinskii Kom	30	50	13.3	*Achillea capillaries, Stipa bungeana*	3.9

[a]means the main vegetation in forest/shrub land.

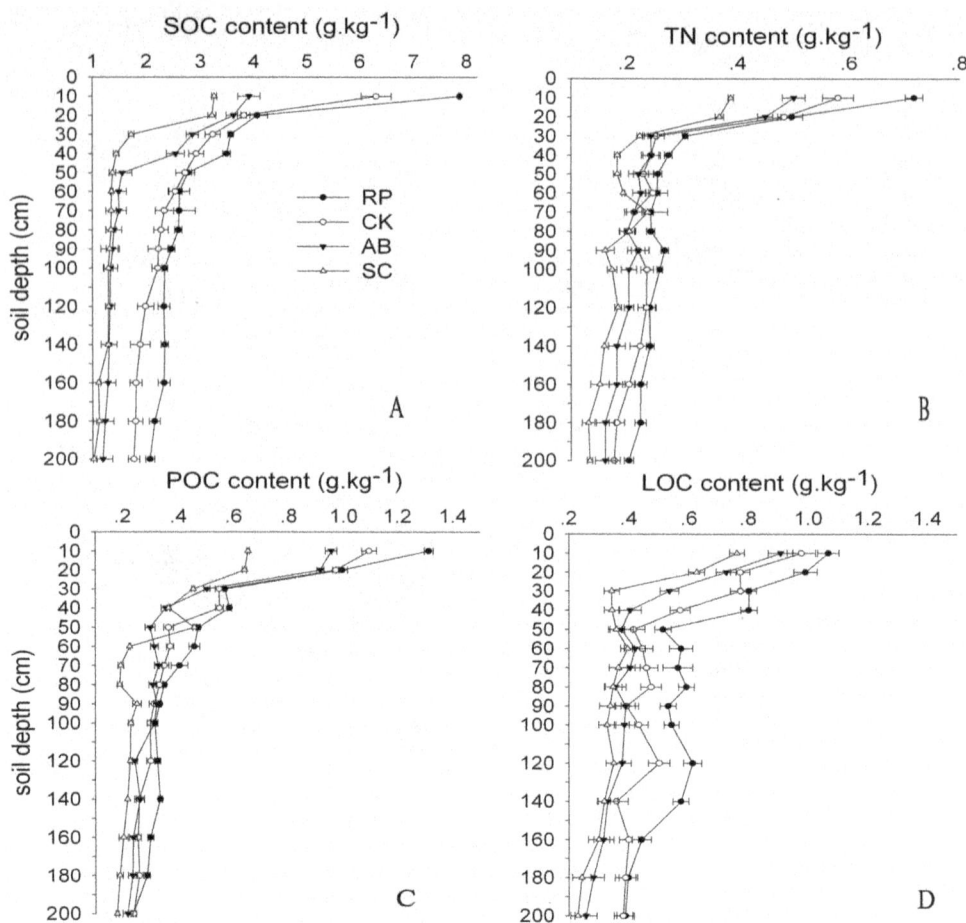

Figure 2. Distribution of soil organic carbon (SOC, A), total nitrogen (TN, B), particulate organic carbon (POC, C), and labile organic carbon (LOC, D) contents of different land used types in soil depth of 0–200 cm. The error bars are the standard errors.

distance from farmers' residences [27–28]. Despite wild grasslands and shrub lands were usually found on steep slopes, these sites were used for firewood collection as well. So the wild vegetation was of limited coverage or even barren for long periods. In 1999, most slope lands were closed for vegetation restoration under the GTGP [29].

Soil sampling

In September 2012, based on land use history, 30 year old *Robinia psendoacacia* (RP), *Caragana Korshinskii Kom* (CK), abandoned land (AB) and slope cropland (SC) in the Zhifanggou catchment were selected. Three 30 m×20 m plots were established for each land use type. All sites were located on the same physiographical units with same slope aspects, same elevation of 1250 m and a spatial distance of 1200 m.

Soil samples were taken at several soil depths using a soil auger (diameter 5 cm) from 10 points within "S" shape at each plot (0–10 cm, 10–20 cm, 20–30 cm, 30–40 cm, 40–50 cm, 50–60 cm, 60–70 cm, 70–80 cm, 80–90 cm, 90–100 cm, 100–120 cm, 120–140 cm, 140–160 cm, 160–180 cm, and 180–200 cm). Then after removing the litter layer, ten soil samples at each depth of each plot were mixed to make one sample. Samples were collected at least 80 cm away from the trees. All samples were sieved through a 2 mm screen, and roots and other debris were removed. Soil samples were air-dried and stored at room temperature for the

determination of soil chemical properties. A ring tube was used to determine the bulk density in each soil depth.

Laboratory analysis

SOC content $(g.kg^{-1})$ and TN content $(g.kg^{-1})$ were determined using $K_2Cr_2O_7$ oxidation method and Kjeldhal method, respectively [30].

To determine POC content, 25 g soil was dispersed with 100 mL of 5 g L^{-1} sodium hexametaphosphate before being. Then, the mixed soil solution was shaken for 1 h at high speed on an end-to-end shaker and screened by a 0.053 mm sieve with several deionized water rinses. The soil remained on the sieve was backwashed into a pre-weighed aluminum box and dried at 60°C for 24 h, then it was grounded for analysis of C [31].

Soil labile organic carbon (LOC) was measured following the method described in Graeme et al. [32]. A 2–6 g air dried soil sample was put into a 50 mL centrifuge tube, and 25 mL of 333 mmolL^{-1} KMnO$_4$ solution was added before being shaken with a rate of 120 rpm for 1 h, and centrifuged for 5 min with a rate of 5,000×g. The upper clear solution was transferred, and diluted by 250 times, and then the absorbance at 565 nm wavelength was determined. The absorbances at values 565 nm with different KMnO$_4$ concentrations were also determined for preparation of standard curve, which was used for the determination of the KMnO$_4$ concentrations. Difference between the

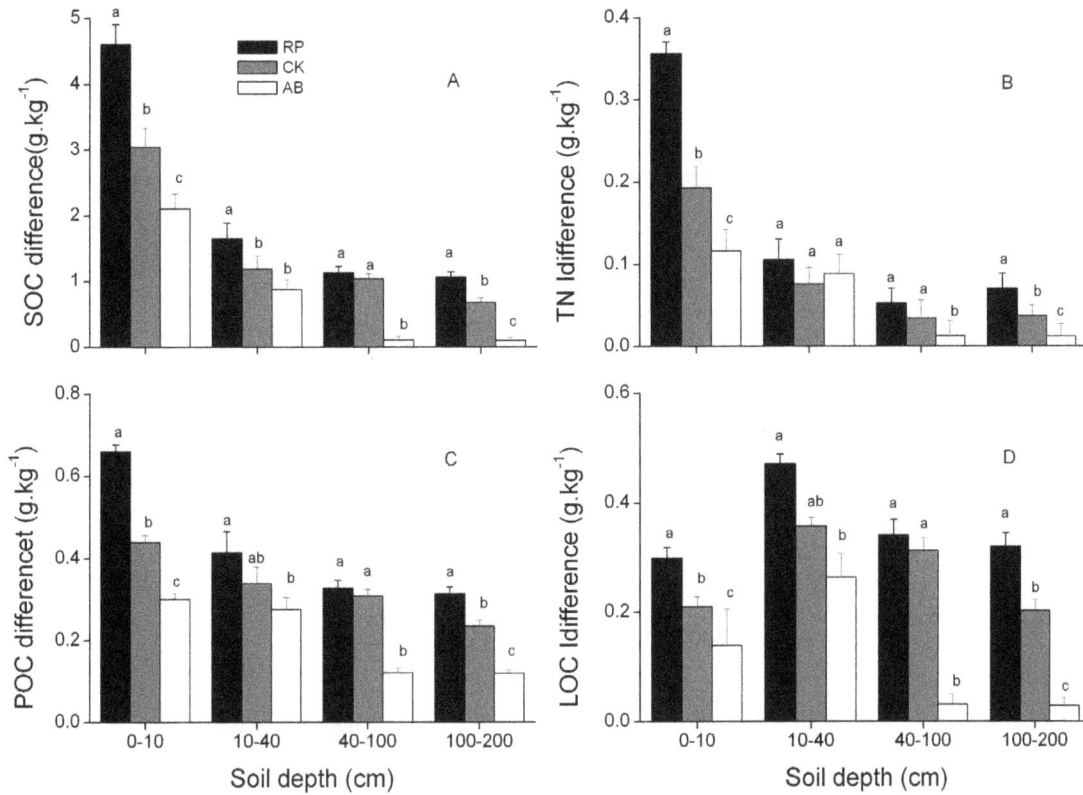

Figure 3. Differences in soil organic carbon (SOC, A), total nitrogen (TN, B), particulate organic carbon (POC, C), labile organic carbon (LOC, D) contents between SC and RP, CK or AB (RP/CK/AB - SC). Error bars are the standard errors. Different lowercase letters indicate significant difference among different land use types within same soil layer (P<0.05). The same for Fig 4

amounts of $KMnO_4$ added and remained was used to calculate labile C concentration in the soil sample.

Calculation of SOC (TN) stocks, SR of SOC (TN, POC, and LOC) and CMI

SOC density (SOCD, TND) represents the total SOC (TN) storage of overall certain sampling depth. SOCD(TND) of different sampling depths were calculated:

$$SOCD(TND) = C_{SOC,TN} \times \rho \times H \times (1 - \delta/100) \times 10^{-1} \quad (1)$$

where SOCD (TND) is the density ($Mg \cdot ha^{-1}$) of SOC (TN) and $C_{SOC,TN}$ is the content ($g \cdot kg^{-1}$) of SOC (TN). ρ is the bulk density ($g \cdot cm^{-3}$), H is the soil horizon thickness (cm), and δ is the fraction (%) of gravels >2 mm in size in soil. Because the soil gravel size of loess in China is mostly below 2 mm, this fraction was assumed to be 0 [33].

SR values (0–10 cm: 10–40 cm, 0–10 cm: 40–100 cm and 0–10 cm: 100–200 cm) were calculated from the contents of SOC, TN, POC and LOC following the method in Franzluebbers (2002).

CMI values were calculated using following procedures:

Firstly, a C pool index (CPI) was calculated:

$$CPI = \frac{sample\ total\ organic\ C\ (g/kg)}{reference\ sample\ total\ C\ (g/kg)} \quad (2)$$

where reference sample is SC soil. Then, a lability index (LI) was calculated:

$$LI = \frac{L\ in\ each\ sampled\ soil}{L\ in\ the\ reference\ soil} \quad (3)$$

where reference soil is SC soil, and L was calculated from the C lability:

$$L = \frac{content\ of\ labile\ C}{content\ of\ non-labile\ C} \quad (4)$$

At last, CMI was calculated:

$$CMI = CPI \times LI \times 100 \quad (5)$$

Statistical analyses

All statistical analyses were carried out with SPSS 17.0. Analysis of variance (ANOVA) and Duncan's Multiple Range Test (DMRT) at 5% level of significance were used to compare the difference in contents and/or stocks of SOC, TN, POC, LOC, CMI, and SR among different land use types or soil depths. A sample linear-regression analysis was used to estimate the relationships between carbon stocks with CMI or SR values.

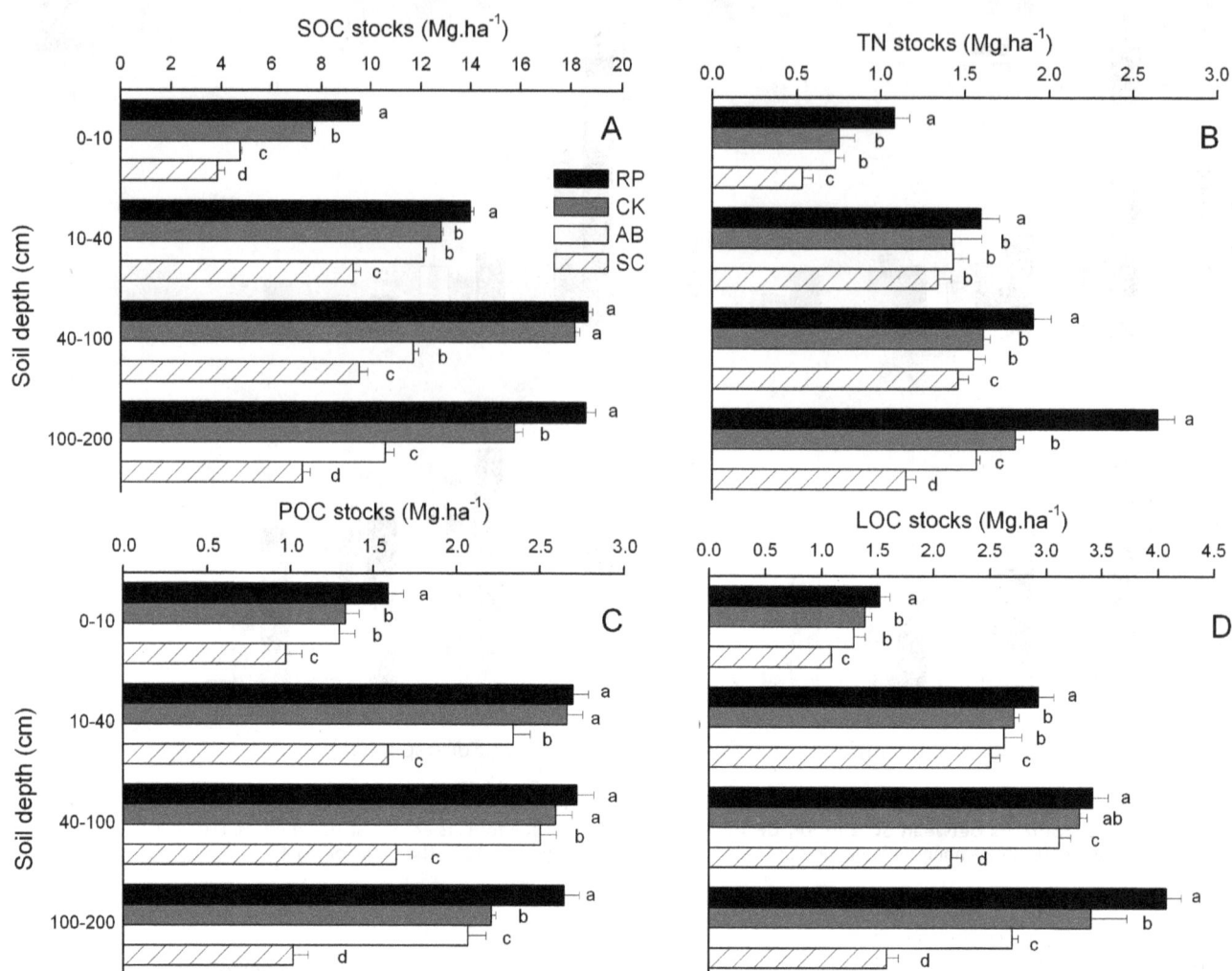

Figure 4. Stocks of soil organic carbon (SOC, A), total nitrogen (TN, B), particulate organic carbon (POC, C), labile organic carbon (LOC, D) of different land use types. The error bars are the standard errors.

Results

Changes in contents of SOC, TN, POC and LOC

The contents of SOC, TN, POC and LOC responded differently as the change of soil depth (Fig. 2). In all land use types, contents of SOC, TN, POC and LOC in top soil (0–10 cm) were 3.26–7.86 g.kg^{-1}, 0.39–0.72 g.kg^{-1}, 0.65–1.31 g.kg^{-1} and 0.76–1.07 g.kg^{-1}, respectively, which were significantly higher than other soil layers (P<0.05). The contents of SOC, TN, POC and LOC decreased significantly in soil depth of 10–40 cm while the decreases trended to be flatter in subsoil (40–100 cm). Additionally, the differences in contents of SOC, TN, POC and LOC in deep subsoil (100–200 cm) were negligible (P<0.05).

The differences in contents of SOC, TN, POC and LOC between three forest/shrub types (RP, CK and AB) and SC are shown in Fig. 3. The differences in SOC, TN, POC and LOC of RP and SC in soil depths of 0–10 cm and 100–200 cm were significantly higher than that between other land use types and SC (P<0.05). The differences in SOC and TN of RP were 33.78% and 45.97% larger than that of CK, and 54.13% and 67.28% larger than that of AB in soil depth of 0–10 cm (P<0.05), while the differences in POC and LOC were 32.8%, 54.0% higher than that of CK, and 23.3% and 45.0% higher than that of AB (P<0.05).

Moreover, the differences in SOC, TN, POC and LOC of RP were 25.05–85.29% higher than that of CK, and 61.78–90.70% higher than that of AB in soil depth of 100–200 cm. Additionally, significant differences in SOC, TN, POC and LOC contents were observed between RP and CK in soil depths of 10–40 cm and 40–100 cm, but there was no difference between CK and AB (P<0.05).

Changes and distribution of SOC, TN, POC and LOC stocks

SOC, TN, POC and LOC stocks of RP, CK and AB were higher than SC in all soil profiles (Fig 4). The SOC, TN, POC and LOC stocks of RP were significantly increased (P<0.05), which were 0.43–5.8 Mg.ha^{-1}, 0.25–4.70 Mg.ha^{-1}, 0.44–9.14 Mg.ha^{-1} and 1.49–11.38 Mg.ha^{-1} higher than that of SC in soil layers of 0–10, 10–40, 40–100 and 100–200 cm, respectively. Moreover, the stocks of SOC, TN, POC and LOC in soil layer of 100–200 cm of RP were higher than that of CK and AB by 15.4–32.1% and 21.8–43.1%, respectively (P<0.05).

The SOC, TN, POC and LOC stocks responded differently as the change of soil depth (Fig. 5). Although the distribution of SOC, TN, POC and LOC stocks in soil depths of 0–10 cm and 10–

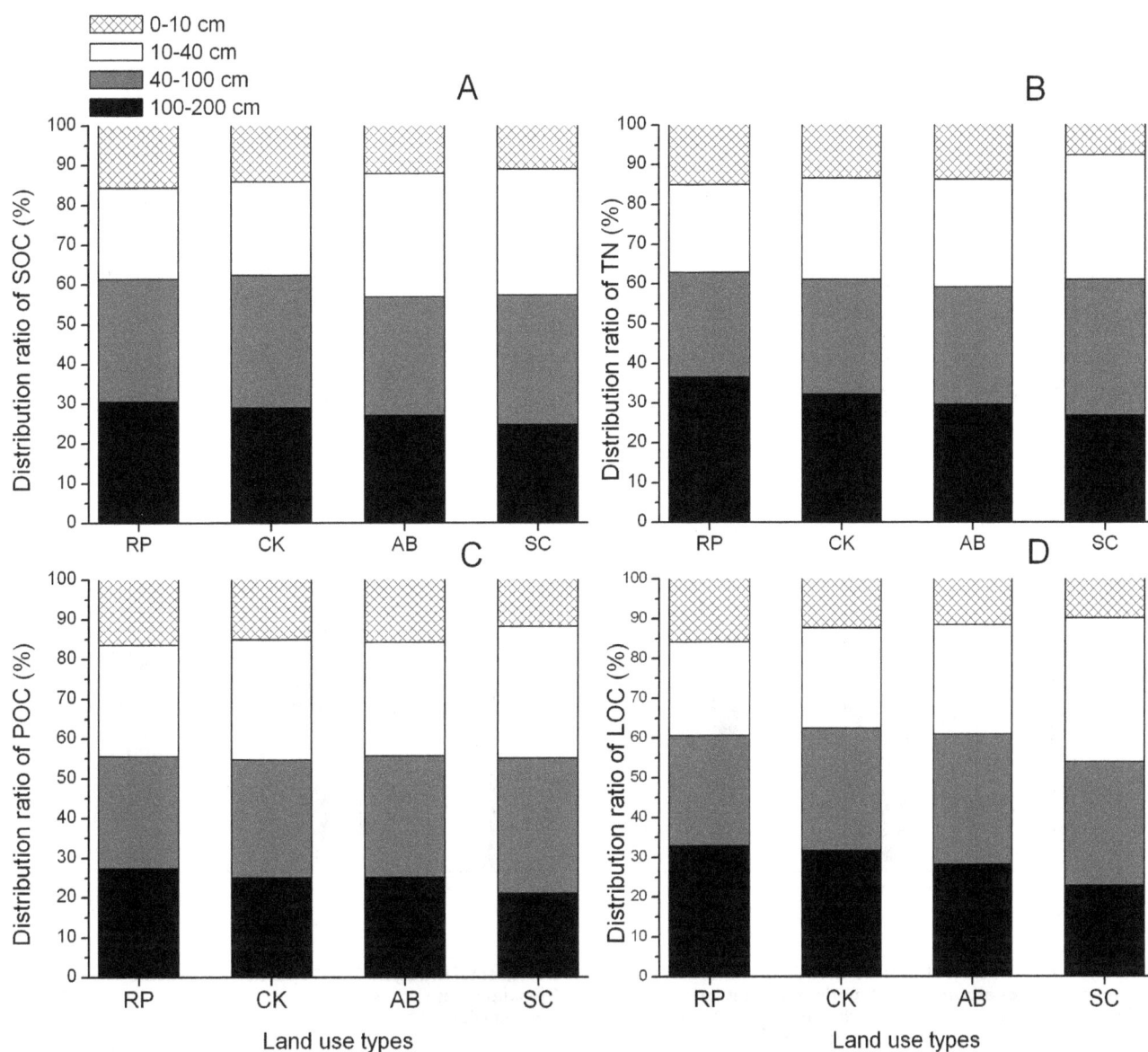

Figure 5. Distribution ratios of soil organic carbon (SOC, A), total nitrogen (TN, B), particulate organic carbon (POC, C), labile organic carbon (LOC, D) in soil depth of 0–200 cm under different land use types.

40 cm accounted for the majority, 26.36–34.06% and 21.08–36.62% were distributed in soil layers of 40–100 cm and 100–200 cm, respectively. Among four land use types, the highest proportion of SOC, TN, POC and LOC stocks were found in RP, while the lowest were in soil depths of SC in 0–10 cm and 100–200 cm of SC. The proportion of SOC, TN, POC and LOC stocks under RP were higher than SC by 4.68%, 7.32%, 4.65% and 5.96% respectively in soil depth of 0–10 cm soil depth, whereas by 5.90%, 9.78%, 6.30% and 10.06% was higher in soil depth of 100–200 cm soil depth respectively.

Change in SR and CMI values

Responses of SR in different land use types to change of soil depth were different (Fig 6). The SR values of SOC, TN and LOC differed significantly among different soil depths (P<0.05), while the SR values of LOC differed only between 0–10:10–40 cm, 0–10:40–100 cm and 0–10:100–200 cm. Among four land use types, the SR values of SOC, TN, POC and LOC of RP were the

highest, but that of SC were the lowest in each soil depth(P<0.05). The SR values of SOC, TN, POC and LOC were in a decreasing order of CK>AB>SC. The SR values differed significantly between CK or AB with SC (P<0.05), while there was no significant difference between CK and AB. Additionally, the ratios of SR values of SOC, TN, POC and LOC in the surface layer (0–10 cm) to that in layer of 10–40 cm were >2.0 in most cases.

The CMI values were significantly affected by land use types. In our study, the CMI values were in a decreasing order of RP>CK>AB>SC in four soil profiles and CMI values were significantly enhanced by RP compared with SC (Fig 7). Averaged CMI values of RP, CK, and AB were 40.60%, 50.54%, 37.81%, and 14.1% higher than that of SC in soil layers of 0–10 cm, 10–40 cm, 40–100 cm and 100–200 cm.

Regression equations to assess CMI/SR values of TN, POC, and LOC (Y) were showed in Table 2. There was a significant

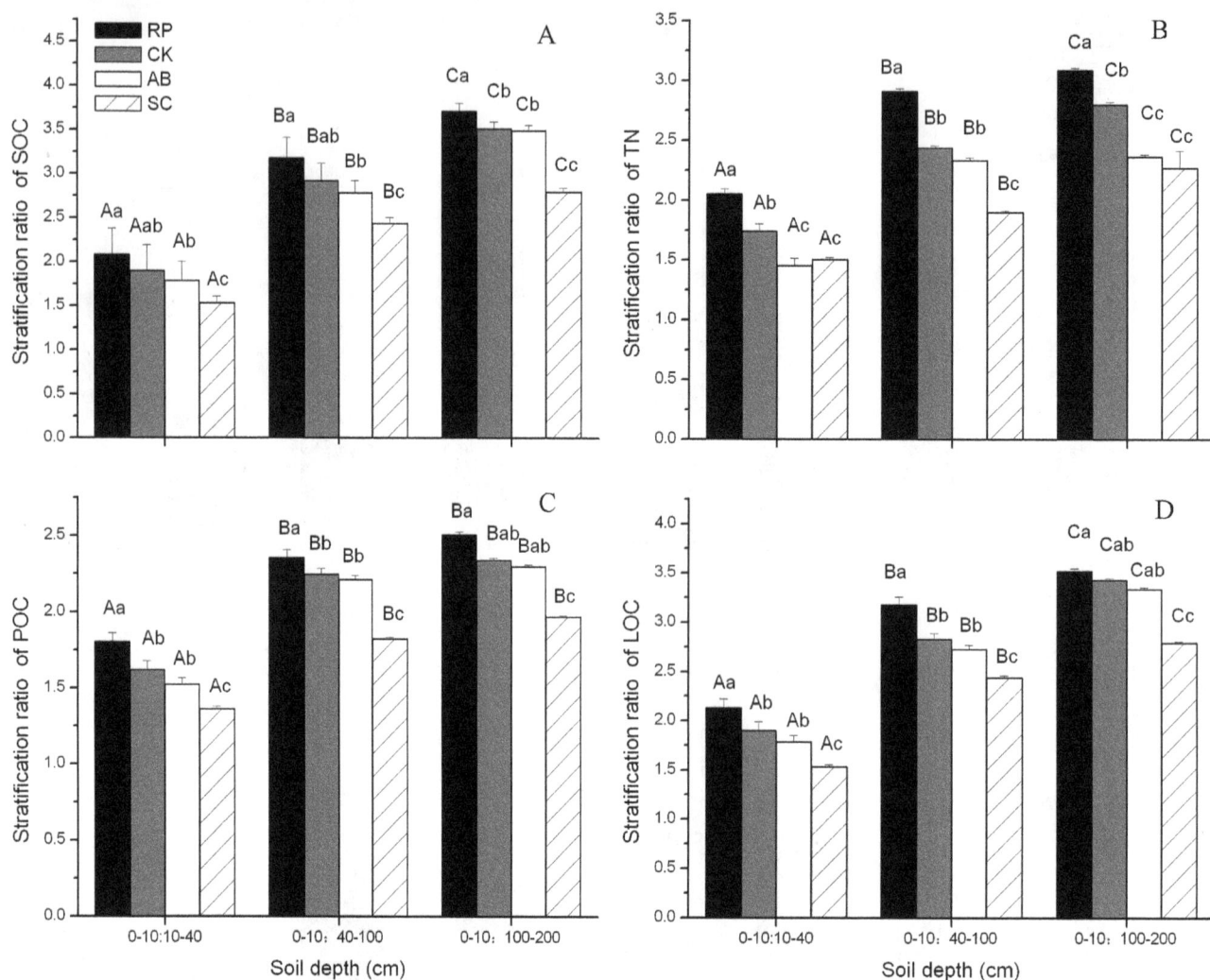

Figure 6. Comparison of stratification ratio of soil organic carbon (SOC, A), total nitrogen (TN, B), particulate organic carbon (POC, C), labile organic carbon (LOC, D) under different land use types. Different uppercase letters indicate significant difference among different soil depths within same land use type while the different lowercase letters indicate significant difference among different land use types within same soil depth. The error bars are the standard errors.

positive correlation between CMI/SR values of TN, POC, and LOC with SOC stocks in surface soil and deep soil.

Discussion

SOC, TN, POC and LOC contents and SOC and TN stocks

Vegetation can greatly influence soil quality, C and N cycling, and regional socioeconomic development [34–35]. It is also reported that converting cropland into land with perennial vegetation would increase the SOC content [36]. Our results showed that land use type and soil depth significantly affected the contents of SOC and TN (Fig. 2). The conclusion that both land use type and soil depth are important factors influencing the soil carbon and nitrogen distribution was consistent with previous studies [35,37]. We also observed that the lowest SOC, TN, POC and LOC contents were found in slope cropland (Fig 4), which essentially agree with a previous study [38], indicating that the conversion of slope cropland to vegetation improves the C and N contents. A possible reason is that the lower residue input into the soil in slope cropland leads to lower SOC and TN contents.

Additionally, our results showed that SOC, TN, POC and LOC contents of RP were greater than that of CK and AB (Fig 4). It infers that the effects of RP on soil C and N play a significant role in land use and ecosystem management. The conclusion was consistent with Qiu et al [39], who reported that RP has potential to improve SOC content in the loessial gully region of the Loess Plateau and the improvements are greater in long-term than middle-term.

Recently it was reported that the depth of sampling is an important factor for the measurement of change in SOC stocks [40], and land use could influence subsoil C pools [41]. We found that SOC, TN, POC and LOC stocks of RP, CK, and AB were higher than SC for different soil profiles, especially in depths of 40–100 cm and 100–200 cm (Fig. 4 and 5). It is demonstrated that converting slope cropland into woodland and shrubland not only affects SOC and TN stocks in surface soil, but also largely influences that in deep soil. The result was consistent with Wang et al [42], who reported that deep layer (50–200 cm) SOC stocks were equivalent to approximately 25% of that in the shallow layer (0–50) in Hilly Loess Plateau. This is mainly due to the fact that

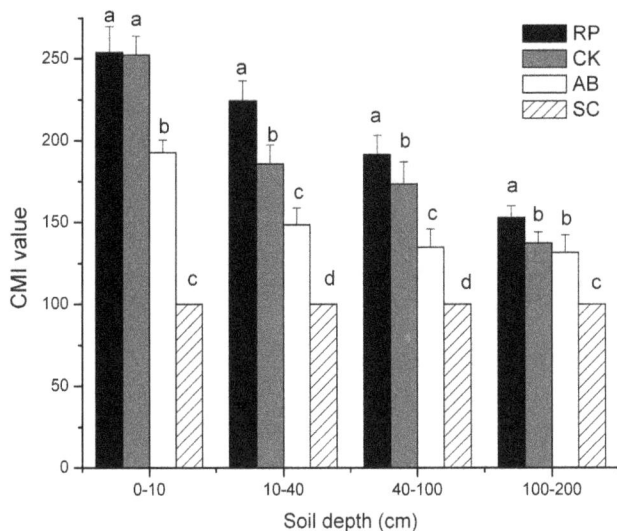

Figure 7. Carbon management index (CMI) values of different land use types at different soil depths. The error bars are the standard errors. Different lowercase letters indicate significant difference among different land use types within same soil depth.

SOC input into subsoil is largely affected by plant roots and root exudates, dissolved organic matter and bioturbation. In addition, most important factors leading to protection of SOC in subsoil include the spatial separation of SOM, microorganisms and extracellular enzyme activity related to the heterogeneity of C input [43]. As a result, stabilized SOC in subsoil is horizontally stratified.

Stratification ratios of SOC, TN, POC, and LOC

According to Franzluebbers [7], SOC SR values >2 in degraded conditions is uncommon, and the SR values of SOC

are generally low and seldom reach 2.0. And SR values of soil organic C and N pools with value of >2 would be an indicator that soil quality might be improved [7]. In our study, the most of SR in SOC, TN, POC and LOC was more than 2 after convert slope cropland to forest or shrub land (Fig. 6). This means soil quality was improved in these afforested soils without disturbance. Greater C stratification ratios could be related to the fact that, during soil recovery by re-vegetation or land abandonment, soil was undisturbed thus reducing oxidation and favoring soil C [44]. The result was consistent with Sá et al [45]. Similar results were also reported by Moreno et al [46] and Franzluebbers [7], who reported that stratification of SOC occurs over time when soil tillage and disturbance is stopped and it is usually greater in undisturbed soils than in disturbed soils. In addition, the stratification may increase with time, and SOC, TN, POC and LOC contents are still aggrading but have not reached soil C saturation yet. That is the reason why SR values of SOC, TN, POC and LOC under CK and AB were higher but no significant differences were observed compared with SC (Fig. 6). Sá et al [45] concluded that the SOC pool stabilization may be attained in about 40 years after long-term no-tillage adoption.

Carbon management index

CMI value was calculated to obtain indications of the C dynamics of the system and provide an integrated measure for quantity and quality of SOC [15]. Soils with higher CMI values are considered as better managed [47]. We found that CMI values were significantly enhanced by RP forest compared with CK, AB and SC in both surface soil and subsoil and deep soil (Fig. 7). Soil management under RP plot was more appropriate to improve the SOC status than other land use types. Similar result was reported by Qiu et al [39], who illustrated that RP forest has significantly increased SOC, total nitrogen, ratio of carbon to nitrogen and ratio of carbon to phosphorus compared to other vegetation types. Our result showed that there were significant positive correlations between SOC stocks and CMI/SR in both surface soil and deep soil (Table 2). These findings showed that SR values of SOC, TN,

Table 2. Regression equations among SOC stocks and CMI/SR for different soil layers.

Axis		Soil depth (cm)	Equations	R^2	Significant level
X, CMI/SR[a]	Y, SOC stocks[b]				
CMI	SOC	0–10	$Y = 53.5 + 23.20X$	0.91	$P = 0.048$
		10–40	$Y = -146 + 25.81X$	0.97	$P = 0.027$
		40–100	$Y = 24.2 + 8.60X$	0.93	$P = 0.023$
		100–200	$Y = 77.5 + 4.05X$	0.93	$P = 0.046$
TN	SOC	0–10:10–40	$Y = 1.12 + 0.08X$	0.93	$P = 0.047$
		0–10:40–100	$Y = 1.26 + 0.07X$	0.96	$P = 0.032$
		0–10:100–200	$Y = 1.67 + 0.07X$	0.98	$P = 0.017$
POC	SOC	0–10:10–40	$Y = 1.18 + 0.06X$	0.97	$P = 0.023$
		0–10:40–100	$Y = 1.55 + 0.04X$	0.93	$P = 0.041$
		0–10:100–200	$Y = 1.74 + 0.04X$	0.92	$P = 0.047$
LOC	SOC	0–10:10–40	$Y = 1.30 + 0.08X$	0.96	$P = 0.031$
		0–10:40–100	$Y = 1.95 + 0.05X$	0.97	$P = 0.025$
		0–10:100–200	$Y = 2.53 + 0.05X$	0.98	$P = 0.013$

[a]CMI= carbon management index, SR= stratification ration, TN= SR of total nitrogen, POC= SR of particulate organic carbon, LOC= SR of labile organic carbon.
[b]For the Y-axis, the SOC stocks (0–10 cm, 40–100 cm, and 100–200 cm) were used to analyze correlations between SOC stocks and SR values of TN, POC, and LOC (0–10:10–40, 0–10:40–100, 0–10:100–200).

POM, and LOC, and CMI are suitable indicators for evaluating soil quality and C changes induced by GTGP in surface soil and deep soil.

Conclusion

In this study, the SOC, TN, POC and LOC contents of RP, CK and AB in soil layer of 100–200 cm were higher than SC, especially for RP plot. Although the SOC, TN, POC and LOC stocks in soil layer of 100–200 cm were lower, there was more than 27.38–36.62%, 25.10–32.91%, 21.59–31.69% and 21.08–26.83% of SOC, TN, POC and LOC stocks were distributed in 100–200 cm soil depth under RP, CK and AB. Meanwhile, the SR of SOC, TN, POC and LOC in the surface to lower depth ratio (i.e., 0–10:10–40 cm) was >2.0 in most of case. And SR and as well CMI values were significantly enhanced by RP compared with SC in deep soil (100–200 cm) (P<0.05). Indicating that soil quality

was improved after converting slope land into perennial vegetation, especially under RP plot from surface soil to deep soil. Moreover, there were significant and positive correlations between SOC stocks and CMI or SR of TN, POC, LOC both surface soil and deep soil indicated that the SR and CMI value are suitable indicators for evaluating soil quality and C changes in surface soil as well as in deep soil. We, therefore, propose deep soil organic carbon should be considered in C cycle induced by Grain-to-Green Program (GTGP) and under RP forest is more appropriate strategy to improve the SOC status than other land use types in surface soils and deep soil.

Author Contributions

Conceived and designed the experiments: GY XH. Analyzed the data: FZ. Contributed reagents/materials/analysis tools: YF GR. Wrote the paper: FZ.

References

1. Zhang CS, McGrath D (2004) Geostatistical and GIS analyses on soil organic carbon concentrations in grassland of southeastern Ireland from two different periods. Geoderma 119: 261–275.
2. Jobbágy EG, Jackson RB (2000) The vertical distribution of soil organic carbon and its relation to climate and vegetation. Ecological Applications 10: 423–436.
3. Cole CV, Duxbury J, Freney J, Heinemeyer O, Minami K, et al. (1997) Global estimates of potential mitigation of greenhouse gas emissions by agriculture. Nutrient Cycling in Agroecosystems 49: 221–228.
4. Amundson R (2001) The soil carbon budget in soils. Annual Reviews of Earth and Planetary Sciences 29: 535–562.
5. Guo L, Gifford RM (2002) Soil carbon stock s and land use change: a meta analysis. Global Change Biology 8: 345–360.
6. IPCC (2007) Climate change 2007: the physical Science basis. In: Solomon, S., Qin, D., Manning, M., Chen, Z., et al. (Eds.), Contribution of Working Group I to the Fourth Assessment Report of the Intergovernmental Panel on Climate Change. Cambridge University Press, Cambridge.
7. Franzluebbers AJ (2002) Soil organic matter stratification ratio as an indicator of soil quality. Soil Tillage Research 66: 95–106.
8. Prescott CE, Weetman GF, DeMontigny LE, Preston CM, Keenan RJ (1995) Carbon chemistry and nutrient supply in cedar–hemlock and hemlock –amabilis fir forest floors. In: McFee, W.W., Kelley, J.M. (Eds.), Carbon Forms and Functions in Forest Soils. Soil Sci. Soc. Am., Madison, WI, pp. 377–396.
9. Van Lear DH, Kapeluck PR, Parker MM (1995) Distribution of carbon in a Piedmont soil as affected by loblolly pine management. In: McFee, W.W., Kelley, J.M. (Eds.), Carbon Forms and Functions in Forest Soils. Soil Sci. Soc. Am., Madison, WI, pp. 489–501.
10. Schnabel RR, Franzluebbers AJ, Stout WL, Sanderson MA, Stuedemann,JA. (2001) The effects of pasture management practices. In: Follett, R.F., Kimble, J.M., Lal, R. (Eds.), The Potential of US Grazing Lands to Sequester Carbon and Mitigate the Greenhouse Effect. Lewis Publishers, Boca Raton, FL, pp. 291–322.
11. Sa JCM, Lal R (2009) Stratification ratio of soil organic matter pools as an indicator of carbon sequestration in a tillage chronosequence on a Brazilian Oxisol. Soil & Tillage Research 103: 46–56.
12. Haynes RJ (2005) Labile organic matter fractions as central components of the quality of agricultural soils: an overview. Adv. Agron 85: 221–268.
13. Franzluebbers AJ, Arshad MA (1992) Particulate organic carbon content and potential mineralisation as affected by tillage and texture. Soil Science Society of America Journal 61: 1382–1386.
14. Sparling GP (1997) Soil microbial biomass activity and nutrient cycling: an indicator of soil health. In: Pankhurst, C.E., Doube, B.M., Gupta, V.V.S.R. (Eds.), Biological Indicators of Soil Health. CAB International, Wallingford, UK, pp. 97–119.
15. Blair GJ, Lefroy RDB, Lisle L (1995) Soil carbon fractions based on their degree of oxidation and the development of a carbon management index for agricultural systems. Australian Journal of Agricultural Research 46: 1459–1466.
16. Bai ZG, Dent D (2009) Recent land degradation and improvement in China. Ambio 38: 150–156.
17. Jia ZB (2009) Investigation Report on Forestry major problem in 2008. Forestry Press in China. 267–273 (In Chinese)
18. LüY H, Fu BJ, Feng XM, Zeng Y, Liu Y, et al. (2012) A Policy-Driven Large Scale Ecological Restoration: Quantifying Ecosystem Services Changes in the Loess Plateau of China. PLoS ONE 7, e31782. doi:10.1371/journal.pone.0031782.
19. Zhao YT (2010) Analysis on the Necessity and Feasibility of Implementing the Project for Conversion of Cropland to Forest. Ecological Economy 7: 81–83.
20. Wei J, Cheng J, Li W, Liu W (2012) Comparing the Effect of Naturally Restored Forest and Grassland on Carbon Sequestration and Its Vertical Distribution in the Chinese Loess Plateau. PLoS ONE 7(7): e40123. doi:10.1371/journal.pone.0040123
21. Wei XR, Qiu LP, Shao MA, Zhang XC, Gale WJ (2012) The Accumulation of Organic Carbon in Mineral Soils by Afforestation of Abandoned Farmland. PLoS ONE 7(3): e32054. doi:10.1371/journal.pone.0032054
22. Gong ZT, Lei WJ, Chen ZC, Gao YX, Zeng SG, et al. (1999) Chinese Soil Taxonomy. Science Press, Beijing 36–38.
23. Liu G (1999) Soil conservation and sustainable agriculture on the Loess Plateau: challenges and prospective. Ambio 28: 663–668.
24. Xue S, Liu GB, Pan YP, Dai QH, Zhang C, et al. (2009) Evolution of Soil Labile Organic Matter and Carbon Management Index in the Artificial Robinia of Loess Hilly Area. Scientia Agricultura Sinica 4: 1458–1464
25. Jiao JY, Zhang ZG, Bai WJ, Jia YF, Wang N (2012) Assessing the Ecological Success of Restoration by Afforestation on the Chinese Loess Plateau. Restoration Ecology 20: 240–249.
26. Zhang F, Zhang SL, Cheng ZJ, Zhao HY (2007) Time structure and dynamics of the insect communities in bush vegetation restoration areas of Zhifanggou watershed in Loess hilly region. Acta Ecologica Sinica 27: 4555–4562. (in Chinese with English abstract)
27. Chen QB, Wang KQ, Qi S, Sun LD (2003) Soil and water erosion in its relation to slope field productivity in hilly gully areas of the Loess Plateau. Aata Ecologica Sinica 23: 1463–1469.
28. Li FM, Song QH, Jjemba PK, Shi YC (2004) Dynamics of soil microbial biomass C and soil fertility in cropland mulched with plastic film in a semiarid agro-ecosystem. Soil Biology and Biochemistry 36: 1893–1902.
29. Wang Z, Liu GB, Liu MX, Zhang J, Wang Y, et al. (2012) Temporal and spatial variations in soil organic carbon sequestration following revegetation in the hilly Loess Plateau, China. Catena 99: 26–33.
30. Bao SD (2000) Soil and Agricultural Chemistry Analysis. China Agriculture Press, Beijing, China (in Chinese).
31. Cambardella CA, Elliot ET (1992) Particulate soil organic matter changes a grassland cultivation sequence. Soil Science Society of America Journal 56: 777–783.
32. Graeme JB, Rod DBL, Leanne L (1995) Soil carbon fractions based on their degree of oxidation, and the development of a carbon management index for agricultural systems. Australian Journal of Agricultural Research 46:1459–1466
33. Wang YF, Fu BJ, Lu YH, Song CJ, Luan Y (2010) Local-scale spatial variability of soil organic carbon and its stock in the hilly area of the Loess Plateau, China. Qua-ternary Research 73: 70–76.
34. Eaton JM, McGoff NM, Byme KA, Leahy P, Kiely G (2008) Land cover change and soil organic carbon stocks in the Republic of Ireland 1851–2000. Climate Change 91: 317–334.
35. Fu XL, Shao MA, Wei XR, Robertm H (2010) Soil organic carbon and total nitrogen as affected by vegetation types in Northern Loess Plateau of China. Geoderma 155: 31–35.
36. Groenendijk FM, Condron LM, Rijkse WC (2002) Effect of afforestation on organic carbon, nitrogen, and sulfur concentration in New Zealand hill country soils. Geoderma 108: 91–100.
37. Davis M, Nordmeyer A, Henley D, Watt M (2007) Ecosystem carbon accretion 10 years after afforestation of depleted subhumid grassland planted with three densities of Pinus nigra. Global Change Biology 13: 1414–1422
38. Chen LD, Gong J, Fu BJ, Huang ZL, Huang YL, et al. (2007) Effect of land use conversion on soil organic carbon sequestration in the loess hilly area, loess plateau of China. Ecology. Research 22: 641–648.
39. Qiu LP, Zhang XC, Cheng JM, Yin XQ (2010) Effects of black locust (Robinia pseudoacacia) on soil properties in the loessial gully region of the Loess Plateau, China. Plant Soil 332: 207–217.
40. VandenBygaart AJ, Bremer E, McConkey BG, Ellert BH, Janzen HH, et al. (2010) Impact of Sampling Depth on Differences in Soil Carbon Stocks in Long-Term Agroecosystem Experiments. Soil Science Society of America Journal 75: 226–234

41. Strahm BD, Harpison RB, TeRPy TA, Harpington TB, Adams AB, et al. (2009) Changes in dissolved organic matter with depth suggest the potential for postharvest organic matter retention to increase subsurface soil carbon pools. Forest Ecology Management 258: 2347–2352

42. Wang Z, Liu GB, Xu MM (2010) Effect of revegetation on soil organic carbon concentration in deep soil layers in the hilly Loess Plateau of China. Acta Ecologica Sinica 14: 3947–3952 (in Chinese with English abstract)

43. Rumpel C, Kögel-Knabner I (2011) Deep soil organic matter-a key but poorly understood component of terrestrial C cycle. Plant Soil 338: 143–158

44. Fayez R (2012) Soil properties and C dynamics in abandoned and cultivated farmlands in a semi-arid ecosystem. Plant Soil 351: 161–175.

45. Sá JCM, Cerpi CC, Dick WA, Lal R, Vesnke-Filho SP, et al. (2001) Organic matter dynamics and carbon sequestration rates for a tillage chronosequence in a Brazilian Oxisol. Soil Science Society of America Journal 5: 1486–1499.

46. Moreno F, Murillo JM, Pelegrín F, Girón IF (2006) Long-term impact of conservation tillage on stratification ratio of soil organic carbon and loss of total and active CaCO3. Soil Tillage Research 85:86–93

47. Diekow J, Mielniczuk J, Knicker H, Bayer C, Dick DP, et al. (2005) Carbon and nitrogen stocks in physical fractions of a subtropical Acrisol as influenced by long-term no-till cropping systems and N fertilization. Plant Soil 268: 319–328.

17

Establishing a Regional Nitrogen Management Approach to Mitigate Greenhouse Gas Emission Intensity from Intensive Smallholder Maize Production

Liang Wu, Xinping Chen, Zhenling Cui*, Weifeng Zhang, Fusuo Zhang

Center for Resources, Environment and Food Security, China Agricultural University, Beijing, People's Republic of China

Abstract

The overuse of Nitrogen (N) fertilizers on smallholder farms in rapidly developing countries has increased greenhouse gas (GHG) emissions and accelerated global N consumption over the past 20 years. In this study, a regional N management approach was developed based on the cost of the agricultural response to N application rates from 1,726 on-farm experiments to optimize N management across 12 agroecological subregions in the intensive Chinese smallholder maize belt. The grain yield and GHG emission intensity of this regional N management approach was investigated and compared to field-specific N management and farmers' practices. The regional N rate ranged from 150 to 219 kg N ha^{-1} for the 12 agroecological subregions. Grain yields and GHG emission intensities were consistent with this regional N management approach compared to field-specific N management, which indicated that this regional N rate was close to the economically optimal N application. This regional N management approach, if widely adopted in China, could reduce N fertilizer use by more than 1.4 MT per year, increase maize production by 31.9 MT annually, and reduce annual GHG emissions by 18.6 MT. This regional N management approach can minimize net N losses and reduce GHG emission intensity from over- and underapplications, and therefore can also be used as a reference point for regional agricultural extension employees where soil and/or plant N monitoring is lacking.

Editor: Shuijin Hu, North Carolina State University, United States of America

Funding: The work has been funded by the National Basic Research Program of China (973, Program: 2009CB118606) (website: http://www.973.gov.cn/AreaAppl. aspx). National Maize Production System in China (CARS-02-24)(website: http://119.253.58.231/). Special Fund for Agro-scientific Research in the Public Interest (201103003)(website: http://www.hymof.net.cn/webapp/login.asp). The funders had no role in study design, data collection and analysis, decision to publish, or preparation of the manuscript.

Competing Interests: The authors have declared that no competing interests exist.

* E-mail: cuizl@cau.edu.cn

Introduction

The need to increase global food production while also increasing nitrogen (N) use efficiency and limiting environmental costs [e.g., greenhouse gas (GHG) emissions] have received increasing public and scientific attention [1–6]. Coordinated global efforts are particularly critical when dealing with N-related GHG emissions because such emissions and their impacts recognize no borders. The most rapidly developing countries, such as China and India, are becoming central to the issue, not only because these countries consume the most chemical N fertilizer [7,8], but they have also become dominating forces in the production of new N fertilizers in recent decades [7,8]. From 2001 to 2010, global N fertilizer consumption increased from 83 to 105 MT, with 83% of this global increase originating from five rapidly developing countries, specifically China (9.9 MT), India (5.2 MT), Pakistan (0.8 MT), Indonesia (1.1 MT), and Brazil (1.1 MT). In comparison, chemical N fertilizer consumption decreased by 6.5% (0.7 MT) in Western Europe and Central Europe, and increased by only 7.1% (0.8 MT) in the United States over this period [8]. Optimizing N management in these rapidly developing countries clearly has important implications worldwide.

In the past 30 years, the N application rate in many developed economies has been optimized based on recommended systems, and have included soil nitrate (NO$_3$) and plant testing [9,10], and more recently, remote sensing [11]. However, in rapidly developing countries, small-scale farming with high variability between fields and poor infrastructure in the extension service makes the use of many advanced N management technologies difficult. Fox example, the average area per farm in China is only 0.6 ha, and individually managed fields are generally 0.1–0.3 ha [12]. Therefore, the challenge is to develop agronomically effective and environmentally friendly practices that are applicable to hundreds of millions of smallholder farmers, while producing high yields and reducing N losses.

Decisions regarding the optimal N fertilizer application rate require knowledge of existing soil N supplies, crop N uptake, and the expected crop yield in response to N application [13]. Optimal N rates often vary depending on soil-specific criteria and/or crop management variables such as soil productivity, producer management level, and geographic location [14]. However, the optimal N rate will become more uniform under geographically similar soil and climatic conditions, and when the main factors causing the variation in optimal N rates are either addressed or removed [14].

Our hypothesis is that a regional N management approach could be adopted to accommodate hundreds of millions of small farmers and reduce variation among farms, increase crop yield, and lower the GHG emission intensity of maize production. In China, maize (*Zea mays* L.) is the largest food crop produced, accounting for 37% of Chinese cereal production and 22% of the global maize output in 2011 [15]. Chinese maize production results in some of the most intensive N applications globally, and the resulting enrichment of N in soil, water, and air has created serious environmental problems.

In the present study, we developed a regional N management approach across major maize agroecological regions in China. We also compared grain yield and GHG emissions between the regional N management approach and site-specific N management, and evaluated the potential for increasing grain yields and mitigating GHG emission intensity using this regional N management approach when compared to farmers' practices across each region.

Materials and Methods

Description of China's agroecological maize regions

In China, maize is grown primarily in 4 main agroecological regions and 12 agroecological subregions, including Northeast China (NE1, NE2, NE3, NE4), North China Plain (NCP1, NCP2), Northwest China (NW1, NW2, NW3), and Southwest China (SW1, SW2, SW3) (Fig. 1) [16]. These agroecological subregions were divided based on climatic conditions, terrains, agricultural management practices (e.g., irrigation), and soil types. Detailed information on each of these subregions is provided in Table S1 and Text S1.

Farmers' survey

A multistage sampling technique was used to select representative farmers for a face-to-face, questionnaire-based household survey conducted once a year between 2007 and 2009 [17]. In this study, 5,406 farmers from 66 counties in 22 provinces were surveyed (Table 1). In each province, three counties were randomly selected, three townships were randomly selected in each county, two to five villages were randomly selected in each township, and 20 farmers from the villages were randomly surveyed to collect information on N fertilizer use and grain yield in each farmer's household. This study was approved by a research ethics review committee at the College of Resources and Environmental Science (CRES), China Agricultural University, Beijing, China. Data was collected through an in-house survey, which was conducted by research staff at the College of Resources and Environmental Science. Before beginning the survey, an informed consent information sheet was given to the farmer to read (or in some cases was read to the farmer), and verbal informed consent was requested. Because this study was considered anonymous and each participating household could not be identified directly or indirectly, the research ethics review committee of CRES waived the need for written informed consent from the participants.

On-farm field experiments

In total, 1,726 on-farm maize N fertilizer experiments in 181 counties of 22 provinces were conducted from 2005 to 2010 in the NE ($n = 397$) and NW ($n = 416$) spring maize areas, and in the NCP ($n = 407$) and SW ($n = 506$) summer maize areas. All 66 counties where farm surveys were conducted were included in these 181 counties.

All experimental fields received four treatments without replication: without N fertilizer (N0), medium N rate (MN), 50% and 150% of MN. The amount of N fertilizer for the MN treatment was recommended by local agricultural extension

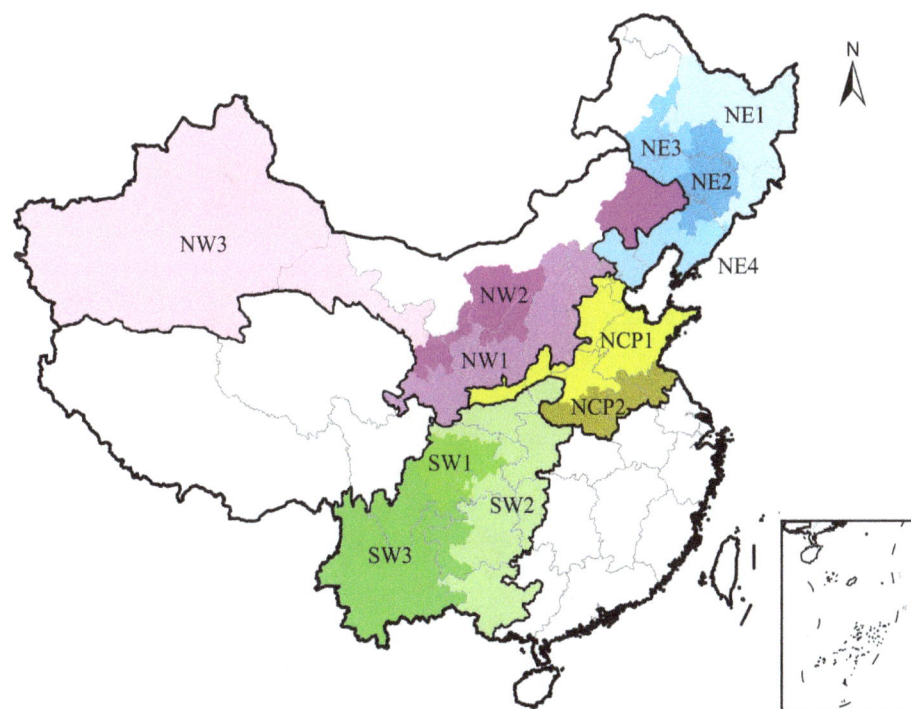

Figure 1. Map showing the four major maize-planting agroecological regions (thick lines, NE, NCP, NW, SW) and their subregions in China (different colors). Northeast China (NE1, NE2, NE3, NE4), North China Plain (NCP1, NCP2), Northwest China (NW1, NW2, NW3), and Southwest China (SW1, SW2, SW3). Here, we show the distribution of maize production in China; the total maize sowing area in the 12 subregions is approximately 32 million hectares, which represents 96% of the total maize production in China.

Table 1. N fertilizer application rate, maize grain yield, N balance, and GHG emission intensity of N fertilizer use, N fertilizer production and other sources in different agro-ecological subregions.

Region & Subregion	n [a]	N rate (kg ha^{-1})	Grain yield (Mg ha^{-1})	N balance (kg N ha^{-1})	GHG emission intensity (kg CO$_2$ eq Mg^{-1} grain)			
					N fertilizer use	N fertilizer production	Other sources	Total
NE	1263	195±61	8.91±1.19	32±30	115±38	182±60	50±17	347±131
NE1	361	156±43	8.59±0.79	3±13	96±27	151±40	51±13	298±85
NE2	411	201±59	9.03±1.16	40±28	116±34	183±54	49±15	348±124
NE3	311	226±76	8.80±1.52	68±41	137±46	213±71	51±17	402±164
NE4	180	205±77	8.68±1.50	49±39	124±51	196±74	52±21	373±183
NCP	1983	208±72	7.42±1.24	61±43	148±46	233±76	55±19	436±178
NCP1	1460	206±71	7.68±1.27	54±38	141±49	223±77	54±19	418±180
NCP2	523	217±66	7.14±1.15	76±46	161±45	252±75	57±17	471±174
NW	882	238±107	7.58±1.91	95±68	170±77	261±119	56±27	487±240
NW1	394	234±103	6.93±1.70	98±73	182±80	280±123	58±26	520±282
NW2	289	246±128	8.22±2.50	91±64	163±86	248±129	54±29	466±171
NW3	199	234±83	7.15±1.48	106±47	176±62	272±96	60±20	508±257
SW	1278	250±91	5.45±1.17	144±86	251±93	381±140	78±30	710±319
SW1	427	257±83	5.41±1.13	151±71	263±85	394±127	75±24	732±225
SW2	447	232±90	5.36±1.08	129±93	232±90	358±134	84±34	675±359
SW3	404	272±101	5.59±1.32	160±103	274±100	403±152	75±28	752±375

[a]n: number of observations.

employees based on experience and target yield (1.1 times the average yield of the past 5 years). The median N application rates for the 1,726 sites are shown in Table 2. Approximately one-third of the granular urea was applied by broadcasting at sowing, while the remainder was applied as a side-dressing at the six-leaf stage. All experimental fields received 30–150 kg P_2O_5 (P) ha^{-1} as triple superphosphate and 30–135 kg K_2O (K) ha^{-1} as potassium chloride, based on experience and target yield. All P and K fertilizers were applied by broadcasting before sowing. No manure was used, which is common for maize production in China. Detailed information regarding the N application rate and selected soil chemical properties before maize planting at 1,726 on-farm experimental sites is provided in Table S2.

Individual plots were approximately 40 m^2 (5 m wide and 8 m long). All experiments were managed (including maize variety, density, planting, harvesting, herbicide and insecticide for pests, diseases, and weeds) by local farmers based on a field manual provided by local agricultural extension employees, whereas for the treatments, local agricultural extension employees conducted fertilizer applications. The time of planting and harvest were determined by farmers and differed among sites. Generally, in NE and NW, maize was planted in early May and harvested in late September. Maize was planted from June to October in NCP and from April to August in SW. Plant densities were 50,000–65,000 plants ha^{-1} in NE, 70,000–75,000 plants ha^{-1} in NCP, 65,000–75,000 plants ha^{-1} in NW, and 45,000–50,000 plants ha^{-1} in SW. The locations of the 1,726 experiments were not privately-owned or protected in any way. No specific permits were required for the field studies. The farming operations employed during the experiment were similar to the operations routinely employed on rural farms and did not involve endangered or protected species. All operations were approved by the CRES, China Agricultural University.

Sampling and laboratory procedures

Prior to the experiments, five chemical soil properties were examined. Values were determined based on soil samples from a combined soil sample of the 10–20 cores from depths of 0–20 cm. Soil samples collected before planting were air-dried and sieved through a 0.2-mm mesh. Soil samples were used to measure organic matter content (OM) [18], alkaline hydrolyzable N (AN) [19], Olsen-P [20], NH_4OAc-K [21], and pH [22]. Upon harvest, approximately 2.5×8-m^2 sections of each plot were assessed, and ears were harvested from all plants by hand. The grain yield was adjusted to a moisture content of 15.5%.

A regional N management approach

A guideline for regional N rate was calculated for each subregion through several steps. First, yield data were collected from a large number of N response trials ($n = 1,726$). Grain yield responses to N fertilizer curves were fit using a quadratic model with PROC NLIN (SAS Institute Inc., Cary, NC, USA) to generate yield function equations (the yield significantly ($P<0.05$) responded to N) [23,24]. Next, from the response curve equation at each experimental site, the yield increase (above the yield in the N0 treatment), gross Chinese yuan return at that yield increase (maize grain price times yield), N fertilizer cost (N fertilizer price times N fertilizer rate), and net return to N ratio (gross yuan return minus N fertilizer cost) were calculated for each 1 kg N fertilizer rate increment from 0 to 270 kg N ha^{-1}. Finally, for each incremental N rate, the net return was averaged across all trials in the subregional data set to generate an estimated ratio of the maximum return to N rate, and the corresponding yield across all trials at an N fertilizer:maize grain price ratio [14,25]. In recent

years, the fertilizer:maize grain price ratio has remained relatively stable, and a value of 2.05 was used in this study.

Field-specific N management

In total, grain yield responses to N fertilizer curves were fit for 1,726 on-farm sites, using a quadratic model with PROC NLIN (SAS Institute Inc.) to generate yield function equations (the yield significantly ($P<0.05$) responded to N) [23,24]. The minimum N rate for the maximum net return was calculated from the selected model based on an N:maize price ratio of 2.05.

Nitrogen use efficiency and N balance

Nitrogen use efficiency for each treatment using the partial factor productivity (PFP_N) indices.

$$PFP_N = \frac{Y_N}{F_N} \qquad (1)$$

Where Y_N = Crop yield with N applied;
F_N = Amount of N applied.
Soil surface N balance was calculated as described in the Organization for Economic Co-operation and Development (OECD) [26].

$$N \text{ balance} = N \text{ input} - N \text{ uptake} \qquad (2)$$

where N input is N applied as chemical fertilizer, and N uptake is N in the harvested yield.

$$N \text{ uptake} = \text{Aboveground N uptake} \times \text{Yield} \qquad (3)$$

The maize aboveground N uptake requirement per million grams (Mg) grain yield in China was determined previously; spring maize grain yield was <7.5 Mg ha^{-1}, 7.5–9.0 Mg ha^{-1}, 9.0–10.5 Mg ha^{-1}, and 10.5–12.0 Mg ha^{-1}, and N uptake requirements per Mg grain yield were 19.8, 18.1, 17.4 and 17.1 kg, respectively [27]. Summer maize N uptake requirements per Mg grain yield were 20 kg [28].

Estimation of GHG emissions and emission intensity

Total GHG emissions during the entire life cycle of maize production, including CO_2, CH_4, and N_2O, consisted of three components: (1) emissions during N fertilizer application, production and transportation, (2) emissions during P and K fertilizer production and transportation, and (3) emissions from pesticide and herbicide production (delivered to the gate) and diesel fuel consumption during sowing, harvesting, and tillaging operations [29].

$$GHG = (GHGm + GHGt) \times N \text{ rate} + \text{total } N_2O \times 44/28 \\ \times 298 + GHGothers \qquad (4)$$

where GHG (kg CO_2 eq ha^{-1}) is the total GHG emission, and GHGm is the GHG emission originating from fossil fuel mining as the industry's energy source to N product manufacturing, and was 8.21 kg CO_2 eq kg N^{-1} (Table S3) [30]. GHGt is the N fertilizer transportation emission factor, and was 0.09 kg CO_2 eq kg N^{-1} (Table S3) [30]. N rate is the N fertilizer application rate (kg N ha^{-1}). GHG$_{others}$ represents GHG emission of P and K fertilizer

Table 2. The number of on-farm experiments, maize yield without N, medium N rate and N rate, grain yield, GHG emission intensity of N fertilizer use, N fertilizer production and other sources for regional N management approach and field-specific N management.

Subregion	n[a]	Yield without N (Mg ha⁻¹)	Medium N rate (kg ha⁻¹)	Yield for medium N rate (Mg ha⁻¹)	Regional N management approach							Field-specific N management					
					N rate (kg ha⁻¹)	Grain yield (Mg ha⁻¹)	GHG emission intensity (kg CO_2 eq Mg⁻¹ grain)					N rate (kg ha⁻¹)	Grain yield (Mg ha⁻¹)	GHG emission intensity (kg CO_2 eq Mg⁻¹ grain)			
							N fertilizer use	N fertilizer production	Other sources	Total				N fertilizer use	N fertilizer production	Other sources	Total
NE1	132	6.40±1.01[b]	153±6	8.98±1.09	150	8.85	91	141	49	280		158±26	8.87±1.11	91±19	149±32	49±7	289±55
NE2	62	6.82±1.23	147±21	9.05±1.51	150	9.18	87	136	49	272		155±25	9.13±1.48	87±14	143±23	51±8	281±42
NE3	126	6.50±1.26	162±15	9.48±1.38	164	9.01	96	151	50	298		165±20	9.10±1.25	92±18	153±29	50±8	295±53
NE4	77	6.92±1.66	204±28	8.93±1.59	188	8.76	113	178	55	346		191±49	8.84±1.48	117±37	183±53	57±9	356±94
NCP1	348	6.58±1.13	194±22	8.23±1.13	178	8.13	115	182	58	355		179±27	8.14±1.19	113±23	185±36	59±9	357±64
NCP2	59	6.91±1.13	213±30	8.67±0.95	177	8.37	111	176	55	342		185±33	7.59±1.12	129±30	208±45	62±9	399±80
NW1	100	6.30±1.06	190±34	8.35±1.03	181	8.13	117	185	59	360		180±44	8.13±1.12	115±29	184±45	59±9	357±77
NW2	309	8.12±1.83	190±20	10.53±1.71	176	10.38	89	141	47	277		182±34	10.48±1.82	91±25	148±37	48±9	288±68
NW3	7	7.23±1.74	221±9	10.33±1.59	219	9.83	118	185	46	349		215±16	9.85±1.56	116±20	185±30	46±7	347±57
SW1	78	5.70±1.30	217±22	7.63±1.20	174	7.46	123	194	63	379		191±39	7.56±1.19	134±34	214±53	63±11	412±93
SW2	368	5.59±1.13	195±22	7.72±1.18	183	7.71	125	197	66	387		184±38	7.77±1.26	125±33	202±52	67±12	394±92
SW3	60	6.00±1.09	207±25	8.29±1.33	186	8.10	121	191	65	376		191±37	8.38±1.33	120±28	192±44	65±11	376±78
National[c]	–	6.60	187	8.69	174	8.56	108	171	56	334		178	8.55	109	185	57	343

[a]n: number of observations.

[b]Mean ± SD.

[c]National values are computed from the regional values weighted by area. The regional weights are as follows:
NE1, 4.5%; NE2, 14.9%; NE3, 4.7%; NE4, 6.4%; NCP1, 25.6%; NCP2, 6.0%; NW1, 10.4%; NW2, 7.3%; NW3, 2.6%; SW1, 3.5%; SW2, 7.9%; SW3, 6.2%.

production and transportation, pesticide and herbicide production and transportation, and diesel fuel consumption (Table S3).

Total N_2O emission included direct and indirect N_2O emissions. Indirect N_2O emissions were estimated with a method used by the International Panel on Climate Change [31], where 1% and 0.75% of ammonia (NH_3) volatilization and nitrate (NO_3^-) leaching, respectively, is lost as N_2O. N_2O emission is calculated based on empirical models. Based on previous reports, the final data set consisted of 10 (30 observations) and 22 (117 observations) studies on direct N_2O emissions for spring maize and summer maize, respectively. Detailed information is provided in Table S4 and Figure S1.

$$\text{Direct } N_2O \text{ emission for spring maize} = 0.576\exp(0.0049 \times N \text{ rate}) \tag{5}$$

$$\text{Direct } N_2O \text{ emission for summer maize} = 0.593\exp(0.0045 \times N \text{ rate}) \tag{6}$$

NH_3 volatilization and N leaching employs the following equation (Cui *et al* 2013, Global Change Biology, main text, Fig. 2) [6].

$$NH_3 \text{ volatilization} = 0.24 \times N \text{ rate} + 1.30 \tag{7}$$

$$N \text{ leaching} = 4.46\exp(0.0094 \times N \text{ rate}) \tag{8}$$

The system boundaries were set as the periods of the life cycle from the production inputs (such as fertilizers, pesticides, and herbicides), delivery of the inputs to the farm gates, and farming operations. We calculated total GHG emissions expressed as kg CO_2 eq ha^{-1} and the GHG emission intensity expressed as kg CO_2 eq Mg^{-1} grain. The change in soil organic carbon content was also not included in our analysis, because it was difficult to detect the small magnitude of the changes that occurred over a short time [32]. The soil CO_2 flux as a contributor to global warming potential (GWP) was also not included in this study because the net flux was estimated to contribute less than 1% to the GWP of agriculture on a global scale [33].

To calculate total GHG emissions and emission intensity, the N rate and corresponding yield of each farm were used for farmers' N practices. The regional N rate and corresponding yield of each subregion were used for the regional N management approach, and the optimal N rate and corresponding yield of each field were used for field-specific N management.

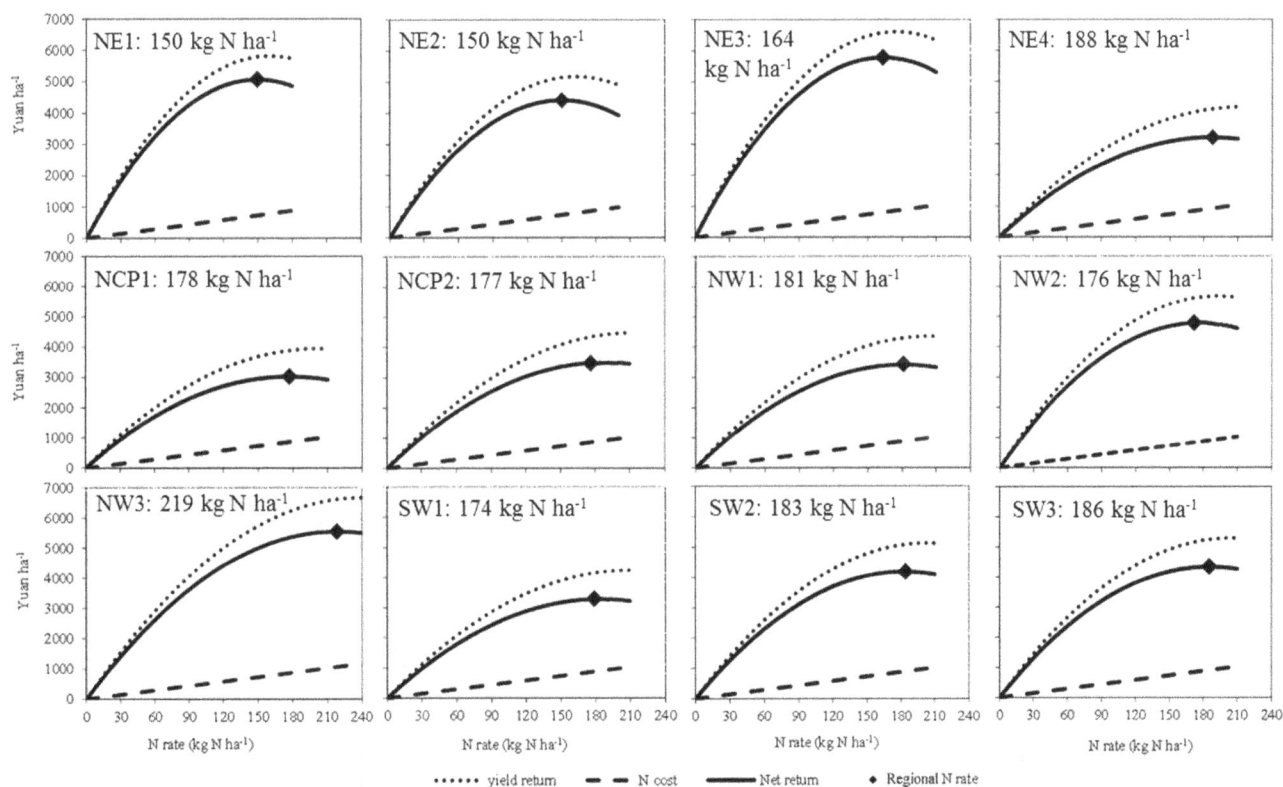

Figure 2. Maize grain yield and fertilizer economic components of calculated net return across N rates using the regional N management approach indicated at the 2.05 price ratio (N price 4.87 yuan kg^{-1} and maize price 2.37 yuan ha^{-1}) in the 12 agroecological subregions. In total, 1,726 N responses trials were used to estimate the regional N rate. The net return is the increase in yield times the grain price at a particular N rate, minus the cost of that amount of N fertilizer. The maximum return is the N rate at which the net return is greatest.

Results

Farmers' Practice

Across all 5,406 farms, maize grain yield averaged 7.56 Mg ha^{-1}, the corresponding N application rate averaged 220 kg ha^{-1}, and the N balance averaged was 69 kg N ha^{-1} (Table S5). Calculated GHG emission intensity averaged 482 kg CO_2 eq Mg^{-1} grain (Table S5), including the contributions of 155, 242, and 85 kg CO_2 eq Mg^{-1} grain from N fertilizer use, N fertilizer production, and other sources, respectively (data not shown).

Large variations were observed in grain yield and N fertilizer application rates across the four main agroecological regions. The N application rates followed the order SW (250 kg N ha^{-1}) ≈ NW (238 kg N ha^{-1}) > NCP (208 kg N ha^{-1}) ≈ NE (195 kg N ha^{-1}). In contrast, the maize grain yields were highest in NE (8.91 Mg ha^{-1}) followed by NW (7.58 Mg ha^{-1}), NCP (7.42 Mg ha^{-1}) and SW (5.45 Mg ha^{-1}). The GHG emission intensity averaged 347, 436, 487, and 710 kg CO_2 eq Mg^{-1} grain for NE, NCP, NW, and SW, respectively (Table 1).

Regional N management approach

Across all 1,726 on-farm experiments, the average grain yield under the N0 treatment, weighted by maize area in each subregion, was 6.60 Mg ha^{-1} and ranged from 5.59 Mg ha^{-1} (SW2) to 8.12 Mg ha^{-1} (NW2) (Table 2). The average medium N rate (MN) recommended by local extension employees, weighted by maize area in each subregion, was 187 kg N ha^{-1} and ranged from 147 kg N ha^{-1} (NE2) to 221 kg N ha^{-1} (NW3). The corresponding grain yield under MN treatment averaged 8.69 Mg ha^{-1} and ranged from 7.63 Mg ha^{-1} (SW1) to 10.53 Mg ha^{-1} (NW2) (Table 2).

Considering all on-farm experiments, the calculated regional N rate based on the cost response to N application rate for the subregions, weighted by maize area in each subregion, averaged 174 kg N ha^{-1} and ranged from 150 kg N ha^{-1} (NE1 & NE2) to 219 kg N ha^{-1} (NW3) (Table 2, Fig 2). The corresponding grain yield averaged 8.56 Mg ha^{-1} and ranged from 7.46 Mg ha^{-1} (SW1) to 10.38 Mg ha^{-1} (NW2) (Table 2). Calculated GHG emission intensity, weighted by maize area in each subregion, averaged 334 kg CO_2 eq Mg^{-1} grain and ranged from 272 kg CO_2 eq Mg^{-1} grain (NE2) to 387 kg CO_2 eq Mg^{-1} grain (SW2).

Based on the maize grain yield response to N application rates in all 1,726 on-farm experiments, the calculated field-specific N rate, weighted by maize area in each subregion, averaged 178 kg N ha^{-1} (Table 2) and ranged from 53 kg N ha^{-1} to 271 kg N ha^{-1} (Table S2), with a coefficient of variation (CV) of 18% (data not shown). The corresponding grain yield averaged 8.63 Mg ha^{-1} (Table 2) and ranged from 4.29 Mg ha^{-1} to 14.91 Mg ha^{-1} (Table S2), with a CV of 19% (data not shown). The calculated GHG emission intensity averaged 343 kg CO_2 eq Mg^{-1} grain (Table 2). The similar N rate, grain yield and GHG emission intensity between the regional N management approach and field-specific N management supported the notion that the regional N rate was close to an economic and environmentally optimal N application (Table 2).

Opportunities to reduce the GHG emission intensity

Compared to farmer's practices, the regional N management approach proposed reducing N fertilizer by 20.9% (220 vs. 174 kg N ha^{-1}). The grain yield would increase by 13.2% (7.56 vs. 8.56 Mg ha^{-1}). The GHG emission intensity would decrease by 30.7%, from 482 to 334 kg CO_2 eq Mg^{-1} grain. The overuse and high variability of N use by farmers has resulted in a high variability in GHG emission intensity, ranging from 364 to

1,399 kg CO_2 eq Mg^{-1} grain (Table S5) with a CV of 43% (data not shown).

Of the 12 agroecological subregions, NE2, NE3, NW1, NW2, SW1, SW2, and SW3 showed the highest potential for N-reduction (>20%), ranging from 21.0% to 31.5% and accounting for 55% of the total maize-sown area. Reduced N rates in other subregions ranged from 3.8% to 18.4% and accounted for 45% of the total maize-sown area. The subregions with a high yield increase potential (>15%; Fig. 3) were NCP2, NW1, NW2, NW3, SW1, SW2, and SW3, with increases ranging from 17.2% to 44.9% and accounting for 44% of the total maize-sown area. Grain yield in other regions ranged from 0.5% to 5.9%, accounting for 56% of the total maize-sown area. Subregions with a high potential to decrease GHG emission intensity (>20%) included NE2, NE3, NCP2, NW1, NW2, NW3, SW1, SW2, and SW3, ranging from 21.8% to 50.0% and accounting for 64% of the total maize-sown area. Reduced GHG emission intensity in other regions ranged from 6.0% to 15.1%, accounting for 36% of the total maize-sown area.

This regional N management approach, if widely adopted in China, regional N fertilizer consumption would be reduced by 1.4 MT (−20.3%), and 91% of this reduction would occur in the NE2, NE3, NCP1, NW1, NW2, SW1, SW2, and SW3 subregions (Table 3). At the same time, Chinese maize production could be increased by 31.9 MT (13.1%), from 244.1 MT to 276.0 MT, when undertaking this regional N management approach (Table 3). Total GHG emissions would be reduced by 18.6 MT eq CO_2 year^{-1} (−16.9%) (from 110.2 to 91.5 MT eq CO_2 year^{-1}) (Table 3), with 91% of this reduction occurring in the NE2, NE3, NCP1, NW1, NW2, SW1, SW2, and SW3 subregions.

Discussion

The current intensive maize system used in farmers' practices in China results in a median yield, high N application, and GHG emission intensity of 7.56 Mg ha^{-1}, 220 kg N ha^{-1}, and 482 kg CO_2 eq Mg^{-1} grain, respectively. These yields and N application rates are higher than the reported global averages (4.81 Mg ha^{-1} and 104.9 kg N ha^{-1}, the N rate calculated based on maize N fertilizer consumption and maize area harvested) for these crops in 2006 [8,15,34] and are similar to the previously reported Chinese averages for maize [35,36]. In comparison, grain yield in central Nebraska, USA, averaged 13.2 Mg ha^{-1} with only 183 kg N ha^{-1}. GHG emission intensity in this region was only 231 kg CO_2 eq Mg^{-1} grain, which was 48% lower than the average for China [37] and 109% lower than the 482 kg CO_2 eq Mg^{-1} grain for individual farmer's practices in China. The median yield and large GHG emission intensity for Chinese maize systems were attributable to the large variation in N application rates among fields. Considering 5,406 farms, N application rates ranged from 46 (only 56% of crop N uptake) to 615 kg N ha^{-1} (414% of crop uptake). Similar results were reported by Wang *et al* (2007), showing that one-third of farmers apply too little N, while another one-third of farmers apply too much ($n = 10,000$) [38].

In small-scale farming, a lack of basic knowledge and information on crop responses to N fertilizer often results in the over- and underapplication of N fertilizer [39,40]. We developed and assessed regional N management approach using large pools of response trial data that have been grouped according to criteria that indicate differing N responses for regions with similar management, climates, and soil. Our guide provides a N application rate that can be used to reduce the potential for N-deficiency or N-surplus, lowers the likelihood of reduced yields and profits, and lessens GHG emissions intensity (particularly N_2O

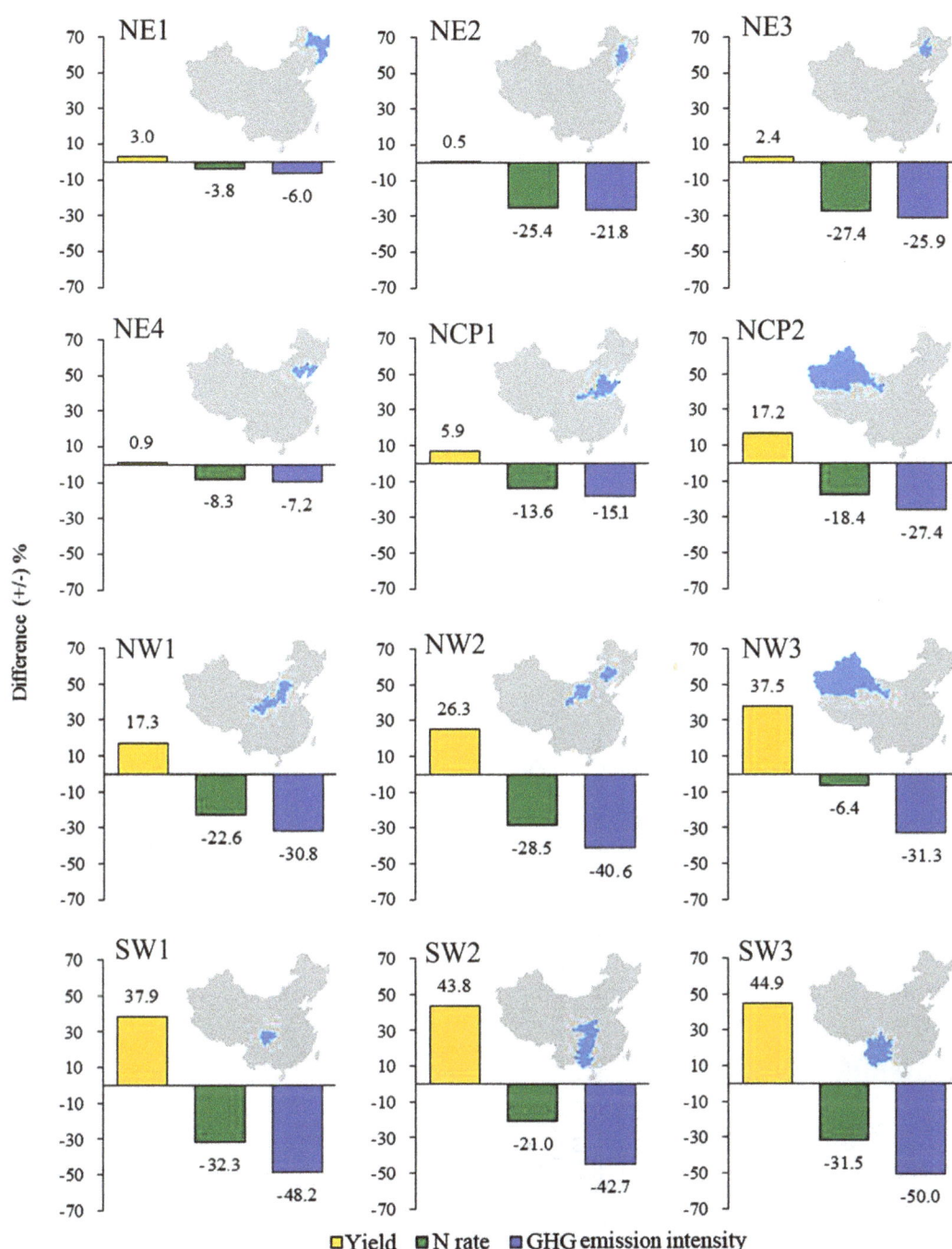

Figure 3. Regional differences (±%) in N application rates, grain yield, and GHG emission intensity between the regional N management approach and farmers' practice in the 12 agroecological subregions. Regional difference (±%) = (regional approach minus farmers' practice)/farmers' practice ×100.

emissions associated with N fertilization). Using a regional N management approach, potential for crop productivity increases and the mitigation of GHG emission intensity are likely to be achieved through a combination of increased N application in regions with a low N input and improved PFP$_N$ in regions where N fertilizer application is already high. Meanwhile, crop N uptake and N use efficiency can improve the ratio split application, with one-third for base dressing and two-thirds for top dressing [36]. Currently, typical farmers' practices apply 50% of the total N fertilizer before planting or at the early growth stage [36,41]. Some

recent practices have indicated that the amount of basal application should be added to the ratio of the top dressing to improve N use efficiency and increase grain yield [36].

The gains in yield and reduced GHG emissions achieved using regional N management approach are significant. Moreover, we believe these benefits can be further improved by applying other best-management strategies to fertilizer (e.g., slow-release N fertilizer, N transformation inhibitors, and fertigation) [42] and related practices that enhance the crop recovery of applied N (e.g., rotation with N fixing crops, precision agriculture management

Table 3. Maize production, N fertilizer consumption and total GHG emission between the regional N rate and farmers' practice in 12 agro-ecological subregions.

Subregion	Area (million ha)	N fertilizer consumption (MT)			Maize production (MT)			Total GHG emission (MT eq CO_2 yr^{-1})		
		Farmers' practice	Regional N rate	Difference [a]	Farmers' practice	Regional N rate	Difference [a]	Farmers' practice	Regional N rate	Difference [a]
NE1	1.45	0.23	0.22	−0.01	12.5	12.8	0.4	3.7	3.6	−0.1
NE2	4.80	0.96	0.72	−0.24	43.8	44.1	0.2	15.2	12.0	−3.3
NE3	1.50	0.34	0.25	−0.09	13.2	13.5	0.3	5.3	4.0	−1.3
NE4	2.06	0.42	0.39	−0.04	17.9	18.0	0.2	6.7	6.2	−0.4
NCP1	8.24	1.70	1.47	−0.23	63.3	67.0	3.7	26.4	23.8	−2.7
NCP2	1.94	0.42	0.34	−0.08	13.9	16.2	2.4	6.5	5.6	−1.0
NW1	3.35	0.78	0.61	−0.18	23.2	27.2	4.0	12.1	9.8	−2.3
NW2	2.36	0.58	0.42	−0.17	19.4	24.5	5.1	9.0	6.8	−2.2
NW3	0.85	0.20	0.19	−0.01	6.1	8.4	2.3	3.1	2.9	−0.2
SW1	1.14	0.29	0.20	−0.09	6.2	8.5	2.3	4.5	3.2	−1.3
SW2	2.54	0.59	0.46	−0.12	13.6	19.6	6.0	9.2	7.6	−1.6
SW3	1.99	0.54	0.37	−0.17	11.1	16.1	5.0	8.4	6.1	−2.3
National [b]	32.23	7.06	5.62	−1.43	244.1	276.0	31.9	110.2	91.5	−18.6

[a]Different mean the different of maize production, N fertilizer consumption, and total GHG emission between regional N rate and farmer's practice.
[b]National values are computed from the regional values weighted by area. The regional weights are as follows:
NE1, 4.5%; NE2, 14.9%; NE3, 4.7%; NE4, 6.4%; NCP1, 25.6%; NCP2, 6.0%; NW1, 10.4%; NW2, 7.3%; NW3, 2.6%; SW1, 3.5%; SW2, 7.9%; SW3, 6.2.

techniques) [42]. While this approach for N fertilizer management should be extended to farmers throughout the entire Chinese cereal production area, it is also relevant to other high-yield cropping systems outside of China. The economic approach to N rate recommendations based on multiple N rate trials has been applied for two to three decades in the U.S. Midwest, and has been more recently "formalized" with the Iowa State MRTN approach for seven Midwestern states [43].

This regional N management approach, if widely adopted in China, could reduce fertilizer N consumption by 20.3%, increase Chinese maize production by 13.1%, and reduce total GHG emissions by 16.9%. Moreover, the recommendations provide reasonable N rates and high net return, and can be easily adopted in rural areas of China where no available soil and/or plant N monitoring facilities exist [44]. The regional N rate can also be used as a reference point for agricultural extension employees without any soil and/or plant N monitoring. In practice, some factors also affect these suggested regional N rates, such as timing of crop rotation, tillage system, and soil productivity [14]. For example, the recommended N rate for soybean following maize rotations is lower than maize following maize rotations [14]. No-till management can delay or reduce residue breakdown, or mineralization, thereby reducing the N supplied from crop residue [14]. Soils where productivity is limited frequently require higher rates of fertilizer N to reach optimum yield. Conversely, lower rates of fertilizer N may be needed to reach optimum yield on highly productive soils [14].

Although this regional N management approach can easily be adopted in rural areas, delivering this technology to millions of farmers is challenging due to the lack of effective advisory systems and knowledgeable farmers. For example, educated young male farmers tend to leave the farming sector for more profitable jobs, leaving farmwork to the older and less educated individuals, especially in low income or remote areas [45]. In addition, adding more N fertilizer based on the regional N rate is difficult for farmers with low incomes or in remote areas. The Chinese central government has been aware of this problem and has attempted to provide agricultural technologies to these areas. For example, China has launched national programs for soil testing and fertilizer recommendations since 2005. In 2009, 2,500 counties in China were involved in the programs, receiving a total of 1.5 billion yuan from the Chinese central government [40].

Although the on-farm trials were conducted by local farmers in the same counties as the farmers' surveys (including experimental counties), the management and environment is not always the same for on-farm trials and farmers' surveys. While gains in grain yield and GHG were achieved by farmers using the trials, we believe that the majority of these gains can be realized in practice in many counties if improved agronomic and N management techniques are adopted. The management and environment differed among four maize regions; thus, N losses may also differ. For example, the annual direct N_2O emission accounted for 0.92% of the applied N with an uncertainty of 29%. The highest N_2O fluxes occurred in East China as compared with the lowest

fluxes in West China [46]. In this study, we use the different exponential relationships of the N application rate and N_2O fluxes for spring maize and summer maize, respectively. However, developing N loss models at the regional or subregional scale is difficult due to insufficient field measurement data in China. Long-term field observations covering all subregions are required to accurately assess farming potential and mitigate GHG emissions.

Supporting Information

Figure S1   Relationships between the N application rate and direct N_2O emissions for spring maize (A) and summer maize (B) production in China based on a meta-analysis. The direct N_2O emission data was taken from Table S4.

Table S1   The criteria and values for the sub-regional divisions.

Table S2   The site, year, soil type, irrigation, crop rotations, soil organic matter (SOM) content, alkaline hydrolyzable N (AN), Olsen-P (AP), NH_4OAc-K (AK), pH, medium N rate (MN), recommended P_2O_5 rate (RP), recommended K_2O rate (RK), grain yield without N fertilizer, yield at 50% MN, yield at 100% MN, yield at 150% MN, economic optimal N rate (EONR), yield at EONR, and GHG emissions intensity at EONR for all 1,726 on-farm experiments.

Table S3   GHG emission factors of agricultural inputs.

Table S4   The site, year, annual mean precipitation, temperature, soil organic matter (SOM), total N content, pH, N rate, grain yield, and direct N_2O emissions at different experimental sites.

Table S5   Descriptive statistics of the surveyed farms N fertilizer application rate, maize grain yield, PFP_N, N balance and GHG emission intensity for 5,406 farmed fields between 2007 and 2009 in China.

Text S1   Detailed information for each of these regions.

Author Contributions

Conceived and designed the experiments: FsZ XpC. Performed the experiments: FsZ XpC. Analyzed the data: LW ZlC WfZ. Contributed reagents/materials/analysis tools: LW ZlC. Wrote the paper: LW. Designed the NH3 volatilization and N leaching models used in analysis: ZlC.

References

1. Tilman D, Fargione J, Wolff B, D'Antonio C, Dobson A, et al. (2001) Forecasting agriculturally driven global environmental change. Science. 292: 281–284.
2. Tilman D, Cassman KG, Matson PA, Naylor R, Polasky S (2002) Agricultural sustainability and intensive production practices. Nature. 418: 671–677.
3. Conley D J, Paerl H W, Howarth R W, Boesch D F, Seitzinger S P, et al. (2009) Controlling eutrophication: nitrogen and phosphorus. Science. 323: 1014–1015.
4. Tilman D, Balzer C, Hill J, Befort B L (2011) Global food demand and the sustainable intensification of agriculture. Proc. Natl. Acad. Sci. USA.108: 20260–20264.
5. Zhang F, Cui Z, Fan M, Zhang W, Chen X, Jiang R (2011) Integrated soil-crop system management: reducing environmental risk while increasing crop productivity and improving nutrient use efficiency in China J Environ. Qual. 40: 1051–1057.
6. Cui Z, Yue S, Wang G, Meng Q, Wu L, Yang Z, et al. (2013) Closing the yield gap could reduce projected greenhouse gas emissions: a case study of maize production in China. Global Change Biol. 19: 2467–2477.
7. Zhang F, Cui Z, Chen Z, Ju X, Shen J, et al. (2012) Chapter one-Integrated nutrient management for food security and environmental quality in China. In Adv. Agron. ed Donald L S (Academic Press) 1–40 p.

8. IFA IFA Statistics (Paris: International Fertilizer Industry Association). Available at: www.fertilizer.org/ifa/HomePage/STATISTICS. Accessed 2013 Sept 6.

9. Soper R, Huang P (1963) The effect of nitrate nitrogen in the soil profile on the response of barley to fertilizer nitrogen Can. J. Soil Sci. 43: 350–358.

10. Wehrmann J, Scharpf HC, Kuhlmann H (1988) The Nmin method – an aid to improve nitrogen efficiency in plant production, In Nitrogen Efficiency in Agricultural Soils ed Jenkinson D S, Smith K A (Netherlands: Elsevier Applied Science) 38–45 p.

11. Gebbers R, Adamchuk VI (2010) Precision agriculture and food security. Science. 327: 828–831.

12. Chen X P, Cu Z L, Vitousek P M, Cassman K G, Matson P A, et al. (2011) Integrated soil-crop system management for food security. Proc. Natl. Acad. Sci. USA 108 6399–6404.

13. Dobermann A, Witt C, Abdulrachman S, Gines H, Nagarajan R, et al. (2003) Estimating indigenous nutrient supplies for site-specific nutrient management in irrigated rice. Agron. J. 95: 924–35

14. Sawyer J, Nafziger E, Randall G, Bundy L, Rehm G, Joern B (2006) Concepts and rationale for regional nitrogen rate guidelines for corn. Iowa: Iowa State University, University Extension. 15–24 p.

15. FAO FAOSTAT–Agriculture Database. Available: http://faostat.fao.org/site/339/default.aspx. Accessed 2013 Sept 6.

16. National Bureau of Statistics of China. China Statistical Yearbook. Available: http://www.stats.gov.cn/tjsj/ndsj/. Accessed 2013 Sept 6.

17. Etimi N A, Solomon VA (2010) Determinants of rural poverty among broiler farmers in Uyo, Nigeria: implications for rural household food security. J. Agric. Soc. Sci. 6: 24–28.

18. Walkley A (1947) A critical examination of a rapid method for determining organic carbon in soils-effect of variations in digestion conditions and of inorganic soil constituents. Soil Sci. 63: 251–264.

19. Khan S, Mulvaney R, Hoeft R (2001) A simple soil test for detecting sites that are nonresponsive to nitrogen fertilization. Soil Sci. Soc. Am. J. 65: 1751–1760.

20. Olsen SR (1954) Estimation of available phosphorus in soils by extraction with sodium bicarbonate (Washington, DC: US Department of Agriculture)

21. van Reeuwijk LP (1993) Procedures for soil analysis (International Soil Reference and Information Centre).

22. Richards LA (ed) (1954) Diagnosis and improvement of saline and alkali soils (Washington, DC: US USDA. U.S. Gov. Print. Office).

23. Wallach D, Loisel P (1949) Effect of parameter estimation on fertilizer optimization Appl. Stat. 641–651.

24. Magee L (1990) R^2 measures based on Wald and likelihood ratio joint significance tests. American Statistician. 44: 250–253.

25. Hoben J, Gehl R, Millar N, Grace P, Robertson G (2011) Nonlinear nitrous oxide (N_2O) response to nitrogen fertilizer in on–farm corn crops of the US Midwest. Global Change Biol. 17: 1140–1152.

26. OECE. Environmental indicators for agriculture: Methods and results (Paris: Organisation for Economic Co-operation and Development). Available: www.oecd.org/greengrowth/sustainable-agriculture/1916629.pdf. Accessed 2013 Sept 6.

27. Hou P, Gao Q, Xie R, Li S, Meng Q, et al. (2012) Grain yields in relation to N requirement: Optimizing nitrogen management for spring maize grown in China. Field Crop Res. 129: 1–6.

28. Meng Q F (2012) Strategies for achieving high yield and high nutrient use efficiency simultaneously for maize (Zea mays L.) and wheat (Triticum aestivum L.), Ph.D. Diss. China Agriculture University.

29. Forster P, Ramaswamy V, Artaxo P, Berntsen T, Betts R, et al. (2007) Changes in atmospheric constituents and in radiative forcing In Climate Change. The Physical Science Basis Contribution of Working Group I to the Fourth Assessment Report of the Intergovernmental Panel on Climate Change. ed Solomon S, Qin D, Manning M, Chen Z, Marquis M, Averyt K B, Tignor M, Miller H L 2007(Cambridge: Cambridge University Press).

30. Zhang WF, Dou ZX, He P, Ju XT, Powlson D, et al. (2013) New technologies reduce greenhouse gas emissions from nitrogenous fertilizer in China Proc. Natl. Acad. Sci. USDA 110: 8375–8380

31. Klein CD, et al. (2006) IPCC Guidelines for National Greenhouse Gas Inventories Chapter 11: N_2O emissions from managed soils, and CO_2 emissions from lime and urea application avaluable at: www.ipcc-nggip.iges.or.jp/public/2006gl/pdf/4_Volume4/V4_11_Ch11_N2O&CO2.pdf

32. Conant RT, Ogle SM, Paul EA, Paustian K (2010) Measuring and monitoring soil organic carbon stocks in agricultural lands for climate mitigation Front. Ecol. Environ 9: 169–173.

33. IPCC 2007Climate Change 2007: Mitigation. Contribution of Working Group III to the Fourth Assessment Report of the Intergovernmental Panel on Climate Change. ed Smith P, et al(Cambridge: Cambridge University Press).

34. Heffer P (2009) Assessment of fertilizer use by crop at the global level 2006/07–2007/08. International Fertilizer Industry Association. (Paris, France).

35. Cui Z (2005) Optimization of the nitrogen fertilizer management for a winter wheat-summer maize rotation system in the North China Plain - from field to regional scale Ph.D. Diss. China Agriculture University. (Chinese with English abstract).

36. Cui Z, Chen X, Miao Y, Zhang F, Sun Q, et al. (2008) On-farm evaluation of the improved soil N-based nitrogen management for summer maize in North China. Plain Agron. J. 100: 517–525.

37. Grassini P, Cassman KG (2012) High-yield maize with large net energy yield and small global warming intensity Proc. Natl. Acad. Sci. USA 109: 1074–1079.

38. Wang JQ (2007) Analysis and evaluation of yield increase of fertilization and nutrient utilization efficiency for major cereal crops in China Ph.D. Diss. China Agriculture University.

39. Huang J, Hu R, Cao J, Rozelle S (2008) Training programs and in-the-field guidance to reduce China's overuse of fertilizer without hurting profitability J. Soil Water Conserv. 63: 165A–167A.

40. Cui Z, Chen X, Zhang F (2010) Current nitrogen management status and measures to improve the intensive wheat–maize system in China AMBIO. 39: 376–384.

41. Chen XP (2003) Optimization of the N fertilizer management of a winter wheat/summer maize rotation system in the Northern China Plain Ph.D. diss. Univ.of Hohenheim.

42. Good AG, Beatty PH (2011) Fertilizing nature: a tragedy of excess in the commons. Plos Biol. 9: e1001124.

43. Iowa State University – Agronomy Extension. Corn Nitrogen Rate Calculator. Available: extension.agron.iastate.edu/soilfertility/nrate.aspx. Accessed 2013 Sept 6.

44. Zhu Z, Chen D (2002) Nitrogen fertilizer use in China – Contributions to food production, impacts on the environment and best management strategies Nutr. Cycl. Agroecos. 63: 117–127.

45. Barning R (2008) Economic evaluation of nitrogen application in the North China Plain Ph.D. diss. Univ. of Hohenheim.

46. Lu Y, Huang Y, Zou J, Zheng X (2006) An inventory of N_2O emissions from agriculture in China using precipitation-rectified emission factor and background emission. Chemosphere. 65: 1915–1924.

18

Evaluation of Soil Contamination Indices in a Mining Area of Jiangxi, China

Jin Wu[1], Yanguo Teng[1]*, Sijin Lu[2], Yeyao Wang[2], Xudong Jiao[1]

1 College of Water Science, Beijing Normal University, Beijing, China, **2** China National Environmental Monitoring Center, Beijing, China

Abstract

There is currently a wide variety of methods used to evaluate soil contamination. We present a discussion of the advantages and limitations of different soil contamination assessment methods. In this study, we analyzed seven trace elements (As, Cd, Cr, Cu, Hg, Pb, and Zn) that are indicators of soil contamination in Dexing, a city in China that is famous for its vast nonferrous mineral resources in China, using enrichment factor (EF), geoaccumulation index (I_{geo}), pollution index (PI), and principal component analysis (PCA). The three contamination indices and PCA were then mapped to understand the status and trends of soil contamination in this region. The entire study area is strongly enriched in Cd, Cu, Pb, and Zn, especially in areas near mine sites. As and Hg were also present in high concentrations in urban areas. Results indicated that Cr in this area originated from both anthropogenic and natural sources. PCA combined with Geographic Information System (GIS) was successfully used to discriminate between natural and anthropogenic trace metals.

Editor: Stephen J. Johnson, University of Kansas, United States of America

Funding: This work was supported by the Specific Research on Public Service of Environmental Protection in China (No. 201509031) and National Natural Science Foundation in China (No.41303069). The funders had no role in study design, data collection and analysis, decision to publish, or preparation of the manuscript.

Competing Interests: The authors have declared that no competing interests exist.

* Email: teng1974@163.com

Introduction

Environmental issues that pose a threat to soil health include erosion, a decline in organic matter content and biodiversity, contamination, sealing, compaction, salinization, and landslides [1]. In China, contamination is recognized as a major threat to soil. In recent years, there have been numerous review and research articles providing assessments of various kinds of soil contamination, including urban soil contamination, agricultural soil contamination, and soil contamination in mining areas [2]. Several studies have also provided a comparison of the results of different methods for the assessment of soil contamination [3–5]. Such studies help to raise public awareness of soil contamination and to facilitate research on contamination and contamination control strategies. However, the status and trends of soil contamination, especially at regional scales, have not been well described. Knowledge of soil geochemistry is fundamental to assessing soil contamination at the regional scale. One of the most efficient tools for studying environmental geochemistry problems is geographical information system (GIS) based on geostatistical analysis. To our knowledge, maps and comparisons of indices derived from different soil contamination methods are not widely available.

The objective of our work was to determine the origin of trace metals in soils using various indices based on geochemistry mapping, including enrichment factor (EF), geoaccumulation index (I_{geo}), and pollution index (PI), along with principal component analysis (PCA); we also aimed to critically evaluate the advantages and limitations of these methods. The data we used were obtained from a regional geochemical survey carried out in

Dexing, a city in China that is famous for its vast nonferrous mineral resources. To better understand the outcome of this work, we first present a brief overview of core issues and problems associated with current soil contamination assessment methods.

Selection of reference values

A major methodological problem associated with correctly assessing soil contamination is the identification of appropriate reference values for uncontaminated soil conditions, since all quantitative assessment methods rely on reference values of background concentrations [6]. The background, the crust, and the regulatory reference values are common reference values used for soil contamination assessment; the background value is the most appropriate reference value to evaluate soil contamination for theoretical considerations alone.

There is some variability in the definition of background. A selection of definitions and relevant terms is presented in Table 1 [7,8,9]. Indiscriminate usage of the term "background" to evaluate soil contamination can result in misinterpretations if several flaws are ignored. Reimann and de Caritat critically discuss the definitions and use of background values in environmental geochemistry [10]. Some characteristics are summarized:

(1) No specific global background levels of elements can be defined. Natural element concentrations can be as high or even higher than any visible anthropogenic contamination, therefore it is difficult to identify anthropogenic additions and contamination in most cases.

(2) Background levels depend on location and scale, and should usually be restricted to the local scale. It has been

Table 1. A selection of definitions of background and relevant term.

Definition	Term	Reference
The normal abundance of an element in barren earth material, and it is more realistic to view background as a range rather than an absolute value	Background	[7]
Geogeneous or pedogeneous average concentration of a substance in an examined soil	Background	[8]
If the atmosphere in a particular area is polluted by some substance from a particular local source, then the background level of pollution is that concentration, which would exist without the local source being present.	Background	[9]
Widely used to infer background levels reflecting natural processes uninfluenced by human activities.	Natural background	[10]
used to describe the unmeasurably perturbed and no longer pristine natural background	Ambient background	[10]
Used when data either come from age-dated materials or are collected from areas believed to represent a survey/study area in its supposed preindustrialization state.	Pre-industrial background	[10]
The outer limit of background variation	Threshold	[11]
A depature from the geochemical patterns that are normal for a given area or geochemical landscape	Anomaly	[7]
Concentrations of substances characterizing variability in the geochemistry of earth's surface materials and are needed for documenting the present state of the surface environment and to provide datum against which any changes can be measured	Baseline	[12]

demonstrated that background levels may vary both within and between regions.

(3) It is more realistic to view background as a range rather than an absolute value. There are a range of values characterizing any particular area or region that reflect the heterogeneity of the environment.

(4) It can be argued that natural background no longer exists on this planet. There is evidence from the world's ice sheets and glaciers that small amounts of elements have been transported on intercontinental scales to remote regions and deposited as a result of being released into the atmosphere from human activities.

Threshold is usually expressed as a single value showing the upper background between anomalous and background concentrations, while the baseline, usually expressed as an observed or 95% expected range, is used mainly in geochemical exploration, and is not appropriate for environmental purposes. The background values derived from different percentiles of trace metal soil concentrations for some countries are summarized in Table 2 [11–16]. The use of percentile as an upper background (threshold) provides a practical approach to continue to use the term "background". This implies the availability of reliable procedures to evaluate soil contamination, but raises the question of data comparability.

When local information is unavailable, and more cannot be obtained, it is necessary to resort to data generated by surveys from different parts of the world covering spatially significant areas (Table 2). The average concentrations of 90 naturally occurring elements in the Earth's crust have been estimated; these are known as "Clarke values" and can be found in Taylor and Wedepohl [17,18]. These two papers summarize published data on the composition of the upper continental crust, which varies slightly because there are hypothetical concentrations based on assumed proportions of various crustal rock types. The concentrations of elements differ so widely from one geologic unit to another, that the use of the Clarke value for an element in a regional or local context does not sufficiently represent variations in element distributions caused by mineralization or contamination in a particular sampling medium [19]. However, such values can give a preliminary indication of whether results from a new investigation are within an expected range and whether they reflect natural variations in concentrations present in different environments [10].

The use of regulatory reference value (RRVs), which are generally based on background values in combination with toxicity levels, is a different approach to evaluating soil contamination. RRV is set by a state authority, and is not always based solely on scientific evidence, but also on economic or political considerations. The RRVs for trace metals in soil of some countries are provided in Table 3 [20–22]. RRVs have been given various names in their original languages that translate in English to maximum admissible concentration values, target values, intervention values, guideline, cut-off values, and many others. Advantages of using screening values have been pointed out by several authors [23,24] and are confirmed in practice by their long term and successful use in many countries. Advantages include their speed and ease of application, their clarity for use by regulators and other non-specialist stakeholders, and their comparability and transparency. The major limitation of screening values is that crucial site-specific considerations cannot be included. Screening values may give rise to a misleading feeling of certainty, knowledge, and confidence, which can lead to reluctance on the part of users to apply them to site-specific risk assessments [25]. A combined approach, using guideline values to streamline the preliminary stages of decision making and site-specific risk assessment to achieve fine-tuning in later stages of an investigation, is generally considered the most appropriate [26].

Table 2. Summary of often used background values of trace metal soil concentrations (mg/kg).

Country	As	Cd	Cr	Cu	Hg	Ni	Pb	Zn
Austria[13]	nd	0.37	50	35	0.19	40	28	111
China[14]	11.2	0.1	61	22.6	0.07	26.9	26	74.2
Estonia[13]	nd	0.52	30	24	0.08	29	21	58
Germany[13]	nd	1.50	45	45	0.45	38	171	225
Japen[14]	9.02	0.41	41.3	36.97	0.28	28.5	20.4	63.8
Jiangxi (China)[15]	10.4	0.10	48.0	20.8	0.08	19	32.1	69.0
Lithuania[13]	nd	nd	23	13	nd	19	15	37
Medium of world[16]	5	0.3	80	25	0.05	20	17	70
Netherlands[13]	nd	0.60	74	27	0.23	38	42	110
Romania[13]	nd	1.23	37	35	nd	50	38	167
Slovakia[13]	Nd	0.33	55	38	0.14	41	23	85
The Continental Crust[16]	1.7	0.1	126	25	0.04	56	14.8	65
United States[14]	7.2	nd	54	25	0.09	19	19	60
United Kingdom[14]	11.3	0.62	84	25.8	0.1	33.7	29.2	59.8
Upper continental crust[16]	2	0.1	35	14.3	0.06	18.6	17	52
Interval	1.7–11.3	0.1–1.5	23–126	13–45	0.04–0.45	18.6–56	14.8–171	37–225
Range	9.6	1.4	103	32	0.41	37.4	156.2	188
Relative range	85%	93%	82%	71%	91%	67%	91%	84%

Table 3. Summary of regulatory reference values of trace metal in soil of some countries (mg/kg).

Countries	Denomination	As	Cd	Cr	Cu	Hg	Ni	Pb	Zn
Austria[6]	Guidelines		0.5–1	100	100	1	60	100	300
Chinese[20]	Guidelines	30	0.3	200	100	0.5	50	300	250
Canada[21]	Residential/parkland guidelines	12	10	64	63	6.6	50	140	200
Canada[21]	Commercial guidelines	12	22	87	91	24	50	260	360
Canada[21]	Industrial guidelines	12	22	87	91	50	50	600	360
Germany[6]	Clay		1.5	100	60	1	70	100	200
	Loam/silt		1	60	40	0.5	50	70	150
	Sand		0.4	30	20	0.1	15	40	60
The Netherlands[22]	Target guidelines	29	0.8	100	36	0.3	35	85	140
The Netherlands[22]	Intervention guidelines	55	12	380	190	10	210	530	720
Switzerland[6]	Guidelines		0.8	50	40	0.5	50	50	150

Based on the location of a reference area in relation to a study site, two types of reference areas can be classified: on-site and off-site. All the statistically derived references mentioned above are off-site references and are easy to compute. Desaules argued that off-site reference methods are obviously not appropriate to assess weakly contaminated sites, while the specific and sensitive on-site reference method could be used to accurately identify soil contamination based on the observed values of investigated trace metals [6]. On-site reference is a value specific to a particular material and to a particular locality.

Deep soil layer values are not affected by contamination and are considered to be the most convenient for use as on-site references of the same soil profile [27]. There is debate about the use of deep soil layer values to evaluate soil contamination. The use of deep soil layers, instead of the continental crust, as a reference value improves the sensitivity of EF to anthropogenic surface enrichments [27,28]. In contrast to other authors who have promoted the use of deep soil layer values, Reimann and de Caritat demonstrate that it does not significantly reduce the shortcomings of the EF approach and may even give spurious results based on results from subcontinental-scale geochemical surveys [10].

Other suggestions for on-site references to identifying contamination are buried fossil topsoils, provided the buried soils have not been contaminated or depleted subsequently by pedogenic processes, and dated peat bog samples, which make it possible to trace the chronology of atmospheric deposition [6,29,30]. However, both these types of bog samples are difficult to obtain.

Indices and methods for the assessment of soil contamination

Popular soil contamination assessment methods can be classified into two categories: quantitative and qualitative. The qualitative methods, such as PCA, factor analysis, and cluster analysis, are inferential and indicative. These multivariate analyses require that each variable shows a normal distribution and that the whole dataset shows a multivariate normal distribution [31]. Some of the most commonly used quantitative methods are the contamination factor (CF), enrichment factor (EF), and geoaccumulation index (I_{geo}). The CF, defined by Hakanson, enables an assessment of soil contamination through the use of concentrations in the surface layer of bottom sediments to preindustrial levels as a reference [32]. In China, the CF was adopted as a pollution index (PI), which is often evaluated by comparing metal concentrations with related environmental guidelines, or with respect to relevant background values. The CF is sometimes used in equivalency to background. The PI will be used in this paper because it has been widely used in soil contamination assessments. EF was introduced in the 1970s, and was initially developed to obtain information on the origin of elements in the atmosphere [33,34]. I_{geo}, a method used for the evaluation of the degree of contamination in aquatic sediments was originally defined by Müller and has been widely used in soil trace metal studies [35]. There are numerous studies which use the abovementioned factors to assess soil contamination at different scales [36,37], while, several studies use a combination of methods [38–40].

Care needs to be taken when using the terms 'contamination' and 'pollution'. Contamination is the presence of a substance where it should not be, or in levels that are above background levels [29]. The term pollution is defined as contamination that results in adverse biological effects [29]. In the context of soil systems, the difference between contamination and pollution is that contamination is presence of the substance in soil adversely affecting the soil, and pollution is the presence of the substance in the soil adversely affecting the usefulness of the soil [41]. The

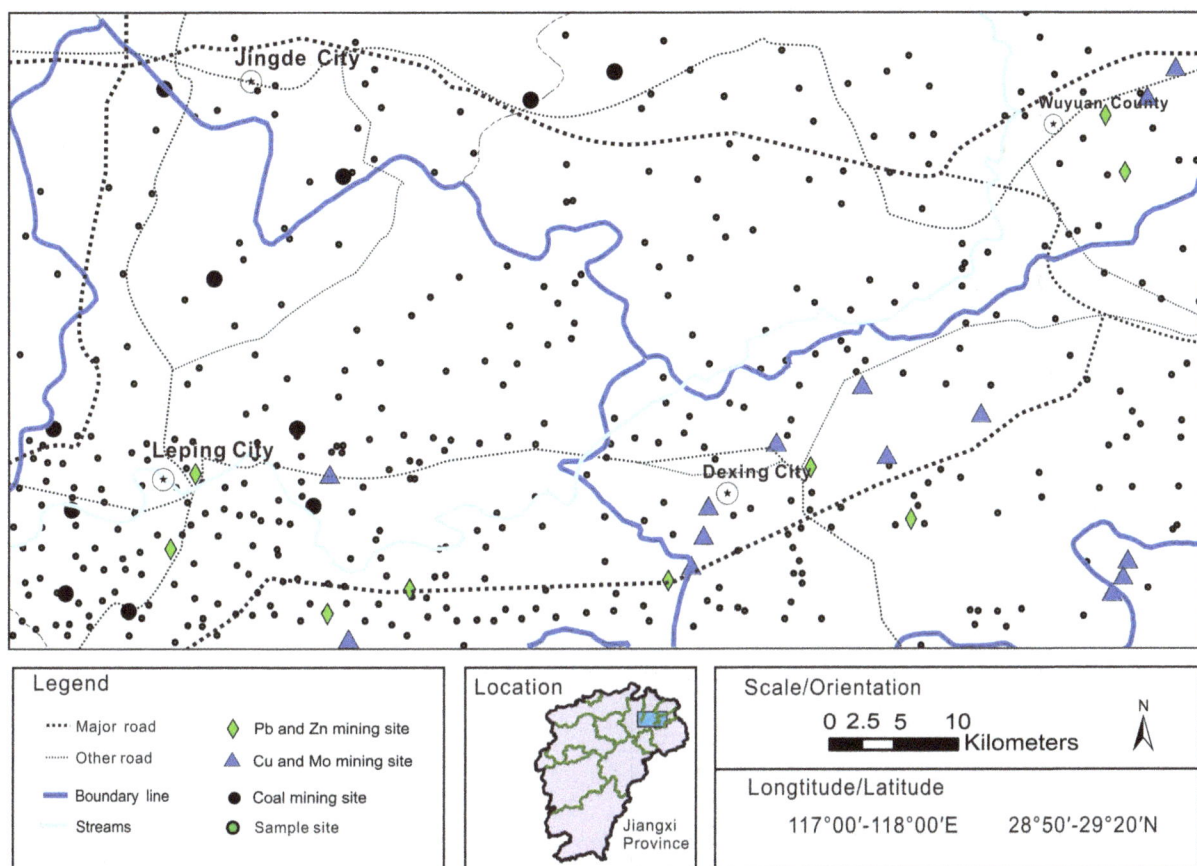

Figure 1. Location of the study area and sampling pattern.

sources of trace metals in soils are manifold, and include natural parent materials and various exogenous pollution sources [42]. Identifying and quantifying anthropogenic trace metals in soil is crucial for the assessment of soil contamination. However, difficulties arise from correctly evaluating the degree of soil contamination, especially at slightly disturbingly area. Generally, local hotspots of soil contamination (such as metal smelters and brownfields) are easier to identify and delimitate than regional contamination by agrochemicals and atmospheric deposition close to urban or industrial sources, or global contamination by long-range transboundary air contamination [6]. There is no soil contamination assessment method available to provide accurate information on the extent of perturbation for a number of reasons.

The formation of soil is a function of climate, soil organisms, landscape, plants, time, and geology. All of these factors can affect the concentration of any one element in a soil system. Because different sample materials will respond differently to the input of an element, it is not appropriate to use a single value (e.g., mean, maximum) to evaluate soil contamination of an entire area. There are two methods to describe characteristics of contamination over an entire area: the calculation of the proportion of contaminated samples in a given area, and geochemical mapping. However, the proportion of contaminated samples does not represent the specific geochemical context of each sample or other relevant information, so that the proportion calculated will not reliably provide a complete picture of soil contamination of a given area. Geochemical mapping, usually performed on GIS, provides a visual representation of the geochemical and contamination processes related to the distribution of trace elements. Additionally, most current soil contamination assessment frameworks are limited to

Table 4. Classification of different soil contamination assessment models.

Index class	I_{geo}	EF	PI	Description of classes
1	$I_{geo}<0$	EF<2	PI<1	Uncontaminated
2	$0\leq I_{geo}<1$	$2\leq EF<5$	$1\leq PI<3$	Moderately contaminated
3	$1\leq I_{geo}<3$	$5\leq EF<20$	$3\leq PI<6$	Considerable contaminated
4	$3\leq I_{geo}<5$	$20\leq EF<40$	$6\leq PI<12$	High contaminated
5	$5\leq I_{geo}$	40<EF	12<PI	Extremely contaminatied

Table 5. Descriptive statistics of metal levels (mg/kg) and selected properties (%) in soil.

Parameter	Range	Mean	S.D.	C.V.(%)	Skewness	Kurtosis	Local background value[46]
As	1.80–52.10	11.63	7.15	0.61	2.05	5.78	19.00
Cd	0.04–1.55	0.24	0.19	0.79	3.55	16.21	0.17
Cr	9.90–659.00	72.09	33.92	0.47	12.87	221.56	92.00
Cu	5.60–629.00	53.48	69.25	1.29	4.13	20.75	48.00
Hg	0.03–0.80	0.10	0.06	0.60	7.68	77.99	0.15
Mn	88.00–916.00	307.28	159.54	0.52	1.51	2.33	854.00
Pb	16.00–244.00	47.02	29.84	0.63	3.85	17.82	47.00
Ti	1082.00–9555.00	5429.92	922.53	0.17	–0.76	5.11	6320.00
Zn	27.20–799.00	87.98	55.75	0.63	7.30	76.34	108.00
Al_2O_3	7.30–20.65	13.51	2.05	0.15	–0.15	0.26	19.08
Fe_2O_3	1.57–9.64	4.31	1.11	0.26	0.91	2.36	6.44
K_2O	0.77–4.80	2.25	0.63	0.28	0.31	1.31	3.06
MgO	0.22–3.64	0.65	0.29	0.45	3.76	31.04	0.90
Na_2O	0.06–3.03	0.37	0.25	0.68	3.79	31.18	0.55
SiO_2	35.93–83.41	69.47	4.97	0.07	–0.71	4.94	74.37

Table 6. Descriptive statistics of soil contamination indices for soil trace metal.

Variable	Min	Max	Mean	Std	CV (%)	skewness	kurtosis
EF(As)	0.23	8.69	1.60	0.99	61.88	2.32	8.86
EF(Cd)	0.72	25.45	3.54	2.78	78.53	3.73	18.85
EF(Cr)	0.26	16.25	2.14	0.87	40.65	10.55	167.87
EF(Cu)	0.32	38.34	3.58	4.36	121.79	3.98	19.06
EF(Hg)	0.59	13.58	18.30	1.12	6.12	5.59	47.21
EF(Pb)	0.93	11.10	2.09	1.33	63.64	3.98	19.30
EF(Zn)	0.63	11.19	1.79	9.61	536.87	5.82	44.96
I_{geo}(As)	−3.12	1.74	−0.64	0.77	120.31	0.29	0.14
I_{geo}(Cd)	−1.8	3.37	0.47	0.75	159.57	0.98	1.77
I_{geo}(Cr)	−2.86	3.19	−0.07	0.45	642.86	−0.93	15.29
I_{geo}(Cu)	−2.47	4.33	−0.29	0.99	341.38	1.54	2.80
I_{geo}(Hg)	−1.82	2.74	−0.37	0.52	140.54	1.38	6.84
I_{geo}(Pb)	−1.60	2.34	−0.19	0.58	305.26	1.76	4.15
I_{geo}(Zn)	−1.92	2.95	−0.36	0.55	152.78	1.10	5.69
PI(As)	0.17	5.01	1.12	0.69	61.61	2.05	5.78
PI(Cd)	0.43	15.50	2.45	1.85	75.51	3.55	16.21
PI(Cr)	0.21	13.73	1.50	0.71	47.33	12.87	221.56
PI(Cu)	0.27	30.24	2.57	3.33	129.57	4.13	20.75
PI(Hg)	0.43	10.00	1.26	0.78	61.90	7.68	77.99
PI(Pb)	0.49	7.60	1.46	0.93	63.70	3.85	17.82
PI(Zn)	0.39	11.58	1.28	0.81	63.28	7.30	76.34
IPI(Ave)	0.66	7.54	1.66	0.83	0.50	2.68	9.78

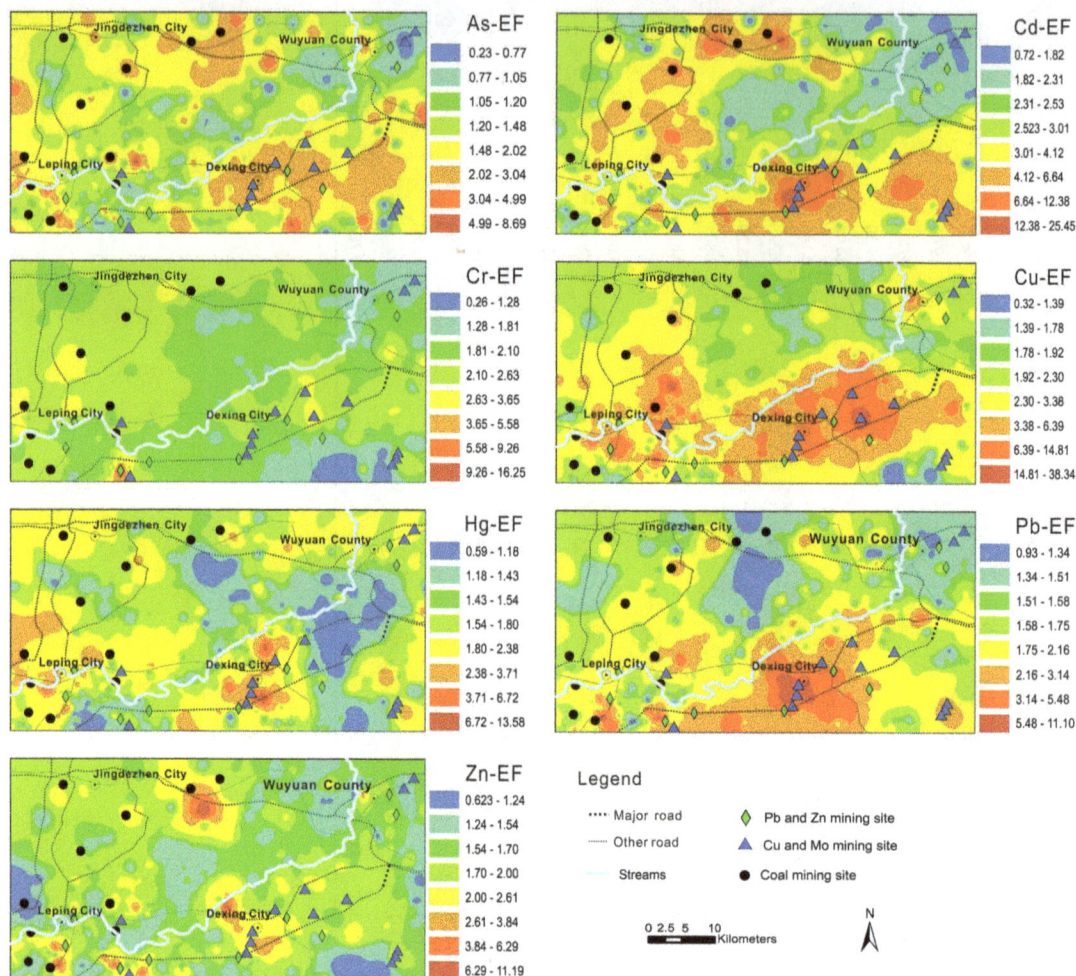

Figure 2. Spatial distribution of EF of trace elements relative to Jiangxi background in soil of the study area.

potentially toxic inorganic trace metals (As, Cd, Cr, Cu, Hg, Mn, Pb, and Zn); it is important to also consider other important inorganic (F, P, and Se) and organic (PAHs, PCBs, and PCDD/Fs) substances.

Materials and Methods

Study site description

The study area is located in the northeast part of Jiangxi province $(117°00'-118°00'E, 28°50'-29°20'N)$, China (Figure 1). No specific permissions were required for these locations. The field studies did not involve endangered or protected species. The altitude ranges from 20–1300 m, and the climate zone is subtropical monsoon, with an annual average temperature of 17°C and rainfall of 1900 mm. The soils are mainly classified as paddy soil in the plains, and yellow soil and red soil in the hilly areas. The stratum is full-fledge and spread across the study area, except for areas containing Silurian, Devonian, and Tertiary strata [43]. The Lean River is the main water body in the study area and has a number of branches, including the Jishui River, the Dawu River, and the Changle River.

Sampling and analyzing

From December 2003 to April 2004, 407 non-agricultural topsoil samples (0–20 cm) were collected using a Global Position-ing System (GPS) to identify sampling locations (Figure 1). The area covered by the sampling sites was approximately 400 km². One sample per 16 km² was collected at sites far from potential contamination sources, and one sample per 4 km² was collected around potential contamination sources, such as the Dexing copper mine, and the Leping coal mine. Each sample represents composite material taken from four points over a 1-km² patch of land; total sample weight was 1–1.5 kg. Samples were air dried at 35–40°C prior to analysis. The soil was passed through a 6-mm sieve to remove stones and plant material, then was milled with a carnelian mortar then passed through a 0.015-mm sieve prior to chemical analysis.

Each soil sample (10–20 mg) was digested in 1 mL of 60% (w/w) HNO_3 and 1 mL of 60% (w/w) $HClO_4$ in a stainless steel high-pressure digestion bomb at 140°C for 6 h. After completely cooling the system, the open vial was transferred to a hot plate (about 190°C) to evaporate the solution until the volume had decreased to several hundred micro-liters, then 0.5 mL of 49.5% (w/w) HF was added and the sample was evaporated again. The HF treatment was repeated several times until the silicate minerals had been completely dissolved. Finally, the residual solution was diluted to 6 mL with 1% (w/w) HNO_3, filtered through a syringe filter (0.45 μm). Total concentrations of Cu, Pb, Zn, and Cr were analyzed by inductively coupled plasma atomic emission spectros-copy, As and Hg were analyzed by atomic fluorescence

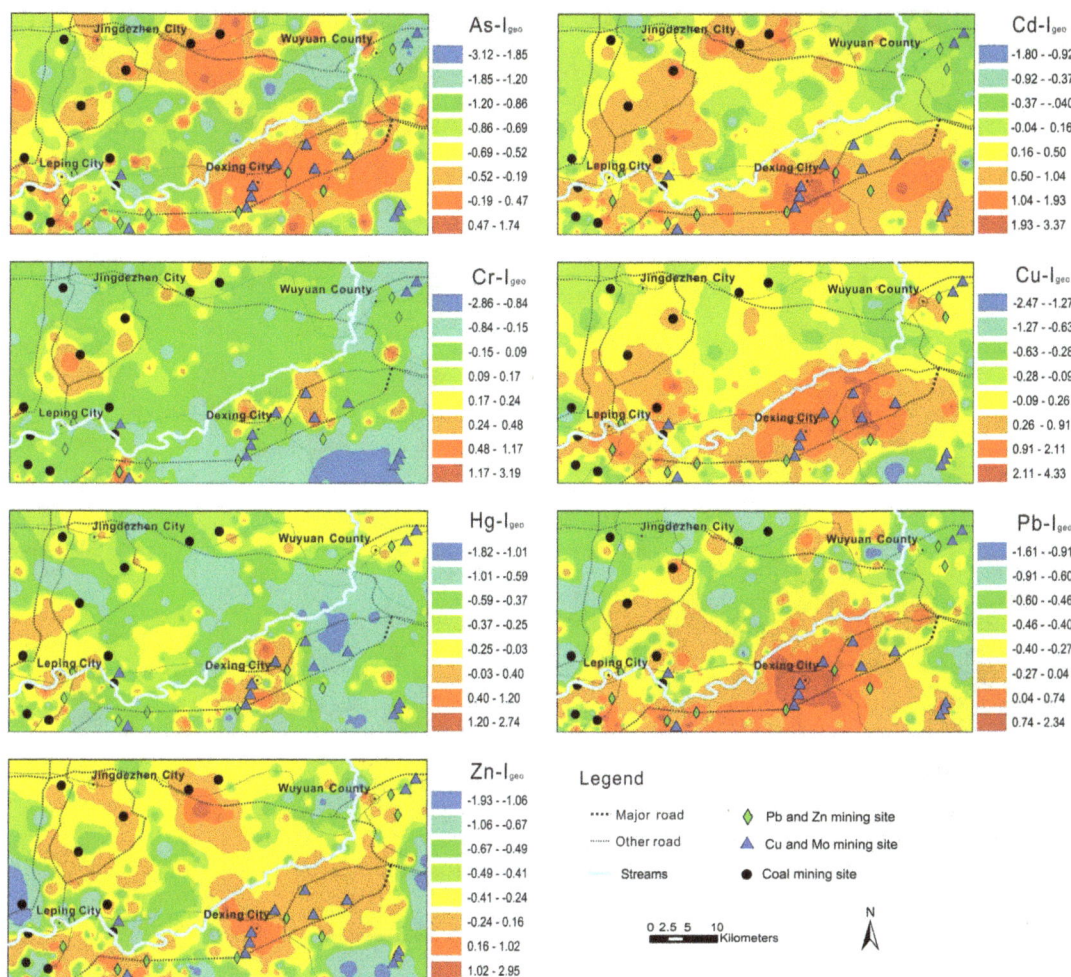

Figure 3. Spatial distribution of I_geo of trace elements in soil of the study area.

spectroscopy, and Cd was analyzed by atomic absorption spectroscopy. The total concentrations of K, Ca, Na, Mg, Si, Al, Mn, Ti, and Fe were determined by wavelength-dispersive X-ray fluorescence spectroscopy. Quality assurance and quality control procedures were performed along with laboratory analyses through the analysis of standard reference materials GSS-1, GSS-2, GSS-3, and GSS-4 soil (National Research Center for Geoanalysis of China). The results showed that the precision and bias of the analysis were generally below 5%. Recoveries of samples spiked with standards ranged from 95 to 105%.

Soil contamination assessment method

The assessment of soil contamination was carried out using EFs, I_{geo}, and PIs. To enable a comparison of the three indices, the value of the EFs, I_{geo}, and PIs were calculated using the modified formula based on the equations suggested by Chester and Stoner, Hakanson, and Müller, respectively [32,33,35].

$$I_{geo} = \log_2 \frac{C_n}{1.5B_n} \quad (1)$$

$$EF = \frac{C_n/X_n}{C_r/X_r} \quad (2)$$

$$PI = \frac{C_n}{B_n} \quad (3)$$

where C_n is the concentration of the element in the soil environment, B_n is the background concentration of soil in Jiangxi, X_n is the concentration of the reference element in the soil environment, and X_r is the concentration of the reference element in the reference environment. For this study, we used Al_2O_3 as the reference element.

For comparison of the degree of contamination, soil contamination indices were divided into five grades according to their classification criteria (Table 4). The classification for I_{geo} and PI were adjusted based on the definitions given by Müller and Hakanson and the classification of EF was done according to Sutherland [32,35,44].

Multivariate statistics analysis

In this study, principal component analysis was conducted to identify the relationship between heavy metals in soil and their potential sources. The common two potential sources were: natural (the biogeochemical processes of parent material and the physicochemical processes of parent material) and anthropogenic (industrial activity, industrial activity, vehicle-related activity and fossil energy activity). PCA is designed to reduce a dataset

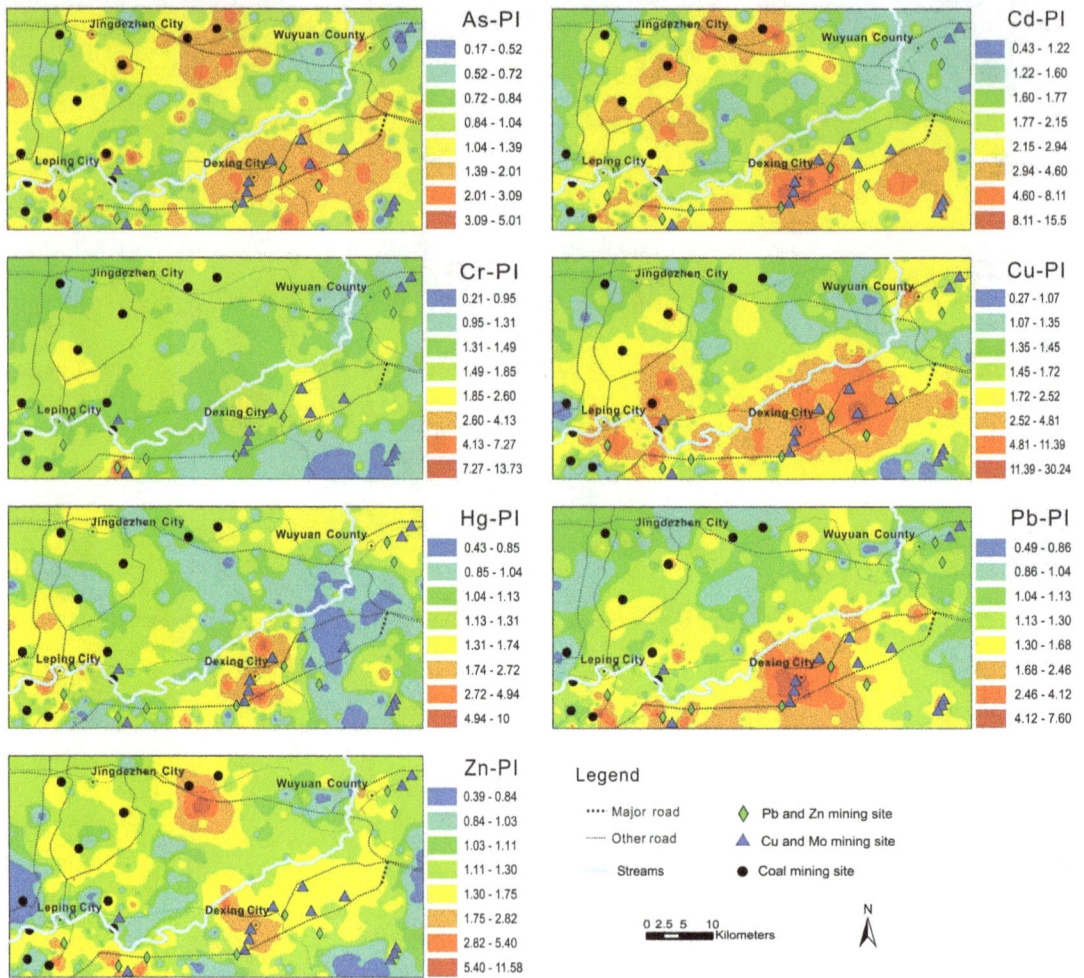

Figure 4. Spatial distribution of PI of trace elements in soil of the study area.

containing a large number of variables to a smaller size by finding a new set of variables called components. Is this study, there are 10 element measurements constituting the variables, and hence 10 components. PCA was conducted using a commercial statistics software package SPSS (version 17) for Windows. The assumption of normality for all variables was checked before multivariate statistical and spatial analyses; when necessary, data transformation was done via a Box-Cox transformation.

Geostatistical Analysis

The Kolmogorov–Smirnovtest (p<0.05) indicated that the various metals had skewed concentration distributions. Only As and Zn fitted a normal distribution after being logarithmically transformed. A log transformation was conducted prior to the analysis because of the skewed distributions of the heavy metal data.

Ordinary kriging is the most commonly used interpolation method to predict the overall trend of soil pollution. However, for the purpose of identifying contaminated areas, inverse distance weighting (IDW) is more appropriate to predict local features of soil pollution, especially local hotspots and cold spots [45]. It is a deterministic spatial interpolation model that is directly related to the values being estimated, and is suited to small datasets for which modeled semi-variograms are very difficult to fit. The interpolating function is:

$$Z(x) = \sum_{i=1}^{n} W_i Z_i / \sum_{i=1}^{n} W_i \qquad (4)$$

$$W_i = d_i^{-u} \qquad (5)$$

Where Z(x) is the predicted value at an interpolated point, W_i is the weight assigned to point i, Z_i is at a known point, d_i is the distance between point i and the prediction point, and n is the number of known points used in the interpolation. Interpolation mapping was conducted using IDW within ArcGIS 9.30 software.

Results and Discussion

Descriptive statistical analysis

Descriptive statistics of heavy metal concentrations of topsoil are presented in Table 5. The arithmetic means concentrations of As, Cd, Cr, Cu, Hg, Pb, and Zn were 11.63, 0.24, 72.09, 53.48, 0.10, 47.02, and 87.98 mg/kg, respectively. Wide concentrations ranges coupled with the relatively high CV values for metal elements demonstrate the anthropogenic contribution in the study area. In this study, the Coefficient of variance was higher for Cu than for the other metals, and their concentrations varied widely. This

Table 7. Total variance explained and rotated component matrix (three principal components selected) for heavy metal contents.

Element	Total	% of variance	Cumulative %	PC1	PC2	PC3
As	3.57	35.713	35.713	**0.58**	0.16	0.26
Cd	1.68	16.78	52.50	**0.79**	−0.044	0.104
Cr	1.19	11.87	64.37	0.11	**0.81**	0.30
Cu	0.86	8.61	72.97	**0.61**	0.37	0.177
Hg	0.81	8.07	81.03	**0.43**	0.24	−0.42
Pb	0.58	5.81	86.84	**0.83**	−0.05	0.41
Zn	0.45	4.51	91.36	**0.71**	0.13	0.45
Fe$_2$O$_3$	0.36	3.62	94.98	0.23	0.54	**0.65**
Al$_2$O$_3$	0.28	2.75	97.73	0.15	0.07	**0.84**
Ti	0.23	2.27	100	0.04	**0.85**	−0.13

suggests that Cu inputs to the soil in the study area may be attributable to anthropogenic sources.

The mean concentrations of all metals, especially Cd and Cu, exceeded the environmental background values for Jiangxi and China [14]. This was probably because of the influence of mining activities in the study area. It was found that the highest concentrations of all heavy metals were higher than their corresponding guidelines for soils, except Pb, based on the Chinese Environmental Quality Standard for Soils [20]. However, the mean concentrations of the metals were lower than the guidelines. The mean concentrations of most metals, except Cd, Cu and Pb, were lower than the background values of local.

Soil contamination assessment based on EF

The descriptive statistics of EF corresponding to the seven trace elements measured in the study area are given in Table 6. Mean values of EF were less than 2 for As and Zn, indicating no contamination by those metals in the soil. The mean EF values of Cd, Cr, Cu, and Pb ranged from 2.09 to 3.58; with respect to those

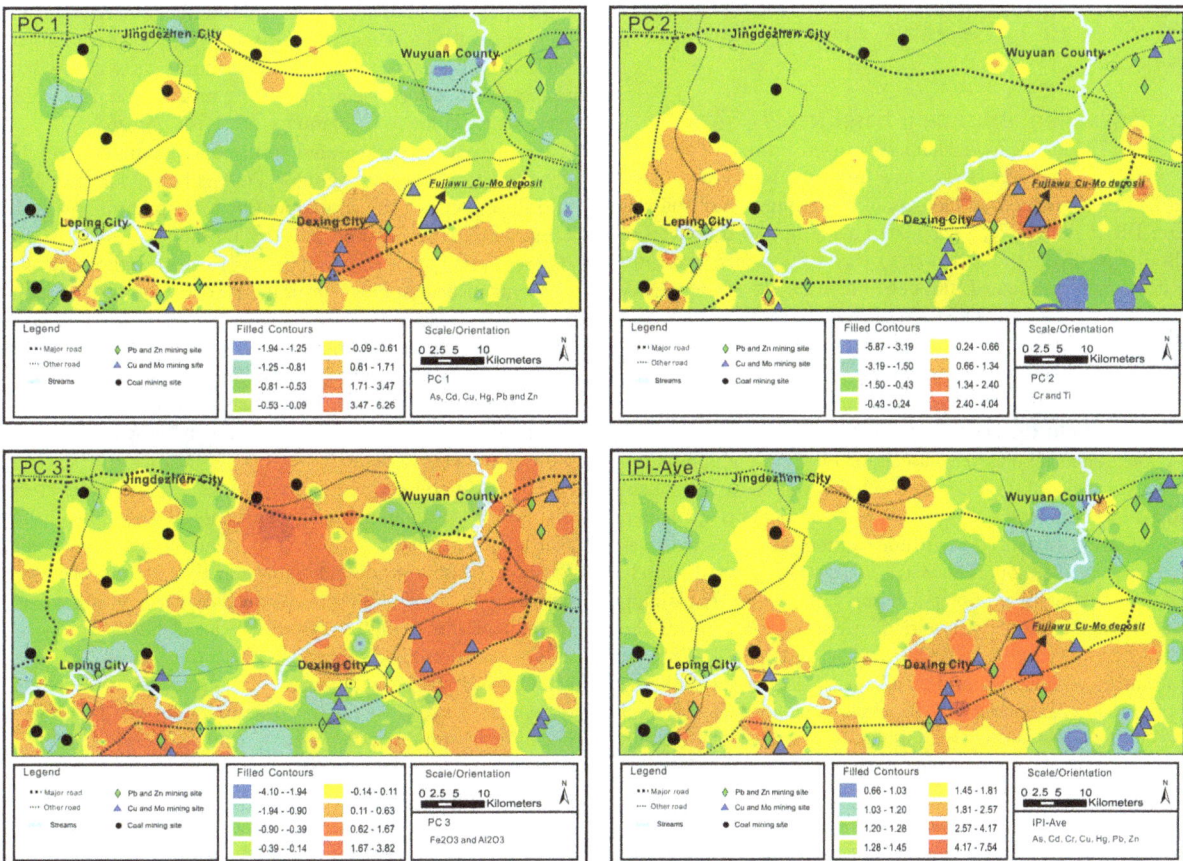

Figure 5. Spatial distribution map of PC scores.

metals, the soil was classified as moderately contaminated. The mean EF value of Hg was approximately 20, which was the highest of all the metals and which indicates considerable soil contamination.

Estimated maps of EF of seven heavy metals in soil are presented in Fig. 2. The EF map of Cu shows higher values in areas surrounding the Dexing and Leping mining areas, which contain many Cu and Mo mining sites. The highest levels of Cd and Pb occurred at the centre of Dexing. Urban vehicular emissions and industrial activity, including incinerator operation and metallurgic activities, have continuously contributed to Cd and Pb contamination of topsoil in this area. The spatial distribution of As was highly heterogeneous in contrast to the other metals, suggesting that As in these samples may originate from point source pollution. In contrast to other heavy metals, the spatial distribution of Cr shows no clear hotspots, suggesting the study area is weakly polluted by Cr.

Soil contamination assessment based on I_{geo}

The mean I_{geo} values for all trace elements were lower than 0 (ranged from -0.07 to -0.64), suggesting a lack of soil contamination, except for Cd (Table 6). The spatial distributions of Cd, Cu, and Pb exhibited similar patterns (Fig. 3), however, the I_{geo} values indicated that the area polluted by Cd and Cu was more extensive than the area polluted by Pb. The spatial distribution of I_{geo} for Cr was similar to the EF for Cr, confirming the lack of Cr contamination. Most soils in the world do not contain elevated concentrations of Hg, which is leached and evaporates after being reduced to the metallic form, although a portion is absorbed by organic matter and clay minerals. The urban areas, including Jingdezhen, Leping, Dexing, and Wuyaun, had the highest Hg I_{geo} values; the remaining area is weakly enriched in Hg.

Soil contamination assessment based on PI

The mean PIs for all trace elements ranged from 1.12 to 2.57, which indicates that the soils were moderately contaminated (Table 6). The assessment of the overall contamination of soil was based on IPI_{Ave}. The IPI_{Ave}, calculated according to the mean of the PIs of the seven trace elements, was 1.66, which indicates moderate contamination. Estimated PI maps of seven heavy metals in soil are presented in Fig. 4. Among these soil contamination indices, the spatial distributions of I_{geo} and PI are remarkably similar across the study area.

Source identification based on PCA

PCA has been extensively used to identify contamination sources. The results of the PCA conducted in this study are shown in Table 7. In this study, three principal components explained 64.36% of total variance, according to the initial eigenvalues (eigenvalues>1). As, Zn, Cd, Cu, Pb, and Hg were closely associated with the first principal component (PC1), explaining 35.71% of total variance; Cr and Ti were associated with the second principal component (PC2), which explained 16.78% of total variance; and Fe_2O_3 and Al_2O_3 were associated with the third principal component (PC3), explaining 11.87% of total variance. The other seven components (eigenvalues<1) explain little of the variability in the dataset and will not be discussed further.

As shown in this Fig. 5, high score areas were distributed in and around some of the Cu-Mo mining sites and along major roads. The areas with high component 1 scores that produced high amounts of Cd, Cu, Pb and Zn, were located around the Fujiawu Cu-Mo deposit (the biggest open store of Cu in Asia). These mining activities represented by PC1 may have be the primary contributors of Cd, Cu, Pb, and Zn contamination in soil. Thus, PC1 was mainly controlled by anthropogenic sources.

Interpolated scores associated with PC2 are displayed in Fig. 5; the scores exhibit a different spatial distribution than PC1 scores. Two high score areas were located in the city of Leping and the Fujiawu Cu-Mo deposit. The high score areas located in Leping are associated more strongly with natural sources. The reasons for the observed high scores of areas located in the Fujiawu Cu-Mo deposit are not clear; in Cu deposits, naturally occurring Cu is often present in higher concentrations than other environment. There are a number of potential causes of high PC2 scores, including the influence of anthropogenic activities. The findings suggest that Ti and Cr in soil originated from both natural and anthropogenic sources.

The spatial distribution of PC3 is presented in Fig. 5. The spatial variability of the score associated with PC3 is different than that of the scores associated with PC1 and PC2. Fe_2O_3 and Al_2O_3 were grouped into PC3, with high factor loading ($Fe_2O_3 = 0.648$; $Al_2O_3 = 0.838$). Fe_2O_3 and Al_2O_3 are ubiquitous components of soil and display some natural soil characters. It is speculated that natural sources may contribute to the Fe_2O_3 and Al_2O_3 present in the soil environment.

Comparative method evaluation

Almost all of the four indices used in this study have been employed previous in soil contamination assessments. However, our assessments based on GIS have some distinct advantages over those done in previously studies: (1) using these maps, soil researchers and managers can visually identify the degree of anthropogenic influence on the environment at a regional scale; (2) all mapping indices incorporate some other relative information, such as land-use type, soil type, and human activities, which lead to increased confidence in the results; and (3) mapping indices can serve as a platform for planning other soil research.

Though similar integrated soil quality evaluation results were obtained from the four indices, PCA is better for than EFs, I_{geo}, and PIs integrated soil contamination assessment in the study area. Using PCA, integrated soil contamination was assessed by differentiating the importance of various indicators. The 10 elements measurements constituting a dataset were included in the statistical analyses to find the influence of anthropogenic components by multivariate analysis. The drawback is that this is a qualitative method, which cannot evaluate the degree of contamination. However, IPI_{ave} treats each trace element as an independent entity and does not consider the specific geochemical context of each element.

The three soil contamination indices we used were dependent on the use of regional background values. Based on the indices, which are calculated according to mean values, Cd was classified to have caused moderate contamination, while the degree of contamination of other heavy metals varied. Using mean trace metal values/regional background ratios of soil on a regional scale is an oversimplified approach and may result in erroneous estimates of soil contamination. Thus, the use of mean values is a reliable way to evaluate contamination of an entire region because different sample material will respond differently to the presence of elements in the soil.

The mapping of the contamination indices we used, which take into account spatial information and human activities, provide an effective way to evaluate the spatial distributions of anthropogenic impact on soil composition. Using EFs, I_{geo}, and PI calculated relative to off-site reference values of an entire region does not improve the sensitivity of the methods to the anthropogenic enrichment and may even give spurious results. This study demonstrates that values of contamination indices can be high relative to off-site values for a number of reasons, and contamination is just one potential cause. The three off-site references methods employed in this study are easy to conduct, and may be used for quantitative analyses to assume consistent effects of geologic and pedogenic processes at regional scale.

Conclusions

The findings of this study suggest that EFs, I_{geo}, and PI calculated according trace metal mean values relative to off-site reference values to assess soil contamination provide different interpretations of the same data. The assessment results are inconsistent, and no conclusions are reliable. However, the mapping of EFs, I_{geo}, PI, and PCA, combined with contamination source analysis, has the potential to differentiate between anthropogenic and natural element sources.

The most plausible results are likely to be obtained from multivariate statistical analysis- methods. In this study, the use of PCA allowed us to discriminate between natural and anthropogenic trace metals in soils of the study area. The results are supported by the resulting EF, I_{geo}, and PI maps. According to the analysis, surface horizons are highly enriched in Cd, Cu, Pb, and Zn. The composition of topsoil is significantly modified by human activity in areas with high population density and areas near mining sites. As and Hg present in the soil were also mainly derived from anthropogenic sources, and occurred in relatively high concentrations in urban areas, in contrast to Cd, Cu, Pb, and Zn. Mapping of the soil contamination assessment indices seems to be an efficient tool for detecting sources of anomalies in the study area.

Acknowledgments

The authors would like to thank the anonymous reviewers for their helpful comments and suggestions.

Author Contributions

Conceived and designed the experiments: YGT YYW. Performed the experiments: JW SJL. Analyzed the data: JW XDJ. Contributed reagents/materials/analysis tools: YGT JW. Contributed to the writing of the manuscript: JW YGT. Edited English grammar and expression: YGT.

References

1. Andrews SS, Carroll CR (2002) Designing a soil quality assessment for sustainable agroecosystem management. Ecol Appl 11: 1573–1585.
2. Teng YG, Wu J, Lu SJ, Wang YY, Jiao XD, et al. (2014) Soil and soil environmental quality monitoring in China: A review. Environ Int 69: 177–199.
3. Li ZY, Ma ZW, de Tsering JVK, Yuan ZW, Huang L (2014) A review of soil heavy metal pollution from mines in China: Pollution and health risk assessment. Sci Total Environ 468–469: 843–853.
4. Khalil A, Hanich L, Bannari A, Zouhri L, Pourret O, et al. (2014) Assessment of soil contamination around an abandoned mine in a semi-arid environment using geochemistry and geostatistics: Pre-work of geochemical process modeling with numerical models. J Geochem Explor 125: 117–129.
5. Ikem A, Campbell M, Nyirakabibi I, Garth J (2008) Baseline concentrations of trace elements in residential soils from Southeastern Missouri. Environ Monit Assess 140: 69–81.
6. Desaules A (2012) Critical evaluation of soil contamination assessment methods for trace metals. Sci Total Environ 426: 120–131.
7. Hawkes HE, Webb JS (1962) Geochemistry in Mineral Exploration. New York: Harper. 409 p.
8. ISO: International Organisation for Standardisation (2005) Soil Quality: Vocabulary. Part 1. Terms and Definitions Relating to the Protection and Pollution of the Soil. Available: http://www.iso.org/iso/home/store/catalogue_ics/catalogue_detail_ics.htm?ics1=13&ics2=80&ics3=1&csnumber=38529.
9. Porteous A (1996) Dictionary of Environmental Science and Technology. 2nd edition. Chichester, NY: John Wiley & Sons. 794 p.
10. Reimann C, Filzmoser P, Garrett RG (2005) Background and threshold: critical comparison of methods of determination. Sci Total Environ 346: 1–16.
11. Garrett RG (1991) The management, analysis and display of exploration geochemical data. Exploration geochemistry workshop. Ottawa: Geological Survey of Canada. 9–1 to 9–41.
12. Darnley AG (1995) International geochemical mapping–a review. J Geochem Explor 55: 5–10.
13. Utermann J, Düwel O, Nagel I (2006) Contents of trace elements and organic matter in European soils. In: Gawlik BM, Bidoglio G, editors. Background values in European soils and sewage sludges. Luxembourg: European Commission. 282 p.
14. CEMS: Chinese environmental monitoring station (1990) Background values of elements in soils of China (in Chinese). Beijing: China Environmental Press. 501 p.
15. He J, Xu G, Zhu H, Peng G (2005) soil background values of Jiangxi Province. Beijing: Chinese Environmental Science Press. 314 p.
16. Reimann C, de Caritat P (1998) Chemical elements in the environment-factsheets for the geochemist and environmental scientist. Berlin, Germany: Springer-Verlag. 398 p.
17. Taylor SR, McLennan SM (1995) The geochemical evolution of the continental crust. Rev Geophys 33: 241–65.
18. Wedepohl KH (1995) The composition of the continental Crust. Geochim Cosmochim Ac 59: 1217–32.
19. Salminen R, Gregorauskiene V (2000) Considerations regarding the definition of a geochemical baseline of elements in the surficial materials in areas differing in basic geology. Appl Geochem 15: 647–653.
20. CEPA: Chinese Environmental Protection Administration (1995) Environmental quality standard for soils (GB 15618-1995) (in Chinese). Available: http://kjs.

mep.gov.cn/hjbhbz/bzwb/trhj/trhjzlbz/199603/W020070313485587994018.pdf.
21. CCME: Canadian Council of Ministers of the Environment (2007) Canadian soil quality guidelines for the protection of environmental and human health. Available: http://ceqg-rcqe.ccme.ca/en/index.html.
22. Li XD, Lee SL, Wong SC, Shi WZ, Thornton L (2004) The study of metal contamination in urban soils of Hong Kong using a GIS-based approach. Sci Total Environ 129: 113–124.
23. Sepulvado JG, Blaine AC, Hundal LS, Higgins CP (2011) Occurrence and Fate of Perfluorochemicals in Soil Following the Land Application of Municipal Biosolids. Environ Sci Technol 45(19): 8106–8112.
24. Nathanail CP, Earl N (2001) Human Health Risk Assessment: Guideline Values and Magic Numbers Issues in Environmental Science and Technology No.16 Assessment and Reclamation of Contaminated Land. The Royal Society of Chemistry. 85–101.
25. Carlon C (2007) Derivation methods of soil screening values in Europe. A review and evaluation of national procedures towards harmonization. European Commission, Joint Research Center, Ispra. 206 p.
26. Ferguson C, Darmendrail D, Freier K, Jensen BK, Jensen J, et al. (1998) Risk Assessment for Contaminated Sites in Europe. Volume 1: Scientific Basis. LQM Press, Nottingham. 165 p.
27. Blaser P, Zimmermann S, Luster J, Shotyk W (2000) Critical examination of trace element enrichments and depletions in soils: As, Cr, Cu, Ni. Pb and Zn in Swiss forest soils. Sci Total Environ 249: 257–280.
28. Hernandez L, Probst A, Probst JL, Ulrich E (2003) Heavy metal distribution in some French forest soils: evidence for atmospheric contamination. Sci Total Environ 312: 195–219.
29. ISO: International Organisation for Standardisation (2005) Soil quality-guidance on the determination of background values. ISO 19258. Available: http://www.iso.org/iso/catalogue_detail.htm?csnumber=33772.
30. Shotyk W, Cherkubin AK, Appleby PG, Fankhauser A, Kramers JD (1997) Lead in three peat bog profiles, Jura mountains, Switzerland: enrichment factors, isotopic composition, and chronology of atmospheric deposition. Water Air Soil Pollut 100: 297–310.
31. Reimann C, de Caritat P (2000) Intrinsic flaws of element enrichment factors (EFs) in environmental geochemistry. Environ Sci Technol 34: 5084–5091.
32. Hakanson L (1980) An ecological risk index for aquatic pollution control. A sedimentological approach. Water Res 14: 975–1001.
33. Chester R, Stoner JH (1973) Pb in particulates from the lower atmosphere of the eastern Atlantic. Nature 245: 27–8.
34. Zoller WH, Gladney ES, Duce RA (1974) Atmospheric concentrations and sources of trace metals at the South Pole. Science 183: 199–201.
35. Müller G (1979) Schwermetalle in den Sedimenten des Rheins-Veränderungen seit. Umschau 24: 773–8.
36. Manta DS, Angelone M, Bellanca A, Neri R, Sprovieri M (2002) Heavy metals in urban soils: a case study from the city of Palermo (Sicily), Italy. Sci Total Environ 300: 229–243.
37. Wang XQ, He MC, Xie J, Xi JH, Lu XF (2010) Heavy metal pollution of the world largest antimony mine-affected agricultural soils in Hunan province (China). J Soil Sediment 10: 827–837.
38. Loska K, Cebula J, Pelczar J, Wiechula D, Kwapilinski J (1997) Use of enrichment and contamination factors together with geoaccumulation indexes to

evaluate the content of Cd, Cu, and Ni in the Rybnik water reservoir in Poland. Water Air Soil Pollut 93: 347–65.

39. Loska K, Wiechula D, Korus I (2004) Metal contamination of farming soils affected by industry. Environ Int 30: 159–165.

40. Gowd SS, Reddy MR, Govil PK (2010) Assessment of heavy metal contamination in soils at Jajmau (Kanpur) and Unnao industrial areas of the Ganga Plain, Uttar Pradesh, India. J Hazard Mater 174: 113–121.

41. USEPA: United States Environmental Protection Agency (1992) Terms of Environment. Communications Education And Public Affairs, USEPA 175-B-92-001. Available: http://iaspub.epa.gov/sor_internet/registry/termreg/searchandretrieve/termsandacronyms/search.do.

42. Luo XS, Yu S, Zhu YG, Li XD (2012) Trace metal contamination in urban soils of China. Sci Total Environ 421–422: 17–30.

43. Teng YG, Ni SJ, Wang JS, Niu LG (2009) Geochemical baseline of trace elements in the sediment in Dexing area, South China. Environ Geol 57: 1646–1660.

44. Sutherland RA (1999) Distribution of organic carbon in bed sediments of Manoa Stream, Oahu, Hawaii. Earth Surf Proc Land 27: 571–583.

45. Xie Y, Chen TB, Lei M, Yang J, Guo QJ, et al. (2011) Spatial distribution of soil heavy metal pollution estimated by different interpolation methods: accuracy and uncertainty analysis. Chemosphere 82: 468–476.

46. Teng YG, Ni SJ, Wang JS, Zuo R, Yang J (2010) A geochemical survey of trace elements in agricultural and non-agricultural topsoil in Dexing area, China. J Geochem Explor 104: 118–127.

19

Water Consumption Characteristics and Water Use Efficiency of Winter Wheat under Long-Term Nitrogen Fertilization Regimes in Northwest China

Yangquanwei Zhong, Zhouping Shangguan*

State Key Laboratory of Soil Erosion and Dryland Farming on the Loess Plateau, Northwest A & F University, Yangling, Shaanxi, P.R. China

Abstract

Water shortage and nitrogen (N) deficiency are the key factors limiting agricultural production in arid and semi-arid regions, and increasing agricultural productivity under rain-fed conditions often requires N management strategies. A field experiment on winter wheat (*Triticum aestivum* L.) was begun in 2004 to investigate effects of long-term N fertilization in the traditional pattern used for wheat in China. Using data collected over three consecutive years, commencing five years after the experiment began, the effects of N fertilization on wheat yield, evapotranspiration (ET) and water use efficiency (WUE, i.e. the ratio of grain yield to total ET in the crop growing season) were examined. In 2010, 2011 and 2012, N increased the yield of wheat cultivar Zhengmai No. 9023 by up to 61.1, 117.9 and 34.7%, respectively, and correspondingly in cultivar Changhan No. 58 by 58.4, 100.8 and 51.7%. N-applied treatments increased water consumption in different layers of 0–200 cm of soil and thus ET was significantly higher in N-applied than in non-N treatments. WUE was in the range of 1.0–2.09 kg/m^3 for 2010, 2011 and 2012. N fertilization significantly increased WUE in 2010 and 2011, but not in 2012. The results indicated the following: (1) in this dryland farming system, increased N fertilization could raise wheat yield, and the drought-tolerant Changhan No. 58 showed a yield advantage in drought environments with high N fertilizer rates; (2) N application affected water consumption in different soil layers, and promoted wheat absorbing deeper soil water and so increased utilization of soil water; and (3) comprehensive consideration of yield and WUE of wheat indicated that the N rate of 270 kg/ha for Changhan No. 58 was better to avoid the risk of reduced production reduction due to lack of precipitation; however, under conditions of better soil moisture, the N rate of 180 kg/ha was more economic.

Editor: Raffaella Balestrini, Institute for Plant Protection (IPP), CNR, Italy

Funding: The study was sponsored by the National Natural Science Foundation of China (41390463, 61273329) and the Important Direction Project of Innovation of CAS (KZCX2-YW-JC408). The funders had no role in study design, data collection and analysis, decision to publish, or preparation of the manuscript.

Competing Interests: The authors have declared that no competing interests exist.

* E-mail: shangguan@ms.iswc.ac.cn

Introduction

Northwest China is a vast semi-arid area with average annual precipitation in the range of 300–600 mm and more than 90% of the land is cropland [1]. This means that water is the primary factor limiting crop yields. In addition, world food demand is expected to double during 2005–2050 [2], thus it is important to increase food production with lower water use [3], particularly in water shortage regions. Currently, water stress and nutrient deficits are the main factors limiting primary production in arid and semi-arid environments [4–8]. Therefore, many rain-fed farming experts have focused on how to increase crop water use efficiency (WUE, i.e. the ratio of grain yield to total ET in the crop growing season) by irrigation and fertilization.

In the 1990 s, many studies on effects of limited irrigation on crop yields and WUE showed that by reducing irrigation volume, crop yield could be generally maintained and product quality improved [9–13],and appropriate irrigation management can increase crop yield and WUE [14–16]. There are several sources of soil water in irrigated or high water-table areas, however, precipitation is the only source of soil water for crop growth in many rain-fed farming systems of arid and semi-arid regions.

Therefore, new methods need to be devised to improve WUE in this non-irrigated farming system.

N fertilization is a common practice to increase grain production, but its performance depends on soil water status [17–19]. The importance of increasing crop yield and improving soil quality through fertilization has been confirmed. The increasing use of N fertilizer could significantly increase maize production [8,20], and already affects a large proportion of the world's food production [21,22]. Fan et al. [1] reported that inorganic N and phosphorus (P) fertilization increased grain yields by 50–60% in China, and reports from Europe showed that N fertilizers can increase crop yield significantly [23]. N fertilization is well known to improve soil fertility [24,25]; however, using excessive N fertilizer can decrease the N utilization rate, which not only causes a huge waste of resources and economic losses, but can also adversely impact the environment [26–28]. Balancing the N rate, WUE and yield is an important problem in dryland farming systems. Better understanding of interactions among precipitation, fertilization and crops production is essential for efficient utilizations of water resources and N fertilizers, and sustainable food productions in rain-fed cropping systems experiencing climate change [1]. Long-term fertilization experiments are

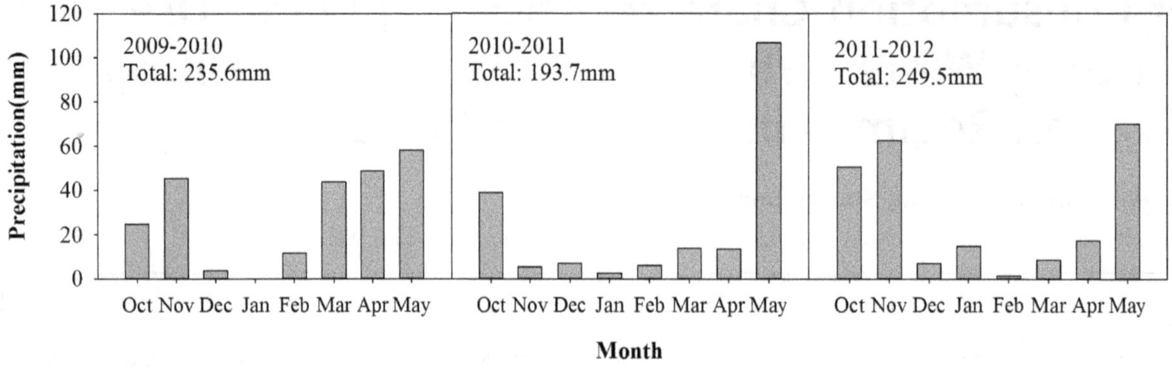

Figure 1. Monthly and total precipitation during the 2009–2012 wheat growing seasons. Monthly precipitation of wheat growing seasons in 2009–2010, 2010–2011, 2011–2012 in Shaanxi, Yangling.

valuable to follow crop yield, soil fertility, WUE and risk management over time [29,30]. Various long-term experiments have examined how to increase yield and WUE of wheat, using irrigation, organic or inorganic fertilizer, soil tillage and crop management [31–33]. However, few experiments have been done on evapotranspiration (ET) and WUE under circumstances with only N fertilizer and without irrigation in northwest China. With China's urbanization, increasing numbers of farmers have abandoned farms to urban construction and this has led to a loss of labor. Thus, most farmland in northwest China region still uses traditional cropping practices that all fertilizer applied once prior to planting [1], lack of careful management of irrigation and other tasks. Kang et al. [14] reported that difference in yield and WUE are also related to regional variability in environment and crop varieties, so information specific to a region is needed for developing and refining the agricultural performance in this region. In these circumstances, it is very important to determine the advantages and disadvantages of long-term N fertilization on yield of different varieties.

This study examined two different water-sensitive cultivars of winter wheat (*Triticum aestivum* L.) to investigate effects of N fertilizers on crop yield, ET and WUE, using the most common management of farmers in northwest China. The objectives were (1) to investigate impacts of traditional long-term N fertilization on yields of two different water-sensitive wheat cultivars; (2) to examine the effect of N fertilizers on total ET, and soil water consumption from different soil layers of the two cultivars; and (3) to establish relationships among crop yield, WUE and ET and determine optimum N fertilizer rates in northwest China. This study may compensate for some of the lack of long-term influence only N fertilizer on crop production, and the results should provide guidelines to farmers in the region on choosing appropriate cultivars and obtaining high yields with appropriate N application.

Materials and Methods

Experiment site and climatic conditions

The study commenced in October 2004 in an experiment field of the Institute of Soil and Water Conservation of the Northwest A & F University, Yangling, Shaanxi (34°17′56″N, 108°04′7″E). Located on the southern boundary of the Loess Plateau, the experiment site has a temperate and semi-humid climate with a mean annual temperature of 13°C and a mean annual precipitation of 632 mm, of which about 60% occurs during July–September.

Experiment design

The study adopted a randomized block design with three replications. Two winter wheat (*Triticum. aestivum* L.) cultivars were used: Zhengmai No. 9023 (ZM) is water sensitive and poorly drought-tolerant and Changhan No. 58 (CH) is drought-tolerant and suitable for drought prone environments. The thousand-kernel weights of ZM and CH were 43.58 and 43.61 g, respectively. N treatments were applied at five rates: 0, 90, 180, 270 and 360 kg/ha (N0, N90, N180, N270 and N360, respectively). Plot size was 2 m×3 m with 20 rows (15-cm spaces) of wheat sown at 90 seeds/row. Wheat was sown in early October and harvested in early June the following year. The seeding rate was 130 kg/ha. Immediately before sowing, the fertilizer was evenly spread on the soil surface and then incorporated into the upper 15 cm soil by chiseling. N was applied as urea and P (75 kg P_2O_5/ha) as super phosphate. No potassium fertilizer was applied, and the site was ploughed to bury weeds before sowing.

Measurements

In all treatments, the volumetric soil water content was measured every 10 cm for 0–100 cm of soil and every 20 cm for 100–300 cm with a neutron moisture meter (CNC100, Super Energy, Nuclear Technology Ltd., Beijing, China). The 3-m-long neutron gauge access tube was buried vertically in the center of each plot at the beginning of the study. Soil water was measured during the first week of every wheat growing month except January and February. If any precipitation occurred just before or during the measurement period, then measurements were postponed for several days until the soil moisture attained a normal degree. The yields and the thousand-kernel weights of wheat in all plots were measured at harvest time in early June. Since this paper aimed to test the cumulative effects of N fertilization, we chose data from three wheat growing years with different precipitation characteristics: 2009–2010, 2010–2011 and 2011–2012.

Calculation and statistics

ET of winter wheat was calculated using the following equation [33]:

$$ET = \Delta S + P + I - R - D$$

Table 1. Nitrogen effects on wheat yield and thousand-kernel weights of two cultivars in three years.

Varieties	Treatments	2010 Yield(kg/ha)	2010 Thousand-kernel (g)	2011 Yield(kg/ha)	2011 Thousand-kernel (g)	2012 Yield(kg/ha)	2012 Thousand-kernel (g)
ZM	N0	4716 c	49.5 a	2974 c	47.0 a	5906 bc	49.1 a
	N90	6355 ab	42.5 b	5499 ab	42.2 ab	7272 a	42.3 bcd
	N180	7597 a	42.3 b	6355 ab	42.1 ab	7953 a	41.6 cd
	N270	7527 a	42.5 b	6482 ab	40.5 b	7923 a	41.1 d
	N360	7519 a	43.9 b	6390 ab	41.2 ab	7512 a	40.4 d
CH	N0	4162 c	46.1 ab	3391 c	42.2 ab	5407 c	46.5 ab
	N90	5862 ab	42.3 b	4886 b	41.1 ab	6926 ab	45.5 abc
	N180	6594 ab	42.3 b	5879 ab	41.9 ab	7777 a	40.4 d
	N270	6334 ab	37.9 c	6748 a	40.5 b	8199 a	40.1 d
	N360	6126 b	37.1 c	6808 a	40.3 b	8018 a	38.6 d

Values are means of three replicates for each treatment. Different letters indicate statistical significance at P<0.05 within the same column.

Where ΔS is soil water storage change, P is precipitation, I is irrigation rate, R is surface runoff and D is deep water percolation (all in mm).

No irrigation was used and so I = 0. Precipitation in the three growing seasons is shown in Fig. 1, and measured surface runoff was negligible during these years. Deep percolation was calculated as the difference between soil moisture content and field moisture capacity when the soil water content at this depth was more than the field water holding capacity. In the study site, deep water percolation did not occur.

WUE was defined as follows:

$$WUE = Y/ET$$

Where Y is grain yield (kg/ha).

All data concerned were analyzed by SPSS 16.0 Statistical software. ANOVA was adopted to determine whether treatments were significantly different at P<0.05. Duncan's multiple range test was used to differentiate treatment means at P<0.05.

Results and Discussion

Wheat yield

The yields and thousand-kernel weights of the two wheat cultivars at the different N rates in the different years are presented in Table 1. Grain yield was in the range of 2.94–7.92 t/ha for ZM and 3.39–8.19 t/ha for CH. Grain yields of the two cultivars both differed significantly between N0 and the other N rates, and increased as N rates increased, but did not differ significantly among the N-applied treatments. The lack of significant differences may due to yield in three replications being affected by other factors in field experiment. Usually the significance of increase yield is hard to attain in agricultural research, Morell et al. [36] reported that grain yield mostly 1000 kg/ha higher than control, but still have no statistical difference. The yields of wheat slightly decreased at N360, except for yield of CH in 2011. In 2010, 2011 and 2012, the wheat yields of ZM increased by up to 61.1, 118.0 and 34.7%, respectively, in the N-applied treatments compared to treatment without N fertilization; and corresponding yields of CH increased by up to 58.4, 100.8 and 51.7%. The highest yields of ZM and CH were both for treatments of N180, N270 and N270 in 2010, 2011 and 2012, respectively. Thus, N application significantly increased yields of wheat, but an excessive N rate had no positive effect on grain yield. Previous research has shown similar results with N fertilizer application significantly increasing maize and wheat yield compared to unfertilized treatments [14,15]. Bassoa et al. [23] examined the long-term wheat response to N in rain-fed Mediterranean environments, and showed that yield response was stronger for 120 than 60 and 90 kg N/ha. Many other studies have demonstrated a parabolic relationship between N and grain yield, i.e. when N rate surpassed a certain threshold, the grain yield greatly decreased. In China, there have been many experiments on different wheat cultivars and fertilizer regimes that have shown the maximum N rate is 150–225 kg/ha. At excessive N rates, the leaf protein and chlorophyll contents decrease, and then photosynthesis also decreases [34]. Tinsina et al. [35] also showed that wheat yield was higher at 120 than 180 kg/ha in Bangladesh. Morell et al. [36] showed no additional wheat yield responses to N fertilizers at N rates >100 N kg/ha. All these studies demonstrated that N application could increase wheat yield, but excessive N had no yield benefit.

At the same N rates, the yields of the two cultivars did not differ significantly. In 2010, the rainfall, which was evenly distributed

Table 2. Water consumption from soil (ΔS) or precipitation and its ratio to total evapotranspiration (ET) and WUE.

Year	Varieties	Treatments	Evapotranspiration (ET)(mm)		Soil water consumption (ΔS) Amount (mm)	Ratio (%)	Precipitation Amount (mm)	Ratio (%)	WUE (kg/m³)	
2009–2010	ZM	N0	325.72	e	90.12	27.7	235.6	72.3	1.45	de
		N90	348.58	de	112.98	32.4		67.6	1.83	abcd
		N180	361.57	bcd	125.97	34.8		65.2	2.09	a
		N270	385.25	abc	149.65	38.9		61.2	1.96	ab
		N360	385.59	abc	149.99	38.9		61.1	1.97	ab
	CH	N0	321.39	e	85.79	26.7		73.3	1.29	e
		N90	352.69	cde	117.09	33.2		66.8	1.66	bcde
		N180	373.65	bcd	138.05	36.9		63.1	1.76	abc
		N270	411.98	a	176.38	42.8		57.2	1.48	cde
		N360	395.49	ab	159.89	40.4		59.6	1.52	cde
2010–2011	ZM	N0	298.40	d	104.70	35.1	193.7	64.9	1.00	d
		N90	323.28	bc	129.58	40.1		59.9	1.70	ab
		N180	344.98	ab	151.28	43.8		56.2	1.84	ab
		N270	345.60	ab	151.90	44.0		56.1	1.88	a
		N360	337.41	ab	143.71	42.6		57.4	1.89	a
	CH	N0	308.48	cd	114.78	37.2		62.8	1.10	cd
		N90	344.01	ab	150.31	43.7		56.3	1.42	bc
		N180	351.92	a	158.22	45.0		55.0	1.67	ab
		N270	351.75	a	158.05	44.9		55.0	1.92	a
		N360	344.63	ab	150.93	43.8		56.2	1.84	ab
2011–2012	ZM	N0	373.92	c	124.42	33.3	249.5	66.7	1.58	a
		N90	412.28	bc	162.78	39.5		60.5	1.77	a
		N180	456.45	a	206.95	45.3		54.7	1.75	a
		N270	442.06	ab	192.56	43.6		56.4	1.80	a
		N360	440.44	ab	190.94	43.3		56.6	1.71	a
	CH	N0	381.35	c	131.85	34.6		65.4	1.43	a
		N90	464.53	a	215.03	46.3		53.7	1.50	a
		N180	469.35	a	219.85	46.8		53.5	1.66	a
		N270	461.05	a	211.55	45.9		54.1	1.79	a
		N360	441.13	ab	191.63	43.4		56.6	1.82	a

Values are means of three replicates for each treatment. Different letters indicate statistical significance at P<0.05 within the same column. ΔS has the same significance as total ET.

through the growing season, provided a more favorable environment for ZM, the water-sensitive cultivar, and so its yields were higher than those of CH for all treatments. In 2011, precipitation was the lowest in the whole growing season of all years, despite 106 mm of rainfall in May when wheat filled its seeds. However, such high rainfall at seed filling was unfavorable for wheat yield. Sheng and Wang [37] found that high soil-water contents at the seed-filling stage of wheat can result in lower thousand-kernel weights and grain yields. So drought during the growing season and too much rainfall from the seed-filling to the ripening stages led to the lower yield of wheat in 2011 compared to 2010 and 2012. Before the 2011–2012 growing season, the summer of 2011 received a lots of rain, 672.7 mm during June–September, larger than 421.6 mm in 2009 and 436.7 mm in 2010 summer. This caused total water consumption in 2011–2012 to be higher than previously and so the yields of wheat were the highest among the three years, although precipitation did not differ greatly from the growing season of 2009–2010. Shangguan et al. [38] reported that the fallow efficiencies, expressed as the ratio of soil water accumulation to precipitation received during the period of fallow, were important for yield in the next growing season. The importance of soil-water storage during the fallow period for increasing grain yields of post-fallow crops are supported by many studies on dryland including the Southern Great Plains in the USA [39–41] and the Loess Plateau [38]. In 2011 and 2012, the yield of ZM was higher than that of CH at N0, N90 and N180, and

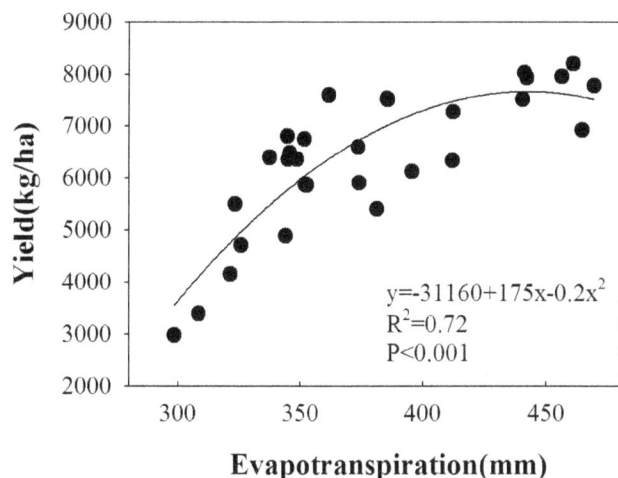

$$y=-31160+175x-0.2x^2$$
$$R^2=0.72$$
$$P<0.001$$

Figure 2. Relationship between wheat evapotranspiration (ET) (mm) and grain yield (Y) (kg/ha) for winter wheat in northwest China. The relationship between ET and yield is shown by the equation.

drought-tolerant CH showed higher yield only at the higher N rate in dry years. This is because CH was developed in recent years and prefers high fertilizer levels – consequently its cultivation has greatly expanded in northwest areas. CH is sensitive to N, and high rates of N result in higher yields; however, in contrast ZM is a poorly drought-tolerant cultivar but can produce higher yield at lower N rates. These cultivar characteristics have been demonstrated by many physiological indices in our previous studies [42]. Overall, the water consumption characteristics differed between the wheat varieties, leading to the different production performance. N180 resulted in higher yields of ZM in 2010 when rainfall was evenly distributed over the growing season. However, when rainfall was unevenly distributed or there was a lack of rainfall in the growing season, appropriate increases in the amount of N for CH could result in higher wheat yield to avoid the risk of reduced production.

The thousand-kernel weights of wheat decreased with increased N rates, consistent with many other research results [34]. There were two reasons for this: first, N can increase numbers of wheat tillers and panicles as well as flag leaf photosynthetic rates, but large and thick leaves would shade one another, affecting starch assimilation and transportation to kernels and resulting in lower thousand-kernel weights; secondly, N application could delay the flowering stage of wheat, thereby shortening the grain-filling stage and leading to lower thousand-kernel weights.

Water consumption characteristics

Total ET. In dryland farming, ET is supplied partly from precipitation in the growing season and partly from soil-water storage before planting. However, the relative contribution between precipitation and crop-consumed soil water to ET differs significantly among crops.

The total ET, ΔS or rainfall and its ratio to total ET at the different N rates are shown in Table 2. Total ET behaved differently between years and cultivars. Total ET was in the range of 298.40–442.46 mm for ZM and 361.57–469.35 mm for CH. Total ET was significantly higher in the N-applied than non-N treatments, except for treatment N90. Total ET were highest in 2011–2012 of CH and ZM. In the N-applied treatments, the ET of ZM increased by up to 18.4, 15.8 and 22.1% in 2009–2010,

2010–2011 and 2011–2012, respectively, and correspondingly for CH by 28.0, 14.1 and 23.1%. The ET slightly decreased for N360, indicating that ET could not increase further if too much N was applied. Zhou et al. [15] showed that N fertilizer application decreased water storage in 0–200 cm of soil and particularly so after wheat harvesting. Hunsaker et al. [43] showed that wheat ET was significantly higher than in low N treatments. One explanation for N increasing ET of wheat is that N promotes wheat to grow more and produce longer roots, enabling more soil water to be absorbed; another explanation is that N fertilization increases the leaf area index and transpiration rates of wheat [44]. However, too much N makes soil environments stressful by increasing N concentration in soil solution, thus preventing roots from absorbing water. The total ET in this study was considerably lower than that reported for the southern high plains of the USA [45,46] and the North China Plain [47], but was close to that for the Loess Plateau [48]. These differences are likely due to different climatic conditions, like temperature and precipitation and also attributed to different field management.

In all the experiment years, CH had higher ET than ZM but not significantly at the same N rates – probably due to different characteristics of the varieties. As a drought-tolerant cultivar with long roots, CH is capable of absorbing deep soil water, thus presenting higher ET than ZM. A deep-growing root system will favor taking up deep soil water under water-limited conditions. Research on dryland crops has shown that deep soil-water utilization is probably limited by root density [11,48,49]. The rainfall was less during 2010–2011 than 2009–2010 and 2011–2012 growing seasons so that the ratio of soil water consumption was higher in the former than the other two years (Table 2). As the N application rates increased, the ratios also increased, indicating that N application helped plants utilize deeper soil water.

The relationship between grain yields and seasonal ET was best described by a quadratic function obtained by regression analysis ($Y = -31160+175x-0.2x^2$; Fig. 2). Grain yield did not increase when ET exceeded a certain critical value, e.g. about 430 mm in the present study. Grain yield required a minimum ET of 244 mm for winter wheat (Fig. 2). This minimum ET value is higher than the 84 mm for wheat in the North China Plain [47] and 156 mm in the Mediterranean region [12], as well as higher than the 206 mm of dryland and irrigated wheat reported by Musick et al. [41] in US southern plains. These differences are likely due to such different climates and crop management. This result may indicate that the crop yield in this area will more relies on precipitation and soil water storage.

Water consumptions in the different soil layers. Water consumption in the different layers of the soil profile in 40-cm increments is plotted with depth in Fig 3. N applications had a significant effect on ΔS, as well as on ET (Table 2), since N application increased water consumption in the different layers above 200 cm, except during the 2011–2012 growing season for both cultivars. In 2011–2012, at 200 cm soil depth, the N treatments still had higher soil water consumption than N0 treatment, likely due to the large amount of rainfall in this year. A similar result was found by Zhou et al. [15], with N fertilizer application decreasing water storage at soil depths of <200 cm after wheat harvesting. N application increased water consumption in the different soil layers; however, in the same soil layers, water consumption did not differ significantly among the different N rates except for some layers of CH.

Soil water consumption clearly changed with the different N rates (Fig. 3). The trends of water consumption were similar in all treatments as soil layers became deeper. N application increased water consumption in all soil layers. Generally, water consumption

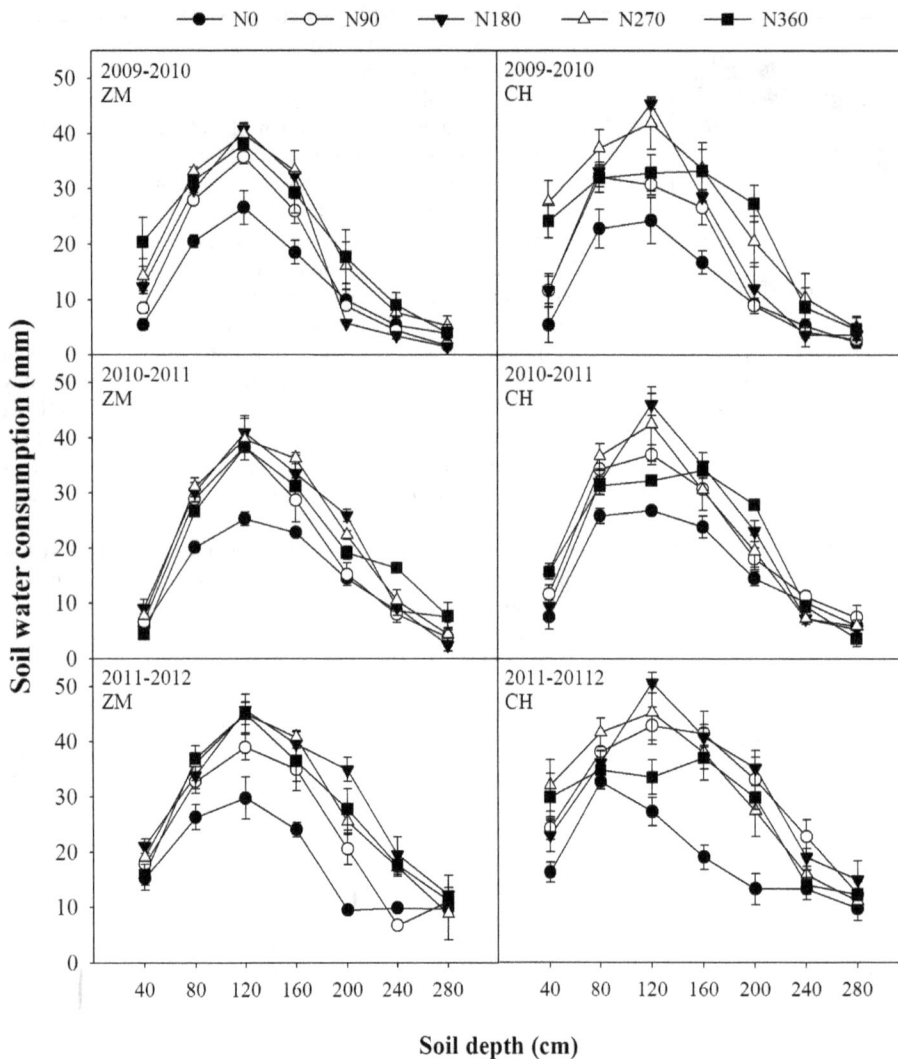

Figure 3. Wheat evapotranspiration (ET) (mm) of two cultivars in different soil layers and different nitrogen (N) treatments in three years. Water ET trends as soil depth increased with influence of N fertilizer for two cultivars in three years. Standard error bars are also shown.

of CH was higher than that of ZM, which was true of the total soil water consumption in all layers (Table 2). Water was mainly consumed in layers of 40–160 cm deep, with the highest water consumption for 100–140 cm. Water was stably absorbed for soil layers <120 cm in the N0 treatment, and <160 cm in the N-applied treatments, shown by ΔS of N treatments at 160 cm being higher than for the non-N treatment at 120 cm. This was probably because N application could promote roots to grow longer and stronger, and which were able to absorb deeper soil water. In addition, the N-applied treatments had greater effects on CH than ZM (Fig. 3), showing that CH was sensitive to low N.

WUE

WUE was in the range of 1–2.09 kg/m^3 for ZM and 1.1–1.92 kg/m^3 in the three years (Table 2). As the N rates increased the WUE increased, and at N360 the WUE increased slightly but not significantly. Zhou et al. [15] reported that grain yields and WUE did not significantly differ between N rates of 120 and 240 kg/ha. The above indicated that excessive N application had no favorable effect on WUE. In 2009–2010 and 2010–2011, the N-applied treatments significantly improved WUE. However, in

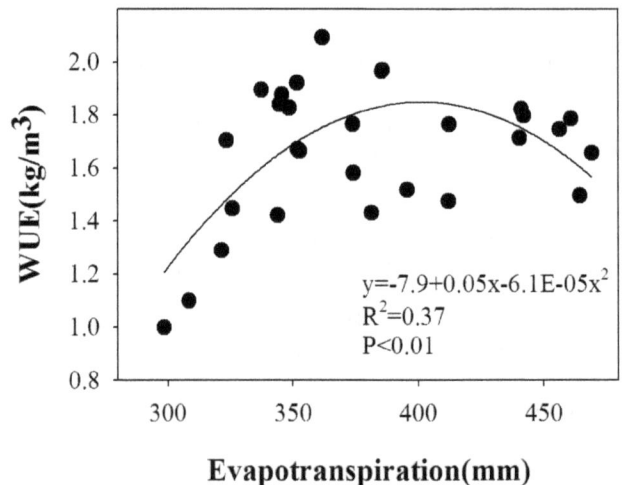

Equation: $y=-7.9+0.05x-6.1E-05x^2$, $R^2=0.37$, $P<0.01$

Figure 4. Relationship between evapotranspiration (ET) (mm) and WUE (Y) for winter wheat in northwest China. The relation between ET and WUE is shown by the equation.

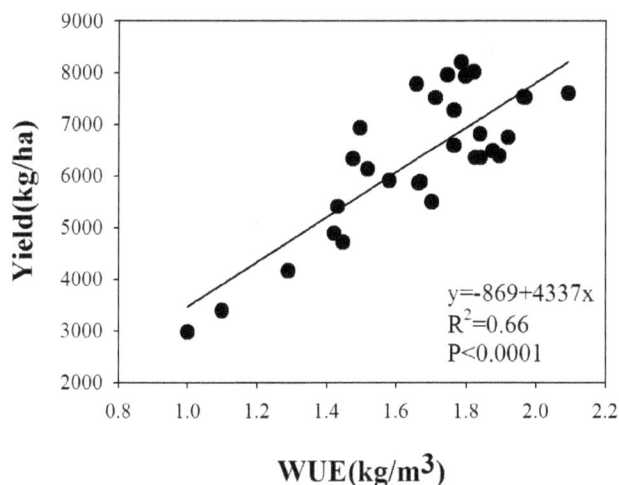

Figure 5. Relationship between WUE and grain yield (Y) for winter wheat in northwest China. The relation between WUE and yield could be deduced from the liner equation.

2011–2012, WUE did not significantly differ between the N-applied and non-N treatments. Compared to 2009–2010 and 2010–2011, the WUE of the non-N treatment was higher in 2011–2012. This may be due to the higher soil water content before sowing and the higher rainfall in the growing season in 2012 that increased the grain yield in the non-N treatment. Consequently higher WUE in non-N treatment reduced the difference between non-N and N-applied treatments.

The WUEs obtained in the present study were higher than those of irrigated winter wheat (0.40–0.88 kg/m³) [45,46] and of irrigated wheat in the US southern plains (0.82 kg/m³) [41], as well as higher than these of irrigated wheat in the Loess Plateau (0.73–0.93 kg/m³) [14], demonstrating that in dryland farming systems N fertilizer can be a useful way to increase WUE. However, the results of the present study were similar to those of winter wheat in the North China Plain (0.84–1.39 kg/m³) [12] and of N-fertilized winter wheat in Yangling, Shaanxi (0.8–1.5 kg/m³) [15]. These differences are caused by different climate or water, fertilizer and crop management. N fertilizer application significantly increases the yield and WUE of both wheat and maize, indicating that N fertilizer application is an effective way to increase grain yield in the study region. Deng et al. [50] reviewed four published research reports and found that N fertilizers increased WUE of wheat and potato by an average of 20% in north central and northwest China.

Regression analysis produced a quadratic relationship between ET and WUE (Fig. 4) and correlations were calculated between WUE and wheat yields (Fig. 5). The yields increased linearly with WUE: WUE reached a maximum value at the ET of 401 mm and then decreased (Fig. 4). However, the maximum WUE did not correspond to the maximum grain yield in the study – the higher WUE means that the crop can gain high yield using less water. This is an important method to obtain a balance between higher yields and lower water supplies of wheat in arid and semi-arid regions by increasing its WUE. In the present study, although there were significant differences between N treatments, both cultivars had relatively higher WUEs at the N rate of 270 kg/ha in a dry year.

Conclusions

N fertilization affected the grain yields, thousand-kernel weights, ET and WUE of the two different water-sensitive wheat cultivars, ZM and CH. The most common pattern of farming in northwest China was used in the present study, with long-term different rates of N fertilization and no irrigation during the wheat growing season. We concluded that (1) in this dryland farming system, increased N fertilization resulted in higher wheat yields in a situation of low precipitation; the drought-tolerant CH showed a yield advantage in a drought environment with high N fertilizer rates; (2) N application affected water consumption in the different soil layers, and promoted absorption and utilization of water from deeper soil layers; and (3) comprehensive consideration of yield and WUE of wheat indicated that the N rate of 270 kg/ha for CH was better to avoid the risk of reduced production due to lack of precipitation; however, under conditions of better soil moisture, the N rate of 180 kg/ha was more economic.

Author Contributions

Conceived and designed the experiments: YZ ZS. Performed the experiments: YZ. Analyzed the data: YZ. Contributed reagents/materials/analysis tools: YZ. Wrote the paper: YZ ZS.

References

1. Fan TL, Stewart BA, Wang YG, Luo JJ, Zhou GY (2005) Long-term fertilization effects on grain yield, water-use efficiency and soil fertility in the dry land of Loess Plateau in China. Agr Ecosyst Environ 106: 313–329.
2. Borlaug NE (2009) Foreword. Food Sec. 1, 1–11
3. Perry C, Steduto P, Allen RG, Burt CM (2009) Increasing productivity in irrigated agriculture: agronomic constraints and hydrological realities. Agr Water Manage 96: 1517–1524
4. Li SX, Wang ZH, Malhi SS, Li SQ, Gao YJ, et al. (2009) Nutrient and water management effects on crop production, and nutrient and water use efficiency in dry land areas of China. Adv Agron 102: 223–265.
5. Hooper DU, Johnson L (1999) Nitrogen limitation in dryland ecosystems: responses to geographical and temporal variation in precipitation. Biogeochemistry 46: 247–293.
6. Rockström J, De Rouw A (1997) Water, nutrients and slope position in on-farm pearl millet cultivation in the Sahel. Plant Soil 195: 311–327.
7. Austin AT (2011) Has water limited our imagination for arid land biogeochemistry? Trends Ecol Evol 26: 229–235.
8. Zand-Parsa S, Sepaskhah A, Ronaghi A (2006) Development and evaluation of integrated water and nitrogen model for maize. Agr Water Manage 81: 227–256.
9. Li YS (1982) Evaluation of field soil moisture condition and the ways to improve crop water use efficiency in Weibei region. Journal of Agronomy Shananxi 2: 1–8. (in Chinese)
10. Shan L (1983) Plant water use efficiency and dryland farming production in North West of China. Newslett Plant Physiology 5: 7–10. (in Chinese)
11. Hamblin AP, Tennant D (1987) Root length density and water uptake in cereals and grain legumes: how well are they correlated. Crop Pasture Sci 38(3): 513–527.
12. Zhang H, Oweis T (1999) Water-yield relations and optimal irrigation scheduling of wheat in the Mediterranean region. Agr Water Manage 38: 195–211.
13. Zhang J, Sui X, Li J, Zhou D (1998) An improved water use efficiency for winter wheat grown under reduced irrigation. Field Crops Res 59: 91–98.
14. Kang SZ, Zhang L, Liang YL, Hu XT, Cai HJ, et al. (2002) Effects of limited irrigation on yield and water use efficiency of winter wheat in the loess plateau of China. Agr Water Manage 55: 203–216.
15. Zhou JB, Wang CY, Zhang H, Dong F, Zheng XF, et al. (2011) Effect of water saving management practices and nitrogen fertilizer rate on crop yield and water use efficiency in a winter wheat–summer maize cropping system. Field Crops Res 122. 157–163.
16. Guo YQ, Wang LM, He XH, Zhang Y, Chen SY, et al. (2008) Water use efficiency and evapotranspiration of winter wheat and its response to irrigation regime in the north China plain. Agr Forest Meteorol 148: 1848–1859.
17. Halvorson AD, Nielsen DC, Reule CA (2004) Nitrogen management nitrogen fertilization and rotation effects on no-till dry land wheat production. Agron J 96: 1196–1201.

18. Turner NC (2004) Agronomic options for improving rain fall-use efficiency of crops in dryland farming systems. J Exp Bot 55: 2413–2415.

19. Turner NC, Asseng S (2005) Productivity, sustainability, and rainfall-use efficiency in Australian rainfed Mediterranean agricultural systems. Aust J Agric Res 56, 1123–1136 56: 1123–1136.

20. Kirda C, Topcu S, Kaman H, Ulger AC, Yazici A, et al. (2005) Grain yield response and N-fertilizer recovery of maize under deficit irrigation. Field Crops Res 93: 132–141.

21. Pimentel D, Hurd L, Bellotti A, Forster M, Oka I, et al. (1973) Food production and the energy crisis. Science 182: 443–449.

22. Erisman JW, Sutton MA, Galloway J, Klimont Z, Winiwarter W (2008) How a century of ammonia synthesis changed the world. Nat Geosci 1: 636–639.

23. Bassoa B, Cammarano D, Troccoli A, Chen DL, Joe T (2010) Long-term wheat response to nitrogen in a rainfed Mediterranean environment: Field data and simulation analysis. Eur J Agron 33: 132–138.

24. Hai L, Li XG, Li FM, Suo DR, Guggenberger G (2010) Long-term fertilization and manuring effects on physically-separated soil organic matter pools under a wheat-wheat-maize cropping system in an arid region of China. Soil Biol Biochem 42: 253–259.

25. Malhi S, Nyborg M, Goddard T, Puurveen D (2011) Long-term tillage, straw management and N fertilization effects on quantity and quality of organic C and N in a Black Chernozem soil. Nutr Cycl Agroecosys 90(2): 227–241.

26. Godfray HCJ, Beddington JR, Crute IR, Haddad L, Lawrence D, et al. (2010) Food security: the challenge of feeding 9 billion people. Science 327: 812–818.

27. Schindler D, Hecky R (2009) Eutrophication: more nitrogen data needed. Science 324: 721–722.

28. Hvistendahl M (2010) China's push to add by subtracting fertilizer. Science 327: 801–801.

29. Dawe D, Dobermann A, Moya P, Abdulrachman S, Bijay S, et al. (2000) How widespread are yield declines in long-term rice experiments in Asia. Field Crops Res 66: 175–193.

30. Regmi AP, Ladha JK, Pathak H, Pasuquin E, Bueno C, et al. (2002) Yield and soil fertility trends in a 20-year rice–rice–wheat experiment in Nepal. Soil Sci Soc Am J 66: 857–867.

31. Huifang H, Jiayin S, Dandan Z, Xuanqi L (2012) Effect of irrigation frequency during the growing season of winter wheat on the water use efficiency of summer maize in a double cropping system. Maydica 56(2).

32. Shen JY, Zhao DD, Han HF, Zhou XB, Li QQ (2012) Effects of straw mulching on water consumption characteristics and yield of different types of summer maize plants. Plant Soil Environ 58(4): 161–166.

33. Zhou XB, Chen YH, Ouyang Z (2011) Effects of row spacing on soil water and water consumption of winter wheat under irrigated and rain-fed conditions. Plant Soil Environ 57(3): 115–121.

34. Shangguan Z, Shao M, Dyckmans J (2000) Effects of nitrogen nutrition and water deficit on net photosynthetic rate and chlorophyll fluorescence in winter wheat. Aust J Plant Physiol 156(1): 46–51.

35. Tinsina J, Singh U, Badaruddin M, Meisiner C, Amin MR (2001) Cultivar, nitrogen, and water effects on productivity and nitrogen-use efficiency and balance for rice-wheat sequences of Bangladesh. Field Corps Res 72: 143–161.

36. Morell FJ, Lampurlane J, Alvaro FJ, Martne C (2011) Yield and water use efficiency of barley in a semiarid Mediterranean agro-ecosystem: Long-term effects of tillage and N fertilization. Soil Till Res 117: 76–84.

37. Sheng HD, Wang PH (1985) The relationship between the weight of 1000-seeds and soil water content in winter wheat season. Acta University of Agricultural Boreali occidentalis 13: 73–79. (in Chinese)

38. Shangguan ZP, Shao MA, Lei TW, Fan TL (2002) Runoff water management technologies for dryland agriculture on the Loess Plateau of China. Int J Sust Dev World 9: 341–350.

39. Johnson WC (1964) Some observations on the contribution of an inch of seeding-time soil moisture to wheat yields in the Great Plains. Agron J 56: 29–35.

40. Unger PW (1972) Dryland winter wheat and grain sorghum cropping systems, northern High Plains of Texas. Texas Agricultural Experiment Station 11–26.

41. Musick JT, Jones OR, Stemart BA, Dusek DA (1994) Water-yield relationships for irrigated and dry land wheat in the US southern plains. Agron J 86: 980–986.

42. Zhang XC, Shangguan ZP (2007) Effects of application nitrogen on photosynthesis and growth of different drought resistance winter wheat cultivars. Chinese Journal of Eco-Agriculture. 15(6). (in Chinese)

43. Hunsaker DJ, Kimball BA, Pinter Jr P, Wall G, LaMorte RL, et al. (2000) CO_2 enrichment and soil nitrogen effects on wheat evapotranspiration and water use efficiency. Agr Forest Meteorol 104: 85–105

44. Rahman MA, Chikushi J, Saifizzaman M, Lauren JG (2005) Rice straw mulching and nitrogen response of no-till wheat following rice in Bangladesh. Field Crops Res. 91: 71–81.

45. Howell TA, Steiner JL, Schneider AD, Evett SR (1995) Evapotranspiration of irrigated winter wheat-southern high plains. Trans ASAE 38: 745–759

46. Schneider AD, Howell TA (1997) Methods, amount, and timing of sprinkler irrigation for winter wheat. Trans ASAE 40: 137–142

47. Zhang H, Wang X, You M, Liu C (1999) Water-yield relations and water use efficiency of winter wheat in the north China plain. Irrigation Sci 19: 37–45

48. Jupp AP, Newman EI (1987) Morphological and anatomical effects of severe drought on the roots of Loium-perene L. New Phytol 105: 393–402

49. McIntyre BD, Riha SJ, Flower DJ (1995) Water uptake by pearl millet in a semiarid environment. Field Crop Res 43: 67–76

50. Deng XP, Shan L, Zhang HP, Turner NC (2006) Improving agricultural water use efficiency in arid and semiarid areas of China. Agr Water Manage 80: 23–40.

20

Long Term Effect of Land Reclamation from Lake on Chemical Composition of Soil Organic Matter and Its Mineralization

Dongmei He, Honghua Ruan*

Faculty of Forest Resources and Environmental Science, and Key Laboratory of Forestry and Ecological Engineering of Jiangsu Province, Nanjing Forestry University, Nanjing, Jiangsu, China

Abstract

Since the late 1950s, land reclamation from lakes has been a common human disturbance to ecosystems in China. It has greatly diminished the lake area, and altered natural ecological succession. However, little is known about its impact on the carbon (C) cycle. We conducted an experiment to examine the variations of chemical properties of dissolved organic matter (DOM) and C mineralization under four land uses, i.e. coniferous forest (CF), evergreen broadleaf forest (EBF), bamboo forest (BF) and cropland (CL) in a reclaimed land area from Taihu Lake. Soils and lake sediments (LS) were incubated for 360 days in the laboratory and the CO_2 evolution from each soil during the incubation was fit to a double exponential model. The DOM was analyzed at the beginning and end of the incubation using UV and fluorescence spectroscopy to understand the relationships between DOM chemistry and C mineralization. The C mineralization in our study was influenced by the land use with different vegetation and management. The greatest cumulative CO_2-C emission was observed in BF soil at 0–10 cm depth. The active C pool in EBF at 10–25 cm had longer (62 days) mean residence time (MRT). LS showed the highest cumulative CO_2-C and shortest MRT comparing with the terrestrial soils. The carbohydrates in DOM were positively correlated with CO_2-C evolution and negatively correlated to phenols in the forest soils. Cropland was consistently an outlier in relationships between DOM chemistry and CO_2-evolution, highlighting the unique effects that this land use on soil C cycling, which may be attributed the tillage practices. Our results suggest that C mineralization is closely related to the chemical composition of DOM and sensitive to its variation. Conversion of an aquatic ecosystem into a terrestrial ecosystem may alter the chemical structure of DOM, and then influences soil C mineralization.

Editor: Han Y.H. Chen, Lakehead University, Canada

Funding: This study was supported by the MOST of China (973 Program No. 2012CB416904), the National Science Foundation of China (No. 31170417) and partially supported by PAPD and Collaborative Innovation plan of Jiangsu higher education, Doctorate Fellowship Foundation of Nanjing Forestry University. The funders had no role in study design, data collection and analysis, decision to publish, or preparation of the manuscript.

Competing Interests: The authors have declared that no competing interests exist.

* E-mail: hhruan@njfu.edu.cn

Introduction

Microbial respiration of soil organic matter (SOM) causes a large flux of CO_2 to the atmosphere, and understanding the controls on this flux is a critical component of society's effort to cope with increasing atmospheric CO_2 and climate change [1–2]. Many abiotic and biotic factors affect the global carbon (C) cycle, especially, human activities, such as the fossil fuel combustion and land use change [3]. Of these factors, land use changes, such as deforestation and afforestation, draining of wetlands, converting grassland to arable cropping and reclaiming lakes, directly affect microbial respiration [4–5]. Many studies on the influence of land use change have shown it results in drastic changes to soil C cycling [6–8]. Land use determines the vegetation type grown on the soil, and therefore the type of organic matter input to the ecosystem [5,9]. Land use change can also increase microbial C mineralization, which may exacerbate the trend of global warming and other effects related to climate change [7].

In the late 1950s, reclaiming land from lakes became a common type of land use in China. China is a traditional agricultural country with a growing population, and it was deemed necessary to expand cultivated land to develop agriculture as well as feed an increasing population. Thus, after several decades more than 1.4×10^4 km^2 of land was reclaimed from lakes [3,10]. Taihu Lake as one of the five largest freshwater lakes in China also experienced land reclamation. Taihu Lake is situated in the south of the Yangtze Delta among 30°55′–31°33′N and 119°55′–120°36′E and with a land area of 2.3×10^3 km^2. From the 1950s to 1980s, more than 160 lakes around the Taihu Lake basin were disappeared and the area of water surface had reduced by 13.6% [11]. In recent years, many studies have focused on the effects of reclamation on ecosystem. They observed that many negative consequences caused by reclaiming land from lakes, such as species decline for habitat loss, eliminating the natural buffers and flood control capacity, water pollution and eutrophication caused by agriculture measures and so on [12]. However, there are few studies about the effect of lake reclamation on the properties of soil organic matter in the terrestrial ecosystem and little is known about its impact on soil C cycling.

Solid soil organic matter must pass through the dissolved phase to be decomposed by microbes [13–14]. Although dissolved organic C (DOC) makes up only a small portion of total soil organic carbon (mostly less than 1%), it represents the active or labile C pool [15–17] and turn over more than 4000 times annually [18]. Thus, the DOM pools is closely correlated to the labile fraction of soil organic matter and is a sensitive indicator of its overall dynamics [16,19]. Also, many studies observed the close relationships between biodegradability of DOC and C mineralization [14,16,20].

Land use is considered as the factor with the greatest influence on soil DOM because it determines the quantity and quality of organic matter input into the soil [5]. However, the mechanism of how land use influences dynamics of DOM is still not clear. Several authors have found that biodegradability of DOM largely depends on its chemical composition [21–23]. It is generally assumed that the easily degradable DOM consists mainly of simple carbohydrate monomers (i.e., glucose, fructose), low molecular organic acids (i.e., citric, oxalic, succinic acid), amino acids, amino sugars and low molecular weight proteins [13,18]. The stable DOM fraction generally contains polyphenolic, aromatic structures and other complex macromolecules [24]. So, the chemical structure of DOM and the complexity of its molecules were considered to correlate with C mineralization. The chemical structure of DOM often described with UV and fluorescence spectroscopy [25–26]. Coupling and understanding of the chemical composition of DOM and C mineralization could be a useful tool for evaluating the influence of land use change on soil C dynamics.

In a previous study, Wang et al. [3] reported the labile SOC concentrations were different from each other under four types of land use from land reclaimed from Taihu Lake, China. Here we endeavored to link the C mineralization potential to the chemical makeup of DOM in soils of land that had been converted from lake sediments to forests and cropland. We hypothesized that this conversion of aquatic ecosystems to terrestrial ecosystems resulted in increased recalcitrance in chemical structure of DOM and ultimately affected potential soil C mineralization, because terrestrial organic matter is usually rich in aromatic and phenolic structures while DOM from aquatic environments has a lower content of aromatic C and more aliphatic C [27–28]. Our objectives were to (1) test for differences in C mineralization in lake sediments and terrestrial soils under different land use on reclaimed land; (2) compare the chemical composition of DOM in lake sediments and soils under different land use; (3) relate C mineralization and chemical properties of DOM in order to explore the role of DOM in C cycling in reclaimed soils.

Materials and Methods

Site description

The study was conducted at the Xiaodian Lake Forest ($31°10'N$, $120°48'E$), located in the northeast of Wujiang City, Jiangsu province, southeast of China (fig. 1) (The Department of Agriculture & Forestry of Jiangsu Province is the authority responsible for it. There were no specific permissions required for the study area. Our field studies did not involve endangered or protected species). The climate in this region is humid north subtropical monsoon with a mean annual temperature of 16°C, annual rainfall of approximately 1100 mm (mainly during the summer months), annual average relative humidity of 78%, and an annual non-frost period of up to 240 days [3]. This area used to be a part of the Taihu Lake, and the lake area was converted to farmland in the early 1960s. The soil properties of the reclaimed land were unsuitable for food crops and so, afforestation projects were carried out in 1969 in much of the region. After more than 40 years of forest management, it has developed into a forest park

with $1.33 km^2$ forest-covered area. The dominant forest types in the park are dominated by coniferous forest (*Metasequoia glyptostroboides*), evergreen broadleaf forest (*Cinnamomum camphora*), bamboo forest (*Phyllostachys heterocycla*) and all of them are the planted tree species monoculture. Additionally, there is approximately 7.0 ha of cropland with rotations of rice and canola in parcels around the park and with the common tillage practices of ploughing, irrigation, mineral and organic fertilization. Soils in this region are all derived from lake sediment and were similar following reclamation. The general characteristics of each site were described by Wang et al. [3].

Soil sampling

We chose four sites within coniferous forest (CF), evergreen broadleaf forest (EBF), bamboo forest (BF) and cropland (CL), respectively. Four 10 m×10 m plots were randomly established within each site. Soil samples were collected randomly using a 3 cm diameter soil sampler at 0–10 cm and 10–25 cm depths from each plot in October 2011. After removal of the litter layer, ten soil cores were taken inside each plot and pooled to form a composite soil sample. Visible roots, residues and stones were immediately removed after sampling and then field-moist soil samples were sieved through a 2 mm mesh. Soil samples were divided into two parts: one stored at 4°C for analysis of soil microbial biomass, DOM and C mineralization incubation experiments. The other part was air dried and analyzed for pH, soil total carbon (TC) and total nitrogen (TN). Lake sediments (LS) were sampled with plexiglass tube (11 cm I.D., 50 cm in length) by cylindrical sampler (Rigo Co. $\Phi110$ mm×500 mm) at four points in the water area of the lake. Sediments in the tube were pushback and sliced at 10 cm for 0–10 cm depth and then 15 cm for 10–25 cm depth using a stainless steel spatula [29–30]. A total of 40 sediment samples were collected and kept in room temperature of 4°C.

Soil physical and chemical characteristics

Soil bulk density was measured with the core method in each layer by using a 5 cm×5 cm metal cylinder [31]. Soil moisture was determined gravimetrically by oven drying at 105°C for 24 h. Soil pH was measured on air-dried soil in a 1:2.5 (w/w) soil-water suspension using a glass electrode. Soil TC and TN were determined via dry combustion and thermal conductivity detection using an elemental analyzer (Vario EL III, Elementar, Germany). Microbial biomass carbon (MBC) was determined by chloroform fumigation as described by Vance et al. [32].

Incubation experiment and analytical methods

Carbon mineralization. C mineralization was measured in each soil sample during incubation in the dark at 25°C for 360 days [33]. From each incubation sample, 100 g fresh moist soil was incubated in a 1000 ml sealed Mason jar. A 50 ml beaker containing 20 ml 1 M NaOH solution was placed in each jar to trap the evolved CO_2. Three additional jars containing 20 ml 1 M NaOH solution but no soil were used as a control. During incubation, the evolved CO_2 trapped in NaOH was determined by titration with 1 M HCl after precipitating the carbonate with 1 M $BaCl_2$ solution. After the CO_2 traps were taken out, the jars were left open for 2 h to maintain aerobic conditions in each jar, and resealed for further incubation. Water loss in the jars was monitored by weighing the jars and replenished by adding ultrapure water after opening [34]. The cumulative C mineralization was expressed as g CO_2-C kg^{-1} soil.

DOM extraction and characterization. DOM was extracted from each soil sample before and after soil incubation. 10 mM

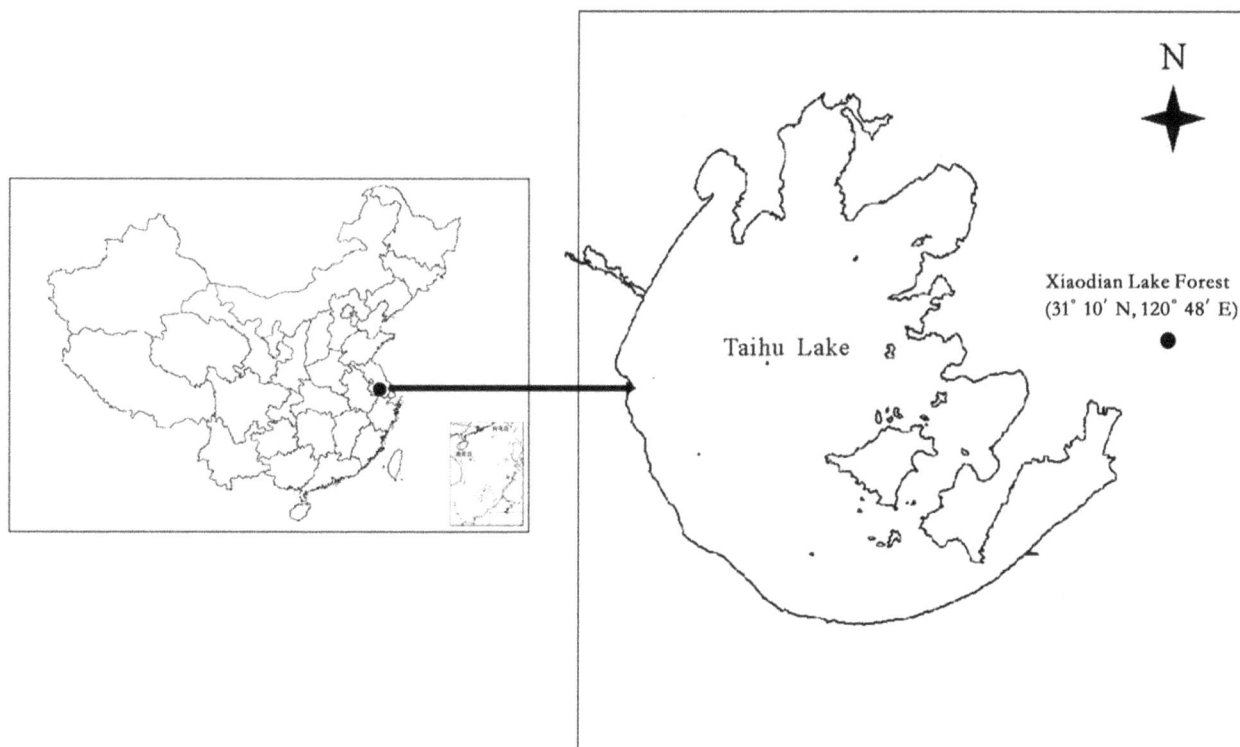

Figure 1. The map of study site in Xiaodian Lake area of the Taihu Lake basin China.

CaCl$_2$ solution was added to an aliquot of soil samples at a soil-solution ratio of 1:2 (w/v) and shaken for 30 min in a horizontal shaker at room temperature. The suspension was spun in a centrifuge for 20 min at 4000 rpm and filtered through a 0.45 μm pore-size cellulose acetate membranes (Schleicher & Schuell, OE 67). The filtered solutions were stored frozen (−20°C) for additional analyses.

DOC concentration was determined by Shimadzu TOC-5050 total organic carbon analyzer. Hydrophilic and hydrophobic fractions in DOM were separated by Amberlite XAD-8 resin (Rohm and Haas, Philadelphia, PA) [35]. Briefly, DOM solutions were acidified to pH 2 with HCl, and then eluted through glass columns filled with Amberlite XAD-8 resin. The column effluent representing the hydrophilic fraction of DOM (Hi) was analyzed for organic C and the amount of organic C in hydrophobic fraction was calculated by the difference between organic C in DOM solution and in hydrophilic fraction. Total carbohydrates in DOM (CH) were measured by phenol-sulfuric acid method [36], using glucose as a standard. Total soluble phenolic compounds (Phe) were analyzed by Folin-Ciocalteau method [37], using tannic acid as a standard.

For spectroscopic measurements, all samples were diluted to 10 mg C L^{-1} to avoid concentration effects and were brought to a constant pH of 7.7 [21] by adding NaOH. UV absorption at 254 nm of DOM was measured using a UV/Vis spectrophotometer (Shimadzu UV-2550, Japan). The specific UV absorbance values (SUVA$_{254}$) were determined as the ratio of the UV absorbance at 254 nm to the DOM concentration and multiplying the value of 100, i.e. (UV$_{254}$/DOC)×100. SUVA$_{254}$ can serve as an indicator of the aromaticity of DOM.

Fluorescence emission spectra were obtained with a Varian Cary Eclipse Fluorescence Spectrophotometer (λ_{ex} 254 nm, slit 10 nm, λ_{em} 300–480 nm, slit 10 nm, and scan speed 1200 nm min^{-1}) using 1 cm cuvettes. Fluorescence efficiency index (FE) proposed by Ewald et al. [38] was considered to be proportional to the quantum efficiency and was defined as the ratio of maximum fluorescence intensity (F$_{max}$) divided by UV absorption at the excitation wavelength of 254 nm (F$_{max}$/Abs). The emission humification index (HIX$_{em}$, dimensionless) was determined as the ratio between the area in the upper quarter (Σ435–480 nm) of the usable fluorescence emission spectrum and the area in the lower usable quarter (Σ300–345 nm) [39–41]. The HIX$_{em}$ can show the degree of complexity and condensation of the DOM.

Statistical analysis

The double exponential model [20,21,42], separating the mineralized organic C into an active C pool and slow C pool can be presented as:

$$C_m(\%) = a \times [1 - \exp(k_1 t)] + (100 - a) \times [1 - \exp(-k_2 t)]$$

where, t is incubation days; C_m is cumulative value of mineralized C at t time presented as percent of initial C in the soil; a is the portion of organic C that is readily decomposed (in % of initial C in the soil = labile C); $(100 - a)$ is the part of organic C that is slowly decomposed (stable C); k_1 and k_2 are mineralization rate constants for active and stable C pools (day^{-1}). Mean residence time (MRT) for each pool is calculated as the reciprocal of the decomposition rate constant in the double exponential model.

This double decomposition equation for two distinct C pools with different mineralization rate constants was fitted with the nonlinear regression that was used in the Marquardt algorithm and an iterative process to find the parameter values that could minimize the residual sum of squares. The model that gave the least squared error was chosen to be the best. For a model to be chosen, all the parameters had to generate plausible results in realistic ranges. For example, the rate constants could not be negative, and the sum of active and stable C pools should not exceed initial C in the soil.

All results were reported as the mean ± SE of the four field replicates. Differences in soil properties, mineralization parameters and DOM chemical properties among land use types for each site were tested using ANOVA followed by least significant difference (LSD). T-test was used to assess the differences between soil depths. Statistical significance was determined at $p < 0.05$ level. Linear regression analysis was used to determine relationships between C mineralization and chemical properties of DOM. We calculated Pearson correlation coefficients among properties of soil, C mineralization parameters and the chemical characteristics of DOM to discuss the influence of selected parameters on C mineralization. All statistical analyses were performed with SPSS 13.0.

Results

Soil properties

We observed variation in several soil characteristics under different land uses and soil depth (Table 1). Soil bulk density was lower at 0–10 cm soil depth than 10–25 cm depth and BF was significantly lower than all other land uses ($p < 0.05$). Soil pH in all soils and sediments were ranging from 4.13 to 6.30 with the highest value in LS. TC and TN concentrations from the terrestrial soils were significantly higher at 0–10 cm soil depth (TC 14.95–35.35 g C/kg soil, TN 1.77–3.41 g N/kg soil), but both of these two factors for LS were lower at upper layer. BF soils were observed remarkably higher TC and TN than other land uses. There was significant difference in MBC and DOM between the land use types and soil depth (Table 1). MBC and DOM were significantly higher at 0–10 cm soil depth in terrestrial land uses. Under the four land uses, MBC showed highest value in CL and lowest value in CF at both soil depths, while DOM was significantly greater in CF at 0–10 cm depth and significantly lower in EBF at 10–25 cm depth. MBC and DOM in LS showed an inverse trend compared with the other land uses, which was greater at 10–25 cm depth.

Chemical characteristics of DOM

The chemical composition of DOM was compared for each sample before and after incubation (Table 2). Before incubation, the fraction of CH at 0–10 cm soil depth ranged from 11.25 to 28.39% of total DOM, with the highest value in BF. The Phe at 0–10 cm soil depth ranging from 1.06 to 3.74% of total DOM was significantly larger than at 10–25 cm depth, with the highest value in CL. The Hi fraction at 0–10 cm depth was varied in the land uses with the highest in CL (61.75%) and lowest in BF (21.22%). At 10–25 cm soil depth, the fraction of CH in BF and LS was significantly larger than the other soils. The largest Phe concentration was observed in LS (3.49%). The proportion of hydrophilic C at 10–25 cm soil depth was no significant difference between each site, but it is significantly higher than at 0–10 cm depth except for CL soil.

From the initial samples before incubation, the spectral indicators were found to be significantly different among the land uses and between soil depths (Table 2). At 0–10 cm depth, $SUVA_{254}$ was significantly greater in CF while FE was significantly greater in CL. At 10–25 cm soil depth, $SUVA_{254}$ and HIX_{em} were both lower in CL. Difference in FE at 10–25 cm was statistically significant between each site. There were no significant differences between soil depths in $SUVA_{254}$ for EBF and LS, but FE for all the

Table 1. Soil total C concentration (TC), total N concentration (TN), dissolved organic M (DOM), microbial biomass C (MBC), pH and the bulk density of different land uses soils collected on reclaimed land from Taihu Lake, China.

Site		TC	TN	DOM	MBC	pH	Bulk density
Depth	Land use	(g C/kg Soil)	(g N/kg Soil)	(mg C/kg Soil)	(mg C/kg Soil)		(g cm^{-3})
0–10 cm	CF	19.12±1.64b*	2.24±0.15b*	126.63±2.75a*	214.50±5.77d*	4.45±0.07d*	1.22±0.03a*
	EBF	16.05±0.69bc*	1.77±0.04b*	93.50±2.56d*	370.40±9.64c*	4.78±0.08c	1.18±0.03a
	BF	35.35±1.79a*	3.41±0.22a*	110.50±2.40b*	423.01±5.13b*	4.13±0.06e*	0.89±0.06b
	CL	14.95±0.96c*	1.84±0.10b*	105.80±2.81bc*	521.98±7.72a*	5.67±0.08b*	1.16±0.03a*
	LS	16.14±1.38bc	2.26±0.23b	99.95±2.11cd*	350.03±7.64c*	6.19±0.02a	Nd
10–25 cm	CF	8.34±0.53c	1.18±0.07b	97.72±4.50b	112.77±4.43a	4.89±0.15bc	1.31±0.01a
	EBF	8.81±0.77c	1.16±0.08b	78.39±3.06c	309.94±3.90c	4.98±0.07b	1.25±0.02a
	BF	23.76±0.94a	2.59±0.25a	95.88±2.46b	232.28±4.65d	4.30±0.02c	0.99±0.06b
	CL	9.30±1.16c	1.23±0.13b	91.40±3.13b	455.99±5.20b	6.18±0.16a	1.31±0.03a
	LS	17.51±1.03c	2.34±0.13c	115.48±4.21a	558.82±10.28e	6.30±0.07a	Nd

Values are mean ± SE (n = 4). Means within a column of the corresponding depth followed by different letters are significantly different and * indicates the significant difference between the soil depth. (Significance at $p < 0.05$). CF: coniferous forest; EBF: evergreen broadleaf forest; BF: bamboo forest; CL: cropland; LS: lake sediment; Nd: Not determined.

Table 2. Chemical properties of soil DOM at two soil depths under different land uses reclaimed from Taihu Lake, China.

Site			SUVA$_{254}$	FE	HIX$_{em}$	CH	Phe	Hi
Depth	Land use	DOM	(l mg C^{-1} m^{-1})			(% of total DOM)		
0-10 cm	CF	Initial	2.30±0.06a*	321.28±2.28b*	1.76±0.11ab	23.86±1.21b	2.50±0.04b*	32.71±1.38c*
		Δ	2.65±0.05a*	29.62±1.95c	4.17±0.11a*	32.87±2.02a*	4.68±0.22a*	12.15±2.20b
	EBF	Initial	1.00±0.09d	301.18±3.40b*	1.64±0.03b	21.83±0.49b*	2.08±0.04c	37.47±1.35b*
		Δ	2.38±0.12b*	95.03±5.61b*	4.50±0.69a*	25.40±0.51b*	2.81±0.17b*	8.72±1.91b
	BF	Initial	2.07±0.07b*	252.31±3.26d*	1.94±0.07a	28.39±1.13a	2.14±0.06c*	29.22±1.14c*
		Δ	1.05±0.06c	216.67±6.46a*	5.59±0.70a*	4.70±0.37c*	1.37±0.10c*	19.90±2.17a
	CL	Initial	1.20±0.04c*	430.40±2.29a*	1.11±0.08c*	18.98±1.02c*	3.74±0.12a*	61.75±0.70a
		Δ	0.55±0.06d*	−59.55±6.02e*	2.78±0.43b	6.09±0.37c*	−2.61±0.11e*	−3.61±1.35c*
	LS	Initial	1.15±0.06c	289.30±17.80c*	1.68±0.11b	11.25±0.72d*	1.06±0.07d*	59.58±1.19a*
		Δ	0.47±0.07d*	−4.65±1.88d	0.79±0.01c	−4.00±0.31d*	0.25±0.07d*	10.87±3.92b*
10-25 cm	CF	Initial	0.98±0.05b	253.32±1.30d	1.69±0.02a	21.37±0.46b	1.33±0.13c	47.77±1.12a
		Δ	0.60±0.07b	24.40±2.49bc	1.84±0.23c	6.74±0.17a	1.72±0.06a	7.31±1.48c
	EBF	Initial	1.20±0.06a	408.93±4.16a	1.07±0.02b	15.35±0.79c	2.18±0.09b	52.48±0.70a
		Δ	0.18±0.06c	11.25±2.46c	2.24±0.15b	2.78±0.38b	0.60±0.03a	6.98±2.83c
	BF	Initial	1.30±0.06a	170.44±2.17e	1.88±0.02a	26.14±0.81a	1.51±0.02c	46.18±0.49a
		Δ	1.10±0.04a	87.96±2.75a	2.85±0.45a	−2.14±0.85c	0.42±0.05a	17.15±0.69b
	CL	Initial	0.78±0.05c	316.91±2.46c	0.48±0.02c	11.28±0.55d	0.92±0.04d	56.06±1.65a
		Δ	0.15±0.05c	25.32±6.10b	3.52±0.16a	3.61±0.50b	−0.08±0.02c	18.33±3.58b
	LS	Initial	1.23±0.05a	350.29±11.95b	1.60±0.27a	24.81±0.37a	3.49±0.12a	40.76±1.08a
		Δ	0.20±0.04c	−48.88±11.18d	0.40±0.23d	−13.39±0.80d	0.38±0.99b	43.26±3.23a

Results shown are characteristics of initial DOM and the variation of each characteristic at the end of 360 days incubation. Values are mean ± SE (n =4). Means within a column of the corresponding depth followed by different letters are significantly different and * indicates the significant difference between the soil depth. (Significance at $p<0.05$). CF: coniferous forest; EBF: evergreen broadleaf forest; BF: bamboo forest; CL: cropland; LS: lake sediment; SUVA$_{254}$: specific UV absorbance at 254 nm; FE: fluorescence efficiency ($F_{max}/A254$); HIX$_{em}$: humification index using emission fluorescence spectra (ratio of areas: 435–480 nm/300–345 nm); CH: carbohydrate C; Phe: phenol C; Hi: hydrophilic C; initial: properties of initial samples (soils before incubation); Δ: variation of DOM chemical properties between the initial and final value during incubation.

sites were observed significantly different between soil depths. Differences in HIX_{em} between soil depths were only found in CL.

After 360 days incubation, there were significant amounts of variation in DOM chemical properties among the land uses and lake sediments and most of the variation between soil depths was also significant (Table 2). The proportion of CH at 0–10 cm soil depth increased 17%–138% in the land uses with significantly greater value in CF. The proportion of Phe and Hi respectively increased 23%–188% and 18%–68% for CF, EBF, BF, and LS at 0–10 cm depth, but the values of these two factors for CL showed decreased trend after incubation. The increased values of Phe were significantly different from each site with the greatest variation in CF, while increased amount of Hi was lager in BF, at 0–10 cm depth. The proportion of CH fraction had the largest increase in CF, but it was observed reducing in BF and LS, at 10–25 cm soil depth. At this layer, both the proportion of Phe and Hi increased after incubation, and the largest increase was found in LS. There was no significant difference in the variation of Phe among the three forest sites. $SUVA_{254}$ and HIX_{em} from the samples at the end of incubation were increased at both 0–10 cm and 10–25 cm soil depth (Table 2). CF had the greatest variation in $SUVA_{254}$ and HIX_{em} at 0–10 cm depth and the variation of FE in BF was larger the other sites. There was also no significant difference in the variation of HIX_{em} among the three forest sites. The variation of FE values was significantly different between each other site and the FE values in CL and LS decreased after incubation. At 10–25 cm depth, the variation of $SUVA_{254}$ and FE in BF was significantly larger than other sites.

Carbon mineralization

The cumulative CO_2 production under different land uses and sediments ranged from 0.88 to 7.72 g CO_2-C kg^{-1} soil and there were significant difference between soil depths (Fig. 2a). At 0–10 cm soil depth, cumulative C mineralization in BF soils (2.87 g CO_2-C kg^{-1} soil) was significantly greater than other land uses and CL had the lowest amounts of cumulative CO_2-C (1.62 g CO_2-C kg^{-1} soil). The cumulative CO_2-C for the three forest sites decreased with the increasing soil depth. At 10–25 cm soil depth, C mineralization was found to be significantly different in the following order of CL>BF>CF>EBF (Table 3). Representing the precursor of the four terrestrial soils, LS had different C mineralization patterns than the four terrestrial soils, and its evolution of cumulative CO_2-C (4.59 and 7.72 g CO_2-C kg^{-1} soil, respectively) was significantly larger than other soils, at 0–10 cm and 10–25 cm soil depth. After 360 days of incubation, the percentage of cumulative mineralized C (as % of TC) ranged from 5.11% to 47.19% with the highest C mineralization in LS at 10–25 cm depth (Fig. 2b). At 0–10 cm soil depth, the percentage of mineralized C was significantly different between each other site with largest value in EBF (15.23%) and lowest value in BF (8.11%). At 10–25 cm soil depth, CL had the highest percentage of cumulative mineralized CO_2-C (21.07%) among the four land uses, approximately four times larger than BF, which had the lowest percentage of mineralized C (5.11%).

The rate of CO_2-C evolution from all land use types were highest at the beginning and then decreased progressively with the advancement of time (Fig. 3). The CO_2-C evolution patterns in all land uses and lake sediments were best described by the double exponential model, with the model fits resulting in r^2 values between 0.96 and 0.99 (Table 3). The size of the labile C pool in different soils was comprised of 0.2%–7.09% with the highest value in LS at 10–25 cm depth. The k_1 values (mineralization rate constant) of labile C at 0–10 cm depth ranged from 0.02 to 0.19 with the mean residence time (MRT_1) of 21 days (average of all

soils), varying from 0.02 to 0.12 at 10–25 cm depth with an average MRT_1 of 33 days. Determination of the size and turnover of the stable C pool showed values at 0–10 cm soil depth were larger than those at 10–25 cm soil depth (Table 3). The mineralization rate constants of the stable C pool (k_2) were 2–3 orders of magnitude lower than k_1, ranging from 0.00022 to 0.00098 in upper soil layer with an average of MRT_2 of 9 years and ranging from 0.00011 to 0.00129 in lower soil layer with an average of MRT_2 of 12 years. At 0–10 cm soil depth, the mean residence time of active and stable C pools in CL sites was larger than the others, whereas at 10–25 cm soil depth, the largest mean residence time of stable C pools was found in BF. The mean residence time of active and stable C pools in LS was lower than all other land use types.

C mineralization in relation to DOM characteristics

C mineralization was significantly correlated with DOM characteristics and its change (the variation of DOM properties between beginning and ending of incubation) under the four land uses (Table 4, Fig. 4). The CO_2-C evolution at the 0–10 cm soil depth had a positive correlation with the proportion of CH, variation of the Hi fraction (ΔHi), fluorescence efficiency variation (ΔFE) and HIX_{em} variation (ΔHIX_{em}), while had a negative correlation with initial Phe, Hi, FE and variation of CH (ΔCH), Phe (ΔPhe) and $SUVA_{254}$ ($\Delta SUVA_{254}$) (Fig. 4). The high proportion of labile C pool (a value of the double exponential model) from upper soil was related to the high proportion of Phe, Hi fraction and high values of FE. In contrast, it was negatively related to most of the parameters including CH fraction, ΔPhe, ΔHi, HIX_{em}, $SUVA_{254}$, ΔFE, ΔHIX_{em} and $\Delta SUVA_{254}$ (Table 4). The k_1 values of upper soils were correlated very well with proportion of Phe and Hi, the values of FE, HIX_{em}, ΔFE and ΔHIX_{em}, whereas the k_2 values were only correlated with Phe, ΔCH, ΔPhe and $\Delta SUVA_{254}$ (Table 4).

In 10–25 cm soil layer, cumulative CO_2-C positively correlated with proportion of CH and the values of HIX_{em}, ΔHi, ΔFE, ΔHIX_{em}, $\Delta SUVA_{254}$, while negatively correlated with proportion of Phe and Hi, as well as the values of FE and $SUVA_{254}$ (Table 4). The proportion of labile C pool showed positive correlations with Phe and FE, and showed the negative relationships with ΔHi, HIX_{em} and ΔFE. The k_1 values highly related to proportion of CH, Phe and Hi, as well as values of $SUVA_{254}$ and ΔHIX_{em}. The relationships between k_2 values and the properties of DOM were similar with the k_1 values (Table 4).

Discussion

Chemical characteristics of DOM

The decomposition rate of organic matter is in part controlled by its chemical fractions and structures [13,21]. The highest portion of CH fraction in BF was due to a lot of litter and root exudate input. For the four land uses, the percentage of CH was observed lower in CL, which may be explained by two reasons: (i) soil microbial community may deplete CH fraction to a greater degree because fertilization can increase the microorganism activity [9] and (ii) the type of crop residue input may influence the chemical composition of DOM [9]. Different from the land uses, the CH for LS was greater at 10–25 cm, probably due to the deeper sediments in anoxic condition would limit the microbial activity [43]. The lowest proportion of phenol C of the DOM from the un-incubated soils was found in LS at 0–10 cm soil depth, possibly because most of the DOM is derived from algae, bacteria and macrophytes, which contain lower phenolic groups [44]. The highest percentage of Hi fraction in CL agreed with Chantigny [5]

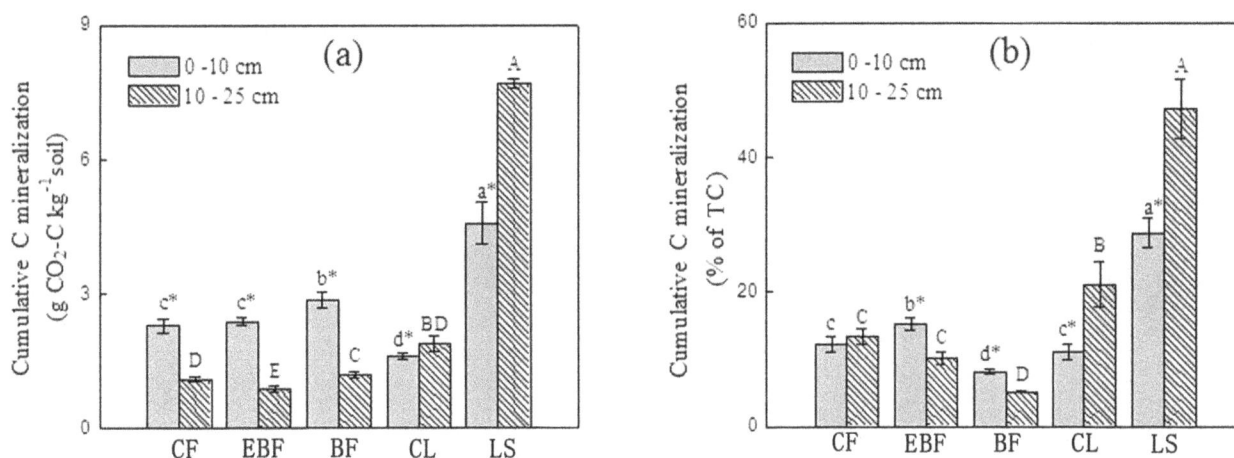

Figure 2. Carbon mineralized at the end of the 360 days incubation period for different land uses and lake sediments. Bars represent standard errors (n = 4). Different letters above bars indicate significant differences ($p<0.05$) of mean values of different sites. * indicates the significant difference ($p<0.05$) between the soil depth.

who reported that DOM from crop residue contained more hydrophilic fractions. During the incubation period, the quality and quantity of organic matter in soils changed, which should lead to variation in the chemical structure of DOM. The proportion of Hi fraction, CH and Phe in DOM under most of the sites were observed to increase at the end of incubation, indicating that some insoluble components were converted to soluble fractions, possibly due to desorption of adsorbed organic matter or other biotic regulatory mechanisms [13,45]. Kalbitz et al. [24] reported that relative increase in polysaccharides after incubation is likely caused by microbial formation and that many bacteria and fungi can release diverse polysaccharides. The relative enrichment in phenols was possibly due to oxidation and fragmentation of lignin in soils [46]. Although both of the percentage of labile and stable fractions in DOM increased at the end of the incubation due to complement of soil organic matter [20], there was a greater increase in the proportion of Phe than CH and Hi fractions. The larger increase in the proportion of Phe at the end of incubation suggested that phenols were the most stable fractions, even though these compounds can experience partial degradation [47]. Our results suggested that CH and Hi fractions were preferentially utilized by microorganisms while phenols were resistant to degradation during incubation. Unlike the other sites, both the proportion of Phe and Hi fractions in CL DOM decreased slightly, probably due to the influence of management practices such as fertilization which changed the soil environment and accelerated both of them degradation [9,48].

The specific UV absorbance at 254 nm (SUVA$_{254}$), humification indices (HIX$_{em}$) and fluorescence efficiency indices (FE) have been used to characterize the content of aromatic structures and complexity degree in DOM [49–51]. In our study, SUVA$_{254}$ and HIX$_{em}$ decreased with soil depth, consistent with Corvasce et al. [52] and Bu et al. [53], suggesting that the aromatic fraction of partially degraded lignin-derived compounds might be gradually adsorbed by upper mineral soil and protected from microbial degradation [54]. CF and BF soils had the highest SUVA$_{254}$ and HIX$_{em}$ values, indicating that there were more complex and condensed polyaromatic structures in the DOM due to the presence of ligninolitic compounds [55]. Our results agree with Khomutova et al. [56] who also found values of specific UV absorbance at 260 nm were larger in coniferous forest soils compared to deciduous forest and pasture soils, suggesting that

DOM under conifer soils is enriched in hydrophobic aromatic compounds [57]. Just like SUVA$_{254}$ and HIX$_{em}$, fluorescence efficiency (FE) was also used to express the degree of condensation of DOM. The low FE in BF soils may be partially attributed to its highly substituted aromatic structural features and its inter- or intra-molecular bonding which could result in self-quenching within macromolecules [26]. Ewald et al. [38] drew a similar conclusion that fluorescence efficiency had an inverse linear correlation with molecular weight of the fulvic acid fractions, which was due to internal quenching in macromolecules caused by energy transfers. The smaller values of SUVA$_{254}$ and HIX$_{em}$ as well as larger value of FE were in CL consistent with Chantigny [5] who reported there were more lignin and other recalcitrant compounds in forest litter compared to crop residues. The increase of SUVA$_{254}$ and HIX$_{em}$ under different sites suggested aromatic structures were accumulating with the decomposition of the labile components of DOM during incubation [21,50]. Hur et al. [27,44] also found the increase of SUVA values for all the DOM samples after microbial incubation and attributed this result to preferential microbial utilization of non-aromatic fractions and/or to microbial transformation of labile compounds into aromatic C structures. At 0–10 cm soil depth, the increase of SUVA$_{254}$ and HIX$_{em}$ in CL were significantly smaller than the forest soils, indicating the same conclusion that DOM in CL contains smaller complex compounds. Although the changes in spectra parameters during incubation exhibited increasing trends under most land use types, the increased DOM for LS was smaller than other land uses. This result suggests that the accumulation of condensed aromatic structures was relatively small during incubation, supporting the previous statement that accumulation of phenols was small compared to hydrophilic fractions for LS DOM. For the soil DOM samples, a pronounced increase in FE value was inconsistent with the changes in HIX$_{em}$ or SUVA$_{254}$, which couldn't be explained by fluorescent quenching. This inconsistency may be attributed to the characterization method of FE influenced by the changes in DOM properties.

Soil carbon mineralization in relation to DOM characteristics

The C mineralization was significantly higher at 0–10 cm soil depth regardless of vegetation types for all land uses except CL.

Table 3. C mineralization kinetics of soil after 360 days incubation at 25°C: cumulative CO_2-C, sizes of the labile and stable C pools, mineralization rate constants and mean residence times of the labile and the stable C pools.

Site	Land use	C_m	Labile C	Stable C	k_1	k_2	MRT_1	MRT_2	r^2
Depth		(%)	(%)	(%)	(day⁻¹)	(day⁻¹)	(days)	(years)	
0–10 cm	CF	12.24	0.75	99.25	0.03	0.00036	30.48	7.68	0.993
	EBF	15.23	0.72	99.28	0.13	0.00048	7.82	5.80	0.979
	BF	8.11	0.20	99.80	0.10	0.00023	10.17	12.13	0.998
	CL	10.95	2.62	97.38	0.02	0.00022	53.13	12.60	0.999
	LS	28.71	1.13	98.87	0.19	0.00098	5.15	2.85	0.983
10–25 cm	CF	13.45	3.07	96.93	0.03	0.00032	33.27	8.67	0.995
	EBF	10.11	4.81	95.19	0.02	0.00016	61.71	17.80	0.999
	BF	5.11	1.42	98.58	0.02	0.00011	48.62	26.11	0.999
	CL	21.07	1.70	98.30	0.07	0.00065	13.53	4.26	0.993
	LS	47.19	7.09	92.91	0.12	0.00128	8.11	2.16	0.957

CF: coniferous forest; EBF: evergreen broadleaf forest; BF: bamboo forest; CL: cropland; LS: lake sediment; C_m: cumulative mineralized C as percentage of initial C in the soil. Labile C: rapidly mineralizable C (calculated using a double exponential model); Stable C: slowly mineralizable C (calculated using a double exponential model); k_1: mineralization rate constant of the labile C pool (double exponential model); k_2: Mineralization rate constant of the stable C pool (double exponential model); MRT_1: Mean residence times of the labile C pool ($MRT_1 = 1/k_1$); MRT_2: Mean residence times of the stable C pool ($MRT_1 = 1/k_2$); r^2: coefficient of determination of the double exponential model.

Values are mean (n = 4).

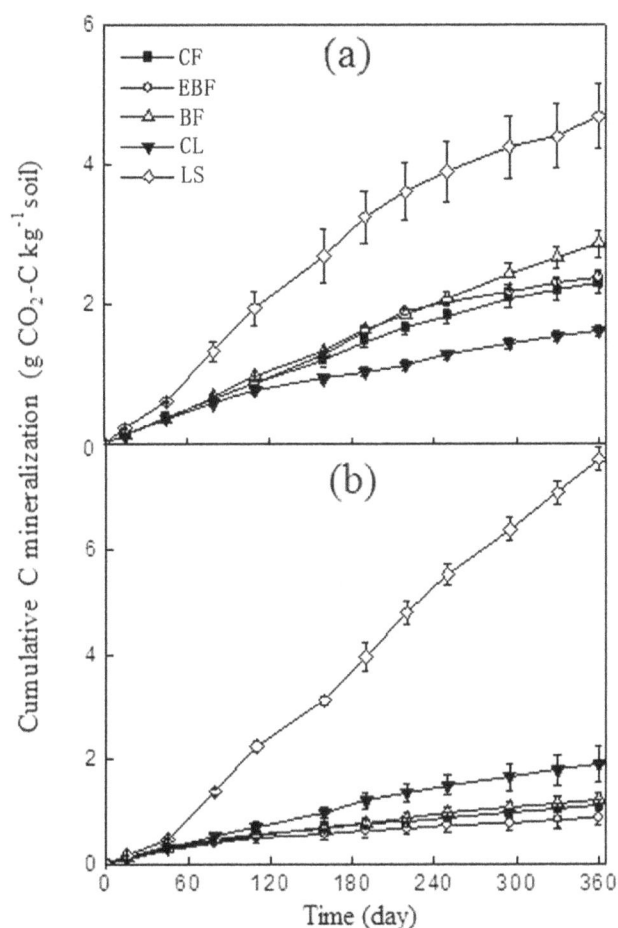

Figure 3. Dynamics of C mineralization of soils under different land use types reclaimed from Taihu Lake, China. (a): C mineralization at 0–10 cm depth; (b): C mineralization at 10–25 cm depth. Bars indicate the standard errors (n = 4).

The higher C mineralization observed at 10–25 cm soil depth in CL was likely due to the tillage practice of ploughing which stimulated the C mineralization in deeper layers [58]. At 0–10 cm soil depth, cumulative C mineralization in BF soils was greater than other land uses, also possibly due to plenty of SOC deriving from degradation of bamboo litter and rhizomatous. While cumulative C emission for CL was significantly lower than the forest soils at 0–10 cm and Zhao et al. [20] observed the similar result that the average cumulative mineralized CO_2-C in forest sites and arable sites was 382 and 279 mg C/kg soil, respectively. The amount of cumulative mineralized CO_2-C during 360 days incubation was significantly higher in lake sediments compared to the terrestrial soils. This result was consistent with our primary hypothesis. Mora et al. [33] observed that during 10 days incubation 0.5% of the sediment C, as well as 0.25% of soil C was mineralized.

The analysis of double exponential model showed that emission of CO_2-C increased as the labile C pool (a) decreased, which was inconsistent with the results of Kalbitz et al. [21] and Bu et al. [59]. During the incubation period, lake sediments had a higher percentage of C mineralized (as % of TC), a larger pool of labile C, and shorter residence times for labile and stable C. This suggested that lake sediments were enriched in less complex, more easily mineralized forms of C [60–62]. The smaller fraction of

mineralized C and lower mineralization rate constants for the stable C pool found in BF soils suggested that organic C in BF soils was composed mainly of polyphenols and aromatic structures deriving from the decomposition of bamboo rhizomes and recalcitrant to biodegradation [63–65]. Compared to forest soils, both labile and stable C pools in CL soils had lower mineralization rate constants, which was different from the result of Kalbitz et al. [21]. This discrepancy was likely explained by DOM representing a small fraction of labile C in CL soils, and mineralization of soil C can't be explained only by consumption of DOM. Zhao et al. [20] also found no correlations between the decrease in quantity of DOM and soil C mineralization. At 0–10 cm depth, mineralization rate constants for both the labile and stable C pool of CF soils were significantly smaller than EBF soils, consistent with previous studies which found C mineralization rate in a conifer forest to be lower than in an evergreen broadleaf forest [66].

In our study, we found the chemical compositions and spectra characteristics of DOM were correlated with C mineralization, but the relationship varied across land uses. The cumulative CO_2-C emission showed a significant positive correlation with proportion of CH at the beginning of the incubation and a significant negative relationship with the proportion of initial Phe. The results suggest that CH and Phe respectively represent the labile and refractory fractions in DOM, and they may regulate the mineralization potential of SOM. Soil carbohydrates are readily available for biodegradation and often have short half-lives due to rapid uptake and assimilation by soil microbial communities [22]. At the end of incubation, the increase in proportion of CH at 0–10 cm soil depth had a negative relationship with cumulative CO_2-C emission, similar to Tian et al. [34], and emphasizing that carbohydrates are utilized preferentially by microorganisms [13,21]. With the consumption of labile fractions, the increasing proportion of Phe was closely related to the less CO_2-C emission, explained by inhibiting the activity of various enzymes [34,46]. The correlation between ΔCH (0–10 cm), ΔPhe (0–10 cm), CH (10–25 cm) and cumulative CO_2-C emission in CL statistically deviated from the regression line (fig. 4a, 4b and 4g), suggesting that the contribution rate to C mineralization made by Phe or CH in CL was different from other land uses. Hi fractions of DOM separated by XAD-8 resin can be characterized as labile soluble organic moieties, particularly carbohydrates, amino sugars and low-molecular-weight organic acids and show a higher degree of biodegradation [13,23,35,57]. In our study, the increase of Hi fraction was strongly tied to increasing C mineralization, which was inconsistent with the general results of previous studies [21,44,67]. This result attributed to the fact that part of the hydrophilic fractions such as hydrophilic neutral (HiN) is composed mainly of carbohydrates that are bound to aromatic compounds and are converted into typical hydrophobic fractions during long term incubation [21]. However, the correlation of Hi and C mineralization at 10–25 cm depth in CL also deviated from the regression line.

Recently, fluorescence and UV spectroscopy has been used successfully to describe the chemical properties of DOM [21,53,68–69]. In many studies, the extent of C mineralization was inversely related to the specific UV absorbance of DOM [13,53,70–71]. This was confirmed by our results observed in 10–25 cm soil depth, indicating that UV-inactive substances were degraded preferentially. The relative increase in $SUVA_{254}$ at the end of incubation was positively related to cumulative CO_2-C emission, which suggests that more aromatic structures were cumulated during the C mineralization. Fluorescence spectroscopy can provide additional information relating to structure, functional groups, conformation, and heterogeneity, as well as dynamic

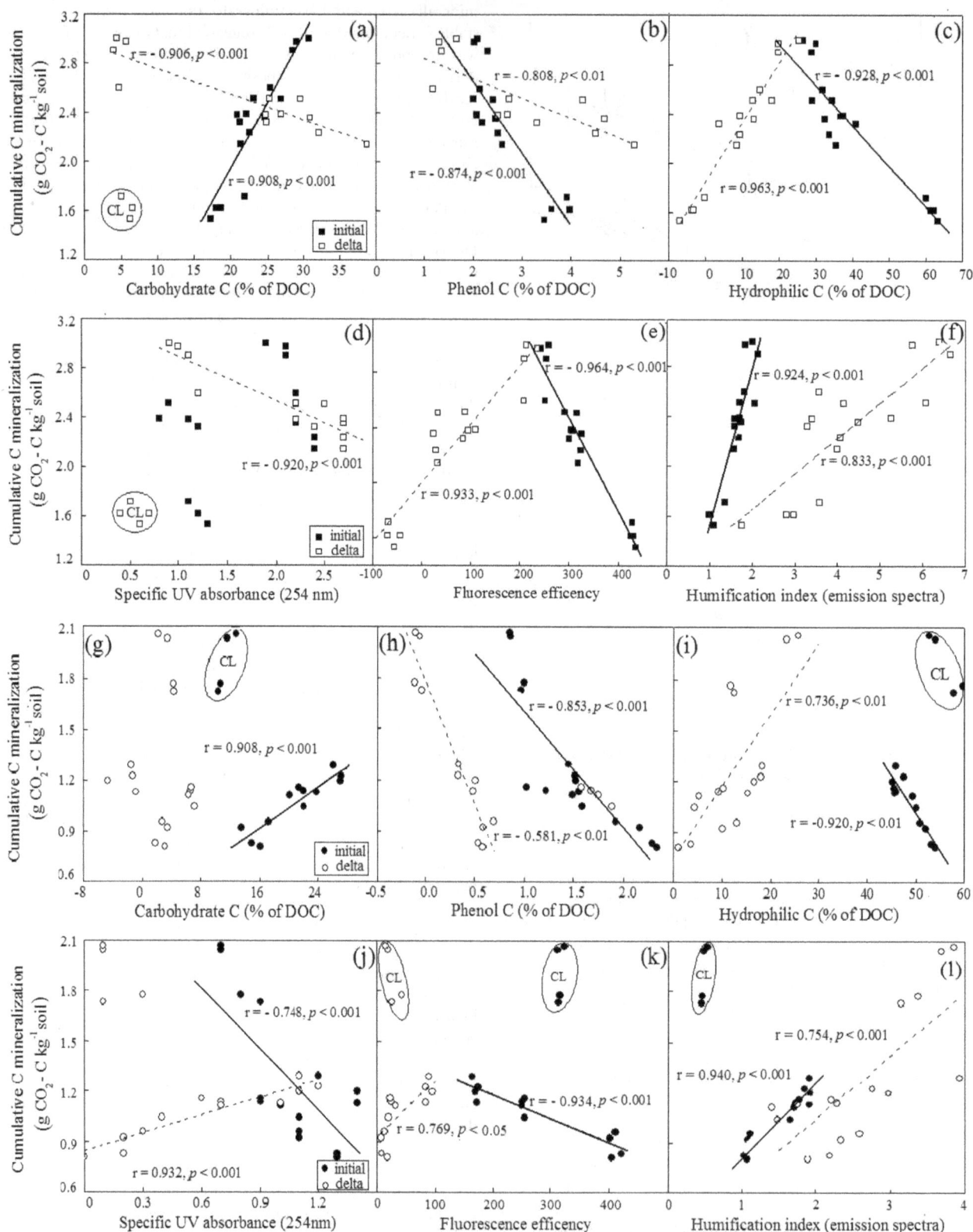

Figure 4. Relationships between cumulative C mineralization and DOM chemical properties of the four land use types. 0–10 cm soil depth: a–f; 10–25 cm soil depth: g–l. (a and g) between carbohydrate carbon percentages and cumulative CO_2-C, (b and h) between phenol carbon percentages and cumulative CO_2-C, (c and i) between hydrophilic carbon and cumulative CO_2-C, (d and j) between $SUVA_{254}$ values and cumulative CO_2-C, (e and k) between the fluorescence efficiency and cumulative CO_2-C, (f and l) between the humification index (emission fluorescence spectra) and cumulative CO_2-C. The curves for the figures are generated with the data of the CL samples excluded.

Table 4. Pearson correlation coefficients (r-value) between C mineralization and soil properties and chemical characteristics of DOM under four terrestrial land use types reclaimed from Taihu Lake, China.

		DOC	CH	ΔCH	Hi	ΔHi	Phe	ΔPhe	$SUVA_{254}$	$\Delta SUVA_{254}$	FE	ΔFE	HIX_{em}	ΔHIX_{em}
0-10 cm	a	-0.140	-0.762**	-0.315	0.976**	-0.893**	0.951**	-0.782**	-0.504*	-0.538*	0.982**	-0.871**	-0.902**	-0.702**
	k_1	-0.586*	0.373	0.048	-0.555*	0.499*	-0.792**	0.339	-0.263	0.275	-0.737**	0.752**	0.510*	0.532*
	k_2	-0.285	-0.148	0.803**	-0.346	0.090	-0.548*	0.666**	-0.289	0.849**	-0.283	0.043	0.210	0.106
	MRT_1	0.270	-0.600*	-0.203	0.822**	-0.742**	0.948**	-0.604*	-0.101	-0.447	0.928**	-0.876**	-0.755**	-0.665**
	MRT_2	0.130	0.099	-0.886**	0.414	-0.145	0.562**	-0.762**	0.142	-0.921**	0.305	-0.031	-0.268	-0.117
	CO_2-C	0.085	0.908**	0.044	-0.928**	0.963**	-0.873**	0.573*	0.432	0.279	-0.964**	0.934**	0.925**	0.833**
10-25 cm	a	-0.600*	-0.263	0.390	0.135	-0.696**	0.755**	0.368	0.200	-0.499	0.776	-0.688**	-0.954**	-0.534*
	k_1	0.170	-0.653**	0.276	0.642**	0.466	-0.796**	-0.483	-0.819**	-0.456	0.090	-0.230	0.575	0.586*
	k_2	-0.420	0.408	-0.383	-0.394	-0.410	0.916**	0.189	0.825**	0.235	0.172	0.125	0.079	0.459
	MRT_1	0.142	-0.710**	0.469	0.662**	0.318	-0.758**	-0.322	-0.886**	-0.560*	0.199	-0.404	-0.700*	-0.436
	MRT_2	-0.088	0.694**	-0.794**	-0.578*	0.028	0.581*	-0.074	0.885**	0.661**	-0.360	0.680**	0.191	-0.122
	CO_2-C	0.310	-0.446	0.014	0.423	0.736**	-0.853**	-0.581*	-0.748**	-0.223	-0.122	0.012	0.940*	0.754**

n=16; abbreviations see Table 2 and Table 3;
*Significance at $p<0.05$;
**Significance at $p<0.01$.

properties of DOM [25–26,49]. In our study, the FE and HIX$_{em}$ deduced from fluorescence emission spectra exhibited strong correlation with CO_2 emission. The FE and HIX$_{em}$ for initial DOM respectively showed negative and positive correlation with C mineralization, suggesting that in the long-term incubation not only labile C but also stable C contributed to the C mineralization. The ΔFE and ΔHIX$_{em}$ during incubation were positively correlated with C mineralization, supporting expectation that with the oxidation or degradation of labile fractions, more aromatic compounds and other complex stable molecules are accumulated [68]. The relationship between part of spectroscopy parameters and cumulative CO_2 emission in CL also deviated from the regression line (fig. 4d, 4k and 4l) possibly due to the effluence of land use and management.

Conclusions

Soils from varied land uses under reclaimed area from Taihu Lake had significant differences in soil DOM chemical properties and C dynamics. The C mineralization of lake sediments was much larger than that in terrestrial ecosystems, confirming our hypothesis. For the terrestrial soils, C mineralization in the upper soil layer was higher in BF and lower in CL. We have found that C mineralization in our study was fitting the double exponential model and the kinetics parameters were closely related to the chemical properties of DOM and sensitive to its variation.

Cumulative CO_2-C was positively correlated with the carbohydrates while negatively correlated with phenols and the aromaticity of DOM, indicating labile compounds are preferentially utilized by microbes. Moreover, the variation of Phe (ΔPhe) and humificaton (ΔHIX$_{em}$) at the end of incubation showed reverse correlation with cumulative CO_2, suggesting in the long-term incubation stable compounds also make contribution to C mineralization.

Overall, our study suggests that in conversion from aquatic ecosystem to terrestrial ecosystem, different land use types with different vegetation cover and varied management practices lead to significant changes in the chemical structure of DOM, which can influence the C dynamics. Inconsistent trend observed in CL soils may be attributed to the tillage practices (e.g. fertilization, crop rotation).

Acknowledgments

We gratefully acknowledge Yiling Luan, Juhua Yu, Jiaojiao Guo, Zilong Ma, Yupeng Zhao and zhiqin Hua for help with field and lab work. Special thanks go to Dr. Jason Vogel for his helpful comments.

Author Contributions

Conceived and designed the experiments: HR DH. Performed the experiments: DH. Analyzed the data: DH. Contributed reagents/materials/analysis tools: DH. Wrote the paper: DH.

References

1. Leinweber P, Jandl G, Baum C, Eckhardt K, Kandeler E (2008) Stability and composition of soil organic matter control respiration and soil enzyme activities. Soil Biology & Biochemistry 40: 1496–1505.
2. Hansson K, Kleja DB, Kalbitz K, Larsson H (2010) Amounts of carbon mineralized and leached as DOC during decomposition of Norway spruce needles and fine roots. Soil Biology & Biochemistry 42: 178–185.
3. Wang Y, Ruan HH, Huang LL, Feng YQ, Zhou JZ, et al. (2010) Soil labile organic carbon with different land uses in reclaimed land area from Taihu lake. Soil Science 175: 624–630.
4. Lal R (2002) Soil carbon dynamics in cropland and rangeland. Environmental Pollution 116: 353–362.
5. Chantigny MH (2003) Dissolved and water-extractable organic matter in soils: a review on the influence of land use and management practices. Geoderma 113: 357–380.
6. Caravaca F, Masciandaro G, Ceccanti B (2002) Land use in relation to soil chemical and biochemical properties in a semiarid Mediterranean environment. Soil & Tillage Research 68: 23–30.
7. Martin D, Lal T, Sachdey CB, Sharma JP (2010) Soil organic carbon storage changes with climate change, landform and land use conditions in Garhwal hills of the Indian Himalayan mountains. Agriculture, Ecosystems and Environment 138: 64–73.
8. Tobiašová E (2011) The effect of organic matter on the structure of soils of different land uses. Soil & Tillage Research 114: 183–192.
9. Kalbitz K, Solinger S, Park JH, Michalzik B, Matzner E (2000) Controls on the dynamics of dissolved organic matter in soils: a review. Soil Science 165: 277–304.
10. Ge QS, Zhao MC, Zheng JY (2000) Land use change of China during the 20th century. Acta Geographica Sinica 55: 698–706.
11. Qin BQ, Xu PZ, Wu QL, Luo LC, Zhang YL (2005) Environmental issues of Lake Taihu, China. Hydrobiologia 581: 3–13.
12. Søndergaard M, Jeppesen E (2007) Anthropogenic impacts on lake and stream ecosystems, and approaches to restoration. Journal of Applied Ecology 44: 1089–1094.
13. Marschner B, Kalbitz K (2003) Controls of bioavailability and biodegradability of dissolved organic matter in soils. Geoderma 113: 211–235.
14. Li ZP, Han CW, Han FX (2010) Organic C and N mineralization as affected by dissolved organic matter in paddy soils of subtropical China. Geoderma 157: 206–213.
15. Cookson WR, Abaye DA, Marschner P, Murphy DV, Stockdale EA, et al. (2005) The contribution of soil organic matter fractions to carbon and nitrogen mineralization and microbial community size and structure. Soil Biology & Biochemistry 37: 1726–1737.
16. Marinari S, Liburdi K, Fliessbach A, Kalbitz K (2010) Effects of organic management on water-extractable organic matter and C mineralization in European arable soils. Soil & Tillage Research 106: 211–217.
17. Kowalczuk P, Durako MJ, Young H, Kahn AE, Cooper WJ, et al. (2009) Characterization of dissolved organic matter fluorescence in the South Atlantic Bight with use of PARAFAC model: Interannual variability. Marine Chemistry 113: 182–196.
18. Boddy E, Hill PW, Farrar J, Jones DL (2007) Fast turnover of low molecular weight components of the dissolved organic carbon pool of temperate grassland field soils. Soil Biology & Biochemistry 39: 827–835.
19. Marschner B, Bredow A (2002) Temperature effects on release and ecologically relevant properties of dissolved organic carbon in sterilised and biologically active soils. Soil Biology & Biochemistry 34: 459–466.
20. Zhao MX, Zhou JB, Kalbitz K (2008) Carbon mineralization and properties of water-extractable organic carbon in soils of the south Loess Plateau in China. European Journal of Soil Biology 44: 158–165.
21. Kalbitz K, Schmerwitz J, Schwesig D, Matzner E (2003) Biodegradation of soil-derived dissolved organic matter as related to its properties. Geoderma 113: 273–291.
22. Don A, Kalbitz K (2005) Amounts and degradability of dissolved organic carbon from foliar litter at different decomposition stages. Soil Biology & Biochemistry 37: 2171–2179.
23. Said-Pullicino D, Kaiser K, Guggenberger G, Gigliotti G (2007) Changes in the chemical composition of water-extractable organic matter during composting: Distribution between stable and labile organic matter pools. Chemosphere 66: 2166–2176.
24. Kalbitz K, Schwesig D, Schmerwitz J, Kaiser K, Haumaier L, et al. (2003) Changes in properties of soil-derived dissolved organic matter induced by biodegradation. Soil Biology & Biochemistry 35: 1129–1142.
25. Chen J, Gu BH, LeBoeuf EJ, Pan HJ, Dai S (2002) Spectroscopic characterization of the structural and functional properties of natural organic matter fractions. Chemsophere 48: 59–68.
26. Chen J, LeBoeuf EJ, Dai S, Gu BH (2003) Fluorescence spectroscopic studies of natural organic matter fractions. Chemosphere 50: 639–647.
27. Hur J, Kim G (2009) Comparison of the heterogeneity within bulk sediment humic substances from a stream and reservoir via selected operational descriptors. Chemosphere 75: 483–490.
28. Teixeira MC, Azevedo JCR, Pagioro TA (2011) Spatial and seasonal distribution of chromophoric dissolved organic matter in the Upper Paraná River floodplain environments (Brazil). Acta Limnologica Brasiliensia 23: 333–343.
29. You BS, Zhong JC, Fan CX, Wang TC, Zhang L, et al. (2007) Effects of hydrodynamics processes on phosphorus fluxes from sediment in large, shallow Taihu Lake. Journal of Environmental Sciences 19: 1055–1060.
30. Yin HB, Fan CX, Ding SM, Zhang L, Zhong JC (2008) Geochemistry of iron, sulfur and related heavy metals in metal-Polluted Taihu Lake sediments. Pedosphere 18: 564–573.
31. Blake GR, Hartge KH (1986) Bulk density. In: Klute A ed. Methods of soil analysis. Part I. Physical and mineralogical methods. Madison, WI: American Society of Agronomy and Soil Science Society of America 363–375.

32. Vance ED, Brookes PC, Jenkinson DS (1987) An extraction method for measuring soil microbial biomass carbon. Soil Biology & Biochemistry 19: 703–707.

33. Mora JL, Guerra JA, Armas CM, Rodríguez-Rodríguez A, Arbelo CD, et al. (2007) Mineralization rate of eroded organic C in Andosols of the Canary Islands. Science of the Total Enviroment 378: 143–146.

34. Tian L, Dell E, Shi W (2010) Chemical composition of dissolved organic matter in agroecosystems: Correlations with soil enzyme activity and carbon and nitrogen mineralization. Applied Soil Ecology 46: 426–435.

35. Simonsson M, Kaiser K, Danielsson R, Andreux F, Ranger J (2005) Estimating nitrate, dissolved organic carbon and DOC fractions in forest floor leachates using ultraviolet absorbance spectra and multivariate analysis. Geoderma 124: 157–168.

36. Chantigny MH, Angers DA, Kaiser K, Kalbitz K (2007) Extraction and characterization of dissolved organic matter. In: Carter MR, Gregorich EG eds. Soil sampling and methods of analysis, chap 48. CRC Press. 617–635.

37. Kalbitz K, Meyer A, Yang R, Gerstberger P (2007) Response of dissolved organic matter in the forest floor to long-term manipulation of litter and throughfall inputs. Biogeochemistry 86: 301–318.

38. Ewald M, Berger P, Visser SA (1988) UV-visible absorption and fluorescence properties of fulvic acids of microbial origin as function of their molecular weights. Geoderma 43: 11–20.

39. Zsolnay Á, Baigar E, Jimenez M, Steinweg B, Saccomandi F (1999) Differentiating with fluorescence spectroscopy the sources of dissolved organic matter in soil subjected to drying. Chemosphere 38: 45–50.

40. Kalbitz K, Geyer W (2001) Humification indices of water-soluble fulvic acids derived from synchronous fluorescence spectra-effects of spectrometer type and concentration. Journal Plant Nutrition and Soil Science 164: 259–265.

41. Kalbitz K, Geyer W, Geyer S (1999) Spectroscopic properties of dissolved humic substances-areflection of land use history in a fen area. Biogeochemistry 47: 219–238.

42. Yang LX, Pan JJ, Yuan SF (2006) Predicting dynamics of soil organic carbon mineralization with a double exponential model in different forest belts of China. Journal of Forestry Research 17: 39–43.

43. Wang XW, Li XZ, Hu YM, Lv JJ, Sun J, et al. (2010) Effect of temperature and moisture on soil organic carbon mineralization of predominantly permafrost peatland in the Great Hing'an Mountains, Northeastern China. Journal of Environmental Sciences 22: 1057–1066.

44. Hur J, Lee B, Shin H (2011) Microbial degradation of dissolved organic matter (DOM) and its influence on phenanthrene–DOM interactions. Chemosphere 85: 1360–1367.

45. Kemmitt SJ, Lanyon CV, Waite IS, Wen Q, Addiscott TM, et al. (2008) Mineralization of native soil organic matter is not regulated by the size, activity or composition of the soil microbial biomass-a new perspective. Soil Biology & Biochemistry 40: 61–73.

46. Rovira P, Vallejo VR (2007) Labile, recalcitrant, and inert organic matter in Mediterranean forest soils. Soil Biology & Biochemistry 39: 202–215.

47. Wieder WR, Cleveland CC, Townsend AR (2008) Tropical tree species composition affects the oxidation of dissolved organic matter from litter. Biogeochemistry 88: 127–38.

48. Majcher EH, Chorover J, Bollag JM, Huang PM (2000) Evolution of CO_2 during birnessite-induced oxidation of ^{14}C-labeled catechol. Soil Science Society of America Journal 64: 157–163.

49. Zsolnay Á (2003) Dissolved organic matter: artefacts, definitions, and functions. Geoderma 113:187–209.

50. Akagi J, Zsolnay Á, Bastida F (2007) Quantity and spectroscopic properties of soil dissolved organic matter (DOM) as a function of soil sample treatments: Air-drying and pre-incubation. Chemosphere 69: 1040–1046.

51. Wang LY, Wu FC, Zhang RY, Li W, Liao HQ (2009) Characterization of dissolved organic matter fractions from Lake Hongfeng, Southwestern China Plateau. Journal of Environmental Sciences 21: 581–588.

52. Corvasce M, Zsolnay A, D'Orazio V, Lopez R, Miano TM (2006) Characterization of water extractable organic matter in a deep soil profile. Chemosphere 62: 1583–1590.

53. Bu XL, Wang LM, Ma WB, Yu XN, McDowell WH, et al. (2010) Spectroscopic characterization of hot-water extractable organic matter from soils under four different vegetation types along an elevation gradient in the Wuyi Mountains. Geoderma 159: 139–146.

54. Traversa A, D'Orazio V, Senesi N (2008) Properties of dissolved organic matter in forest soils: influence of different plant covering. Forest Ecology and Management 256: 2018–2028.

55. Fuentes M, González-Gaitano G., García-Mina JM (2006) The usefulness of uv-visible and fluorescence sepectroscopies to study the chemical nature of humic substances from soils and composts. Organic Geochemistry 37: 1949–1959.

56. Khomutova TE, Shirshova LT, Tinz S, Rolland W, Richter J (2000) Mobilization of DOC from sandy loamy soils under different land use (Lower Saxony, Germany). Plant and Soil 219: 13–19.

57. Kiikkilä O, Kitunen V, Smolander A (2006) Dissolved soil organic matter form surface organic horizons under birch and conifers: Degradation in relation to chemical characteristics. Soil Biology & Biochemistry 38: 737–746.

58. Cookson WR, Murphy DV, Roper MM (2008) Characterizing the relationships between soil organic matter components and microbial function and composition. Soil Biology & Biochemistry 40: 763–777.

59. Bu XL, Ding JM, Wang LM, Yu XN, Huang W, et al. (2011) Biodegradation and chemical characteristics of hot-water extractable organic matter from soils under four different vegetation types in the Wuyi Mountains, southeastern China. European Journal of Soil Biology 47: 102–107.

60. Jacinthe PA, Lal R, Owens LB, Hothem DL (2004) Transport of labile carbon in runoff as affected by land use and rainfall characteristics. Soil and Tillage Research 77: 111–123.

61. Rodríguez-Rodríguez A, Guerra A, Arbelo C, Mora JL, Gorrín SP, et al. (2004) Forms of eroded soil organic carbon in andosols of the Canary Islands. Geoderma 121: 205–219.

62. Juarez S, Rumpel C, Mchnu C, Vincent C (2011) Carbon mineralization and lignin content of eroded sediments from a grazed watershed of South-Africa. Geoderma 167–168: 247–253.

63. Faikd HA (2000) Primary studies on lactone in bamboo leaf. MS degree Thesis. Zhejiang University, China.

64. Zhang LY (2005) The application and Study of Flavone in Bamboo Leaf. Modern Food Science and Technology 22: 247–249 (in Chinese).

65. Zhou Y (2009) Soil organic carbon pools and the characteristics of mineralization along an elevation gradient in Wuyi Mountain, China. Ph.D. Dissertation. Nanjing Forestry University, China.

66. Rey A, Jarvis P (2006) Modelling the effect of temperature on carbon mineralization rates across a network of European forest sites (FORCAST). Global Change Biology 12: 1894–1908.

67. Nguyen HV, Hur J (2011) Tracing the sources of refractory dissolved organic matter in a large artificial lake using multiple analytical tools. Chemosphere 85: 782–789.

68. Glatzel S, Kalbitz K, Dalva M, Moore T (2003) Dissolved organic matter properties and their relationship to carbon dioxide efflux from restored peat bogs. Geoderma 113: 397–411.

69. Matilainen A, Gjessing ET, Lahtinen T, Hed L, Bhatnagar A, et al. (2011) An overview of the methods used in the characterisation of natural organic matter (NOM) in relation to drinking water treatment. Chemosphere 83: 1431–1442.

70. Embacher A, Zsolnay A, Gattinger A, Munch JC (2007) The dynamics of water extractable organic matter (WEOM) in common arable topsoils: I. Quantity, quality and function over a three year period. Geoderma 139: 11–22.

71. Embacher A, Zsolnay A, Gattinger A, Munch JC (2008) The dynamics of water extractable organic matter (WEOM) in common arable topsoils: II. Influence of mineral and combined mineral and manure fertilization in a Haplic Chernozem. Geoderma 148: 63–69.

21

Applications of Low Altitude Remote Sensing in Agriculture upon Farmers' Requests– A Case Study in Northeastern Ontario, Canada

Chunhua Zhang[1]*, Dan Walters[2], John M. Kovacs[2]

1 Department of Geography and Geology, Algoma University, Sault Ste. Marie, Ontario, Canada, **2** Department of Geography, Nipissing University, North Bay, Ontario, Canada

Abstract

With the growth of the low altitude remote sensing (LARS) industry in recent years, their practical application in precision agriculture seems all the more possible. However, only a few scientists have reported using LARS to monitor crop conditions. Moreover, there have been concerns regarding the feasibility of such systems for producers given the issues related to the post-processing of images, technical expertise, and timely delivery of information. The purpose of this study is to showcase actual requests by farmers to monitor crop conditions in their fields using an unmanned aerial vehicle (UAV). Working in collaboration with farmers in northeastern Ontario, we use optical and near-infrared imagery to monitor fertilizer trials, conduct crop scouting and map field tile drainage. We demonstrate that LARS imagery has many practical applications. However, several obstacles remain, including the costs associated with both the LARS system and the image processing software, the extent of professional training required to operate the LARS and to process the imagery, and the influence from local weather conditions (e.g. clouds, wind) on image acquisition all need to be considered. Consequently, at present a feasible solution for producers might be the use of LARS service provided by private consultants or in collaboration with LARS scientific research teams.

Editor: Quazi K. Hassan, University of Calgary, Canada

Funding: Funding for this project was provided by a grant (project #920161) awarded to John M. Kovacs and Dan Walters from the Northern Ontario Heritage Fund Corporation of Canada (http://nohfc.ca/en). The funders had no role in study design, data collection and analysis, decision to publish, or preparation of the manuscript.

Competing Interests: The authors have declared that no competing interests exist.

* Email: chunhua.zhang@algomau.ca

Introduction

With the primary objective of matching agricultural practice with crop and soil conditions, the use of Precision Agriculture (PA) technologies is considered one of the key directions in modern agriculture development. Some of the perceived benefits of PA include increasing crop yield and efficiency by lowering the costs associated with fertilizer, pesticides, herbicides, and fungicides. An additional socio-economic benefit of PA is reducing the transport of agriculture inputs on the air, soil and water. To date, considerable progress has been made in reducing the application of fertilizer, insecticides and fungicides using Variable Rate Technologies (VRT) and Global Positioning Systems (GPS). However, one main remaining challenge is the ability to obtain up-to-date crop/soil condition data (e.g., nutrient deficiency, water stress, pests, disease) for VRT. Historically, yield maps from yield monitors had been applied to create zonal maps for VRT machines [1–3]. However, these maps are normally obtained once a year and often the large variation observed make the reliability of the zonal maps limited [4]. Moreover, these types of yield maps are only available after the season, and many harvesters are still not equipped with yield monitors [5].

Alternatively, remotely sensed imagery obtained during the growing season could be utilized to extract crop condition

information for management purposes in a timely fashion. In addition, yield maps derived from these data could be used as an alternative for yield maps from harvesters [5]. In particular, high spatial resolution satellite imagery can provide crop and soil condition information for management adjustment. For example, a variety of satellite data, including IKONOS, QuickBird, GeoEye-1 and WorldView-2, have been successfully applied in crop yield predictions [6–16]. However, image availability is highly restricted for these sensors due to weather condition and the satellites' poor temporal resolution. Moreover, the spatial resolution of these satellite images is limited with the highest resolution for commercial satellite data (WorldView-2 and GeoEye-1) at approximately 50 cm for the panchromatic band. Although quite good, this spatial resolution along with the limited spectral resolution of the panchromatic band might be not sufficient for examining within-field variations of crop condition and yield.

With finer spatial resolution and real-time monitoring capability [5], airborne multispectral [8,17–19] and hyperspectral [11,20] sensors had been applied to monitor crop conditions and yield. Aerial imagery has been shown to be as effective as high resolution satellite imagery in monitoring spatial variation of crop condition and yield. Furthermore, the rapid development of Low Altitude Remote Sensing Systems (LARS) over the past decade makes its application for PA possible. In 2000, Inoue et al. [21] collected crop images using a Charge-Coupled Device (CCD) camera on-board a blimp to measure biomass and Leaf Area Index (LAI) variation within rice and soybean fields. The results from their study showed that it might be plausible to apply LARS images in studying crop biological parameters. More recently, research scientists at the US Department of Agriculture have been conducting experiments using a fixed wing UAV to monitor various crop characteristics. Specifically, Hunt et al. [22] used a color digital camera on-board a radio controlled model aircraft to collect images of a corn field in order to examine the relationships between Normalized Green Ratio Difference Index (NGRDI), biomass and corn nitrogen status. Similarly, Hunt et al. [23] assessed the relationships between LAI and Green Normalized Difference Vegetation Index (GNDVI) for a wheat field. More recently, Hunt et al. [24] used a modified digital camera on-board a LARS to take high-resolution (i.e. 2.7 and 5.1 cm) color-infrared pictures of two winter wheat fields. They assessed the spectral information with ground collected biophysical data to demonstrate the scientific feasibility of applying LARS to monitor within-field crop variations. Most recently, Primicerio et al. [25] employed an ADC-lite camera on-board a UAV to acquire photos of a vineyard. They were able to convert digital numbers to reflectance and then calculated NDVI to display vineyard vigor. Peña et al. [26] and Torres-Sánchez et al. [27] both applied UAV images to map weeds in corn and sunflower fields, respectively.

While a number of sensors/cameras are available for LARS, optical (either metric or commercial scale) or infrared (metric or commercial with modified filter to record near infrared radiation [22,23]) are the most commonly used for crop monitoring. Thermal infrared sensors have been shown to be useful for monitoring soil moisture or stress [28–30] and, most recently, hyperspectral sensors on board a UAV were used to examine leaf carotenoid content [31]. From the aforementioned studies, the number of crop types examined using LARS is still limited, mainly rice [21,32,33], soybean [21], wheat [24,34], sunflower [27] and corn [22,26,35].

The studies to date demonstrate the scientific feasibility of LARS applications for monitoring crops. LARS appears capable of resolving the spatial resolution restrictions of satellite imagery. However, there are several key limitations apparent in such studies including the small spatial coverage and the image processing of the LARS data. For example, in Canada transportation regulations restrict the operating height of LARS, which means a large number of images need to be collected for each field. Depending on the percent of front- and side-lap of the images, a 30-acre field may require over 300 images. Moreover, because of the relative homogeneity of crops in the field, it is difficult to mosaic the images [24,36]. Hunt et al. [24] reported that calculating NDVI or other vegetation indices from LARS image mosaics is challenging. For example, the same crop feature in several images could have different digital numbers due to changes in the incident angles and/or the atmospheric transmittance [24]. Consequently, most published LARS investigations focus on each image separately and not as an image mosaic [21–23,25].

There appear to be mixed messages about the practical applications of LARS for PA. On one hand, the scientific research demonstrates the ability to quantify relationships between crop biomass [33] and water stress [29] with the digital numbers (or reflectance values) acquired from LARS imagery, which would suggest a very practical use for crop monitoring. On the other hand, the analyses are most often carried out on each image separately, which would not be practical for producers who may require hundreds, if not thousands of images to monitor their

fields. In addition, there are very few examples of applied applications of LARS for crop monitoring in the literature and none so far based on actual requests from producers. Working in collaboration with cash crop producers, we use case studies based in northeastern Ontario, Canada, to explore and describe some applications of LARS mosaic imagery for crop monitoring: scouting, emergency response, and field trials.

Study Area

This research takes place in the clay belt area within in the West Nipissing District of northeastern Ontario, Canada. The main cash crops grown in this region are soybean (*Glycine max*), wheat (*Triticum* spp.), barley (*Hordeum vulgare*), oat (*Avena sativa*) and canola (*Brassica napus*). The annual mean temperature is 3.8°C and the annual mean length of the growing season is 180 days, with a frost-free period of only 120 days. On average, the last spring frost is May 15, and the first fall frost is September 15. The annual precipitation is 1008 mm, in which 273 mm is snow. Agricultural production is influenced by acidic soil, which requires limestone to neutralize the soil pH. Even though the growing season is relatively short, fast growing crop varieties have shown to be successful for this region [37]. With such a short growing season it is necessary to monitor the field crop conditions in a timely fashion. Moreover, the large acreage and scattered distribution of the fields, typical of this region, makes personal visits and scouting of the fields a challenge [38,39].

Producer requests were drawn from members of the North Eastern Ontario Soil and Crop Improvement Association (NEOSCIA). During the 2013 growing season and the spring of 2014, the research team was contacted several times by Steve Roberge of Ferme Roberge and Mitch DesChatelets of Leisure Farms. They requested that we analyze fertilizer field trials, field tile drainage conditions, crop damage from an armyworm [*Spodoptera frugiperda*] infestation, and lodging following a storm event. These farmers gave us permission to fly over their respective farms. Federal permission to fly the LARS over this region of Ontario was granted by Transport Canada (Special Flight Operations Certificate (SFOC) # 5812-15-33-2012-1).

Equipment and Methods

For this study, the UAV system, developed by *Aeryon Labs Inc.*, Canada, consisted of a graphical, touch-screen control station (Figure 1), an aerial vehicle (Aeryon Scout) (Figure 2), and a radio repeater station to extend the control station's transmission range. This aerial vehicle is a commercially available quadrocopter UAV that can be equipped with both an optical and infrared camera. The Aeryon Scout has a maximum flight time of 25 minutes with a communication range of 3 km. The flyer has a maximum ground speed of approximately 50 km/hr and can remain stable in gusts exceeding 60 km/hr. Rechargeable lithium polymer batteries power the flyer and base station. The control station allows the user to create flight plans that can be reused at a later date. This aerial vehicle collects GPS/INS data for each photo, which are later used to orthorectify and create a mosaic image.

Optical images were captured using the Photo3S optical camera (*Aeryon Labs Inc.*, Canada) and near infrared images using an ADC-lite camera (*Tetracam*, United States) that affix to the Aeryon Scout. Both cameras use a Bayer filter to record the radiance from the ground targets. The Photo3S optical camera has three bands: blue, green and red. The images captured with this camera are stored in the flyer and then transferred to a hard drive after landing. The ADC-lite has three bands: near infrared, red, and green. The images captured with the ADC-lite are stored

Figure 1. The touch-screen control station for the Aeryon Scout UAV.

Figure 2. The Aeryon Scout quadrocopter.

directly on a flash card located in the camera. The flight altitude was set at 120 meters, as per our SFOC. Consequently, the spatial resolution for the Photo3S optical and NIR images were 3.5 cm and 5 cm, respectively. Given the homogeneity of the crop fields the front overlap and side lap were 85% and 65% respectively. The high overlapping flight path helps to improve the efficacy of post-flight mosaic processing. The Aeryon Scout can cover close to 0.1 km^2 (~25 ac) per battery charge, flying at 12 km/hr. Moreover, although the rechargeable batteries can be replaced quite quickly, the Aeryon Scout needs to return to the landing/takeoff location during battery replacement.

Ground Control Points (GCP) were set up to help in the orthorectification and georeferencing of the final mosaic images. Each GCP was made of a 30 by 30 cm foam pad placed on the top of a wood stake at a height of 1.5 m. Based on the size of the field, 6 to 8 GCPs were dispersed throughout the field prior to each flight. Locations of the GCPs were recorded using a Trimble GeoXH GPS (*Trimble*, United States). Coordinates recorded from this GPS unit had a positional accuracy of less than 10 cm after real time differential analysis.

For each mission, a team of three was required. One person operated the control unit for the planning and operation of the LARS, while the two others were responsible for flight observation (i.e. spotting aircraft or other potential hazards) and the distribution and collection of the GCPs. When using the ADC-lite infrared camera, a photo of a white Teflon calibration plate was taken upon takeoff for calibrating images taken. *Pixelwrench2* (*Tetracam, USA*) software was used to convert each raw image to a jpeg file and to calibrate the image. Field validation was done at the time the images were being taken. The optical and infrared imagery were then orthorectified and mosaicked using Pix4d Mapper (*Pix4D, Switzerland*) software. Pix4D was also used to generate NDVI images of fields. For the Leisure Farms soybean field a stratified random sample based on 1 m radius sample plots was used to statistically examine the differences in NDVI between the three fertilizer treatments. Specifically, a One-Way Analysis of Variance (ANOVA) was applied to mean values of 18 sample plots generated from treatment areas A and B and 27 sample plots generated from treatment area C (Figures 3&4).

Results and Discussions

The application of UAV imagery to assess fertilizer treatments

There have been several studies demonstrating the benefits of organic manure on soil quality and crop production. For example, adding compost has shown to increase crop production and improve soil fertility [40–43]. However, it is often necessary for producers to conduct their own trials in order to determine the economic feasibility of such products. In 2013 a local producer (Leisure Farms) conducted a trial test of an organic fertilizer on a soybean field to test the economic feasibility. He requested that the research team image of the field prior to an annual mid-season crop tour conducted by the members of the Nipissing District branch of the Ontario Soil and Crop Improvement Association (OSCIA). The three distinct fertilizer treatments were observed in the field (Figure 3). The producer had applied only organic fertilizer (9.37 L/ha or 1 gallon/acre) in section A of the field, whereas in section C a conventional chemical fertilizer (3–14–45, 371.25 kg/ha or 330 lb/acre) was applied. Section B (i.e. middle strip) was treated with a mix of organic (9.37 L/ha or 1 gallon/acre) and chemical fertilizer (185.53 kg/ha or 165 lb/acre). The research team flew his soybean field on a clear day (July 12, 2013) 42 days after seeding. The crop height was approximately 30 cm at the time image acquisition. The mosaicked image (Figure 3) shows a large contrast between the organic treatment and chemical fertilizer treatment. The section treated with only organic fertilizer had the weakest vegetation vigor and consequently appears much darker in the infrared image. The NDVI values are significantly lower than those of the chemical fertilizer treatment (P<0.001, Figure 3). However, there is no statistical difference between the strips of half organic/half chemical (B) and normal chemical fertilizer (C) application (P = 0.59). The observed variability within each treatment area could have been due to soil types, soil moisture content, or other factors. The large patch of high vigor (section D) in the southern section of the field was the result of operator forgetting to turn off the fertilizer spreader. The NDVI difference between treatments B and C were not detectible at the early stages of growth in July 12th imagery. The images taken at later growth stages (August 29, 2013, 90 days after seeding, Figure 4) show greater variability among the three treatments. Significant differences (P<0.001) were observed between treatments A and C, and B and C. While the differences between treatment areas A and B were not statistically significant (P = 0.07), the P values is really close to the critical value of 0.05. Consequently, it is possible that a flight between these two dates would have provided better discrimination of the treatment areas.

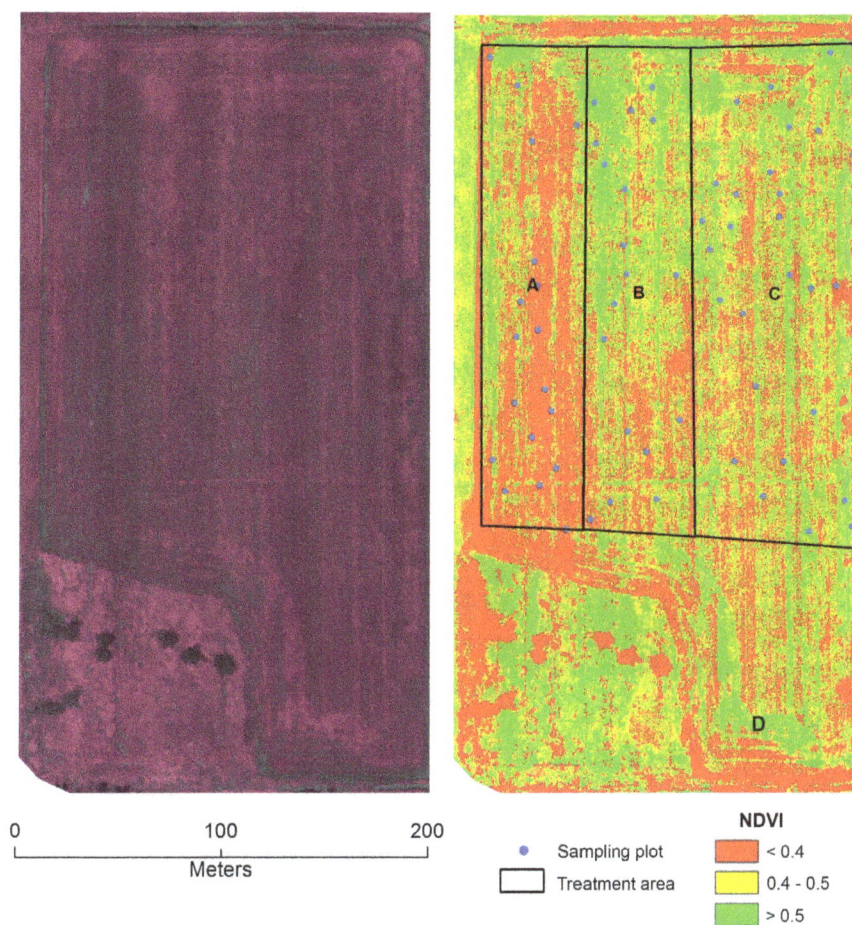

Figure 3. Mosaicked image map based on UAV images of a soybean field in Sturgeon Falls, ON, Canada (79°56′51″E, 46°20′14″N) taken on July 12, 2013. The image map on the left is a mosaicked infrared color composite image (NIR, red, green-no enhancement applied) and the image map on the right a mosaicked NDVI image. The A, B, and C represent treatment areas of organic only, organic and chemical fertilizer and chemical fertilizer only applications, respectively. D indicates a fertilizer application error. The final yields for the treatment areas A, B and C were calculated at 1.73, 2.27 and 2.97 tons/ha, respectively.

The application of UAV images in identifying area of lodging and insect infestation

Fall armyworm is an agricultural pest more typical of tropical and subtropical regions. However, a cool, wet spring followed by warm, humid weather and heavy rainfall favor the propagation of fall armyworm in more temperate regions [44]. On average, one caterpillar needs 140 cm^2 of leaf area to develop through 6 instars [44]. However, the 6th instar itself requires 77.2% of that leaf area. Consequently, the producer in this study only recognized and reported the armyworm infestation at this stage growth. For many crops, including wheat, the fall armyworms tend to consume only the succulent parts of the leaves with the main midribs intact following the infestation (Figure 5). As a result the leaf area of the field or parts of the field drops significantly in a relatively short period of time. Given this type of damage it is believed that armyworm movement/impacts could be assessed using high resolution remotely sensed imagery. For areas infested within a wheat field, the reflectance in the NIR band should decrease whereas that of the red band should increase due to the loss of flag leaves and increased exposure of the soil surface and shadows. During the 2013 growing season the producer notified the research group that his wheat field was hit by the fall armyworm on July 31, 2013 when the wheat crop was at BBCH stage of 83.

Consequently, a mission over the field was taken the following day under somewhat cloud covered conditions.

Lodging, or stem breakage, is a very common type of cereal crop damage which results from stormy weather events and the inadequate standing power of the crop during certain growth stages (i.e. heavy seed heads). Consequently, high nitrogen fertilization may cause plants to be more susceptible to lodging. For the Nipissing district a relatively strong storm event occurred on July 19, 2013 resulting in significant lodging within the same armyworm infested field (Figure 6). In the infrared image, the lodged areas appear as a bright red tone (Figure 7). The lodged wheat covers the bare soil and consequently there is stronger reflectance from wheat leaves and stalks in the IR band which results in the large contrast between the lodged and non-lodged areas.

In addition, it is also quite easy to identify stressed areas on and around the rock outcrop area (Figures 6 and 7). During the field trip the crops on the shallow soils were dead and are shown by a dark tone in the NIR images. The research team provided the producer with a mosaicked hard copy image and knowledge regarding how to interpret the data. The information was actually used to determine whether the producer should invest in equipment required to lift the lodged heads during harvesting.

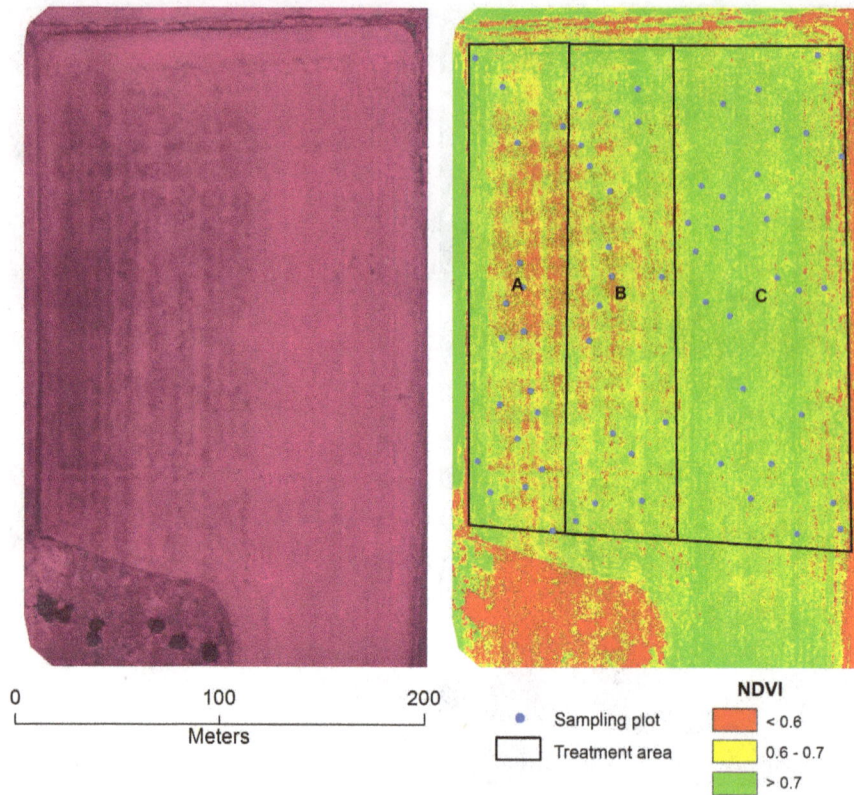

Figure 4. Mosaicked image map based on UAV images of a soybean field in Sturgeon Falls, ON, Canada (79°56′51″E, 46°20′14″N) taken on August 29ᵗʰ, 2013. The image map on the left is a mosaicked infrared color composite image (NIR, red, green-no enhancement applied) and the image map on the right a mosaicked NDVI image. The A, B, and C represent treatment areas of organic only, organic and chemical fertilizer and chemical fertilizer only applications, respectively. D indicates a fertilizer application error. The final yields for the treatment areas A, B and C were calculated at 1.73, 2.27 and 2.97 tons/ha, respectively.

Using UAV images to identify a field tile drainage network

The main soil type for this area of northeastern Ontario is clay and the topography is nearly level with gentle slopes of 1–2%. The combination of clay and flat terrain has led to drainage problems for local producers. Consequently, field tile drainage systems are commonly installed to reduce the risk of crop loss from excess water and provide more uniform crop production amidst climate variability [45]. In addition, producers have higher flexibility in field operations (e.g., planting, drier harvest conditions, less soil compaction, and a wider choice of crops and crop varieties) [45,46]. Good drainage can also reduce the frequency of pests and disease outbreak [45]. On tiled land, producers are able to obtain a modest return [47]. Once installed these systems need to be monitored and maintained and thus it is important for the farmer to know the exact location of their tiles. However, such information is not always available to farmers particularly when ownership of the field changes. In the Nipissing district it is common that the contractors only provide the producers with hand-drawn maps of the drainage system. In Ontario, the Ministry

Figure 5. A comparison of the result of an armyworm attack. The arrows indicate the difference in the flag leaf of the infested (left) versus the healthy (right) wheat plants. Only the mid-rib of the flag leaves remains on the infested plants.

Figure 6. A picture showing lodging and crop stress in a wheat field on Roberge Farms. The photograph was taken between locations C and D in Figure 5, pointing south.

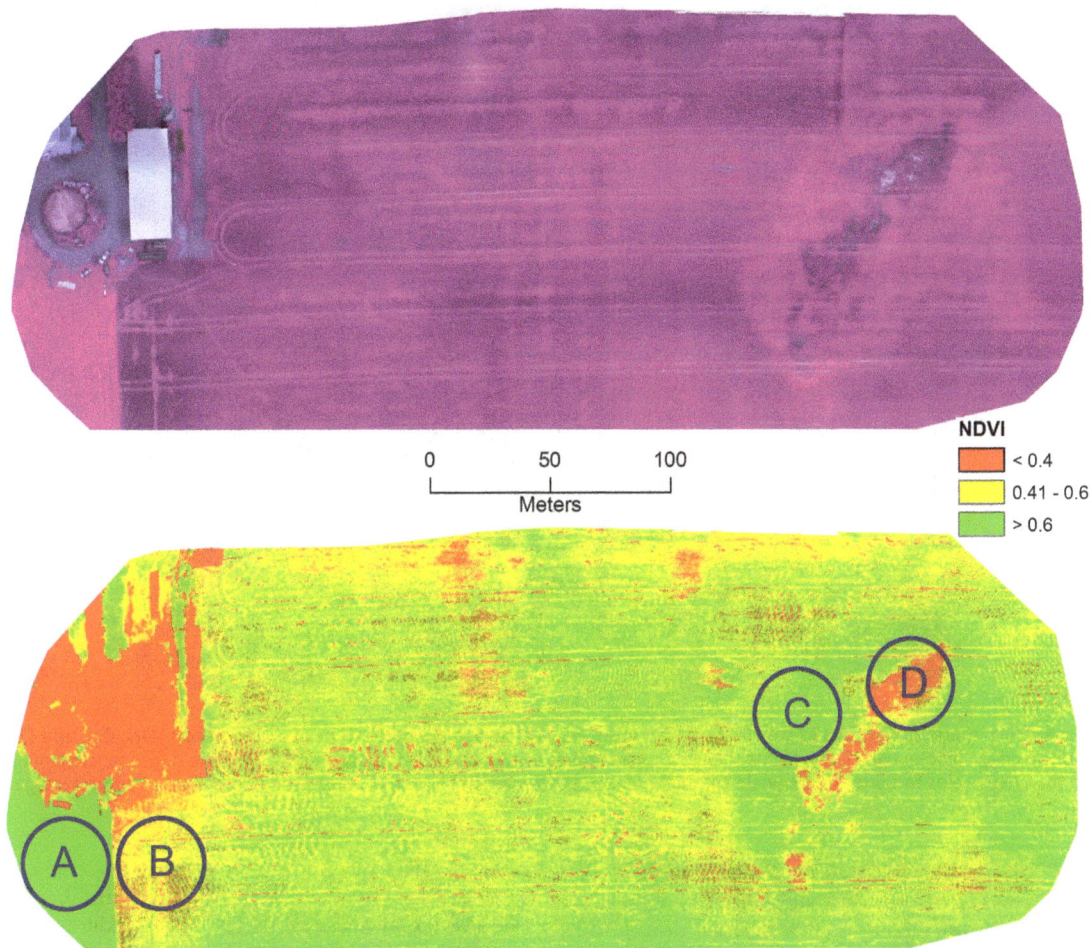

Figure 7. The top image is a mosaicked infrared color composite map (NIR, red, green-no enhancement) of a wheat field located in Verner, ON, Canada (80°5′50″E, 46°22′35″N) that was stricken by army worms and lodging taken on July 31, 2013. The bottom image is the corresponding NDVI derived map. The A indicates a healthy non-infested alfalfa field, the B indicates a section of the wheat crop hit by army worms, the C shows an area of lodging and D indicates a rock outcrop.

of Agriculture and Food normally maintains tile drainage information but access to their GIS database revealed very little coverage for this region of Ontario.

The owner of Leisure Farms had two fields tiled in 2012 but was not able to obtain maps of the location of the tiles from the contractor. Consequently, he requested the research team identify the tile locations prior to seeding. On April 28, 2014 images were collected using the UAV system, processed and mosaicked. The mosaic image was then converted from a tiff file format to a KMZ and e-mailed to the producer for interpretation on Google Earth on the same day. In a follow up phone conversation with the producer we were able to identify locations of some of the tiles in the image which depicted a brighter tone with the expected linear feature. Areas well drained were drier and consequently look brighter (Figure 8). The interpretation is also validated by the fact that the field tile drainage network was located at the expected 50 feet (15 m) on centre interval. Further, we were also able to identify some drainage problems (i.e. excessive wetness) in the field possibly resulting from a poor grade during installation (Figure 8). Interpretation was possible in part to the bare soil present at this time. We were not able to identify the tile drainage system for another of his fields due to the presence of a residual straw cover. Based on the positive results, the producer suggested that we might receive more requests from other local producers to identify field tile locations and drainage problems.

Weather conditions and UAV image acquisition

Weather conditions are critical for remote sensing acquisitions and unfortunately the growing seasons are typically the rainy seasons for many parts of the world. For example, the City of North Bay, located only 50 km east of the study area, experienced 30 rainy days in the 77 days from June 1st to August 16th of 2013. In addition, there were ten days with trace amounts of precipitation. Consequently, sky conditions can considerably hinder the availability of satellite imagery during the peak-growing season. In fact, our research group had requested a WorldView-2 image for the study area during the growing season for two consecutive years (2012–2013) without any success due to persistent cloudy conditions.

In comparison to satellite and high altitude aerial remote sensing, LARS has a higher degree of flexibility with regards to image acquisition. Although it is best to take imagery during a cloud free period, LARS images were successfully collected under full cloud cover. However, the impacts of varying solar radiation on the image should be considered for each task particularly when creating large mosaics based on hundreds of individual images.

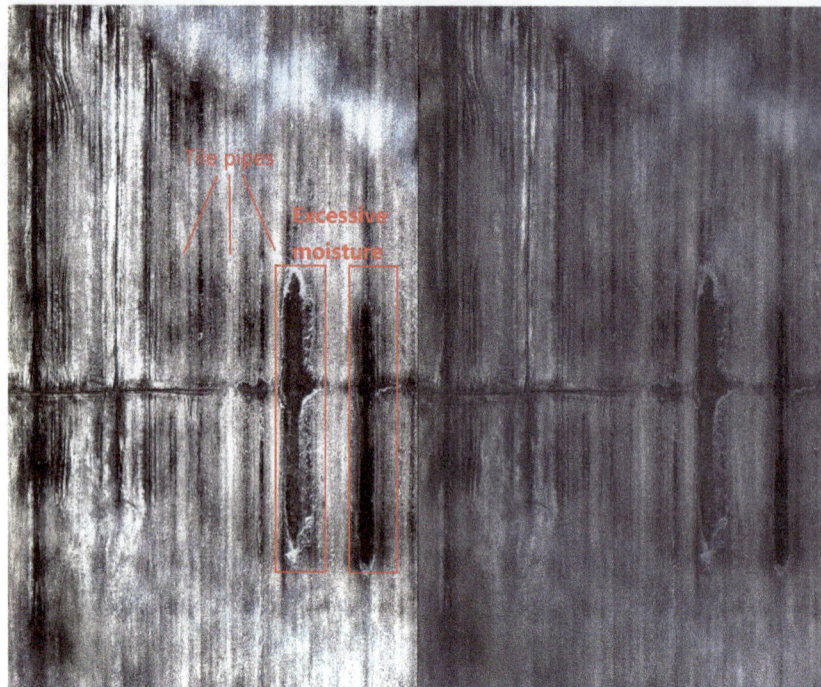

Figure 8. Mosaicked image maps, based on UAV optical images, of a bare field located in Sturgeon Falls, ON, Canada (79°56′51″W, 46°20′44″N). The image maps were used to locate the tile drainage pipes and to identify faulty drainage pipes. A linear enhancement applied to the optical image map (left) helps to better discriminate the tile drainage configuration in comparison to the non-enhanced optical image map (right).

Factors impacting LARS adoption for PA

At present one of the major factors impeding the adoption of LARS is the cost. Currently it is expensive to own LARS equipment with the flyers alone costing between US$20,000 and US$70,000 [48]. The camera cost range from several hundred dollars for a standard commercial camera to upwards of US$7,000 for a metric infrared camera. The maintenance cost should also be considered. For example, the quadrocopter propellers can be damaged and rechargeable batteries have a limited number of cycles. Consequently, an additional US$5,000 should be budgeted for LARS maintenance. For this study an additional software expense was incurred to mosaic the imagery. Finally the costs of liability insurance must also be considered if the LARS is to be flown for commercial or research purposes in Canada and other jurisdictions. Fortunately, insurance was covered through our institution at no additional cost.

Another current limitation for the wide use of LARS for PA is the personnel required. A team of two individuals, one operator and one spotter, were required as part of the SFOC. However, it is recommended that at least three people be present. Specifically, a trained and qualified person needs to be responsible for the assembling, operation and disassembly of the UAV, a spotter is required and it is suggested that a third person set up the GCPs, measure the reference targets spectral responses, and act as a second spotter. In Canada a SFOC is also required for all commercial and research uses of UAV. The UAV team should also be able to mosaic images and georeferencing them shortly after image acquisition. In this investigation we determined that one of the team members needs to spend roughly one hour to download all of files associated with the UAV images, the GPS coordinates, and the spectral measurements. Moreover, another two to four hours, depending on the number of pictures collected, are required to orthorectify these images.

Besides the requirement of short image processing, certain skills in image interpretation or classification are also necessary for effective use of a LARS for PA. Most producers would require image interpretation training. The collection and processing of UAV images in a timely fashion is a key obstacle for their practical application. For our tile drainage example, we received the request from the producer on April 21, 2014. The fight was possible only due to ideal weather conditions and other logistical issues. The travel time from our institute to the study area is roughly 40 minutes. We went to the field on the morning of April 28, 2014 and finish four flights in just under two hours. We had to then travel back to our lab, download all the images and process on the same date. Two KMZ files were sent to the farmer that same day. In this tile drainage case, skills of image interpretation are very important [48]. The farmer was not able to visually identify tiles from the image mosaic until we sent him a KMZ file with our digitization of the tiles.

Feedback from farmers on the application of UAV image in PA

During the process of image acquisition, we had many discussions with the farmers regarding the issues of LARS applications in agriculture. The owner of Leisure Farms was extremely satisfied with the mosaicked imagery we provided to him for the soybean fertilizer trials and tile drainage maps. Based on the results he decided not to bother including his soybean treatment in the annual crop tour of 2013. Moreover, he was impressed with the ability to identify the field tile drainage runs. He anticipated we would receive request for this service from other farmers in the area. Based on the extent of lodging in Roberge Farms wheat fields (approximately 13%) Steve Roberge decided to purchase a lift fork to harvest lodged wheat. According to these farmers, a fast response is the key for the application of UAV imagery in PA, especially for the requests of insect damage and other crop stress scenarios. Consequently it is critical to set up a routine procedure for image capture and processing. As a result, image processing (mosaic, georeferencing and interpretation) should be completed in one or two days with feedback directed to the farmer as quickly as possible. Considering the operating and processing costs of a LARS, it is more practical for a third-party to own the LARS and to provide the service.

Conclusion

This paper examined the feasibility of applying UAV acquired images for monitoring crop conditions based on the actual requests from producers. The results suggest that it is plausible to obtain images and process them in a timely fashion for PA applications. However, due to current costs and operational logistics the

application is still in its infancy stage. A fast adoption of UAV systems should occur as the costs of LARS decrease and more experienced personnel, possibly a service industry, are available to acquire and process these data in a timely fashion.

Acknowledgments

The authors would like to acknowledge the assistance of Ferme Roberge and Leisure Farms of the West Nipissing Agricultural District and the North Eastern Ontario Soil and Crop Improvement Association.

Author Contributions

Conceived and designed the experiments: DW JMK. Performed the experiments: DW JMK CZ. Analyzed the data: CZ DW JMK. Wrote the paper: CZ DW JMK.

References

1. Blackmore S (2000) The interpretation of trends from multiple yield maps. Comput Electron Agr 26: 37–51.
2. Diker K, Heermann DF, Bordahl MK (2004) Frequency analysis of yield for delineating yield response zones. Precis Agric 5: 435–444.
3. Flowers M, Weisz R, White JG (2005) Yield-based management zones and grid sampling strategies: Describing soil test and nutrient variability. Agron J 97: 968–982.
4. Blackmore S, Godwin RJ, Fountas S (2003) The analysis of spatial and temporal trends in yield map data over six years. Biosys Eng 84: 455–466.
5. Yang C, Everitt J, Qian D, Luo B, Chanussot J (2013) Using High-Resolution Airborne and Satellite Imagery to Assess Crop Growth and Yield Variability for Precision Agriculture. P IEEE 101: 582–592
6. Wiegand CL, Rhoades JD, Escobar DE, Everitt JH (1994) Photographic and videographic observations for determining and mapping the response of cotton to soil salinity. Remote Sens Environ 49: 212–223.
7. Lelong CCD, Pinet PC, Poilvé H (1998) Hyperspectral imaging and stress mapping in agriculture: A Case Study on Wheat in Beauce (France). Remote Sens Environ 66: 179–191.
8. Yang C, Anderson GL (1999) Airborne videography to identify spatial plant growth variability for grain sorghum. Precis Agric 1: 67–79.
9. Shanahan J, Schepers J, Francis D, Varvel G, Wilhelm W, et al. (2001) Use of remote sensing imagery to estimate corn grain yield. Agron J93: 583–589.
10. Chang J, Clay DE, Dalsted K, Clay S, O'Neill M (2003) Corn (Zea mays L.) yield prediction using multispectral and multidate reflectance. Agron J 95: 1447–1453.
11. Godwin RJ, Richards TE, Wood GA, Welsh JP, Knight SM (2003) An economic analysis of the potential for precision farming in UK cereal production. Biosyst Eng 84: 533–545.
12. Seelan SK, Laguette S, Casady GM, Seielstad GA (2003) Remote sensing applications for precision agriculture: a learning community approach. Remote Sens Environ 88: 157–169.
13. Yang C, Everitt JH, Bradford JM (2006) Comparison of QuickBird satellite imagery and airborne imagery for mapping grain sorghum yield patterns. Precis Agric 7: 33–44.
14. Yang C, Everitt JH, Bradford JM (2006) Evaluating high resolution QuickBird satellite imagery for estimating cotton yield. T ASAE 49: 1599–1606.
15. Inman D, Khosla R, Reich R, Westfall DG (2008) Normalized difference vegetation index and soil color-based management zones in irrigated maize. Agron J 100: 60–66.
16. Lopez-Lozano R, Baret F, de Cortazar-Atauri IG, Bertrand N, Casterad MA (2009) Optimal geometric configuration and algorithms for LAI indirect estimates under row canopies: The case of vineyards. Agr For Meteorol 149: 1307–1316.
17. Yang C, Everitt JH, Bradford JM, Escobar DE (2000) Mapping grain sorghum growth and yield variations using airborne multispectral digital imagery. T ASAE 43: 1927–1938.
18. Yang C, Everitt JH, Bradford JM (2004) Airborne hyperspectral imagery and yield monitor data for estimating grain sorghum yield variability. TASAE 47: 915–924.
19. Yang C, Everitt JH, Bradford JM, Murden D (2004) Airborne hyperspectral imagery and yield monitor data for mapping cotton yield variability. Precis Agric 5: 445–461.
20. De Tar WR, Chesson JH, Penner JV, Ojala JC (2008) Detection of soil properties with airborne hyperspectral measurements of bare fields. T ASABE 51: 463–470.
21. Inoue Y, Morinaga S, Tomita A (2000) A blimp-based remote sensing system for low-altitude monitoring of plant variables: a preliminary experiment for agricultural and ecological applications. Int J Remote Sens 21: 379–385.
22. Hunt ER, Cavigelli M, Daughtry CST, McMurtrey J, Walthall CL (2005) Evaluation of digital photography from model aircraft for remote sensing of crop biomass and nitrogen status. Precis Agric 6: 359–378.
23. Hunt ER, Hively WD, Daughtry CST, McCarty GW, Fujikawa SJ, et al. (2008) Remote sensing of crop leaf area index using unmanned airborne vehicles. In: Proceedings of the Pecora 17 Symposium, Denver, CO, November 18, 2008. Bethesda, MD: American Society for Photogrammetry and Remote Sensing.
24. Hunt ER, Hively WD, Fujikawa SJ, Linden DS, Daughtry CST, et al. (2010) Acquisition of NIR-Green-Blue digital photographs from unmanned aircraft for crop monitoring. Remote Sens 2: 290–305.
25. Primicerio J, Gennaro SF, Fiorillo E, Genesio L, Lugato E, et al. (2012) A flexible unmanned aerial vehicle for precision agriculture. Precis Agric 13: 517–523.
26. Peña JM, Torres-Sánchez J, de Castro AI, Kelly M, López-Granados F (2013) Weed Mapping in Early-Season Maize Fields Using Object-Based Analysis of Unmanned Aerial Vehicle (UAV) Images. PLoS ONE 8(10): e77151
27. Torres-Sánchez J, López-Granados F, De Castro AI, Peña-Barragán JM (2013) Configuration and specifications of an unmanned aerial vehicle (UAV) for early site specific weed management. PLoS One 8: e58210.
28. Ryo S, Noguchi N, Ishii K (2007) Correction of low-altitude thermal images applied to estimating of soil water status. Biosyst Eng 96: 301–313.
29. Berni JAJ, Zarco-Tejada PJ, Suarez L, Fereres E (2009) Thermal and narrowband multispectral remote sensing for vegetation monitoring from an unmanned aerial vehicle. IEEE Geosci Remote S 47: 722–738.
30. Zarco-Tejada PJ, Gonzalez-Dugo V, Berni JAJ (2012) Fluorescence, temperature and narrowband indices acquired from a UAV platform for water stress detection using a micro-hyperspectral imager and a thermal camera. Remote Sens Environ 117: 322–337.
31. Zarco-Tejada PJ, Guillén-Climent ML, Hernández-Clemente R, Catalina A, González MR, et al. (2013) Estimating leaf carotenoid content in vineyards using high resolution hyperspectral imagery acquired from an unmanned aerial vehicle (UAV). Agr For Meteorol 171: 281–294.
32. Swain KC, Jayasuriya HPW, Salokhe VM (2007) Suitability of low-altitude remote sensing images for estimating nitrogen treatment variations in rice cropping for precision agriculture adoption. J Applied Remote Sens 1: 013547.
33. Swain KC, Thomson SJ, Jayasuriya HPW (2010) Adoption of an unmanned helicopter for low-altitude remote sensing to estimate yield and total biomass of a rice crop. T ASABE 53: 21–27.
34. Honkavaara E, Saari H, Kaivosoja J, Polonen I, Hakala T, et al. (2013) Processing and Assessment of Spectrometric, Stereoscopic Imagery Collected Using a Lightweight UAV Spectral Camera for Precision Agriculture. Remote Sens 5: 5006–5039.
35. Link J, Senner D, Claupein W (2013) Developing and evaluating an aerial sensor platform (ASP) to collect multispectral data for deriving management decisions in precision farming. Comput Electron Agr 94: 20–28
36. Gomez-Candon D, Lopez-Granados F, Caballero-Novella JJ, Gomez-Casero M, Jurado-Exposito M, et al. (2011) Geo-referencing remote images for precision agriculture using artificial terrestrial targets. Precis Agric 12: 876–891.
37. Wilson J, Zhang C, Kovacs JM (2014) Separating crop species in northeastern Ontario using hyperspectral data. Remote Sens 6: 925–945.
38. Cable JW, Kovacs JM, Jiao X, Shang J (2014) Agricultural monitoring in northeastern Ontario, Canada, using multi-temporal polarimetric RADARSAT-2 data. Remote Sens 6: 2343–2371.
39. Cable JW, Kovacs JM, Shang J, Jiao X (2014) Multi-temporal polarimetric RADARSAT-2 for land cover monitoring in northeastern Ontario, Canada. Remote Sens 6: 2372–2392.
40. McSorley R, Gallaher RN (1996) Effect of yard waste compost on nematode densities and maize yield. J Nematol 28: 655–660.

41. Mamo M, Rosen CJ, Hallbach TR, Moncrief JF (1998) Corn yield and nitrogen uptake in sandy soils amended with municipal solid waste compost. J Production Agric 11: 469–475.

42. Stratton ML, Rechcigl JE (1998) Organic mulches, wood products and composts as soil amendments and conditioners. In: Wallace, A., Terry, R.E. (Eds.), Handbook of Soil Conditioners. Marcel Dekker, New York, NY pp: 43–95.

43. Keener HM, Dick WA, Hoitink HAJ (2000) Composting and beneficial utilization of composted by-product materials. In: Power, J.F., Dick, W.A. (Eds.), Land Application of Agricultural, Industrial, and Municipal By-products. Soil Science Society of America, Madison, WI, pp: 315–341.

44. Sparks AN (1979) A review of the biology of the fall armyworm. Florida Entomologist, 82–87.

45. Blann KL, Anderson JL, Sands GR, Vondracek B (2009) Effects of agricultural drainage on aquatic ecosystems: a review. Crit Rev Env Sci Tec 39: 909–1001.

46. Spaling H, Smit B (1995) Conceptual model of cumulative environmental effects of agricultural land drainage. Agr Ecosyst Environ 53: 299–308.

47. Zucker LA, Brown LC (1998) Agricultural drainage: Water quality impacts and subsurface drainage studies in the Midwest. Bulletin, 871–98. University of Minnesota Extension, St. Paul, Minn.

48. Zhang C, Kovacs JM (2012) The application of small unmanned aerial systems for precision agriculture: A review. Precis Agric 13: 693–712.

22

Biochar from Sugarcane Filtercake Reduces Soil CO$_2$ Emissions Relative to Raw Residue and Improves Water Retention and Nutrient Availability in a Highly-Weathered Tropical Soil

Angela Joy Eykelbosh[1]*, **Mark S. Johnson**[1,2], **Edmar Santos de Queiroz**[3], **Higo José Dalmagro**[4], **Eduardo Guimarães Couto**[3]

1 Institute for Resources, Environment, and Sustainability, University of British Columbia (UBC), Vancouver, British Columbia, Canada, **2** Department of Earth, Ocean, and Atmospheric Sciences, University of British Columbia (UBC), Vancouver, British Columbia, Canada, **3** Faculdade de Agronomia, Medicina Veterinária e Zootecnia (FAMEV), Universidade Federal de Mato Grosso, Cuiabaá, Mato Grosso, Brazil, **4** Instituto de Física, Universidade Federal de Mato Grosso, Cuiabá, Mato Grosso, Brazil

Abstract

In Brazil, the degradation of nutrient-poor Ferralsols limits productivity and drives agricultural expansion into pristine areas. However, returning agricultural residues to the soil in a stabilized form may offer opportunities for maintaining or improving soil quality, even under conditions that typically promote carbon loss. We examined the use of biochar made from filtercake (a byproduct of sugarcane processing) on the physicochemical properties of a cultivated tropical soil. Filtercake was pyrolyzed at 575°C for 3 h yielding a biochar with increased surface area and porosity compared to the raw filtercake. Filtercake biochar was primarily composed of aromatic carbon, with some residual cellulose and hemicellulose. In a three-week laboratory incubation, CO$_2$ effluxes from a highly weathered Ferralsol soil amended with 5% biochar (dry weight, d.w.) were roughly four-fold higher than the soil-only control, but 23-fold lower than CO$_2$ effluxes from soil amended with 5% (d.w.) raw filtercake. We also applied vinasse, a carbon-rich liquid waste from bioethanol production typically utilized as a fertilizer on sugarcane soils, to filtercake- and biochar-amended soils. Total CO$_2$ efflux from the biochar-amended soil in response to vinasse application was only 5% of the efflux when vinasse was applied to soil amended with raw filtercake. Furthermore, mixtures of 5 or 10% biochar (d.w.) in this highly weathered tropical soil significantly increased water retention within the plant-available range and also improved nutrient availability. Accordingly, application of sugarcane filtercake as biochar, with or without vinasse application, may better satisfy soil management objectives than filtercake applied to soils in its raw form, and may help to build soil carbon stocks in sugarcane-cultivating regions.

Editor: Ben Bond-Lamberty, DOE Pacific Northwest National Laboratory, United States of America

Funding: Funding was obtained from the UBC Bridge Program and the National Sciences and Engineering Research Council (NSERC) of Canada (awarded to AJE), and from the Conselho Nacional de Desenvolvimento Cientifico e Tecnologico (CNPq) of Brazil (awarded to AJE, MSJ, and EGC). The funders had no role in study design, data collection and analysis, decision to publish, or preparation of the manuscript.

Competing Interests: The authors have declared that no competing interests exist.

* E-mail: a.eykelbosh@alumni.ubc.ca

Introduction

The principal challenge for agricultural sustainability and feeding an ever-growing population is avoiding the degradation of soil, a vital but slowly renewable resource. In Brazil, economic and social drivers have come together to push agricultural expansion into pristine regions [1]. In addition to land use change in the Amazon and Atlantic forests, the savannah or Cerrado of Brazil's center-west has come under intense conversion, with an estimated 50% of this highly biodiverse region already converted to cropping and pasture, primarily within the last 40 years [2]. Conversion has increased the risk of soil degradation through soil organic carbon (SOC) loss, deteriorating soil structure, and increased risk of erosion [3], with knock-on effects for other ecosystem services.

Given that agriculture is central to Brazil's economic development, and that agricultural products from this region will help to feed and fuel the world, innovative agricultural technologies and practices are critically required to prevent soil degradation. Practices such as conservation agriculture have and will continue to play an important role [4], but supplementary strategies have also been proposed. One of these is to enhance soil storage of black carbon, which in natural soils has been associated with increased fertility, water retention, and more rapid incorporation or stabilization of new carbon inputs [5–7]. Black carbon in soil can be enhanced through amendment with man-made charcoal or *biochar* derived from the pyrolysis of waste biomass, through which relatively labile plant-fixed carbon is transformed into a highly recalcitrant char. Although most work to date has focused on temperate soils, studies incorporating biochar into tropical soils have shown evidence of enhanced soil quality, nutrient availability, and productivity, and decreased leaching losses and acidification [8–11].

Sugarcane cultivation in particular may benefit from this approach because the industry produces several pyrolyzable residues. These include bagasse (crushed cane stalks), cane trash (leaves and stalk tips removed during harvest), and filtercake, a sludge that is removed via filtration after the juice clarification step. Previous biochar studies on sugarcane residues have focused exclusively on bagasse, with positive results in terms of soil quality improvements and productivity [8,9]. However, bagasse is a valuable by-product, both under current operations (*e.g.*, for cogeneration of heat and electricity in distilleries) and as a lignocellulosic feedstock for second-generation biofuels [12]. Thus, diverting bagasse for biochar production and soil amendment would come at the cost of other gains, and as such bagasse biochar may be less likely to be implemented as a strategy for soil and carbon management.

Filtercake, in contrast, is a heavy, nutrient-dense residue that is sometimes spread on fields as a fertilizer [13], although the high biological availability of its components likely leads to the rapid mineralization and potential leaching losses common in tropical soils [14]. Filtercake management is hampered by its high water content, which makes it costly to transport and difficult to apply, as well as its potential contribution to nutrient run-off and eutrophication when over-applied [12]. Thus, conversion of this highly labile residue into biochar may be an opportunity to turn a nuisance waste into a valuable soil amendment.

In this study, we developed a benchtop reactor to produce biochar from sugarcane filtercake for research purposes. Filtercake biochar was then characterized and its effects on nutrient availability and water retention analyzed. We also examined the effects of raw and pyrolyzed filtercake amendment on CO_2 efflux in combination with vinasse, a carbon-rich effluent resulting from distillation that is similarly disposed of via soil application [13] and may itself affect carbon cycling through priming of native soil C [15]. Our analyses revealed several ways in which filtercake biochar is distinct from more commonly used woody and herbaceous biochars. Soil quality assays indicated that filtercake biochar may be useful as a soil carbon management tool with agronomic benefits for nutrient and water availability.

Materials and Methods

Material collection and biochar production

Permission to collect soil samples and organic residues (filtercake and vinasse) from private property was granted by the management of Usina Pantanal de Açúcar e Álcool Ltda, Jaciara, Mato Grosso, Brazil (15°55'28.11"S, 55°13'38.94"W, elevation 690 m). The soil was a red-yellow Ferralsol (FAO taxonomy) with a sandy clay loam texture (64% sand, 9% silt, 27% clay) and total carbon (C) and nitrogen (N) contents of 1.5% and 0.07% dry weight (d.w.), respectively. The mineral soil of the top 0–10 cm was collected and sieved to 2 mm. Vinasse was collected from canals close to the point of application and aliquots were frozen at −20°C until use. Vinasse total carbon concentration (1291 mg C L^{-1}) was determined via combustion of the lyophilized solid residue to obtain % C (CHN-1110 elemental analyzer, Carlo Erba Instruments, Italy) and then multiplied by total solids (mg L^{-1}).

Filtercake was collected fresh from the distillery and dried at 45°C for four days in a forced-air oven. The dried material was gently crushed and sieved to collect size fractions of 2–4 and 4–10 mm. Approximately 850 g of this material (50% from each size class) was pyrolyzed in a custom-made benchtop biochar reactor.

The reactor consisted of a 10-×50-cm steel cylinder (diameter × length) closed on one end with a circular steel plate. The closed end was perforated with a 1/4″-brass male compression fitting to serve as an inlet for N_2, which was used to purge the reactor of oxygen. The flanged open end was closed using a gasket made of high-temperature fibreglass cord and a steel plate held in place with a grooved circular clamp. The removable steel plate was perforated with two additional brass fittings, one for exhaust and a second to allow placement of a type-K thermocouple (Omega Engineering Inc., Stamford, CT) for monitoring internal temperature. These apertures were protected by a 1-mm stainless steel mesh at either end of the cylinder.

This assembly was mounted inside a large Linn Elektric muffle furnace with a programmable controller (KK 260 SO 1060; Linn High Therm GmbH, Germany). The slow pyrolysis program was as follows: slow heating to 575°C at a rate of 5–6°C min^{-1}, holding at 575°C for 3 h, followed by slow cooling overnight to room temperature. The entire program was carried out under oxygen-limited conditions (N_2 purge, 0.5–1 L min^{-1}). The program was selected based on previous studies showing that this heating rate, pyrolysis temperature, and time were conducive to creating a biochar (from bagasse) retaining >500 mg C g^{-1} biochar [8,9,16–19].

Biochar characterization

Biochar pH was determined by mixing 3 g of biochar in 27 g of water [17], followed by 30 min of intermittent shaking, 30 min of settling, and measurement with a handheld meter (HI 98121, Hanna Instruments, USA). Total C and total N contents were determined on a CHN analyzer (628 Series, LECO Corp., St. Joseph, MI).

In the carbohydrate analysis, filtercake and biochar samples were finely ground and extracted in hot acetone for 12 h to remove extractives (fats, resins, etc.). Next, 0.2 g of the dried extracted sample were incubated for 2 h in 3 mL of 72% H_2SO_4 (at 20°C), and then diluted to a final concentration of 4% H_2SO_4 and autoclaved at 121°C for 1 hour, followed by filtration. The filtered hydrolysate was then analyzed for carbohydrates via high-performance liquid chromatography, as described in detail by Huntley *et al.* [20].

Total and micropore surface area for both the initial feedstock (raw filtercake) and filtercake biochar were determined using the Brunauer–Emmett–Teller (BET) method [21], calculated from the N_2 adsorption isotherm captured using an ASAP 2020 Physisorption analyzer (Micromeritics, Norcross, GA). Before analysis, samples were degassed at 90°C for 1 hour under a vacuum. For scanning electron microscopy (SEM), samples were sputter-coated with gold and analyzed using a Shimadzu SSX-550 microscope with an accelerating voltage of 15 kV. Finally, Raman spectroscopy was performed using a LabRAM HR system (HORIBA Jobin Yvon S.A.S., France). Spectra were captured using a 442-nm laser source, with a laser power on the sample surface of 1 mW and a 100× objective lens. Samples were acquired over an integration time of 60 s with two scans per sample. Data were analyzed using LabSpec 5 software (HORIBA Jobin Yvon S.A.S., France).

Nutrient availability in soil mixtures

Macro- and micronutrient availability in biochar–soil mixtures were analyzed in solutions prepared at 1:2.5 ratios (biochar–soil mixture:distilled water). Analyses were carried out according to the standard soil methodologies published by the Empresa Brasileira de Pesquisa Agropecuaária (EMBRAPA) [22]. These included the following determinations: available P, Mehlich I extraction with spectrophotometric determination; K^+ and Na^+, Mehlich I extraction with flame photometry; Zn, Cu, Mn, diethylenetriaminepentaacetic (DTPA) extraction followed by atomic absorption; Fe, Mehlich I extraction with atomic absorption; Ca^{2+} and Mg^{2+}, KCl (0.1 M) extraction followed by the complexometric titration method (EDTA method); Al^{3+}, KCl (0.1 M) extraction followed by titration with NaOH; S, Ca_2PO_4 extraction followed by spectrophotometric determination; B, hot water extraction followed by colorimetric determination. Cation-exchange capacity (CEC) was calculated as the sum of exchangeable bases (Ca^{2+}, Mg^{2+}, and K^+) plus potential acidity ($H^+ + Al^{3+}$).

Water retention in soil-biochar mixtures

Water potential in soil–biochar mixtures (0, 5, or 10% biochar on a dry weight basis) were analyzed using a WP4C Dewpoint Potentiameter (Decagon Devices Inc., Pullman, WA). Briefly, each of the oven-dried soil–biochar treatments ($n = 3$ for each treatment) were divided into 5 sub-samples, which were then moistened with 0, 0.5, 1.0, 1.5, 2.0, or 3.0 mL of ultrapure water. These were allowed to equilibrate for at least 16 h, and then ~0.5 g of each sub-sample were analyzed for soil water potential. Immediately afterward, each sub-sample was weighed, dried at 105°C overnight, and weighed again to determine soil water content. To facilitate the statistical analysis of replicates, water potential values (in pF) were plotted against binned soil water contents.

CO2 efflux assay

The mineralization of raw vs. pyrolyzed filtercake was analyzed in the laboratory. Treatments were established as follows: soil alone (S); soil +5% filtercake (SF); soil +5% biochar (SB). To minimize disturbance, fresh, field-moist soil was sieved to 4 mm, roots were removed by hand, and then the soil was gently mixed to ensure a uniform composition; soil for all treatments was handled to an equal degree. Equal weights of these treatments were packed into incubation columns (10×15 cm, D×L; $n = 6$ each) and left open to the atmosphere in a temperature-monitored environment. CO2 efflux was determined on a daily basis for three weeks using a LI-COR 6400 XT apparatus (LI-COR, Lincoln, NE), which was fitted over the incubation column using a PVC sleeve and a foam rubber gasket. CO2 measurements were collected over three weeks under three conditions: week 1, field-moist (no added water); week 2, 30 mL water or vinasse (low dose); week 3, 60 mL water or vinasse (high dose). Vinasse and water were applied by slowly and uniformly dripping liquid over the soil surface using a pipette. CO2 fluxes were monitored over time and total CO2 flux and % C released was determined. Incubations were carried out at ambient temperature, ranging from 25–29°C during the night and 29–35°C during the day.

Statistical analyses

Treatment effects were analyzed using one-way analysis of variance (ANOVA) with post-hoc Tukey tests to detect significant differences among means. All data are presented as the mean ± the standard error (SE). A p value <0.05 was considered significant. All statistical analyses were carried out in R v.3.0.1 [23] using Rcmdr [24].

Results

Filtercake biochar production and characterization

The pyrolysis of raw filtercake yielded what appeared to be a fully pyrolyzed product (biochar yield = 36% dry feedstock weight), as well as an unquantified amount of bio-oil. Filtercake

biochar was alkaline in water (pH = 9.85±0.08) and contained 36.7% C dry weight, and 1.3% N dry weight (C:N, 28), whereas raw filtercake contained 37.4% C and 1.2% N by weight (C:N, 32). Pyrolysis greatly depleted but did not wholly eliminate carbohydrates. Raw filtercake contained 18% cellulose (d.w.) and 11.9% hemicellulose, as well as 32% lignin; filtercake biochar retained 2.7% cellulose and 3.1% hemicellulose (lignin not determined).

The biochar was analyzed by Raman spectroscopy, which revealed a characteristic two-peak spectrum (**Fig. 1**) typically displayed by carbonaceous materials with a core aromatic structure, including biochars [e.g., 25]. The first peak, originally designated the D or *defect* band (1360 cm^{-1}), was initially associated with disordered graphite structure, but has since been associated with benzene rings [26]. The G or *graphite* band (1590 cm^{-1}) is characteristic of graphite; however, in the case of chars, previous work showing the lack of detectable graphitic C via X-ray diffraction led authors to attribute this peak instead to quadrant aromatic ring breathing [27]. These results suggest that the filtercake biochar produced is primarily composed of aromatic C.

Finally, N_2-BET surface analysis revealed that slow pyrolysis increased the specific surface area of raw filtercake from 0.38 m^2 g^{-1} to 26.3 m^2 g^{-1} in the charred product. Analysis of deBoer t-plot micropore area revealed that only approximately 5.5 m^2 g^{-1} of this surface area was contained within pores <2 nm in diameter. Scanning electron microscopy of both the raw filtercake and pyrolyzed product revealed differences in particle size due to fracturing during pyrolysis (**Fig. 2A,B**). Biochar particles were angular with large irregular macropores. At high magnification (2400×), some biochar particles showed scant formation of pores <1 μm in diameter (**Fig. 2C**).

Effect on soil quality parameters and nutrient extractability

Mixing an air-dried dystrophic red-yellow Ferralsol with increasing amounts of filtercake biochar led to increases in soil pH, cation exchange capacity, and carbon, nitrogen, and nutrient availability for increasing levels of biochar (**Table 1**). Even the low dose of 1.25% biochar markedly increased the availability of P, K, and Ca. Among micronutrients, the addition of biochar most strongly affected the extractability of Fe, Mn, and Zn, although these effects were only apparent at a treatment of 5%. This

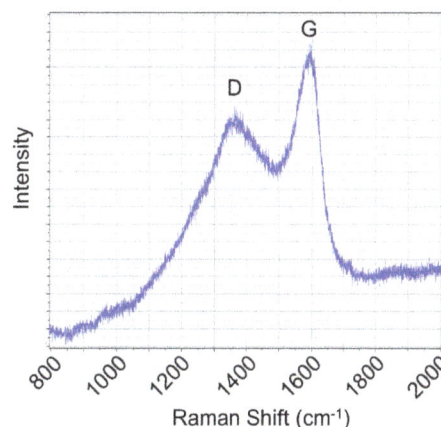

Figure 1. Raman spectrograph of filtercake biochar. D, defect band (1360 cm^{-1}); G, graphite band (1590 cm^{-1}).

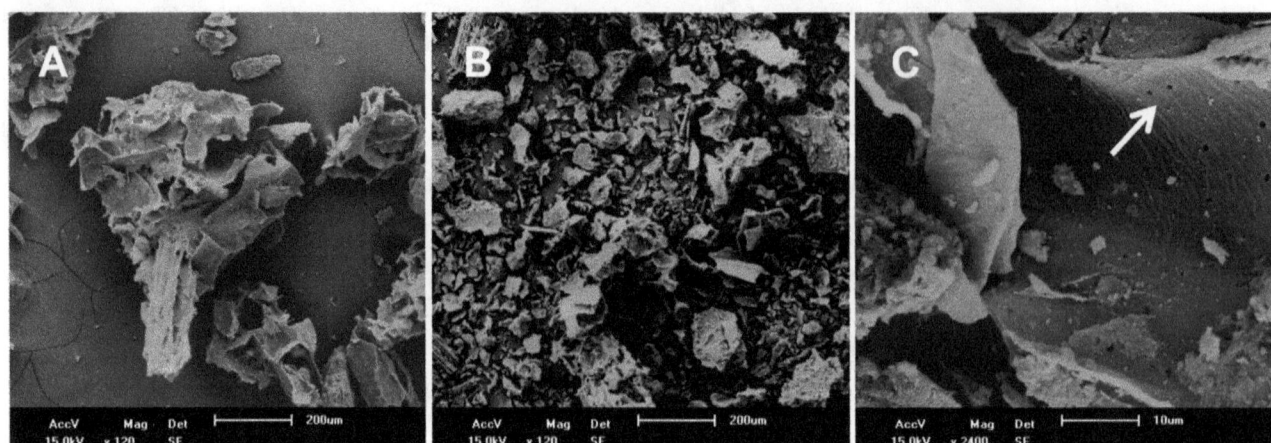

Figure 2. Scanning electron micrographs of raw filtercake and biochar prepared via slow pyrolysis at 575°C. (A) Raw sugarcane filtercake at a magnification of 120×. (B,C) Filtercake biochar at magnifications of 120× and 2400×, respectively. White arrow indicates micropore formation.

increase in extractable nutrients may be due to a direct contribution from the biochar amendment, as well as increases in nutrient availability due to the pH change resulting from the addition of biochar.

Water retention

The effect of biochar on soil water retention was examined by generating matric potential–soil moisture curves for soil alone and soil mixed with 5 or 10% biochar. The addition of 5 or 10% biochar led to a dose-dependent increase in soil matric potential at a given soil moisture content; importantly, this effect was only observed within the range of soil water potentials coinciding with plant-available water (pF<4.18), and grew more pronounced at the wetter end of the curve (**Fig. 3**). These data suggest that

filtercake biochar increased the availability of pores that release water at potentials less than pF = 4.18 (pores >0.2 μm in diameter) [28]. This is consistent with our N_2-BET surface area analysis, which found that this filtercake biochar showed very little surface area in pores <0.2 μm in diameter, which is desirable in terms of retaining plant-available water.

Short-term CO_2 effluxes

To examine the effect of sugarcane residues on soil respiration and carbon utilization, field-moist soil was mixed with filtercake or biochar and incubated for three weeks. During the first 7 days, a large initial efflux of CO_2 was observed for soil amended with filtercake (SF), which peaked after two days; the soil plus biochar (SB) and soil-only treatment showed a more muted response that

Table 1. Physicochemical characterization of soil–biochar mixtures.

Parameters	% Biochar (by weight)			
	0%	1.25%	2.5%	5%
pH (H2O)	6.13±0.03, d	6.70±0.06, c	7.03±0.03, b	7.40±0.06, a
CEC	8.0±0.2, b	9.0±0.2, b	10.5±0.2, a	11.6±0.4, a
% C	1.48±0.06, b	—	—	2.65±0.03, a
% N	0.07±0.01, a	—	—	0.08±0.01, a
Macronutrients				
P	2.5±1.0, b	107.8±5.8, a	146.0±22.8, a	151.6±17.8, a
K	46.7±0.7, d	114.3±3.2, c	164.3±3.5, b	243.0±10.8, a
Ca	3.1±0.1, c	4.9±0.1, b	6.1±0.1, a	6.7±0.3, a
Mg	1.5±0.1, c	2.0±0.1, c	2.7±0.2, b	3.5±0.2, a
S	13.1±0.7, a	13.9±0.1, a	13.1±0.1, a	13.2±0.3, a
Micronutrients				
B	0.26±0.02, a	0.27±0.02, a	0.29±0.01, a	0.28±0.01, a
Fe	115.0±12.8, b	173.7±14.5, ab	195.7±18.8, ab	223.7±29.6, a
Mn	7.1±0.9, b	8.0±0.9, b	10.1±0.7, ab	12.9±0.5, a
Zn	2.8±0.6, b	3.7±0.6, b	4.3±0.2, b	6.4±0.1, a

All statistical comparisons were performed using one-way analysis of variance with a *post hoc* Tukey test. Data represent the mean ± SE. P and K values are given in mg dm^{-3}. Remaining nutrient values and CEC and all other nutrients are given in cmol$_c$ dm^{-3}.

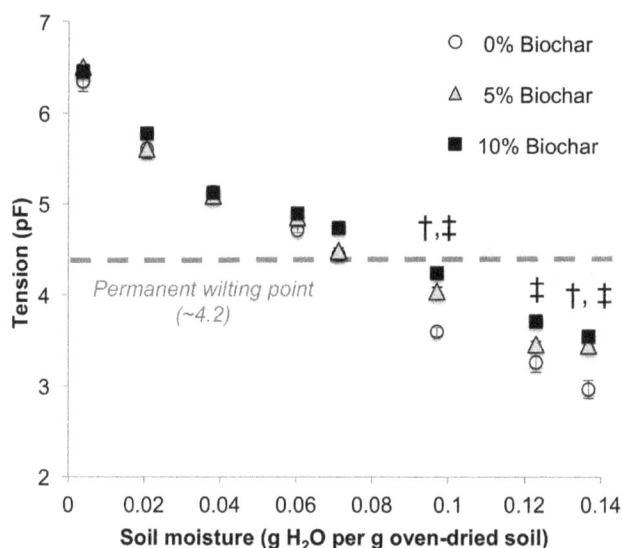

Figure 3. Soil matrix potential and moisture content in soils containing 0, 5, or 10% filtercake biochar. Treatments were compared using one-way ANOVA with a *post hoc* Tukey test for samples in each soil moisture bin. Data represent the mean ± SE. Note that significant differences were observed only below the permanent wilting point (dashed line). †, significant with respect to 5% biochar treatment; ‡, significant with respect to 10% biochar, $p = 0.05$.

began to decrease immediately (**Fig. 4**). Respiration was also moisture-limited, as evidenced by brief increases in CO_2 efflux that accompanied water or vinasse addition (gray arrows in **Fig. 4**). Among treatment groups, soil-only (S) incubations showed the lowest mean efflux, whereas biochar incubations (SB) were slightly but consistently higher over time. Vinasse application increased CO_2 efflux in the SV and SVB incubations relative to the S and SB controls, respectively. After calculating total g CO_2 emissions over the entire experimental period, we found that the addition of raw filtercake to soil (SF) led to a 100-fold increase in g CO_2 emitted relative to the unamended soil (S) control. However, this large increase was greatly ameliorated by applying the filtercake as a pyrolyzed product (**Fig. 5**).

To further investigate the effects of biochar and vinasse on soil C emissions, data were analyzed with respect to percent C lost from the treatment, based on a calculated estimate of the amount of soil carbon plus amendment carbon added. This revealed that filtercake treatments lost a much greater percentage of total C (SF, 16.3±0.4% C), most likely due to a much larger proportion of filtercake feedstock that was enriched in cellulose and hemicellulose (approx. 55% of total C). Vinasse application with filtercake had no additional effect (SVF 16.2±0.2%), perhaps because the soil had reached its respiratory maximum for the available amount of water. Because the powerful effect of filtercake addition greatly skewed the data set, we examined the differences amongst the S, SV, SVB, and SB treatments separately using oneway ANOVA. This revealed that although total carbon loss from soil-only incubations was low (S, 0.35±0.02% of total C present), vinasse addition did significantly increase overall % C lost (SV, 0.70±0.02%, $p = 0.003$; **Table 2**), as would be expected given this highly labile substrate. Biochar addition alone (SB) also led to a minor increase in the total C lost (0.69±0.15%, $p = 0.003$ with respect to S), perhaps due to increased aeration as a result of a slight decrease in soil bulk density (S, 1.08 g cm^{-3}; SB, 1.04 g cm^{-3}, $p<0.001$). Vinasse addition did not augment this effect

(SVB, 0.78±0.04%) compared to biochar alone (SB, 0.69±0.15%, $p = 0.54$).

Discussion

It is frequently noted in the biochar literature that feedstock and pyrolysis conditions are the determining factors in final char characteristics [9,16,17,29]. Here, we compare filtercake biochar with its related product, bagasse biochar as reported in the literature, as well as other feedstocks, and compare these in terms of their agronomic benefits.

Biochars are often proposed as a means to enhance soil black carbon stocks. In natural soils, black carbon plays an important role as a geosorbent and microhabitat [30], a modifier of soil chemistry and carbon cycling [6,7], and as a slowly available C source for microbial activity [31]. In undisturbed Cerrado ecosystems, black carbon produced through natural fires makes up a small but stable pool within the total soil carbon stock; however, this stock is depleted after disturbance [32].

In this study, filtercake biochar produced under the described pyrolysis conditions falls under the category of a low-carbon char, according to the biochar classification proposed by Joseph *et al.* [29]. This is somewhat surprising given that our feedstock was relatively rich in lignin (32%) compared to softwoods and other non-woody residues [33]. Previous work has shown that lignin has a higher char yield (C in feedstock retained in char, 49%) compared to cellulose (19%) or hemicellulose (23.5%) after pyrolysis at 800°C [34], and feedstocks rich in lignin often produce chars with higher % C [17,35]. For example, bagasse with an initial lignin content of approximately 20% [12,33], produces chars in the range of 63–84% C at 600°C [8,9,18]. Thus, the filtercake char described here had less C than might be expected for a lignin-rich material pyrolyzed near to 600°C. Because filtercake has been subjected to fermentation, the depletion of biodegradable non-lignin C may account for both the overall low C yield in the biochar product as well as lignin enrichment in the filtercake.

Nevertheless, filtercake biochar may be very valuable as an agricultural amendment. Regarding its effects on nutrients, treatments as a low as 1.25% biochar (d.w.) significantly increased P, K, and Ca availability and treatment at 5% increased micronutrient availability (**Table 2**) in a dystrophic red-yellow Ferralsol. Furthermore, in an agricultural context, the presence of a small amount of non-pyrolyzed carbohydrates can be viewed as a benefit as these compounds represent a source of bioavailable nutrients and energy for the soil microbial community.

Surface area is a key physical characteristic of biochars because it indirectly indicates a char's ability to retain water as well as dissolved nutrients and low-molecular weight carbon compounds through pore filling [36]. In this study, we noted that filtercake biochar pyrolyzed at 575°C had a high N$_2$-BET surface area (26 m^2 g^{-1}) compared to the raw filtercake feedstock (0.38 m^2 g^{-1}), but a relatively low N$_2$-BET surface area compared to bagasse pyrolyzed at close to the same temperature (218 m^2 g^{-1} at 600°C) [9]. Further SEM analyses revealed highly angular, macroporous particles (Fig. 2b), demonstrating why biochar addition decreased bulk density and increased porosity in treated soils. Pore formation depends largely on production parameters and original biomass structure [37]. Although some remnants of the original plant vasculature were observed through SEM (data not shown), filtercake biochar overall lacked the regular, highly porous structure observed in bagasse biochar made at near the same temperature [17,18]. Because filtercake is subject to

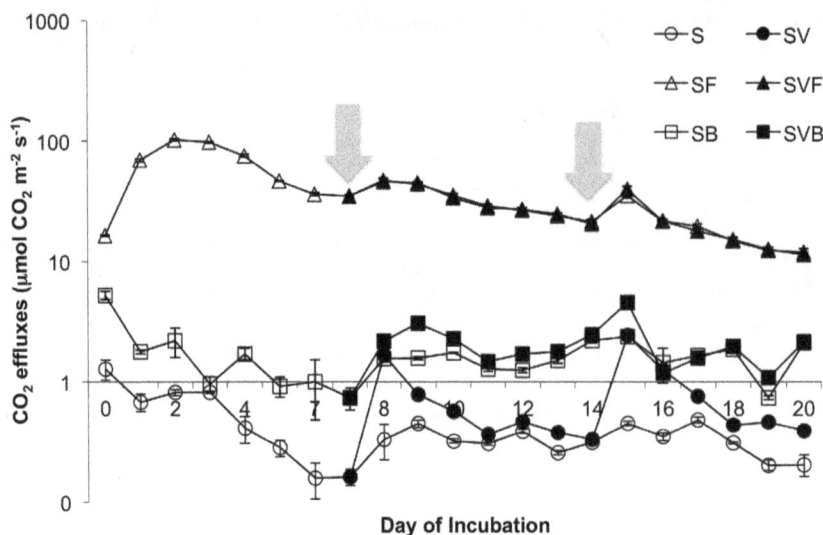

Figure 4. CO_2 effluxes from soil treated with 5% raw filtercake or 5% biochar under field-moist conditions. Data points represent the mean treatment value ± SE at each time point. Note that data are plotted on a logarithmic scale. Water (open symbols) or vinasse (solid symbols) was added on the days indicated by a gray arrow; 30 mL were added at the beginning of week 2 and 60 mL at the beginning of week 3. S, soil; SV, soil with vinasse; SF, soil with 5% filtercake (d.w.); SVF, soil with 5% filtercake and vinasse; SVB, soil with 5% biochar and vinasse; SB, soil with 5% biochar.

maceration and fermentation, destruction of the plant's vascular structure may have contributed to relatively low total surface area.

Nevertheless, even this small increase in total surface area and porosity over the raw feedstock may underlie the increase in soil matrix potential in soil-biochar mixtures. Our results showed that biochar dose-dependently increased the soil matric potential for a given quantity of water added, indicating that water was held more tightly, and this effect was significant within the range of plant available water (pF<4.18). This work is consistent with numerous previous studies showing that biochar increases plant-available water, especially in sandy soils similar to the one examined here [9,38–40]. Given that modeling studies have predicted a warmer, drier future for this region of Brazil [41], utilizing filtercake as

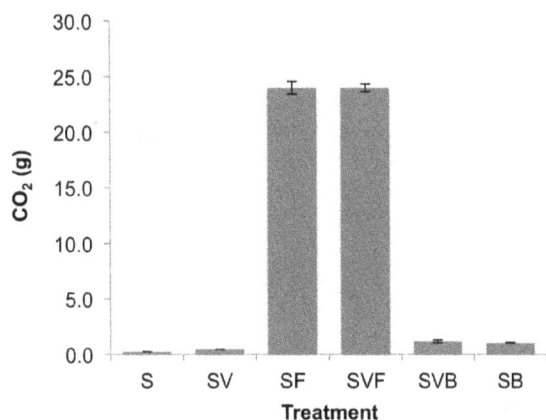

Figure 5. Total CO_2 effluxes from soil and soil amended with filtercake, biochar and/or vinasse. Treatments included the control (soil, S), soil amended vinasse (SV), soil with 5% filtercake (d.w.) (SF), soil amended with filtercake and vinasse (SVF), soil amended with 5% biochar (d.w.) produced from filtercake (SB), and soil amended with biochar and vinasse (SVB). These data show the strong effect of applying filtercake as biochar on reducing CO_2 effluxes (e.g., SF vs. SB and SFV vs. SBV).

biochar to increase plant-available water may be a useful climate adaptation strategy.

Biochar is also posited as a possible climate mitigation strategy, based on the assumption that aromatic carbon within biochar is highly resistant to microbial attack, and may therefore increase carbon sequestration [42]. However, the effect of biochar on soil respiration and carbon turnover is complex. Although some studies have reported that biochar suppresses or has no effect on soil respiration [16,43], others report a transient or sustained increase in CO_2 efflux [16] and microbial biomass [44] over the study periods, suggesting a biological response. We also noted a small but sustained increased in CO_2 efflux from biochar-amended soils relative to the untreated control, and that biochar-treated soils lost a minor, though overall greater percentage of total C compared to the unamended soil (**Table 2**). This may indicate a priming effect, which is defined as an increase in the turnover of existing soil organic carbon in response to carbon addition that is mediated by soil microbes [15]; these priming effects occur in many systems and their ecological consequences are not well understood. Alternatively, increased CO_2 efflux in the presence of biochar may occur through an abiotic mechanism, such as oxidation or off-gassing of CO_2, which in some studies have accounted for >50% of CO_2 emitted from amended soils [45,46]. Regardless, the amount of C lost via abiotic or biotic mechanisms seems to be dependent on biochar production temperature; chars produced at higher temperature with lower remnant labile or volatile C produced a smaller or suppressed efflux [47]. Thus, although the filtercake biochar used here was produced at a relatively high temperature, the residual cellulose and hemicellulose detected in the carbohydrate analysis may have been responsible for elevated soil CO_2 effluxes.

Use of raw vs. pyrolyzed filtercake in sugarcane cultivation

Recent lifecycle assessments examining the contribution of soil organic carbon to total carbon emissions from biofuel systems have generated new appreciation for crop residue management and its role in keeping carbon in the soil [48]. Returning sugarcane

Table 2. Total CO_2 emitted and percentage of total carbon lost after three weeks of incubation.

Treatments	CO$_2$ emitted	Total C content	Total C lost
	(g CO$_2$)	(g C)	(%)
S	0.24±0.01, a	18.3	0.35±0.01, a
SV	0.47±0.01, a	18.4	0.70±0.01, b
SVB	1.17±0.13, b	40.7	0.78±0.08, b
SB	1.04±0.03, b	40.6	0.69±0.02, b

To further probe differences among the S, SV, SVB, and SB treatments, total CO_2 emitted and % C lost were compared via one-way ANOVA. Data represent the mean ± SE. Total C content refers to the sum of initial carbon present in the soil plus C added through amendments. Letters indicate significant differences between treatments as determined using Tukey HSD test.
S, soil; SV, soil with vinasse; SVB, soil with 5% biochar and vinasse; SB, soil with 5% biochar. S and SB received water at the beginning of week 2 (30 mL) and week 3 (60 mL). SV and SVB received the same amount of vinasse.

residues to the field is promoted as a means to re-build C stocks and re-capture valuable nutrients [13,49,50]. However, residue application also provokes transient increases in N_2O and CO_2 emissions [51,52], suggesting that this biomass undergoes relatively rapid mineralization and may contribute to further negative effects. Perhaps because of this, sugarcane cultivation areas in Brazil often show low to no recovery of soil C relative to the pre-cultivated baseline, despite large annual inputs from these residues [53–55].

To mitigate SOC degradation, several previous authors have called for the integration of biochar science into biofuel cultivation and crop management [42,56]. The objective of this paper was to reconsider the waste product, filtercake, and gauge its potential merits as a biochar soil amendment through chemical characterization and a panel of assays assessing the most critical agronomic benefits of a biochar. Filtercake biochar showed benefits in terms of increasing soil pH, CEC, nutrient availability, and water retention. These parameters are directly relevant to exploratory studies investigating the effect of biochar amendment on sugarcane productivity via the use of biophysical models (e.g., APSIM-sugarcane [57]). Furthermore, raw filtercake led to a large increase in soil respiration and a greater percentage of initial carbon lost compared to biochar treatment, likely due to the immediate utilization of labile sugars present in the filtercake. Although this was a very short-term study, from which it is not possible to extrapolate long-term biochar stability, it demonstrates the large difference in biodegradability between filtercake in its raw vs. pyrolyzed forms.

As such, the use of raw filtercake may represent a lost opportunity for soil improvement in this region, where a hot, semi-humid climate and nutrient-poor, leachable, acidic soils severely limit plant productivity and increase management costs [58]. In contrast, applying filtercake as biochar may slow the loss of organic carbon from the field, and contribute to a liming effect, increased CEC, water retention, nutrient availability, and activation of the soil microbial community, all of which are positive indicators of soil quality. Thus, our data suggest that soil improvement objectives on sugarcane plantations might be better served by applying filtercake as biochar, rather than the raw material. Biochar implementation may also have economic benefits, as high-temperature pyrolysis is exothermic and self-sustaining through the production of syn-gas; it also generates saleable bioenergy products (bio-oil and biochar itself) and shows potential as a tradable GHG emission-reducing soil amendment [59]. However, further studies are required to determine the long-term stability of this biochar and how this product might influence lifecycle C emissions, as well as its effects on plant productivity and

the feasibility of incorporating pyrolysis bioenergy and biochar into plant operations.

In summary, physical and chemical characterization revealed that the novel filtercake biochar described here has significant benefits in terms of increased CEC, pH, nutrient availability, and water retention, indicating that this product may be beneficial as a soil amendment. Importantly, we also found that filtercake applied as biochar greatly reduced soil CO_2 effluxes compared to the application of raw filtercake, without wholly eliminating microbial activity. These findings open the way to further studies examining the use and stability of filtercake biochar in the field, and demonstrate in principle how organic waste management on sugarcane plantations could be modified to improve soil quality and reduce CO_2 effluxes from cultivated areas.

Supporting Information

Table S1 Raw data for physicochemical characterization of soil-biochar mixtures presented in Table 1.

Table S2 Raw data for calculating soil matrix potential against moisture content in soils containing 0, 5, or 10% filtercake biochar, as presented in Figure 3. Binned water content values represent the average true water content determined gravimetrically for all samples (0, 5, and 10% biochar) for a set of samples to which a specific amount of water was added before analysis. This allowed statistical comparison of tension (pF) values for the treatment groups for a given moisture content.

Table S3 Raw data for daily and total CO_2 effluxes presented. The first data set represents daily CO_2 fluxes obtained from incubation columns over a three-week period, as presented in Figure 4. The second data set represents CO_2 emitted (μmol m^{-2}) in the preceding 24 hours. Where values were missing in the first data set, CO_2 emitted was calculated as the area under the trace based on the next available timepoint assuming a linear increase or decrease between the two time points. Daily CO_2 effluxes were summed to obtain total efflux, which was then converted into g CO_2, for which treatment means were compared in Figure 5. Next, to further probe differences among the low-emitting treatments, the S, SV, SVB, and SB data were analyzed via oneway ANOVA for both g CO_2 emitted and total percent C lost. Total percent C lost refers to g CO_2-C emitted divided by the total amount of carbon present in the incubation (soil carbon plus added biochar/vinasse/filtercake carbon). The treatment means for g CO_2 emitted and % C lost for the S, SV, SVB, and SB treatments are presented in Table 2.

Acknowledgments

We thank Dr. F. Lobo for providing laboratory space at UFMT, Dr. L. Lavkulich for providing Raman spectrographs, Dr. F. Unda for performing the carbohydrate analysis, and Dr. H. McLaughlin for advice on benchtop biochar production. We would also like to thank the faculty and staff of the Programa de Pós-Graduação em Física Ambiental at UFMT for assistance in sourcing materials and equipment loans and the Laboratório de Caracterização em Novos Materiais (LACANM) for SEM analysis.

Author Contributions

Conceived and designed the experiments: AJE MSJ EGC. Performed the experiments: AJE ESQ HJD. Analyzed the data: AJE MSJ EGC. Contributed reagents/materials/analysis tools: AJE MSJ HJD EGC. Contributed to the writing of the manuscript: AJE MSJ. Revised the manuscript for important intellectual content: AJE MSJ ESQ HJD EGC.

References

1. Martinelli LA, Naylor R, Vitousek PM, Moutinho P (2010) Agriculture in Brazil: impacts, costs, and opportunities for a sustainable future. Curr Opin Environ Sustain 2: 431–438.

2. Klink CA, Machado RB (2005) Conservation of the Brazilian Cerrado. Conserv Biol 19: 707–713. doi:10.1111/j.1523-1739.2005.00702.x.

3. Sparovek G, Correchel V, Barretto AGOP (2007) The risk of erosion in Brazilian cultivated pastures. Sci Agric 64: 77–82. doi:10.1590/S0103-90162007000100012.

4. Machado PLO de A, Silva CA (2001) Soil management under no-tillage systems in the tropics with special reference to Brazil. Nutr Cycl Agroecosystems 61: 119–130. doi:10.1023/A:1013331805519.

5. Glaser B, Haumaier L, Guggenberger G, Zech W (2001) The "Terra Preta" phenomenon: a model for sustainable agriculture in the humid tropics. Naturwissenschaften 88: 37–41. doi:10.1007/s001140000193.

6. Liang B, Lehmann J, Solomon D, Kinyangi J, Grossman J, et al. (2006) Black carbon increases cation exchange capacity in soils. Soil Sci Soc Am J 70: 1719. doi:10.2136/sssaj2005.0383.

7. Liang B, Lehmann J, Sohi SP, Thies JE, O'Neill B, et al. (2010) Black carbon affects the cycling of non-black carbon in soil. Org Geochem 41: 206–213. doi:doi:10.1016/j.orggeochem.2009.09.007.

8. Chen Y, Shinogi Y, Taira M (2010) Influence of biochar use on sugarcane growth, soil parameters, and groundwater quality. Aust J Soil Res 48: 526. doi:10.1071/SR10011.

9. Kameyama K, Miyamoto T, Shiono T, Shinogi Y (2012) Influence of sugarcane bagasse-derived biochar application on nitrate leaching in calcaric dark red soil. J Environ Qual 41: 1131–1137. doi:10.2134/jeq2010.0453.

10. Major J, Rondon M, Molina D, Riha SJ, Lehmann J (2012) Nutrient leaching in a Colombian savanna Oxisol amended with biochar. J Environ Qual 41: 1076–1086. doi:10.2134/jeq2011.0128.

11. Steiner C, Teixeira WG, Lehmann J, Nehls T, Macêdo JLV, et al. (2007) Long term effects of manure, charcoal and mineral fertilization on crop production and fertility on a highly weathered Central Amazonian upland soil. Plant Soil 291: 275–290. doi:10.1007/s11104-007-9193-9.

12. George PAO, Eras JJC, Gutierrez AS, Hens L, Vandecasteele C (2010) Residue from Sugarcane Juice Filtration (Filter Cake): Energy Use at the Sugar Factory. Waste Biomass Valor 1: 407–413. doi:10.1007/s12649-010-9046-2.

13. Prado R de M, Caione G, Campos CNS (2013) Filter Cake and Vinasse as Fertilizers Contributing to Conservation Agriculture. Appl Environ Soil Sci 2013: 1–8. doi:10.1155/2013/581984.

14. Khalil MI, Hossain MB, Schmidhalter U (2005) Carbon and nitrogen mineralization in different upland soils of the subtropics treated with organic materials. Soil Biol Biochem 37: 1507–1518. doi:10.1016/j.soilbio.2005.01.014.

15. Kuzyakov Y (2010) Priming effects: Interactions between living and dead organic matter. Soil Biol Biochem 42: 1363–1371. doi:10.1016/j.soilbio.2010.04.003.

16. Cross A, Sohi SP (2011) The priming potential of biochar products in relation to labile carbon contents and soil organic matter status. Soil Biol Biochem 43: 2134–2127. doi:10.1016/j.soilbio.2011.06.016.

17. Lee Y, Park J, Ryu C, Gang KS, Yang W, et al. (2013) Comparison of biochar properties from biomass residues produced by slow pyrolysis at 500°C. Bioresour Technol 148: 196–201. doi:10.1016/j.biortech.2013.08.135.

18. Inyang M, Gao B, Pullammanappallil P, Ding W, Zimmerman AR (2010) Biochar from anaerobically digested sugarcane bagasse. Bioresour Technol 101: 8868–8872. doi:10.1016/j.biortech.2010.06.088.

19. Yao Y, Gao B, Zhang M, Inyang M, Zimmerman AR (2012) Effect of biochar amendment on sorption and leaching of nitrate, ammonium, and phosphate in a sandy soil. Chemosphere 89: 1467–1471. doi:10.1016/j.chemosphere.2012.06.002.

20. Huntley SK, Ellis D, Gilbert M, Chapple C, Mansfield SD (2003) Significant increases in pulping efficiency in C4H-F5H-transformed poplars: improved chemical savings and reduced environmental toxins. J Agric Food Chem 51: 6178–6183. doi:10.1021/jf034320o.

21. Brunauer S, Emmett P, Teller E (1938) Adsorption of gases in multimolecular layers. J Am Chem Soc 60: 309–319. doi:10.1021/ja01269a023.

22. EMBRAPA (2009) Manual de anaálises qui???micas de solos, plantas e fertilizantes. 2 Edition. Silva FC da, editor Brasi???lia, DF: Empresa Brasileira de Pesquisa Agropecuaária (EMBRAPA).

23. R Core Team (2013) R: A language and environment for statistical computing.

24. Fox J (2005) The R Commander: A Basic Statistics Graphical User Interface to R. J Stat Softw 14: 1–42.

25. Fuertes A, Arbestain M, Sevilla M (2010) Chemical and structural properties of carbonaceous products obtained by pyrolysis and hydrothermal carbonisation of corn stover. Soil Res 48: 618–626. doi:10.1071/SR10010.

26. Kim P, Johnson A, Edmunds C (2011) Surface functionality and carbon structures in lignocellulosic-derived biochars produced by fast pyrolysis. Energy & Fuels 25: 4693–4703. doi:10.1021/ef200915s.

27. Asadullah M, Zhang S, Min Z, Yimsiri P, Li C-Z (2010) Effects of biomass char structure on its gasification reactivity. Bioresour Technol 101: 7935–7943. doi:10.1016/j.biortech.2010.05.048.

28. Hamblin AP (1985) The influence of soil structure on water movement, crop root growth, and water uptake. Adv Agron 38: 95–158.

29. Joseph S, Peacocke C, Lehmann J, Munroe P (2009) Developing a Biochar Classification and Test Methods. In: Lehmann J, Joseph S, editors. Biochar for Environmental Management: Science and Technology. London, UK: Earthscan. pp. 107–126.

30. Cornelissen G, Gustafsson Ö, Bucheli TD, Jonker MTO, Koelmans AA, et al. (2005) Extensive Sorption of Organic Compounds to Black Carbon, Coal, and Kerogen in Sediments and Soils: Mechanisms and Consequences for Distribution, Bioaccumulation, and Biodegradation. Environ Sci Technol 39: 6881–6895. doi:10.1021/es050191b.

31. Czimczik C, Masiello C (2007) Controls on black carbon storage in soils. Global Biogeochem Cycles 21: 1–8. doi:10.1029/2006GB002798.

32. Roscoe R, Buurman P, Velhorst E (2001) Soil organic matter dynamics in density and particle size fractions as revealed by the 13C/12C isotopic ratio in Cerrado's Oxisol. Geoderma 104: 185–202. doi:10.1016/S0016-7061(01)00080-5.

33. Arsene M-A, Bilba K, Savastano Junior H, Ghavami K (2013) Treatments of non-wood plant fibres used as reinforcement in composite materials. Mater Res 16: 903–923. doi:10.1590/S1516-14392013005000084.

34. Cagnon B, Py X, Guillot A, Stoeckli F, Chambat G (2009) Contributions of hemicellulose, cellulose and lignin to the mass and the porous properties of chars and steam activated carbons from various lignocellulosic precursors. Bioresour Technol 100: 292–298. doi:10.1016/j.biortech.2008.06.009.

35. Demirbas A (2004) Effects of temperature and particle size on bio-char yield from pyrolysis of agricultural residues. J Anal Appl Pyrolysis 72: 243–248. doi:10.1016/j.jaap.2004.07.003.

36. Kasozi GN, Zimmerman AR, Nkedi-Kizza P, Gao B (2010) Catechol and humic acid sorption onto a range of laboratory-produced black carbons (biochars). Environ Sci Technol 44: 6189–6195. doi:10.1021/es1014423.

37. Brewer CE, Schmidt-Rohr K, Satrio JA, Brown RC (2009) Characterization of biochar from fast pyrolysis and gasification systems. Environ Prog Sustain Energy 28: 386–396. doi:10.1002/ep.10378.

38. Abel S, Peters A, Trinks S, Schonsky H, Facklam M, et al. (2013) Impact of biochar and hydrochar addition on water retention and water repellency of sandy soil. Geoderma 202: 183–191. doi:10.1016/j.geoderma.2013.03.003.

39. Ulyett J, Sakrabani R, Kibblewhite M, Hann M (2014) Impact of biochar addition on water retention, nitrification and carbon dioxide evolution from two sandy loam soils. Eur J Soil Sci 65: 96–104. doi:10.1111/ejss.12081.

40. Basso AS, Miguez FE, Laird DA, Horton R, Westgate M (2013) Assessing potential of biochar for increasing water-holding capacity of sandy soils. GCB Bioenergy 5: 132–143. doi:10.1111/gcbb.12026.

41. Hoffmann WA, Jackson RB (2000) Vegetation–Climate Feedbacks in the Conversion of Tropical Savanna to Grassland. J Clim 13: 1593–1602. doi:10.1175/1520-0442(2000)013<1593:VCFITC>2.0.CO;2.

42. Mathews JA (2008) Carbon-negative biofuels. Energy Policy 36: 940–945. doi:10.1016/j.enpol.2007.11.029.

43. Spokas KA, Reicosky DC (2009) Impacts of sixteen different biochars on soil greenhouse gas production. Ann Environ Sci 3: 179–193.

44. Steinbeiss S, Gleixner G, Antonietti M (2009) Effect of biochar amendment on soil carbon balance and soil microbial activity. Soil Biol Biochem 41: 1301–1310. doi:10.1016/j.soilbio.2009.03.016.

45. Zimmerman AR (2010) Abiotic and microbial oxidation of laboratory-produced black carbon (biochar). Environ Sci Technol 44: 1295–1301. doi:10.1021/es903140c.

46. Jones D, Murphy D, Khalid M, Ahmad W, Edwards-Jones G, et al. (2011) Short-term biochar-induced increase in soil CO2 release is both biotically and abiotically mediated. Soil Biol Biochem 43: 1723–1731. doi:10.1016/j.soilbio.2011.04.018.

47. Zimmerman AR, Gao B, Ahn M-Y (2011) Positive and negative carbon mineralization priming effects among a variety of biochar-amended soils. Soil Biol Biochem 43: 1169–1179. doi:10.1016/j.soilbio.2011.02.005.

48. Liska AJ, Yang H, Milner M, Goddard S, Blanco-Canqui H, et al. (2014) Biofuels from crop residue can reduce soil carbon and increase CO_2 emissions. Nat Clim Chang 4: 398–401. doi:10.1038/nclimate2187.

49. Elsayed MT, Babiker MH, Abdelmalik ME, Mukhtar ON, Montange D (2008) Impact of filter mud applications on the germination of sugarcane and small-seeded plants and on soil and sugarcane nitrogen contents. Bioresour Technol 99: 4164–4168. doi:10.1016/j.biortech.2007.08.079.

50. Anderson-Teixeira KJ, Davis SC, Masters MD, Delucia EH (2009) Changes in soil organic carbon under biofuel crops. GCB Bioenergy 1: 75–96. doi:10.1111/j.1757-1707.2008.01001.x.

51. Carmo JB do, Filoso S, Zotelli LC, de Sousa Neto ER, Pitombo LM, et al. (2013) Infield greenhouse gas emissions from sugarcane soils in Brazil: effects from synthetic and organic fertilizer application and crop trash accumulation. GCB Bioenergy 5: 267–280. doi:10.1111/j.1757-1707.2012.01199.x.

52. De Oliveira BG, Carvalho JLN, Cerri CEP, Cerri CC, Feigl BJ (2013) Soil greenhouse gas fluxes from vinasse application in Brazilian sugarcane areas. Geoderma 200-201: 77–84. doi:10.1016/j.geoderma.2013.02.005.

53. Vasconcelos RFB de, Cantalice JRB, Oliveira VS de, Costa YDJ da, Cavalcante DM (2010) Estabilidade de agregados de um latossolo amarelo distrocoeso de tabuleiro costeiro sob diferentes aportes de resíduos orgânicos da cana-de-açúcar. Rev Bras Ciência do Solo 34: 309–316. doi:10.1590/S0100-06832010000200004.

54. Resende ASAS de, Xavier RP, Oliveira OC, Urquiaga S, Alves BJR, et al. (2006) Long-term Effects of Pre-harvest Burning and Nitrogen and Vinasse Applications on Yield of Sugar Cane and Soil Carbon and Nitrogen Stocks on a Plantation in Pernambuco, N.E. Brazil. Plant Soil 281: 339–351. doi:10.1007/s11104-005-4640-y.

55. Oliveira VS, Rolim MM, Vasconcelos RFB, Pedrosa EMR (2010) Distribuição de agregados e carbono orgânico em um Argissolo Amarelo distrocoeso em diferentes manejos. Rev Bras Eng Agrícola e Ambient 14: 907–913. doi:10.1590/S1415-43662010000900001.

56. Abiven S, Schmidt MWI, Lehmann J (2014) Biochar by design. Nat Geosci 7: 326–327. doi:10.1038/ngeo2154.

57. McCown RL, Hammer GL, Hargreaves JNG, Holzworth DP, Freebairn DM (1996) APSIM: a novel software system for model development, model testing and simulation in agricultural systems research. Agric Syst 50: 255–271. doi:10.1016/0308-521X(94)00055-V.

58. Goedert WJ (1983) Management of the Cerrado soils of Brazil: a review. J Soil Sci 34: 405–428. doi:10.1111/j.1365-2389.1983.tb01045.x.

59. Gaunt JL, Lehmann J (2008) Energy Balance and Emissions Associated with Biochar Sequestration and Pyrolysis Bioenergy Production. Environ Sci Technol 42: 4152–4158. doi:10.1021/es071361i.

23

Assessment of Bacterial *bph* Gene in Amazonian Dark Earth and Their Adjacent Soils

Maria Julia de Lima Brossi[1]*****, **Lucas William Mendes**[1], **Mariana Gomes Germano**[2], **Amanda Barbosa Lima**[1], **Siu Mui Tsai**[1]

1 Cellular and Molecular Biology Laboratory, Center for Nuclear Energy in Agriculture, University of São Paulo, Piracicaba, SP, Brazil, **2** Brazilian Agricultural Research Corporation, Embrapa Soybean, Londrina, PR, Brazil

Abstract

Amazonian Anthrosols are known to harbour distinct and highly diverse microbial communities. As most of the current assessments of these communities are based on taxonomic profiles, the functional gene structure of these communities, such as those responsible for key steps in the carbon cycle, mostly remain elusive. To gain insights into the diversity of catabolic genes involved in the degradation of hydrocarbons in anthropogenic horizons, we analysed the bacterial *bph* gene community structure, composition and abundance using T-RFLP, 454-pyrosequencing and quantitative PCR essays, respectively. Soil samples were collected in two Brazilian Amazon Dark Earth (ADE) sites and at their corresponding non-anthropogenic adjacent soils (ADJ), under two different land use systems, secondary forest (SF) and manioc cultivation (M). Redundancy analysis of T-RFLP data revealed differences in *bph* gene structure according to both soil type and land use. Chemical properties of ADE soils, such as high organic carbon and organic matter, as well as effective cation exchange capacity and pH, were significantly correlated with the structure of *bph* communities. Also, the taxonomic affiliation of *bph* gene sequences revealed the segregation of community composition according to the soil type. Sequences at ADE sites were mostly affiliated to aromatic hydrocarbon degraders belonging to the genera *Streptomyces*, *Sphingomonas*, *Rhodococcus*, *Mycobacterium*, *Conexibacter* and *Burkholderia*. In both land use sites, shannon's diversity indices based on the *bph* gene data were higher in ADE than ADJ soils. Collectively, our findings provide evidence that specific properties in ADE soils shape the structure and composition of *bph* communities. These results provide a basis for further investigations focusing on the bio-exploration of novel enzymes with potential use in the biotechnology/biodegradation industry.

Editor: Niyaz Ahmed, University of Hyderabad, India

Funding: Funding provided by 'Conselho Nacional de Desenvolvimento Científico e Tecnológico' and the 'Fundação de Amparo à Pesquisa do Estado de São Paulo' (FAPESP/Biota 2011/50914-3). The funders had no role in study design, data collection and analysis, decision to publish, or preparation of the manuscript.

Competing Interests: The authors have declared that no competing interests exist.

* E-mail: majubrossi@gmail.com

Introduction

Amazonian Dark Earth (ADE), locally termed *'Terra Preta de Índio'*, are anthropogenic soil horizons built-up by the Pre-Colombian Indians between 500 and 8,700 years ago. These soil sites were formed by the progressive deposit of materials and organic compounds, such as charcoal, bone, and pottery sheds, which gradually shifted the natural physical and chemical properties of the soil. As a result, relatively infertile Amazon soils were progressively converted into highly fertile spots through processes like increasing the cation exchange capacity and the nutrient content, as well as promoting the stabilization of the soil physical structure [1,2]. Substantial increments of organic material in these sites gradually increased the carbon content, yielding to the formation of soil spots with a high proportion of incompletely combusted biomass (biochar). These spots have been reported to reach up to a 70-fold higher amount of carbon than native soils at adjacent locations (ADJ) [3].

The existence of ADE sites close to their natural ADJ soil locations, which present the same geological history, provides a unique opportunity to investigate the role of biotic and abiotic factors influencing the microbial community assembly and dynamics at these sites. Previous studies revealed that ADE and ADJ sites present differences in microbial community composition, and bacterial diversity has been reported to be higher at ADE sites [4,5,6,7,8]. Most of these studies rely on comparisons between the taxonomic profiles of these communities (i.e., based on the taxonomic bacterial 16S rRNA gene). In this sense, the extent to which the local environment shapes the functional profiles of these communities, and influences their performance, remains mostly elusive [9,10,11,12,13].

Soil is one of the most biodiverse ecosystems on Earth, being able to support communities from multiple trophic levels, which are constantly performing the metabolism of diverse and complex substrates. The extreme spatial and temporal heterogeneity of the soil matrix, paired with the myriad of internal and external feedbacks, are known to determine the structure and function of these communities [14]. Microbes are involved in many ecosystem processes, including biodegradation, decomposition and mineralization, inorganic nutrient cycling, disease causation and suppression, and pollutant removal. Soil disturbances are known to cause shifts in microbial activities, shifting the rate of these processes and triggering impacts on the ecosystem performance [15]. Several environmental factors are known to affect microbial community composition in the soil, including soil temperature, moisture, texture, carbon content, nutrient availability, pH, land use history, seasonality, and the content of incompletely combusted biomass, such as biochar [13,16,17,18,19].

The biochar content is the major physical distinction between ADE and their ADJ soils, which is also known to play an important role in global carbon biogeochemistry [20]. In anthrosols, such as ADE, it is predicted that distinct microbial communities can perform unique processes, such as the retention of high-labile carbon [5]. Despite the unique and specialized capabilities of these soils, functional assessments of their microbial community, particularly those of genes encoding important steps in the carbon cycle, are still scarce. Biodegradation through bacterial activity is one of the most important processes occurring in soils regarding organic matter recycling. This process involves genes acting on key steps in the carbon cycle, for the turnover of more recalcitrant organic carbon, as well as for pollutant degradation in the ecosystem [21]. This process, along with biosynthesis, largely governs the carbon cycle in the environment, which is dependent on microbial enzymatic activities that most often use organic compounds as a primary energy source [22].

The primary step involved in the aerobic microbial degradation of aromatic hydrocarbons is an oxidative attack [23,24], where enzymes named oxygenases are responsible for the insertion of molecular oxygen into aromatic benzene rings [25]. Genes encoding these enzymes have been characterized in *Rhodococcus*, *Acinetobacter*, *Pseudomonas*, *Mycobacterium*, *Burkholderia*, *inter alia* [25,26]. The α-subunit of oxigenases is known as the catalytic domain involved in the transfer of electrons to oxygen molecules. Due its DNA sequence conservation, this subunit has been currently used as a target gene for the detection of such enzymes in complex communities [21,27,28,29].

In this study, we evaluated the structure, composition and abundance of the bacterial catabolic gene Biphenyl Dioxygenase (*bph*) involved in aromatic hydrocarbon degradation in Amazonian Dark Earth and their adjacent soil locations. We aimed to determine the role of anthropogenic action in the diversity of the *bph* gene in soil bacterial communities. Understanding the diversity of specific bacterial genes in ADE should led to future studies that investigate the microbial ecology of anthropogenic altered soils, especially in regards to their potential source of novel enzymes. Collectively, this study characterized this catabolic gene occurring in Amazonian Dark Earth, and compared the profiles with those obtained from adjacent sites, as well as under distinct land use systems.

Materials and Methods

Ethic statement

No specific permits were required for the described field studies. The locations are not protected. The field studies did not involve endangered or protected species.

Study sites, sample collection and soil chemical analyses

Studied sites are located at the Caldeirão Experimental Station of Amazon Brazilian Agricultural Research Corporation (Embrapa) in Iranduba County in the Brazilian Central Amazon (03°26′00″ S, 60°23′00″ W). A detailed description of the soil sampling locations is given by Taketani et al [8]. Briefly, the four sites sampled are composed of two Amazonian Dark Earth (ADE) and their two correspondent adjacent soils (Haplic Acrisol, ADJ). These sites are under a ~35 year-old secondary forest (SF) or under manioc (*Manihot esculenta*) cultivation (M). Hereafter these sites are termed as ADE-SF, ADJ-SF, ADE-M and ADJ-M. Soil samples were taken in triplicate from the topsoil layer (ca. top

10 cm), and the overlaying litter was discarded. Each sample contained approximately 300 g of soil and was transported to the laboratory at 4°C to further processing (<24 h). A portion of the samples were frozen (−20°C) for total DNA extraction while the other portion was kept at 4°C for chemical measurements.

Chemical analyses were performed at Amazon Embrapa (Manaus, Brazil), according to instructions provided by the Embrapa protocol [30]. Briefly, soil samples were analysed in triplicate for pH (H_2O, 1:2.5); H+Al (calcium extractor 0.5 mol L^{-1}, pH 7.0); sum of bases (SB); soil organic matter (SOM); soil organic carbon (SOC; Walkely-Black method); extractable fraction of Al, Ca, and Mg (1 M KCl); extractable fraction of P and K (double acid solution of 0.025 M sulphuric acid and 0.05 M hydrochloric acid Mehlich 1); and effective cation exchange capacity (eCEC).

Total soil DNA extraction and PCR amplifications for T-RFLP

Total DNA was extracted using 0.25 g of soil as an initial material. Extractions were carried out in triplicate for each site, using the PowerSoil DNA isolation kit (MoBio Laboratories, Carlsbad, CA, USA), according to the manufacturer's protocol. DNA quality and quantity were measured spectrophotometrically using NanoDrop 1000 (Thermo Scientific, Waltham, EUA).

For T-RFLP analyses the bacterial 16S rRNA gene was amplified with the primer set 27F - FAM labelled (5′ AGA GTT TGA TCC TGG CTC AG 3′) and 1492r (5′ ACC TTG TTA CGA CTT 3′) [31]. The *bph* gene was amplified with the primer set BPHD F1 – FAM labelled (5′ TAY ATG GGB GAR GAY CCI GT 3′) and BPHD R0 (5′ ACC CAG TTY TCI CCR TCG TC 3′) [21]. For 16S rRNA gene amplification, PCR reactions were carried out in a volume of 25 µL containing 2.5 µL reaction buffer 10× (Invitrogen, Carslbad, CA, USA), 1.5 µL $MgCl_2$ (50 mM), 1 µL of each primer (5 pmol $µL^{-1}$), 0.2 µL (5 U) of Platinum Taq DNA polymerase (Invitrogen), 0.5 mL of deoxyribonucleotide triphosphate mixture (2.5 mM), 0.25 µL of bovine serum albumin (1 ng mL^{-1}), 1 µL of DNA template (ca. 10 ng) and 18.05 µL of sterilized ultrapure water. Amplifications were performed in the GeneAmp PCR System 9700 thermal cycler (Applied Biosystems, Foster City, CA, USA). Reaction conditions were 94°C for 3 min, followed by 35 cycles of 94°C for 30 s, 59°C for 45 s, and 72°C for 1 min with a final extension step at 72°C for 15 min. For the *bph* gene amplification, PCR reactions were carried out in a volume of 25 µL containing 2.5 µL reaction buffer 10× (Invitrogen, Carslbad, CA, USA), 1.5 µL $MgCl_2$ (50 mM), 1.25 µL of each primer (5 pmol $µL^{-1}$), 0.5 µL (5 U) of Platinum Taq DNA polymerase (Invitrogen), 0.3 µL of deoxyribonucleotide triphosphate mixture (2.5 mM), 1 µL of DNA template (ca. 10 ng) and 16.7 µL of sterilized ultrapure water. For *bph* gene, similar PCR cycling conditions were used, except the annealing temperature that was set at 60°C. Negative PCR controls (without DNA template) and positive controls (using the *Escherichia coli* ATCC 25922 DNA for the 16S rRNA gene and the DSM 6899 *Pseudomonas putida* DNA for *bph*) were run in parallel for both amplifications. After the amplifications, 5 µL of obtained products (ca. 60 ng) was digested with the endonuclease *Hha*I (Invitrogen) in 15 µL reaction for 3 h at 37°C. Obtained fragments were further purified using sodium acetate/EDTA precipitation and then mixed with 0.25 µL of the Genescan 500 ROX size standard (Applied Biosystems) and 9.75 µL of deionized formamide. Prior to fragment analysis, samples were denatured at 95°C for 5 min and chilled on ice. Analysis of terminal restriction fragment (T-RF) sizes and quantities was performed on an ABI PRISM 3100 genetic analyzer (Applied Biosystems).

T-RFLP profiles were analysed using PeakScanner v1.0 software (Applied Biosystems, Foster City, CA, USA). Terminal restriction fragments (T-RFs) of less than 25 bp were excluded prior to the analysis. The total values of T-RFs for each soil sample were pulled together to construct a Venn's diagram showing shared T-RFs among samples. The relative abundance of a single T-RF was calculated as percent fluorescence intensity relative to total fluorescence intensity of the peaks [32]. Data from individual samples were combined to soil chemical parameters and subjected to multivariate analysis using Canoco 4.5 (Biometris, Wageningen, The Netherlands) and Primer6 (PrimerE, Ivybridge, United Kingdom). All matrices were initially analysed using de-trended correspondence analysis (DCA) to evaluate the length of the gradient of the species distribution; this analysis indicated linearly distributed data (length of gradient <3), revealing that the best-fit mathematical model for the data was the redundancy analysis (RDA). Forward selection (FS) and the Monte Carlo permutation test were applied with 1,000 random permutations to verify the significance of soil chemical properties upon the microbial community. In addition to P values for the significance of each soil chemical property, RDA and Monte Carlo permutation test supplied information about the marginal effects of environmental variables, quantifying the amount of variance explained by each factor. We used ANOSIM based on relative abundance of T-RFs to test for statistical differences between samples.

454-Pyrosequencing analyses of the bacterial *bph* gene

A partial region of the *bph* gene was amplified for the 454-pyrosequencing using the primer set BPHD F3 (5′ ACT GGA ART TYG CIG CVG A 3′) and BPHD R0 (5′ ACC CAG TTY TCI CCR TCG TC 3′) [20] containing specific Roche 454-pyrosequencing adaptors and barcodes of 8 bp. The expected fragment size was ca. 520 bp. Three independent amplifications were performed for each sample. The 20 μL PCR mixture contained 1× FastStart High Fidelity Reaction Buffer (Roche Diagnostics, Basel, Switzerland), 1.25 mM of each primer, 150 ng mL^{-1} of bovine serum albumin (New England BioLabs, Ipswich, MA, USA), 0.2 mM of dNTPs, 0.5 mL (2.5 U) of FastStart High Fidelity PCR System Enzyme Blend (Roche Diagnostics) and 4 ng of template DNA. The PCR conditions were optimized using the genomic DNA of *Burkholderia xenovorans* LB400 [33], which carries one of the target dioxygenase genes. Amplifications were performed as follows: 95°C for 3 min, 30 cycles of 95°C for 45 s, 60°C for 45 s and 72°C for 40 s, with the final extension of 72°C for 4 min. Triplicate PCR products containing the expected fragment size were purified using the QIAquick Gel Extraction Kit (Qiagen, Hilden, Germany) and QIAquick PCR Purification Kit (Qiagen). DNA concentrations were determined using the NanoDrop ND-1000 spectrophotometer (NanoDrop Technologies, Wilmington, DE, USA). Purified PCR products were pooled and subjected to pyrosequencing using the FLX sequencing system (454 Life Sciences, Branford, CT, USA).

Raw data was filtered for valid sequences using the FunGene Pipeline Repository (http://fungene.cme.msu.edu/FunGenePipeline/). Quality sequences were translated in the correct frame of aminoacids using the RDP FrameBot tool. The RDP pipeline extracted a set of representative sequences from known *bph* sequences to use as subject sequences with FrameBot. The FrameBot produces an optimal alignment between the query and the subject sequences in the presence of frameshifts. Only the protein pairwise alignment with the best score was reported. Protein sequences passing FrameBot were aligned with HMMER using a model trained on the same set of representative sequences used by FrameBot. The aligned protein sequences were chopped

at position 351 of the reference sequence of *Pseudomonas putida* F1 (YP_001268196). The total valid sequences were rarefied to the smallest number of sequences per sample in order to minimize effects of sampling effort upon analysis.

Distance matrices were constructed using the MOTHUR software [34]. The resulting matrices were used to estimate the number of operational protein families (OPF) (i.e. group of proteins that share a common evolutionary origin) and to estimate richness (i.e. Chao1, Jackknife, and ACE indices) and diversity (i.e. Shannon and Simpson indices). Rarefaction curves were constructed at a cutoff level of 94% of amino acid identity. MOTHUR was also used to perform ∫-Libshuff comparisons between the four studied sites. The Good's coverage estimator was used to calculate the sample coverage using the formula C = 1-(n$_i$/N), where N is the total number of sequences analysed and n_i is the number of reads that occurs only once among the total number of reads analysed at a cutoff value of 94% of amino acid identity [35]. Unweighted UniFrac distances among communities were estimated using a tree constructed *de novo* using FastTree. One representative sequence per OPF was selected and subjected to taxonomic affiliation by the comparison tool of NCBI Tblastx (GenBank) using Blast2Go [36].

Sequence data generated by 454-pyrosequencing are available at the MG-RAST server (http://metagenomics.anl.gov) under the project 'Diversity of bhp gene in Amazon soils' (ID 8489) and accession numbers 4557319.3 (ADE_SF), 4557318.3 (ADE_M), 4557321.3 (ADJ_SF) and 4557320.3 (ADJ_M).

Quantitative PCR (qPCR) of the bacterial 16S rRNA and *bph* genes

The bacterial 16S rRNA gene was amplified with the primer set U968F (5' AAC GCG AAG AAC CTT AC 3') and R1387 (5' CGG TGT GTA CAA GGC CCG GGA ACG 3') [37], which amplify a fragment of approximately 400 bp. The 520 bp fragment of the *bph* gene was amplified with the primer set BPHD F3 (5′ ACT GGA ART TYG CIG CVG A 3′) and BPHD R0 (5′ ACC CAG TTY TCI CCR TCG TC 3′) [21]. qPCR reactions were performed in 10 μL containing 5 μL of SYBR green PCR master mix (Fermentas, Brazil), 1 μL of each primer (5 pmol μL^{-1}), 1 μL of DNA template (ca. 10 ng) and 2 μL of sterilized ultrapure water. Thermocycling conditions for the 16S rRNA gene were set as follows: 94°C for 10 min; 40 cycles of 94°C for 30 s, 56°C for 30 s and 72°C for 40 s. Amplification specificity was checked by a melting curve and the data collection was performed at every 0.7°C. qPCR reactions for the *bph* gene were performed at similar conditions, except for the annealing temperature set at 60°C. Reactions were performed in a StepOnePlus system (Applied Biosytems). The Cts values (cycle threshold) were used as standers for determining the amount of DNA template in each sample. Standard curves were produced for the 16S rRNA and *bph* genes using specific cloned fragments. Gene fragments were quantified in a spectrophotometer (190 a 840 nm - NanoDrop ND-1000) and diluted (10^7 to 10^3 genes μL^{-1} for the 16S rRNA gene and 10^6 to 10^2 genes μL^{-1} for *bph*) to generate each specific standard curves. The gene copy numbers in different soil samples were expressed as log copy numbers of the gene per gram of soil. Statistical comparisons were performed using one-way ANOVA (Tukey's test).

Results

Variation in soil chemical properties

Soil chemical properties were measured for each individual sample collected in ADE and ADJ sites (for a detailed description see Table S1). Statistical differences were observed using Tukey's test. Overall, soil chemical properties of ADE-SF were chemically similar to ADE-M. Likewise, ADJ-SF chemical properties were also very similar to ADJ-M. As expected, major differences were attributed mostly to soil type rather than the land use history.

Higher soil pH values were observed in ADE rather than ADJ sites. While ADJ soils were very acidic with a pH of 3.53 (ADJ-SF) and 3.74 (ADJ-M), ADE sites were only weakly acidic with a pH of 5.51 (ADE-SF) and 5.41 (ADE-M). Sites at ADE showed lower total and exchangeable Al (H+Al), a phenomenon that is likely directly connected to observed variations in soil pH.

Soil organic carbon (SOC), soil organic matter (SOM) and effective cation exchange capacity (eCEC) were higher in ADE sites. Different land uses did not influence these properties in ADJ soils; however, the same properties showed significantly higher values in the site under secondary forest rather than in manioc cultivation, in ADE soil locations. In detail, SOC, SOM and eCEC values were approximately 30% higher in the ADE-SF site when compared to the ADE-M.

Assessment of community structures based on the bacterial 16S rRNA and bph genes

T-RFLP analyses for the bacterial 16S rRNA and bph genes were performed for the four sites. The obtained profiles were used to determine the richness of terminal restriction fragments (T-RFs) and to perform the multivariate analyses. A total of 152, 144, 147 and 141 T-RFs were obtained for the analysis of the bacterial 16S rRNA gene in ADE-SF, ADE-M, ADJ-SF and ADJ-M sites, respectively. There were 14 T-RFs detected as dominant throughout all sites, accounting for >50% of the total fluorescence detected for the 16S rRNA gene analyses. Tukey's test (P>0.05) indicated no difference between sites in the richness of T-RFs for the obtained profiles of bacterial 16S rRNA gene.

For the bph gene analysis no dominant T-RFs were found, possibly due to the high heterogeneity of this gene. There were 90, 78, 73 and 69 T-RFs in ADE-SF, ADE-M, ADJ-SF and ADJ-M sites, respectively. Samples from ADE sites showed statistically higher richness of T-RFs (Tukey's test) than the observed at ADJ sites (P>0.05).

The Venn's diagram according to soil type showed that ADE and ADJ soils shared more common T-RFs for the bacterial 16S rRNA rather than for bph gene. Also, the number of unique T-RFs was higher in ADE sites for both assessed genes (Figure 1a). Conversely, Venn's diagram combining the four sites showed a core containing 122 T-RFs for the bacterial 16S rRNA gene, while the distribution of T-RFs for the bph gene was more site specific, and only 6 T-RFs comprised a common core (Figure 1b).

Clustering analysis of T-RFLP data for the bph gene segregated samples according to soil type and land use (Figure 2). This analysis revealed the formation of two main clusters: the first cluster (a) included samples from ADE soils (SF and M) and the second (b) samples from ADJ soils (SF and M). This analysis also revealed that ADE and ADJ sites segregated at 12% of similarity. Concerning the ADE sites, land use systems separated different land uses at 27% of similarity. Conversely, for ADJ sites, land use systems differed at 20% similarity (Figure 2).

Redundancy analysis based on T-RFLP data explained 78.1% of the variation in the first two axes, thus confirming the segregation of sites primary according to soil type, and further in relation to different land use types (Figure 3). Replicates within each soil site were very consistent, evidenced by the formation of concise clusters. We also observed that different soil types also correlated differently to measured chemical parameters. More precisely, the bph community structure from ADE-SF correlated

mostly with pH, eCEC, SOM and SOC, while sites at ADJ-FS presented a significant correlation to H+Al.

ANOSIM analysis indicated statistical differences between the two soil types and land use systems (Table 1). R-values revealed a clear segregation of bph gene structures (R>0.75) in ADJ soils, while the 16S rRNA gene differed to a lower extent across sites (R<0.2).

Diversity of the bacterial bph gene across soil types and land uses

To access the composition of bph gene, samples were sequenced using 454-pyrosequencing. A total of 7,710 reads matched the barcodes, of which 6,877 reads passed the initial filtering (89.2%) and 5,965 (86.7%) were effectively translated into amino acid sequences using the FrameBot. A total of 4,690 valid amino acid sequences were further rarified to the depth of 750 sequences per sample (the minimum in a single sample) for comparative analysis.

The diversity indices for the bph gene (Table 2) revealed that ADE sites (SF and M) ($H' = 4.24$ and 4.05, $L' = 0.024$ and 0.028, respectively) were more diverse than ADJ sites (SF and M) ($H' = 3.19$ and 3.17, $L' = 0.098$ and 0.107, respectively). These indices also showed that sites under SF were more diverse than sites under M. Richness estimators (i.e. Chao1, ACE and Jackknife) also revealed ADE sites (SF and M) (Chao 1 = 238 and 229, ACE = 248 and 286, Jackknife = 271 and 255, respectively) to present higher values than ADJ (Chao 1 = 151 and 127, ACE = 210 and 166, Jackknife = 170 and 130, respectively), with SF sites being also higher than sites under M.

ADE sites presented a total of 159 and 129 OPFs for SF and M sites, respectively. Conversely, these values for ADJ were lower (95 and 90, for SF and M, respectively). These sites also presented a different number of singletons (number of unique reads per OPF): 66, 57, 43 and 36 for ADE-SF, ADE-M, ADJ-SF and ADJ-M, respectively.

Statistical differences among sites for the composition of the bph gene were confirmed by ∫-Libshuff (P<0.001). Venn's diagrams highlight the number of shared OPFs among samples (Figure 4). The number of shared OPFs between ADE soils under different land uses systems (SF and M) was 84 (41% of the total OPFs presented in ADE sites). Conversely, the number of shared OPFs between ADJ soils under different land uses was 55 (42% of the total OPFs in ADJ sites). Soils under SF presented higher numbers of unique OPFs for both soils (ADE and ADJ) (75 and 40 unique OPFs for SF sites and 45 and 35 unique OPFs for ADE sites, respectively).

The estimation of Good's coverage revealed higher values for ADJ sites (0.86 for ADJ-SF and 0.87 for ADJ-M) than for ADE sites (0.79 for ADE-SF and 0.80 for ADE-M), suggesting a highest number of unique sequences in ADE sites. Rarefaction curves (Figure S1) indicate that ADE (SF and M sites) presented a more diverse community than ADJ (SF and M). Sites under SF (ADE and ADJ) were also comparatively more diverse than sites under M (ADE and ADJ). For all comparative analysis the sampling effort did not covered the richness of bph gene. The exception was observed for samples from ADJ sites, where a trend towards a "plateau" was observed.

The Principal Coordinate Analysis (PCoA) based on Unweighted UniFrac distances revealed distinct patterns in phylogenetic community composition (Figure 5). The first axis explained 59.59% of the data variation, and this axis separated samples according to soil type. The second axis explained 22.63% of the data variation, and this axis segregated samples according to land use system.

(a)

(b)

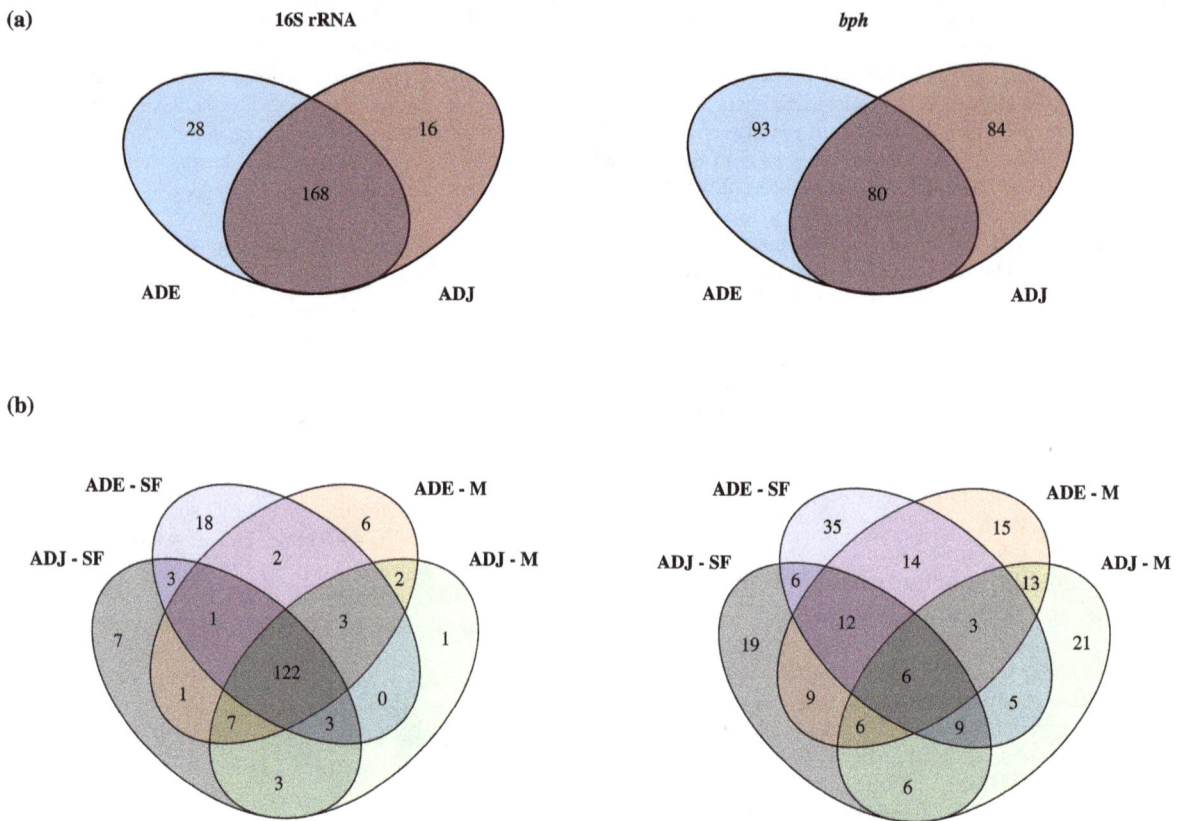

Figure 1. Venn's diagram of T-RFs for 16S rRNA and *bph* genes according to (a) soil types and (b) soil types and land uses. ADE = Amazon Dark Earth; ADJ = Adjacent soils; SF = Secondary Forest; M = Manioc cultivation.

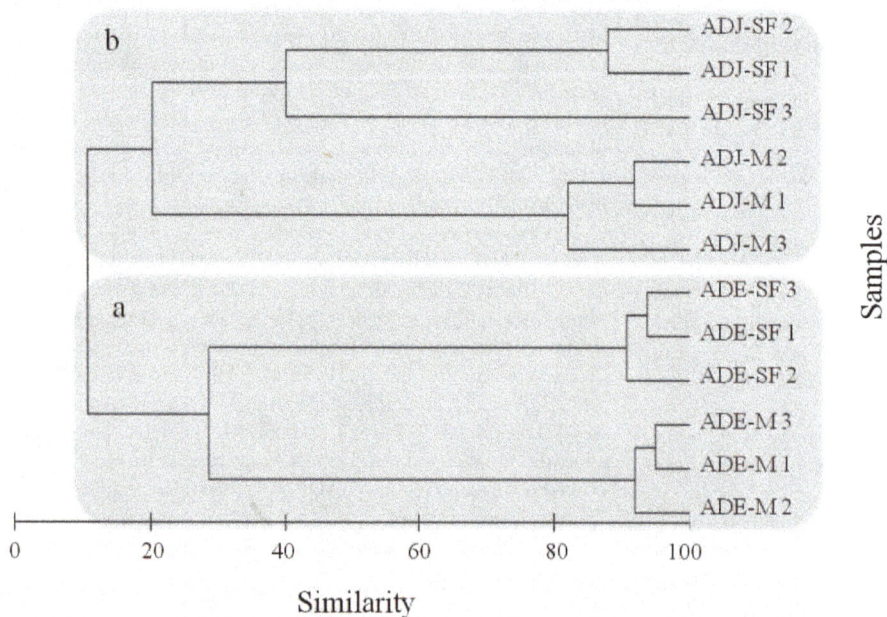

Figure 2. Clustering analysis of T-RFLP data based on Bray-Curtis similarity for the *bph* gene. 'a' and 'b' indicate the segregation patterns according to soil type. ADE = Amazon Dark Earth; ADJ = Adjacent soils; SF = Secondary Forest; M = Manioc cultivation.

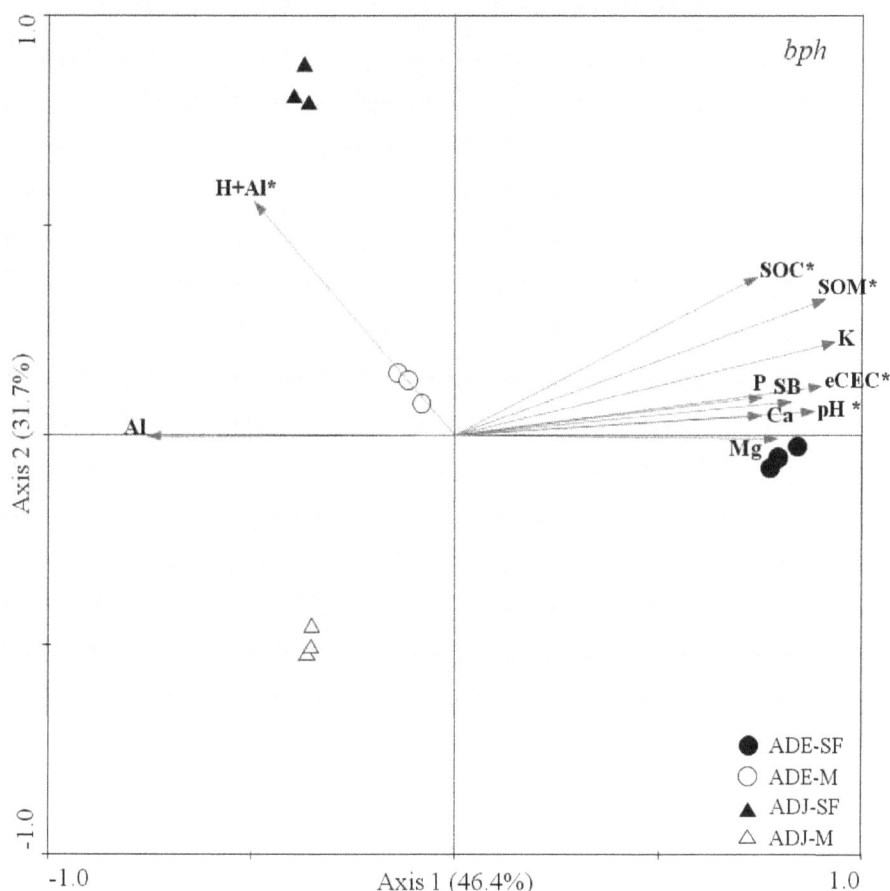

Figure 3. Redundancy analysis (RDA) based on T-RFLP data obtained for the *bph* gene, and soil properties, at the four studied sites. Arrows indicate correlation between the chemical parameters and community structure of samples. The significance of correlations was evaluated via Monte Carlo permutation test and it is indicated as follows: * $p<0.05$. ADE = Amazon Dark Earth; ADJ = Adjacent soils; SF = Secondary Forest; M = Manioc cultivation.

Taxonomic composition of the *bph* gene

Each obtained OPF were further compared to sequences from the GenBank database for taxonomic assignment. All analyzed sequences matched translated proteins described as dioxygenases or putative dioxygenases, with E-values $<10^{-3}$. The most abundant differences in dioxygenases (established as dioxygenases at least ten-fold higher in one site than another) are shown in Figure 6. Sequence matches were associated with aromatic hydrocarbon degradation genes belonging mostly to the genera *Streptomyces*, *Sphingomonas*, *Rhodococcus*, *Mycobacterium*, *Conexibacter* and *Burkholderia*, and uncultured bacterial clones. The taxonomic

affiliation of the reads also revealed the predominance of the dioxygenase sequence belonging to unculturable organisms (rdh cf33) in ADE sites, and the predominance of the dioxygenase sequences similar to those previously described at Australian soil (od16) in ADJ sites.

Quantification of the bacterial 16S rRNA and *bph* genes

Variations in bacterial 16S rRNA gene abundances were clear among the evaluated sites. Bacterial 16S rRNA gene ranged from 2.7×10^7 (ADJ-M) up to 9.7×10^7 (ADE-SF) copies per gram of soil. The results indicated that, soil type and land use influenced the

Table 1. ANOSIM test for the bacterial 16S rRNA and *bph* gene based on T-RFLP data of the Amazonian Dark Earth (ADE) and Adjacent soil (ADJ) under secondary forest (SF) and under manioc cultivation (M).

	16S rRNA[a]		*bph*[a]	
	R values	p values	*R* values	p values
ADE × ADJ	0.17	<0.001	1.00	<0.001
ADE-SF × ADE-CULT	0.11	<0.001	1.00	<0.001
ADJ-SF × ADJ-CULT	0.03	<0.001	1.00	<0.001

[a]Samples were compared using T-RF peak height as a measure of abundance.

Table 2. Comparison of diversity indices and richness estimators for the *bph* gene.

Site	OPFs [a]	Richness			Diversity			ESC [b]
		Chao 1	ACE	Jackknife	Shannon (H')	Simpson (L')		
ADE-SF	159	238	248	271	4.24	0.024		0.79
ADE-M	129	229	286	255	4.05	0.028		0.80
ADJ-SF	95	151	210	170	3.19	0.098		0.86
ADJ-M	90	127	166	130	3.17	0.107		0.87

[a]The operational protein family (OPFs), richness estimators (ACE, Chao1 and Jackknife), diversity indices (Shannon and Simpson) and [b]estimated sample coverage were calculated at a cutoff value of 94% of sequence identity. ADE = Amazon Dark Earth soils; ADJ = Adjacent soils; SF = Secondary forest; M= Manioc cultivation.

abundance of bacteria in the analyzed samples, and higher values were found at ADE-SF site (Figure 7a).

The abundance of the *bph* gene ranged from 1.6×10^6 (ADJ-M) to 2.9×10^6 (ADE SF) copies per gram of soil. Conversely to the data obtained for the total bacteria (i.e. 16S rRNA gene data), the abundance of the *bph* gene was also higher at ADE-SF site (Figure 7b).

Discussion

The aim of this study was to assess bacterial hydrocarbon degrading genes in anthropogenic sites from Brazilian Amazon comparatively to their adjacent locations, under two different land use systems. Recent studies have described the high fertility of ADE soils when compared to adjacent soils in the same area (ADJ), mostly because of their increased pH, higher cation exchange capacity, nutrient content and incompletely combusted biomass [3,6,8,28,38]. Although it is well-known that the taxonomic composition of bacterial communities is strongly influenced by pH [39,40], the pH variation in our data indicate that this is also a strong predictor of the composition and diversity of the bacterial *bph* gene. Variations in soil pH, together with eCEC, SOC and SOM, collectively accounted for 78.1% of the total variation explained in the RDA plot based on T-RFLP data. Higher P values in ADE sites are likely to be an effect of pH, seeing that low acidity soils are known to increase the P solubility [28]. Also, the historical formation of ADE sites (constantly amended with bones and vegetation burning activities) is an intrinsic characteristic of this system, which could possibly explain the high content of phosphorus. There is also a direct relationship between soil P content and eCEC values in ADE sites mainly because the P adsorption decreases due to the formation of complex compounds between P and the organic matter present in the upper layer of the soil[41]. The values of eCEC in our samples were mostly correlated to the organic matter concentration, which was 2-fold higher in ADE than ADJ sites. These findings are likely explained by the higher amount of biochar found in these anthropogenic sites. Biochar is known for retaining soil nutrients due to its specific surface and negative charge density per unit of surface area [20,42,43]. In short, the collective chemical properties intrinsic of ADE sites may play an important role in their high levels of nutrient availability and, ultimately, their fertility. In this context, we hypothesize that such fertility has caused the ADE soils to harbor a higher microbial diversity and functionality than their adjacent soils. Agricultural practices can also alter soil properties, mostly by interfering on the biochemistry of the organic matter available in the system [44]. Our data revealed such an influence in the measurements of SOC, SOM and eCEC, which were significantly higher in ADE sites under secondary forest than under manioc cultivation system. Other chemical properties such as pH, SB and amount of P, Mg and H+Al, did not statistically differ between different land uses in ADE and ADJ sites.

Soil chemical data was also used in regression analyses to understand patterns in the microbial community structure of the *bph* gene. Microbial community structure can be defined by patterns of species abundance and population within a given community [45], which are mostly regulated by the ability of the microorganisms to interact among them and with local conditions [46]. In this study, T-RFLP results did not reveal significant differences among the richness of T-RFs between ADE and ADJ samples for the total bacterial community (16S rRNA). However, ANOSIM revealed significant differences in total bacterial community structure between ADJ and ADE soils types (R = 0.17) and between land uses (R = 0.11 for ADE-SF versus

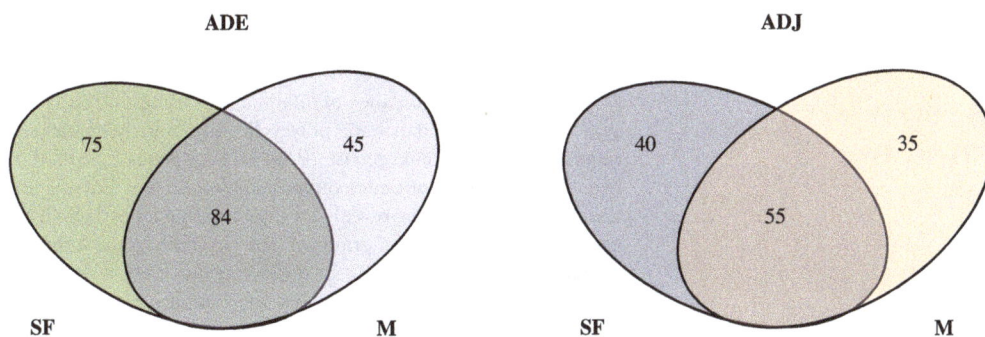

Figure 4. Venn's diagram of _bph_ data belonging to operational protein families (OPFs) for different soil types under different land uses. ADE = Amazon Dark Earth; ADJ = Adjacent soils; SF = Secondary Forest; M = Manioc cultivation. Sequences were grouped into OPFs based on sequence identity of 94%.

ADE-M; and R = 0.03 for ADJ-SF versus ADJ-M; Table 1). Lima [47], who similarly assessed the variation of ADE-SF, ADE-M, ADJ-SF and ADJ-M sites, and Cannavan [48], who investigated changes across different ADE sites, also found differences in 16S rRNA community structure between ADE and ADJ sites. Collectively, these studies support the idea that archaeological sites with a long history of anthropogenic activity influence the inhabiting microbiota of their respective soils.

Analysis of the community patterns for the bacterial _bph_ gene revealed higher richness of T-RFs in ADE than in ADJ sites. Our results show that _bph_ community changes are clear among sites (Figure 2), which can be observed by changes in the abundance of specific T-RFs across sites. In the same way, ANOSIM revealed significant differences in _bph_ gene community structure between ADJ and ADE soils types (R = 1.00) and between land uses (R = 1.00 for ADE-SF versus ADE-M; and R = 1.00 for ADJ-SF versus ADJ-M; Table 1). We suspect, despite all variables present in our system, that the high presence of biochar, which chemically

is formed by an aromatic polycyclic structure, could be a factor influencing the observed differences. Táncsics et al [49], by analyzing the structure community of catechol 2,3-dioxygenase genes in aromatic hydrocarbon contaminated environments by T-RFLP, observed that T-RFLP chromatograms obtained from contaminated samples had entirely different T-RFs compared to the control non-contaminated sample.

Redundancy analysis revealed _bph_ gene structure to be correlated with different soil parameters (Figure 4). Briefly, ADE-SF correlated with pH, eCEC, SOM and SOC. ADJ-SF showed a significant correlation with H+Al. Several studies have suggested that variations in soil pH and properties related to soil acidity (e.g., K, Al, base saturation) are stronger predictors of the richness and diversity of inhabiting microbial communities [6,18,50,51,52,53]. We extend this concept by advocating that other factors also might play a role in structuring the _bph_ gene communities. For instance, the quantity and/or quality of soil organic matter (SOM) and its fractions are likely to regulate

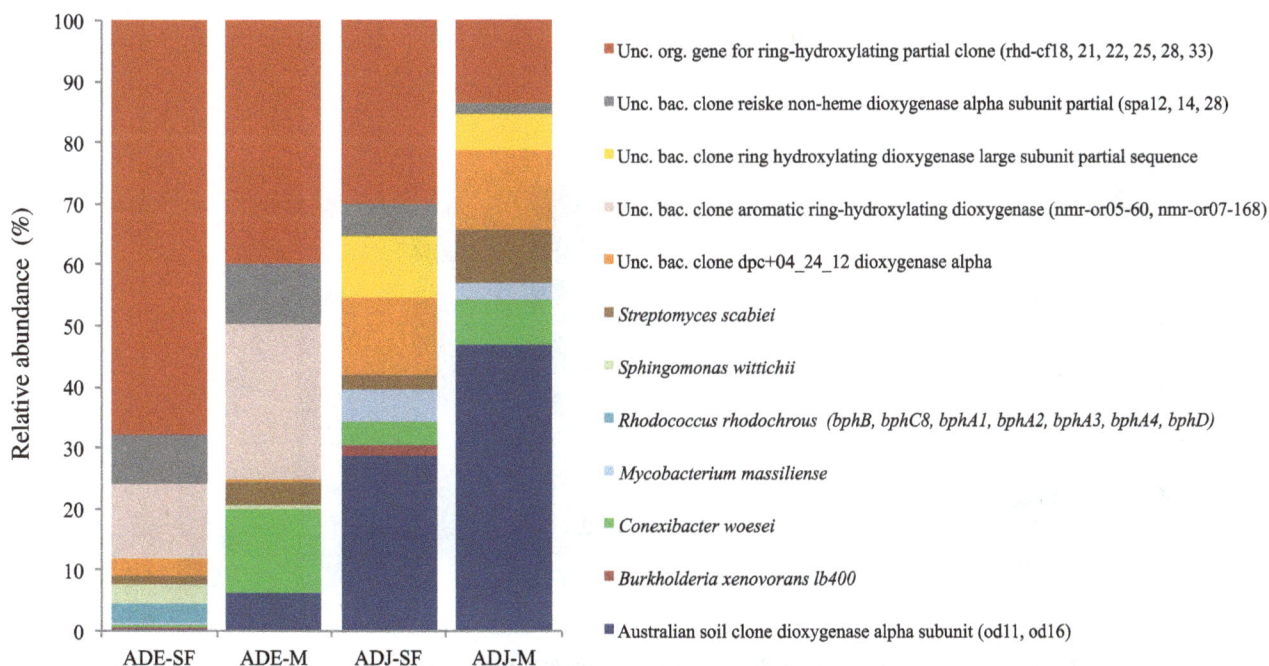

Figure 5. Bar charts representing the taxonomic affiliation of _bph_ gene sequences. Sequences were affiliated using the TBlastX tool available in the GenBank database. ADE = Amazon Dark Earth; ADJ = Adjacent soils; SF = Secondary Forest; M = Manioc cultivation.

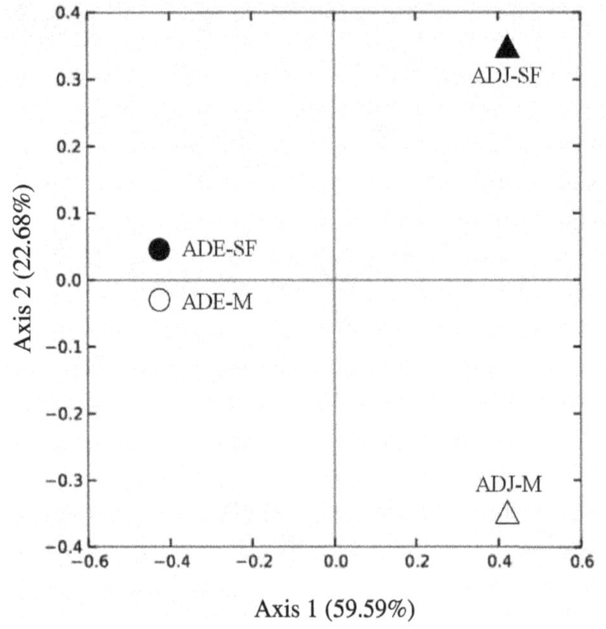

Figure 6. Principal Coordinate Analysis (PCoA) performed with the sequences obtained for the *bph* gene (based on Unweighted Unifrac distance). ADE = Amazon Dark Earth; ADJ = Adjacent soils; SF = Secondary Forest; M = Manioc cultivation.

genes in ADE sites, peaking at ADE-SF. These results indicate that soil characteristics of ADE, as well as their land uses has an influence on the abundance of both analyzed genes. Ding et al [55] analyzed the abundance of total bacteria using the 16S rRNA gene and the abundance of polycyclic aromatic hydrocarbons ring-hydroxylating dioxygenase (PAH-RHD$_\alpha$) genes by qPCR in two different phenanthrene-contaminated soils (i.e. Luvisol and Cambisol). Their results revealed a significantly higher 16S rRNA gene copy number per gram of soil in both phenanthrene-contaminated soils compared to their controls. PAH-RHD genes were detected only in contaminated soil samples, and the values ranged from 2.0×10^7 (Luvisol phenanthrene-contaminated soil) to 1.7×10^6 (Cambisol phenanthrene-contaminated soil) copies per gram of soil, showing similar values to those found in our study. Similarly, a study about the diversity of naphthalene dioxygenase genes (*nahAc*) in soil environments from the Maritime Antarctic revealed that the quantities detected in bulk and rhizospheric soils from PAH-affected sites ranged from 6.4×10^4 to 1.7×10^6 *nahAc* gene copies per gram of soil and presented significantly higher abundance compared with the corresponding counterparts of bulk and rhizospheric soils from non-polluted sites [56].

We used 454-pyrosequencing to investigate the community composition of the *bph* gene. The use of a high-throughput culture-independent approach enabled an overview of the taxonomic groups occurring in the sampled sites. Taxonomic analyses of 3,000 partial sequences of the *bph* gene revealed that this gene was most represented by bacteria belonging to the genera *Streptomyces*, *Sphingomonas*, *Rhodococcus*, *Mycobacterium*, *Conexibacter* and *Burkholderia*, in addition to sequences matching uncultured bacteria (Figure 5). Overall, these groups encompass mixed sequences of toluene/biphenyl dioxygenases, such as *bph*B, *bph*C8, *bph*A1, *bph*A2, *bph*A3, *bph*A4 and *bph*D, and enzymes encoding the alpha subunit of dioxygenases for the degradation of Polycyclic Aromatic Hydrocarbons (PAHs), which are described as environmental widely distributed dioxygenases. Other studies targeting dioxy-

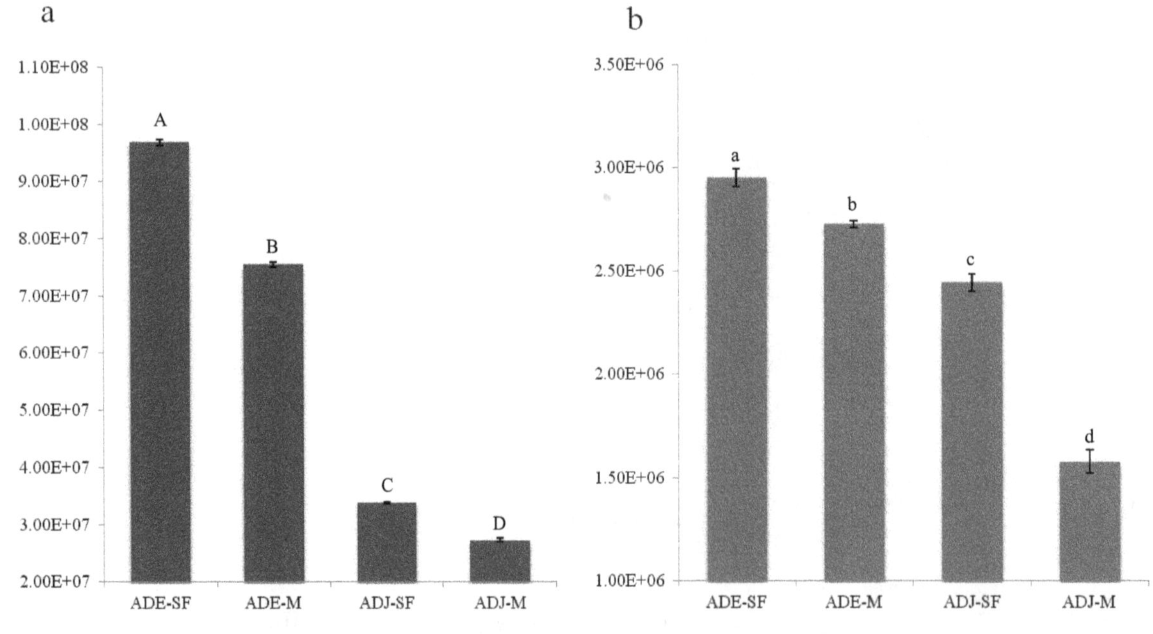

microbial community composition and associate function [54]. Since hydrocarbon degradation is one step into the carbon cycle, *bph* degraders might have an advantage in ADE sites by harnessing energy from elevated levels of SOC and SOM.

Quantification of total bacterial (16S rRNA) and *bph* gene abundances (Figure 7) revealed higher copy numbers of these

Figure 7. Log gene copy number (*y* axis) of (a) 16S rRNA and (b) *bph* genes across the studied sites. Sites are indicated in the *x* axis. Error bars represent the standard deviation of three independent replicates. ADE = Amazon Dark Earth; ADJ = Adjacent soils; SF = Secondary Forest; M = Manioc cultivation. Different upper case letters refer to differences for the 16S rRNA across sites; while different lower case letters refer to differences for the *bph* gene (*P*<0.01, Tukey test).

genases have also reported sequences belonging to *Sphingomonas*, *Rhodococcus*, *Mycobacterium*, *Conexibacter* and *Burkholderia* as major taxonomic groups [21,28,57]. Zhou et al [58] were also able to isolate aromatic hydrocarbon degrading bacteria from mangrove sediments, which were classified as the genera *Mycobacterium* and *Sphingomonas*. Members of these genera have commonly been isolated from diverse sediments and soils [59,60], and they play an important role in hydrocarbon biodegradation [61,62,63]. For instance, Ding et al [55] who studied the diversity of dioxygenases using clone libraries for aromatic ring hydroxylating dioxygenases (ARHD) genes identified gene sequences corresponding to the *phnAc* gene belonging to the *Burkholderia* sp. strain Eh1-1 and PAH-RHD genes of the *Mycobacterium* sp. strain JLS.

Confirming the results of T-RFLP analysis, differences among *bph* communities for ADE and ADJ sites and between SF and M land uses were also observed by differences in their Shannon and Simpson diversity indices (Table 2). Similarly, Germano et al [28] reported a higher diversity of ARHD genes in ADE sites under secondary forest rather than under agricultural cultivation, suggesting that deforestation in these sites has an influence on the diversity of these catabolic genes. We observed that in both soil types, the number of OPF (Table 2) was higher in SF samples, also indicating an influence of the land use system on the richness of this gene. According to Jesus et al [51], the land conversion of tropical forest to agricultural use modifies the size, activity and composition of soil microbial communities. This deeply influences specific bacterial functions, including those acting on organic matter decomposition and nutrient cycling in soils.

The intensive land use by agricultural practices and the conversion of Amazon soil into agricultural areas has been reported to cause significant shifts in the chemical properties of the soil, such as variations in SOC, SOM and eCEC, leading towards a homogenization of the inhabiting bacterial community [64]. However, we hypothesized that, despite an effect of land use, greater differences in community composition would be observed according to soil type. We also expected communities in these different soils to respond differently to agricultural practices. In this context, the literature describes the resilience phenomenon as the ability of the soil to cope with external disturbances and to retain its functional capacity upon the imposition of a stress [65,66,67]. We observed that higher differences in the diversity of *bph* occurred between the different land uses in ADJ sites rather than in ADE sites. Thus, land use type appeared to have a stronger effect on the *bph* community in ADJ sites, maybe due to the higher resilience of the ADE soil against agricultural practices. Principal Coordinate Analysis (PCoA) performed for the *bph* gene also supports these results (Figure 6), revealing that ADE sites (SF and M) clustered closer to each other than observed for ADJ sites (SF and M).

The taxonomic analyses of sequences also revealed that ADE sites harbored distinct *bph* phylogenetic structure from ADJ. These results suggest that the heterogeneity of bacterial *bph* communities could be related to their ability to respond to differences of land use type and soil chemical properties. In a previous study Germano et al. [28] compared the phylogenetic structure of *bph* sequences from ADE sites and revealed that most of the protein clusters from these sites group apart from the main well-known dioxygenase groups previously proposed by Kweon et al [27].

In conclusion, we have taken a distinct and highly diverse soil to elucidate the ecological properties and taxonomic affiliation of bacterial communities characterized by the presence of the *bph* gene, which is crucial to the biodegradation of aromatic compounds. These results enable us to understand differences in the structure, abundance and composition of the main active organisms in ADE soils when compared to their adjacent locations. Further studies focusing on the catabolic activities of these communities are needed to enable a collective view of the formation, dynamics and maintenance of functional properties in ADE soils.

Supporting Information

Figure S1 Rarefaction curves of bacterial *bph* gene sequences were grouped into OPF based on distance sequence of 0.06. ADE = Amazonian Dark Earth soils; ADJ = Adjacent soils; SF = Secondary forest; M = Manioc Cultivation.

Table S1 Soil chemical properties of the studied sites. Amazonian Dark Earth (ADE) and Adjacent soil (ADJ). Sites were located under two different land uses: Secondary forest (SF) and under Manioc (*Manihot esculenta*) cultivation (M). Significant differences between sites are followed by different letters (*P*<0.05, Tukey test).

Acknowledgments

We thank Wenceslau Geraldes Teixeira and Western Amazon Embrapa for technical support. We are grateful to ACG Souza, AK Silveira, RS Macedo and TT Souza for helping with sample collection. We thank J Quensen and the Michigan State University Genomics Technology Support Facility team for the 454-pyrosequencing support; and José Elias Gomes and Fábio Duarte for their assistance in molecular analysis. We also thank Cyrus A. Mallon for English revision and helpful comments on the manuscript.

Author Contributions

Conceived and designed the experiments: MJLB MGG ABL SMT. Performed the experiments: MJLB MGG ABL. Analyzed the data: MJLB LWM. Contributed reagents/materials/analysis tools: SMT. Wrote the paper: MJLB SMT.

References

1. Lehmann J, da Silva JP Jr, Steiner C, Nehls T, Zech W, et al. (2003) Nutrient availability and leaching in an archaeological Anthrosol and a Ferralsol of the Central Amazon basin: fertilizer, manure and charcoal amendments. Plant Soil 249: 343–357.
2. Teixeira WG, Martins GC (2003) Soil physical characterization. In: Lehmann J, Kern DC, Glaser B, Woods WI, editors.Amazonian Dark Earths: Origin, properties, management.Dordrecht: Kluwer Academic. pp. 271–286.
3. Glaser B (2007) Prehistorically modified soils of central Amazonia: a model for sustainable agriculture in the twenty-first century. Philos T Roy Soc B 362: 187–196.
4. Kim JS, Sparovek G, Longo RM, De Melo WJ, Crowley D (2007) Bacterial diversity of terra preta and pristine forest soil from the Western Amazon. Soil Biol Biochem 39: 684–690.
5. O'neill B, Grossman J, Tsai SM, Gomes JE, Lehmann J, et al. (2009) Bacterial community composition in Brazilian anthrosols and adjacent soils characterized using culturing and molecular identification. Microbial Ecol 58: 23–35.
6. Taketani RG, Tsai SM (2010) The influence of different land uses on the structure of archaeal communities in Amazonian anthrosols based on 16S rRNA and amoA genes. Microbial Ecol 59: 734–743.
7. Navarrete AA, Cannavan FS, Taketani RG, Tsai SM (2010) A molecular survey of the diversity of microbial communities in different Amazonian agricultural model systems. Diversity 2: 787–809.
8. Taketani RG, Lima AB, Jesus EC, Teixeira WG, Tiedje JM, et al. (2013) Bacterial community composition of anthropogenic biochar and Amazonian anthrosols assessed by 16S rRNA gene 454 pyrosequencing. A van Leeuw J Microb 104: 233–242.

9. Thies J, Suzuki K (2003) Amazonian dark earths: biological measurements. In: Lehmann J, Kern D, Glaser B, Woods W, editors.Amazonian Dark Earths: Origin, Properties, Management.Dordrecht: Kluwer Academic. pp 287–332.

10. Fitter AH, Gilligan CA, Hollingworth K, Kleczkowski A, Twyman RM, et al. (2005) Biodiversity and ecosystem function in soil. Funct Ecol 19: 369–377.

11. Horner-Devine M, Carney K, Bohannan B (2004) An ecological perspective on bacterial biodiversity. Proc R Soc Lond B Biol Sci 271: 113–122.

12. Wawrik B, Kerkhof L, Kukor J, Zylstra G (2005) Effect of different carbon sources on community composition of bacterial enrichments from soil.Appl Environ Microbiol 71: 6776– 6783.

13. Zhou JZ, Xia BC, Treves DS, Wu LY, Marsh TL, et al. (2002) Spatial and resource factors influencing high microbial diversity in soil. Appl Environ Microbiol 68: 326–334.

14. Vogel TM, Simonet P, Jansson JK, Hirsch PR, Tiedje JM, et al. (2009) TerraGenome: a consortium for the sequencing of a soil metagenome. Nat Rev Microbiol 7: 252.

15. Bissett A, Brown MV, Siciliano SD, Thrall PH (2013) Microbial community responses to anthropogenically induced environmental change: towards a systems approach. Ecol Lett 16: 128–139.

16. Buckley DH, Schmidt TM (2001) The structure of microbial communities in soil and the lasting impact of cultivation. Microbial Ecol 42: 11–21.

17. Buckley DH, Schmidt TM (2003) Diversity and dynamics of microbial communities in soils from agro-ecosystems. Environ Microbiol 5: 441–452.

18. Fierer N, Jackson RB (2006) The diversity and biogeography of soil bacterial communities. Proc Natl Acad Sci USA 103: 626– 631.

19. Anderson CR, Condron LM, Clough TJ, Fiers M, Stewart A, et al. (2011) Biochar induced soil microbial community change: Implications for biogeo-chemical cycling of carbon, nitrogen and phosphorus. Pedobiologia 54: 309–220.

20. Liang B, Lehmann J, Sohi SP, Thies JE, O'neill B, et al. (2010) Black carbon affects the cycling of non-black carbon in soil. Org Geochem 41: 206–213.

21. Iwai S, Chai B, Sul WJ, Cole JR, Hashsham SA, et al. (2010) Gene-targeted-metagenomics reveals extensive diversity of aromatic dioxygenase genes in the environment. ISME J 4: 279–285.

22. Wackett LP (2004) Evolution of enzymes for the metabolism of new chemical inputs into the environment. J Biol Chem 279: 41259–41262.

23. Mason JR, Cammack R (1992) The electron-transport proteins of hydroxylating bacterial dioxygenases. Annu Rev Microbiol 46: 277–305.

24. Butler CS, Mason JR (1997) Structure-function analysis of the bacterial aromatic ring-hydroxylating dioxygenases. Adv Microb Physiol 38: 47–84.

25. Hayaishi O (1962) History and scope. In Hayaishi O, editor. Oxygenases. New York: Academic Press.pp. 1–29.

26. Luz AP, Pellizari VH, Whyte LG, Greer C (2004) A survey of indigenous microbial hydrocarbon degradation genes in soils from Antarctica and Brazil. Can J Microbiol 50: 323–333.

27. Kweon O, Kim SJ, Baek S, Chae JC, Adjei MD, et al. (2008) A new classification system for bacterial Rieske non-heme iron aromatic ring-hydroxylating oxygenases. BMC Biochem 9: 11.

28. Germano MG, Cannavan FS, Mendes LW, Lima AB, Teixeira WG, et al. (2012) Functional diversity of bacterial genes associated with aromatic hydrocarbon degradation in anthropogenic dark earth of Amazonia. Pesq Agropec Bras 47: 654–664.

29. Gibson DT, Parales RE (2000) Aromatic hydrocarbon dioxygenases in environmental biotechnology. Curr Opin Biotechnol 11: 236–243.

30. Claessen MEC, Barreto WDO, Paula JL, Duarte MN (1997) Análises químicas para avaliação da fertilidade do solo. Rio de Janeiro: Brazilian Agricultural Research Corporation Press. 212p

31. Amann RI, Ludwig W, Schleifer KH (1995) Phylogenetic identification and in situ detection of individual microbial cells without cultivation. Microbiol Rev 59: 143–169.

32. Culman SW, Gauch HG, Blackwood CB, Thies JE (2008) Analysis of T-RFLP data using analysis of variance and ordination methods: a comparative study. J Microbiol Methods 75: 55–63.

33. Goris J, De Vos P, Caballero-Mellado J, Park JH, Falsen E, et al. (2004) Classification of the PCB and biphenyl-degrading strain LB400 and relatives as Burkholderia xenovorans sp. nov. Int J Syst Evol Microbiol 54: 1677–1681.

34. Schloss PD, Westcott SL, Ryabin T, Hall JR, Hartmann M, et al. (2009) Introducing mothur: open-source, platform-independent, community supported software for describing and comparing microbial communities. Appl Environ Microbiol 75: 7537–7541.

35. Good IJ (1953) The population frequencies of species and the estimation of the population parameters. Biometrika 40: 237–264.

36. Conesa A, Götz S, Garcia-Gomez JM, Terol J, Talon M, et al. (2005) Blast2GO: a universal tool for annotation, visualization and analysis in functional genomics research. Bioinformatics 21: 3674–3676.

37. Heuer H, Krsek M, Baker P, Smalla K, Wellington EMH (1997) Analysis of actinomycete communities by specific amplification of genes encoding 16S rRNA and gelelectrophoresis separation in denaturing gradients. Appl Environ Microbiol 63: 3233–3241.

38. Grossman JM, O'neill BE, Tsai SM, Liang B, Neves E, et al. (2010) Amazonian anthrosols support similar microbial communities that differ distinctly from those extant in adjacent, unmodified soils of the same mineralogy. Microb Ecolol 60: 192–205.

39. Fierer N, Bradford MA, Jackson RB (2007) Toward an ecological classification of soil bacteria. Ecology 88: 1354–1364.

40. Lauber CL, Hamady M, Knight R, Fierer N (2009) Pyrosequencing-based assessment of soil pH as a predictor of soil bacterial community structure at the continental scale. Appl Environ Microbiol 75: 5111–5120.

41. Falcão N, Moreira A, Comenford NB (2009) A fertilidade dos solos de Terra Preta de Índio da Amazônia Central. In: Teixeira WG, Kern DC, Madari BE, Lima HN, Woods W, editors.As Terras Pretas de Índio da Amazônia: sua caracterização e uso deste conhecimento na criação de novas áreas.Manaus: Embrapa Amazônia Ocidental. pp. 189–200.

42. Glaser B, Haumaier L, Guggenberger G, Zech W (2001) The 'Terra Preta' phenomenon: a model for sustainable agriculture in the humid tropics. Naturwissenschaften 88: 37–41.

43. Cunha TJF, Novotny EH, Madari BE, Benites VM, Martin-Neto L, et al. (2009) O Carbono Pirogênico. In: Teixeira WG, Kern DC, Madari BE, Lima HN, Woods W, editors.As Terras Pretas de Índio da Amazônia: sua caracterização e uso deste conhecimento na criação de novas áreas.Manaus: Embrapa Amazônia Ocidental. pp. 264–285.

44. Bünemann EK, Marschner P, Smernik RJ, Conyers M, Mcneill AM (2008) Soil organic phosphorus and microbial community composition as affected by 26 years of different management strategies. Biol Fert Soils 44: 717–726.

45. Ricklefs RE, Miller G (1999) Ecology. New York: W H Freeman. 822p.

46. Bernhard AE, Colbert D, McManus J, Field KG (2005) Microbial community dynamics based on 16S rRNA gene profiles in a Pacific Northwest estuary and its tributaries. FEMS Microbiol Ecol 52: 115–128.

47. Lima AB (2012) Influência da cobertura vegetal nas comunidades de bactérias em Terra Preta de Índio na Amazônia Central brasileira. Piracicaba: University of São Paulo. 116 p.

48. Cannavan FS (2012) A estrutura e composição de comunidades microbianas (Bacteria e Archaea) em fragmentos de carvão pirogênico de Terra Preta de Índio da Amazônia Central. Piracicaba: University of São Paulo. 116 p.

49. Táncsics A, Szabóc I, Bakab E, Szoboszlayc S, Kukolyad J, et al. (2010) Investigation of catechol 2,3-dioxygenase and 16S rRNA gene diversity in hypoxic, petroleum hydrocarbon contaminated groundwater. Syst Appl Microbiol 33: 398–406.

50. Nicol GW, Tscherko D, Chang L, Hammesfahr U, Prosser JI (2006) Crenarchaeal community assembly and microdiversity in developing soils at two sites associated with deglaciation. Environ Microbiol 8: 1382–1393.

51. Jesus EC, Marsh TL, Tiedje JM, Moreira FMS (2009) Changes in land use alter the structure of bacterial communities in Western Amazon soils. ISME J 3: 1004–1011.

52. Nielsen UN, Osler GHR, Campbell CD, Burslem DFRP, van der Wal R (2010) The influence of vegetation type, soil properties and precipitation on the composition of soil mite and microbial communities at the landscape scale. J Biogeogr 37: 1317–1328

53. Wessen E, Hallin S, Philippot L (2010) Differential responses of bacterial and archaeal groups at high taxonomical ranks to soil management. Soil Biol Biochem 42: 1759–1765.

54. Murphy DV, Cookson WR, Braimbridge M, Marschner P, Jones DL, et al. (2011) Relationships between soil organic matter and the soil microbial biomass (size, functional diversity, and community structure) in crop and pasture systems in a semi-arid environment. Soil Res 49: 582–594.

55. Ding GC, Heuer H, Zuhlke S, Spiteller M, Pronk JG, et al. (2010) Soil Type-Dependent Responses to Phenanthrene as Revealed by Determining the Diversity and Abundance of Polycyclic Aromatic Hydrocarbon Ring-Hydrox-ylating Dioxygenase Genes by Using a Novel PCR Detection System. Appl Environ Microbiol 76: 4765–4771.

56. Flocco CG, Gomes NC, Mac CW, Smalla K (2009) Occurrence and diversity of naphthalene dioxygenase genes in soil microbial communities from the maritime Antarctic. Environ Microbiol 11: 700–714.

57. Leigh MB, Pellizari VH, Uhlik O, Sutka R, Rodrigues J, et al. (2007) Biphenyl-utilizing bacteria and their functional genes in a pine root zone contaminated with polychlorinated biphenyls (PCBs). ISME J 1: 134–148.

58. Zhou HW, Guo CL, Wong YS, Tam NFY (2006) Genetic diversityof dioxygenase genes in polycyclic aromatic hydrocarbon-degrading bacteria isolated from mangrove sediments. FEMS Microbiol Lett 262: 148–157.

59. Leys NMEJ, Ryngaert A, Bastiaens L, Verstraete W, Top EM, et al. (2004) Occurrence and phylogenetic diversity of Sphingomonas strains in soils contaminated with polycyclic aromatic hydrocarbons. Appl Environ Microbiol 70: 1944–1955.

60. Miller CD, Hall K, Liang YN, Nieman K, Sorensen D, et al. (2004) Isolation and characterization of polycyclic aromatic hydrocarbon-degrading Mycobac-terium isolates from soil. Microb Ecol 48: 230–238.

61. Khan AA, Wang RF, Cao WW, Doerge DR, Wennerstrom D, et al. (2001) Molecular cloning, nucleotide sequence, and expression of genes encoding a polcyclic aromatic ring dioxygenase fromMycobacterium sp strain PYR-1. Appl Environ Microbiol 67: 3577–3585.

62. Krivobok S, Kuony S, Meyer C, Louwagie M, Willison JC, et al. (2003) Identification of pyrene-induced proteins in Mycobacterium sp strain 6PY1: evidence for two ringhydroxylating dioxygenases. J Bacteriol 185: 3828–3841.

63. Demaneche S, Meyer C, Micoud J, Louwagie M, Willison JC, et al. (2004) Identification and functional analysis of two aromatic-ring-hydroxylating dioxygenases from a Sphingomonas strain that degrades various polycyclic aromatic hydrocarbons. Appl Environ Microbiol 70: 6714–6725.

64. Rodrigues JLM, Pellizari VH, Mueller R, Baek K, Jesus EC, et al. (2013) Conversion of the Amazon rainforest to agriculture results in biotic homogenization of soil bacterial communities. Proc Natl Acad Sci USA 110: 988–993.

65. Arthur E, Schjønning P, Moldrup P, de Jonge LW (2012) Soil resistance and resilience to mechanical stresses for three differently managed sandy loam soils. Geoderma 173–174: 50–60.

66. Gregory AS, Watts CW, Whalley WR, Kuan HL, Griffiths S, et al. (2007) Physical resilience of soil to field compaction and the interactions with plant growth and microbial community structure. Eur J Soil Sci 58: 1221–1232.

67. Schjønning P, Elmholt S, Christensen BT (2004) Soil quality management: concepts and terms. In: Schjønning P, Elmholt S, Christensen BT, editors.Challenges in Modern Agriculture.Wallingford: CABI Publishing. pp. 1–15.

24

Carbon Dioxide Flux from Rice Paddy Soils in Central China: Effects of Intermittent Flooding and Draining Cycles

Yi Liu[1], Kai-yuan Wan[1], Yong Tao[1], Zhi-guo Li[1], Guo-shi Zhang[1], Shuang-lai Li[2], Fang Chen[1]*

1 Laboratory of Aquatic Botany and Watershed Ecology, Wuhan Botanical Garden, Chinese Academy of Sciences China, Wuhan, China, **2** Institute of Plant Protection and Soil Fertilizer, Hubei Academy of Agricultural Sciences, Wuhan, China

Abstract

A field experiment was conducted to (i) examine the diurnal and seasonal soil carbon dioxide (CO_2) fluxes pattern in rice paddy fields in central China and (ii) assess the role of floodwater in controlling the emissions of CO_2 from soil and floodwater in intermittently draining rice paddy soil. The soil CO_2 flux rates ranged from -0.45 to 8.62 $\mu mol.m^{-2}.s^{-1}$ during the rice-growing season. The net effluxes of CO_2 from the paddy soil were lower when the paddy was flooded than when it was drained. The CO_2 emissions for the drained conditions showed distinct diurnal variation with a maximum efflux observed in the afternoon. When the paddy was flooded, daytime soil CO_2 fluxes reversed with a peak negative efflux just after midday. In draining/flooding alternating periods, a sudden pulse-like event of rapidly increasing CO_2 efflux occured in response to re-flooding after draining. Correlation analysis showed a negative relation between soil CO_2 flux and temperature under flooded conditions, but a positive relation was found under drained conditions. The results showed that draining and flooding cycles play a vital role in controlling CO_2 emissions from paddy soils.

Editor: Dorian Q. Fuller, University College London, United Kingdom

Funding: The study was supported by the National Natural Science Foundation of China (31100386), and the Cooperated Program with International Plant Nutrition Institute (IPNI-HB-33). The funders had no role in study design, data collection and analysis, decision to publish, or preparation of the manuscript.

Competing Interests: The authors have declared that no competing interests exist.

* E-mail: fchenipni@126.com

Introduction

Increases in the emission of greenhouse gases such as carbon dioxide (CO_2), methane (CH_4), and nitrous oxide (N_2O) from soil surface to the atmosphere have been a worldwide concern for several decades [1–3]. CO_2 is recognized as a significant contributor to global warming and climatic change, accounting for 60% of global warming or total greenhouse effect [4]. Measuring the soil CO_2 efflux is crucial for accurately evaluating the effects of soil management practices on global warming and carbon cycling. Temporal variations in soil CO_2 flux have been observed in almost all ecosystems [5,6]. Soil CO_2 fluxes are usually higher during warm seasons and lower during cold seasons [7,8]. The seasonal variation is driven largely by changes in temperature, moisture, and photosynthate production [5,9,10]. The main factors controlling seasonal variations in soil CO_2 flux may depend on the type of ecosystems and the climate.

The increase in population in areas where rice is the main cultivated crop has led to the increase in worldwide area under rice cultivation by approximately 40% over the last 50 years [11]. In particular, Asian countries (China, India, Indonesia, etc.) have accounted for approximately 90% of the total global area under rice cultivation for the last 50 years [11]. Rice paddies in monsoonal Asia play an important role in the global budget of greenhouse gases such as CH_4 and CO_2 [12,13]. Carbon emissons (esp. CH_4) from rice paddies are expected to be a long-term contributor to greenhouse gases, perhaps

increasingly over the past 5000 years [14]. Efforts have been made recently to model carbon emissions based on the history and archaeology of rice cultivation in Asia. However, since these emissions from rice cultivation vary a great deal, this poses a major challenge in modeling this phenomenon [15]. As a result, experimental research from rice paddies assumes greater importance. Many of the factors controlling gas exchange between rice paddies and the atmosphere are different from those in dryland agriculture and other ecosystems because rice is flooded during most of its cultivation period. The dynamics of soil CO_2 fluxes in a paddy field differs significantly from that in fields with upland crop cultivation in which aerobic decomposition process is dominant [6,16,17]. Field studies designed to measure soil CO_2 fluxes and improve our understanding of the factors controlling the fluxes are thus needed.

Intermittent draining and flooding, which is one of the most important water management practices in rice production, was found to be the most promising option for CH_4 mitigation also [18,19]. Mid-season aeration was also found to be one of the basic techniques for raising rice yields in China [20] and was widely adopted in rice cultivation where irrigation/drainage system was well managed. The management induced change of anaerobic and aerobic conditions results in temporal and spatial (vertical, horizontal) variations in reduction and oxidation (redox) reactions affecting the dynamics of organic and mineral soil constituents [21,22]. Thus, intermittent drainage with increased impacts can strongly affect soil CO_2 emissions [6,16]. However, the mechanism of CO_2 exchange between rice paddies and the atmosphere is not fully understood. For example, using eddy covariance measurements, Miyata et al. [16] found a significantly larger net CO_2 flux from the rice paddy soil to atmosphere when the field was drained compared to when it was flooded. These differences in the CO_2 flux were mainly due to increased CO_2 emissions from the soil surface under drained conditions resulting from the removal of diffusion barrier caused by the floodwater. The existence of floodwater, anaerobic soil, or changes in the micrometeorological environment with flooding influences root activity, photosynthesis, and respiration of rice plants [23]. Activity of aquatic plants such as algae in the floodwater may also affect CO_2 exchange between rice paddies and the atmosphere [22]. Most of the data obtained so far were not sufficiently detailed to examine the influence of these factors on the CO_2 exchange in rice paddies.

The scale and dynamics of growing-season CO_2 emissions from paddy fields have been documented mostly through flux measurements made with low time resolution using manual chambers [6,16,17]. In this study, we report a data set that extends hourly CO_2 flux measurements during the rice-growing season in 2011 to improve the understanding of the process controlling CO_2 exchanges in rice paddy soils. The measurements were used to assess the role of floodwater in controlling the exchanges of CO_2 from the paddy soil. The objectives of this study were to: (i) analyze seasonal and diurnal variation of soil CO_2 fluxes in rice paddy fields in the Yangtze River valley; and (ii) determine the effects of related environmental factors associated with flooding and draining cycles in paddy soils on CO_2 flux from the soil surface.

Materials and Methods

Site Description

Field experiments were conducted over one rice growing season, i.e. from June to October 2011, at Nanhu Agricultural Research Station (30°28′N, 114°25′E, altitude 20 m). The research site is owned by Hubei Academy of Agricultural Sciences. The field studies did not involve endangered or protected species and no specific permits were required for the described field studies. The site lies in a typical area of the humid mid-subtropical monsoon climate in the Yangtze River valley of China. The mean annual temperature of the site is 17°C, the cumulative temperature above 10°C is 5,190°C, and the average annual frost-free period is 276 d. The average annual precipitation is 1,300 mm, with most of the rainfall occurring between April and August. The paddy field soil is a Hydromorphic paddy soil, which is a silty clay loam derived from Quaternary yellow sediment. Some physical and chemical properties of the experimental soil (0–20 cm depth) were: pH, 6.3; organic matter, 30.23 g.kg^{-1}; total N, 2.05 g.kg^{-1}; available P, 5 mg.kg^{-1}; available K, 101 mg.kg^{-1}; soil bulk density, 1.26 g.cm^{-3}. The experimental site has been under rice-wheat cultivation since last 30 years, where rice is planted from June to October each year and wheat is planted from November to May the following year. Daily meteorological information (including rainfall and temperature) during the 2011 rice-growing season is presented in Fig. 1.

Field Management

In 2011, rice was transplanted to the paddy field on 15 June with a plant to plant spacing of 20 cm and a row spacing of 27 cm. Irrigation started on 13 June and the field was flooded continuously until 17 July. This was followed by five intermittent flooding and draining cycles, with 3–7 days of flooding and 2–8

days of draining. The field was not irrigated and drained about a month before harvesting. The number of flooded days were 55, while the number of drained days were 53 during the 2011 rice-growing seasons. The depth of standing water during flooding periods was, on average, 10 cm. Before transplanting, base fertilizer – consisting of 36 kg N ha^{-1} in the form of urea (N 46%), 45 kg P_2O_5 ha^{-1} in the form of calcium superphosphate (P_2O_5 12%), and 90 kg K_2O ha^{-1} in the form of potassium sulfate (K_2O 45%) – was broadcast over the soil, which was then turned over by plowing to transfer the fertilizer to the subsurface (i.e., beyond 20 cm soil depth). Additional nitrogen, in the form of urea, was applied at tillering and heading stages of rice growth at rates of 36 and 18 kg N ha^{-1}, respectively. Rice grain was harvested from 1 to 3 October, 2011.

Measurement of Soil CO_2 Flux

The soil CO_2 flux was measured using the soil respiration method, where a cylinder static chamber of 22.5 cm diameter and 30 cm height was placed on the soil. The rate of increase in CO_2 concentration within the chamber was monitored with an ACE (ADC BioScientific Ltd) automated soil CO_2 flux system. The automated design means that during analysis cycles, the soil can be exposed to ambient conditions before the chamber closes to take measurements. This means the ACE will continue to collect data without any human intervention for as long as permitted by its battery life. This makes the ACE an ideal research instrument for continuous assessment of below-ground respiration and carbon stores in on-going experiments. Static chambers were inserted to a depth of approximately 7 cm, extending 23 cm above the soil surface to allow placing of the chamber. During the flooding period, the water remained in situ. The time span between chamber contact with the soil and the start of measurements (the deadband) was 20 s; this has previously been determined to be sufficient for pressure equilibration. The measurement time was set to 180 s. The ACE has a highly accurate CO_2 infrared gas analyzer housed directly inside the soil chamber, with no long gas tubing connecting the soil chamber and no separate analyzer. This ensures accurate and robust measurements, and the fastest possible response times to fluxes in gas exchange. During the soil CO_2 flux measurements, air temperature within the canopy and soil temperature at 2 cm depth were also recorded by the ACE analyzer unit. And the measurements were made at 1-hour intervals during the rice-growing season. During a 24-hour period, the values were averaged to give the mean daily soil CO_2 flux. Survey sites of three replications were taken from the experiment plot. Survey sites were located in the space between two rows, and the two sites were located 5–7 m apart. Three ACE stations were connected via an ACE Master control unit. Each CO_2 flux measurement from the experiment plot was thus an average of three individual measurements.

In order to examine the diurnal soil CO_2 flux pattern in a paddy field, soil CO_2 flux as well as canopy air temperature, soil temperature and PAR were also measured simultaneously at 1 hour intervals for 24 hours under both flooded (6/28~6/29 and 8/14~8/15) and drained (7/20~7/21 and 9/4~9/5) conditions. During these 24 hour periods, the sky was clear and with no clouds.

To study the soil CO_2 emissions in relation to draining and flooding cycle system, two draining/flooding alternation and circulation periods (7/23~7/28 and 8/29~9/4) were tested. We continuously monitored soil CO_2 fluxes along with air temperature within the canopy and soil temperature before, during, and after each flooding and draining cycle in the experiment paddy soil. Clear days continued during the experiment, but temperature

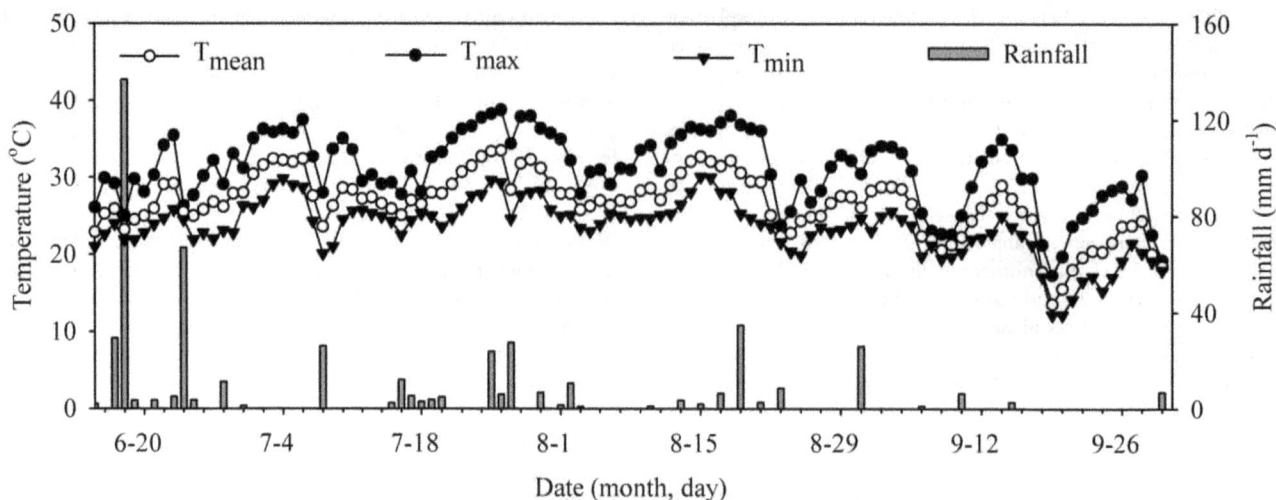

Figure 1. Air temperature records and rainfall events at the study site during the experimental period. (T_{mean}: mean temperature; T_{max}: maximum temperature; T_{min}: minimum temperature).

conditions were a little different from day to day. Flooding started at 9 am (09:00 h) and water depth reached 10 cm around midday. The water level was gradually decreased with cessation of irrigation.

Results

Seasonal Variations in Soil CO_2 Fluxes from Paddy Fields

The daily course of soil CO_2 flux rate is shown in Fig. 2A, while Fig. 2B shows the air temperature within the canopy and soil temperature (2 cm). The soil CO_2 flux rates ranged from -0.45 to 8.62 $\mu mol.m^{-2}.s^{-1}$, exhibiting a wide seasonal fluctuation during the rice-growing season. The soil CO_2 fluxes were generally low at the rice seedling stage, when it remained at about $0\sim1\mu mol.m^{-2}.s^{-1}$ until the first mid-summer drainage. Then the fluxes increased gradually until the tillering stage, with a midway peak near the end of the first mid-summer drainage. From the tillering stage to the physiological maturity stage (i.e, from July to September), the daily average soil CO_2 flux rates had a magnitude ranging between 0 and 9 $\mu mol.m^{-2}.s^{-1}$, which then settled at around $1\sim3$ $\mu mol.m^{-2}.s^{-1}$ until the end of the season. The differences in the rates of soil CO_2 fluxes between drained and flooded conditions are also shown in Fig. 2a. Mean soil CO_2 fluxes under flooded conditions was 0.72 (with standard deviation of 0.48) $\mu mol.m^{-2}.s^{-1}$ (n = 55), whereas under drained conditions, the corresponding value was 2.79 (with standard deviation of 1.73) $\mu mol.m^{-2}.s^{-1}$ (n = 53). It is likely that floodwater decreased topsoil diffusivity, and may thus have decreased soil CO_2 effluxes [24]. Reduction of biological activity under anoxic condition may be another reason for low soil CO_2 fluxes during the flooding period [22].

The air temperature within the canopy and soil temperature (2 cm) exhibited seasonal patterns similar to soil CO_2 fluxes. The temperature varied from 15 to 33°C during the whole growing period of rice in 2011. From June to September, the temperature ranged from 21 to 33°C, and several peaks occurred. From the mid of September (9/18) to the day before harvesting (about 15 days), the average temperature of 19.7°C for air temperature within the canopy and 19.8°C for soil temperature (0–2 cm) are shown in Fig. 2.

Diurnal Patterns of Soil CO_2 Fluxes in Paddy Fields

The diurnal variations in soil CO_2 fluxes and incident PAR, air temperature within the canopy, and soil temperature under both flooding (6/28~6/29 and 8/14~8/15) and draining (7/20~7/21 and 9/4~9/5) conditions are shown in Fig. 3. These experiments began in the early evening, running for just under 24 h. Under flooding conditions, fluxes of CO_2 were, as expected, lower because the diffusivity and biological activity of the topsoil was substantially reduced by floodwater. Initially, there was a slow release of CO_2 into the atmosphere as a positive efflux settled at around $0–1$ $\mu mol.m^{-2}.s^{-1}$ throughout the night. At sunrise the fluxes decreased, even negatively peaked at around 16:00 (negative values indicate carbon sequestration). This may have been because some aquatic plants, such as algae, inside the floodwater began to photosynthesize again. In contrast, CO_2 flux under draining conditions was positive and settled around $2–4$ $\mu mol.m^{-2}.s^{-1}$ throughout the night, despite falling temperatures (Fig. 3). After sunrise, CO_2 fluxes remained positive and increased with temperature, reaching a peak at 2 pm (14:00 h) before falling again as temperatures declined.

Soil CO_2 Fluxes Related to Conversion Processes of Draining and Flooding Cycles

Fig. 4 shows soil CO_2 fluxes, canopy air temperature, and soil temperature before, during, and after the flooding and draining cycle. Soil CO_2 fluxes increased immediately after flooding, and exceeded pre-flooding values by two-thirds. This increase was abrupt and pulselike. Replacement of soil air by water should thus cause an enriched CO_2 pulse. And then, the soil CO_2 flux rate subsequently decreased by 70~90% within only one hour after the water pulse. Within the following days, the CO_2 fluxes remained at minimum levels (about $-2\sim2$ $\mu mol.m^{-2}.s^{-1}$) during flooding. As standing water declined and eventually disappeared, the CO_2 fluxes gradually increased and finally reached to maximum levels (about $6\sim8$ $\mu mol.m^{-2}.s^{-1}$). This indicates that draining and flooding cycles play vital roles in controlling CO_2 emissions in a paddy soil.

Variability of Soil CO_2 Fluxes Related to Temperature

Temperature has a marked effect on CO_2 emissions from the soil surface. To study the relationship between soil CO_2 flux rates

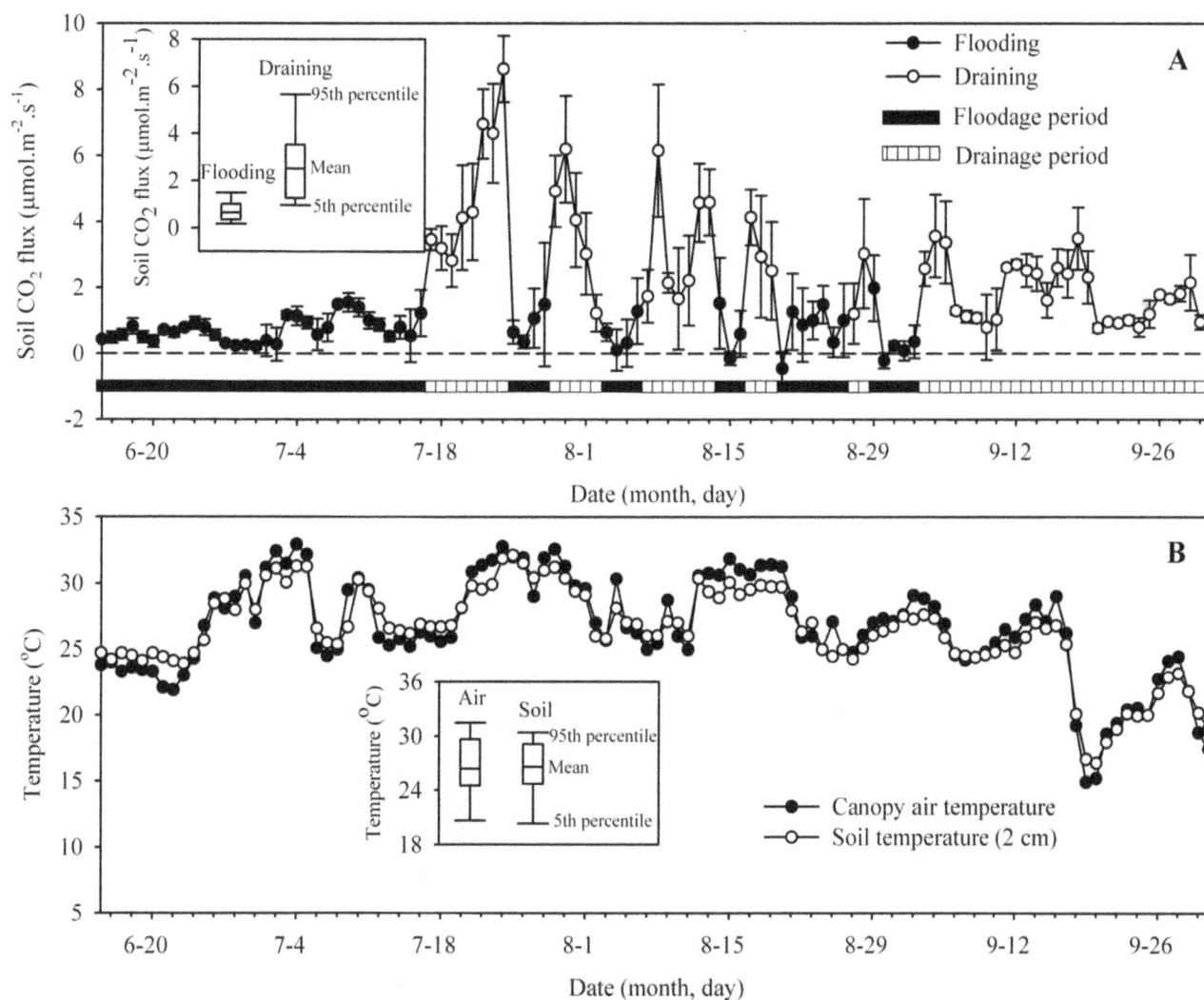

Figure 2. Seasonal variations of soil CO_2 flux, soil temperature (2 cm) and crop canopy temperature in an intermittent draining and flooding rice paddy. The insets indicate box-plot of soil CO_2 fluxes under flooded and drained conditions, and box-plot of canopy air and soil temperature during the 2011 rice-growing season.

and temperature, two environmental temperatures (air temperature within the canopy and soil temperature) were tested in this study (Fig. 5). Linear and exponential regression analysis were used to model the influence of temperature on soil CO_2 flux rates under both flooded and drained conditions. Negative linear correlations between temperature and soil CO_2 fluxes were found under flooded conditions ($R^2 = 0.1524$, $P<0.001$ and $R^2 = 0.0535$, $P<0.001$ for canopy air and soil temperatures, respectively), presumably because standing water limited soil CO_2 emissions. On the contrary, soil CO_2 flux rates increased as an exponential function of temperature under drained conditions ($R^2 = 0.1963$, $P<0.001$ and $R^2 = 0.2382$, $P<0.001$ for canopy air and soil temperatures, respectively).

Discussion

Previous research had revealed that water management systems show the highest potential in controlling CH_4 emissions [25]. CH_4 emissions were higher under continuous flooding than intermittent draining practices [26,27], while they declined during the drainage period to near zero and increased after re-flooding [28]. Drainage

during the rice cultivation period significantly increased CO_2 emissions in our study, while CH_4 emissions were clearly reduced and has been shown by other research [18,29]. Miyata et al. [16] also found that flooded or drainage conditions of paddy soils had strong effects not only on CH_4 emissions but also on CO_2 emissions. Lower CH_4 emissions due to water drainage may increase CO_2 emission. However, during the submerged period of paddy rice cultivation, CO_2 production in the soil is severely restricted under flooding condition. This effect can be explained with two basic mechanisms [8], which could be observed in a paddy soil (Fig. 6). First, flooding a field for subsequent rice cultivation cuts off the oxygen supply from the atmosphere and the microbial activities switch from aerobic (i.e. oxic condition) to facultative (i.e. hypoxic condition) and to anaerobic (i.e. anoxic condition) conditions [22]. As a consequence, biological activity reduction under anoxic condition, rather than completely, inhibits CO_2 production. At the same time, water replaces the gaseous phase in the soil pores. Since CO_2 diffusion rates in water are four orders of magnitude lower than those in air, a part of the produced CO_2 is stored in the soil. Hence, the soil CO_2 fluxes can be dramatically reduced by flooding during the paddy rice cultivation

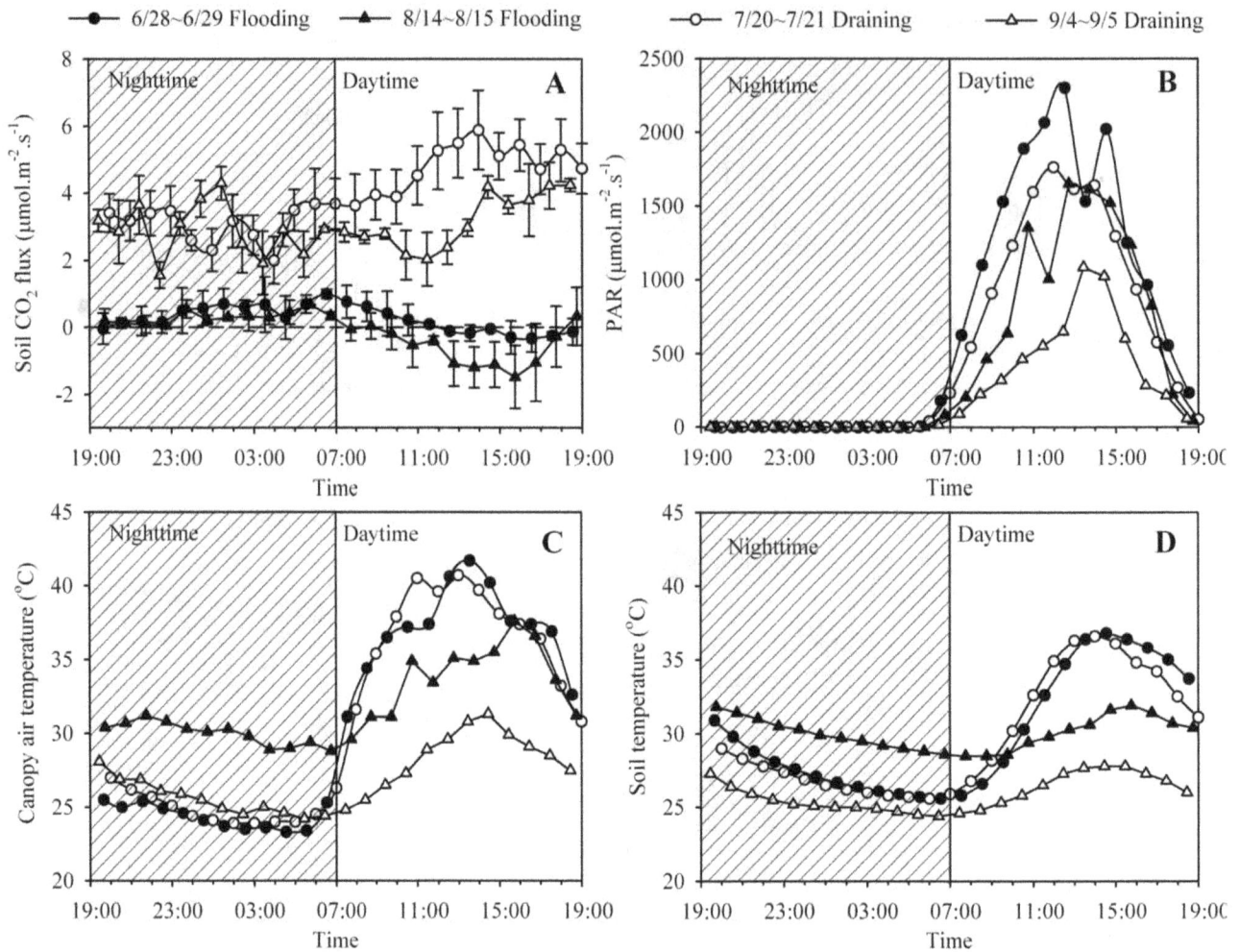

Figure 3. Diurnal patterns of soil CO₂ flux related to PAR, canopy air temperature and soil temperature (2 cm) under both drained and flooded condtions in a rice paddy.

[6,16,23]. Results from the present study provide indirect support for this conclusion, since the soil CO_2 flux rates under flooded conditions were significantly lower than those observed under drained conditions (Fig. 2).

Our study also demonstrated that, in rice fields exposed to intermittent flooding and draining cycles, environmental factors regulating diurnal fluctuations in CO_2 flux are quite different from those governing seasonal variations. Under drainage conditions, soil CO_2 flux showed a single peak at 2 pm (14:00 h), and was lowest in the wee hours. This is in agreement with patterns recorded in forests [5], grassland [30] and dryland areas [31]. Furthermore, correlation analysis revealed that canopy air temperature and soil temperature explained most of the diurnal fluctuations in soil CO_2 flux. In contrast, soil CO_2 flux during the flooding period fluctuated within ± 2 $\mu mol.m^{-2}.s^{-1}$ and soil CO_2 flux rates had small negative values in the daytime (i.e., the paddy soil was obviously a net CO_2 sink.), although soil CO_2 fluxes were positive throughout the night. This occurred primarily because of the layer of standing water, which is the habitat of bacteria, phytoplankton, macrophytes and small fauna. The photosynthesis process of these aquatic organisms affects ecosystem respiration [22].

Sudden pulse-like events of rapidly increasing CO_2 efflux occur in soils under paddy fields in response to re-flooding after draining.

Similarly, an abrupt rise in near-surface soil moisture due to precipitation can cause an instantaneous soil respiration pulse [24,32]. Soil respiration is shown to respond rapidly and instantaneously to the onset of rain and return to the pre-rain rate shortly after the rain stops [32]. The likely reason for this is that CO_2 is heavier than air and accumulates by gravitation within the air spaces of the soil. Replacement of this gaseous carbon by dilution will not occur without water and, unstirred by turbulent mixing, accumulation of CO_2 within the soil will increase. A sudden flooding might simply seal the soil pores, replace the captured CO_2 by water, and release it back into the air [33]. These occurrences, termed "Birch effect", can have a marked influence on the ecosystem carbon balance [34,35]. Indeed, this transient effect was observed in several studies at the ecosystem [36] and soil [37] scales. **On the other hand**, our analysis indicates that soil CO_2 flux was gradually increased during flooding to draining conversion processes. Response of soil CO_2 flux rates to these processes can be viewed in terms of increased diffusivity due to decrease in water filled pore space. Besides this general effect of soil aeration on soil CO_2 flux, the higher soil respiration rates during the drainage periods may have resulted from the higher physiological activity of microorganismsin not limiting soil oxic conditions [22].

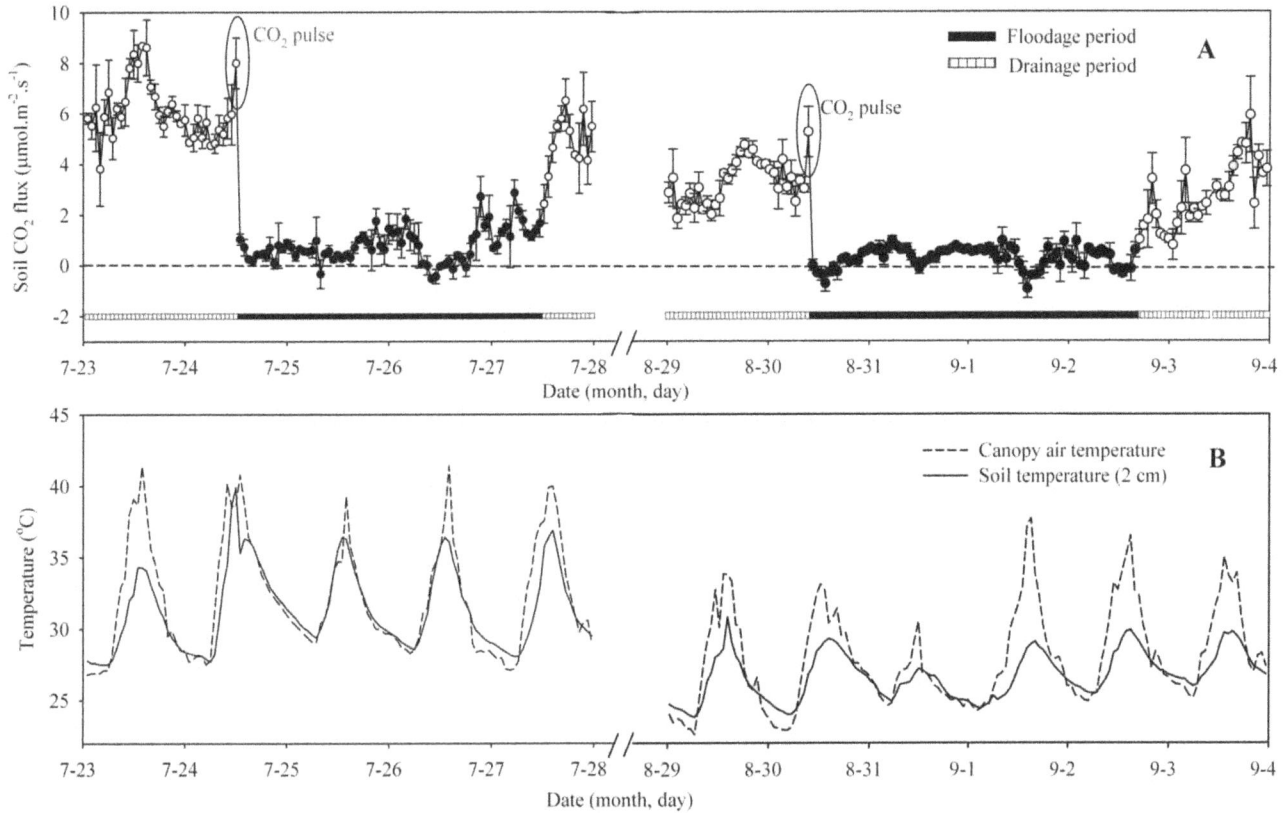

Figure 4. Soil CO$_2$ fluxes, soil temperature (2 cm) and canopy temperature before, during, and after the flooding and draining cycle in a rice paddy field.

We examined possible seasonal effects of temperature on soil CO$_2$ flux and found significant relation between the two under both flooded and drained conditions, but with widely differing mechanisms. In the present study, we found a negative relation between temperature and soil CO$_2$ flux, as long as soil CO$_2$ diffusivity is limiting as is the case during flooding period. An alternative explanation is based on the photosynthetic activity of the aquatic botany. The periods with the high photosynthetic

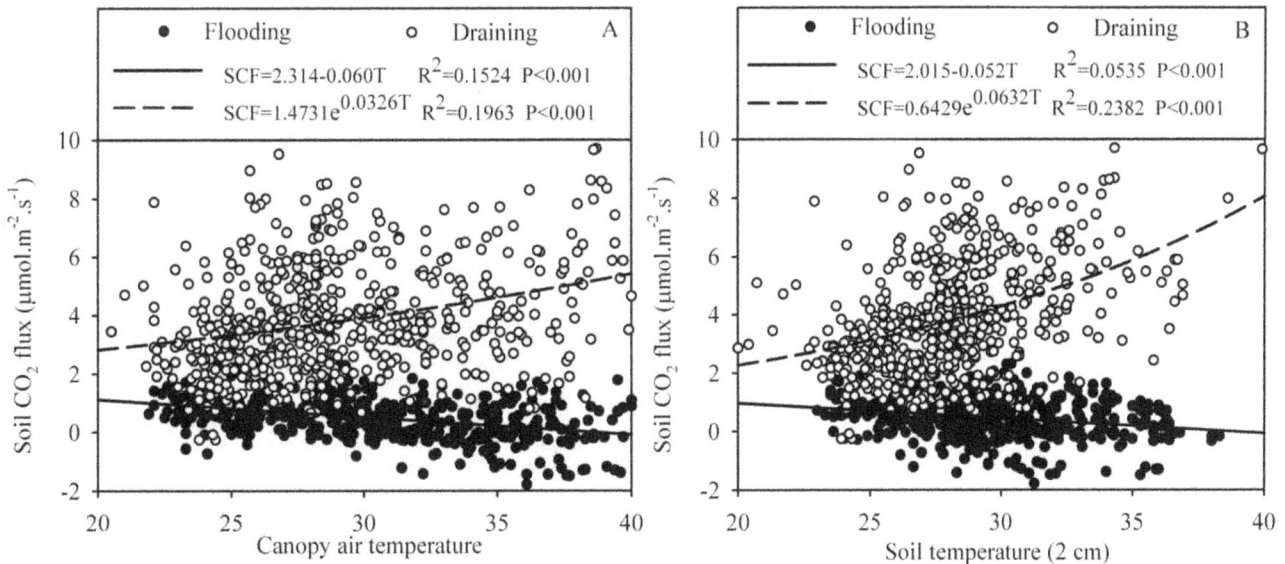

Figure 5. Relationship between soil CO$_2$ fluxes and temperature under both flooded and drained conditions. The solid lines represent the regression functions under flooded conditions, and the dashed lines represent the regression functions under drained conditions. (SCF: soil CO$_2$ fluxes; T: temperature).

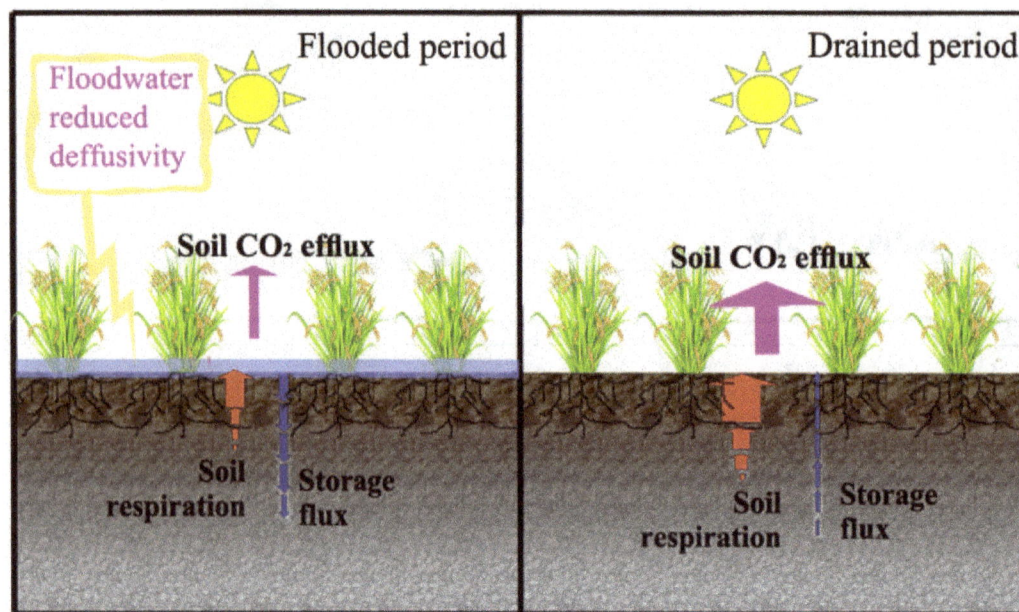

Figure 6. Schematic comparison of soil CO₂ flux processes under the flooded and the drained conditons in rice paddies.

active radiation are associated with conditions of high temperature in daytime (Fig. 3). Under drainage conditions, when soil aeration is assumed to be almost constant, soil temperature is considered to be a major control of soil CO_2 flux. Also the positive exponential relationship between soil CO_2 flux and temperature were observed during drainage period (Fig. 3). The results under drained conditions are similar to those of previous studies of CO_2 flux. For example, Chang et al. [38] found strong relationships between CO_2 flux and soil temperature and indicated that the rates of CO_2 emission increased exponentially with increases in soil temperature. Liu et al. [7], on the other hand, reported a significantly ($P<0.01$) linear relationship between soil CO_2 flux and soil temperature at a depth of 5 cm.

Conclusions

From the comparison of soil CO_2 fluxes under draining and flooding conditions we conclude that: (1) the net effluxes of CO_2 from the paddy soil were lower when the paddy was flooding than

when it was draining, (2) the enhanced fluxes of CO_2 from the draining soil were due to removal of the barrier to gas transport from the soil surface to the air caused by the floodwater, and (3) there was a negative relation between soil CO_2 flux and temperature under flooding condition, whereas a positive relation under draining condition. The present study also showed how flooding and draining cycles affect the exchanges of CO_2 during the rice cultivation period. We need more measurements for multiple years to assess the long-term effect of an intermittent flooding and draining practice on the exchanges of CO_2 in rice paddy fields.

Author Contributions

Conceived and designed the experiments: FC YL. Performed the experiments: KYW YT SLL. Analyzed the data: ZGL. Contributed reagents/materials/analysis tools: GSZ. Wrote the paper: YL.

References

1. Robertson GP, Paul EA, Harwood RR (2000) Greenhouse gases in intensive agriculture: Contributions of individual gases to the radiative forcing of the atmosphere. Science 289: 1922–1925.
2. Li CF, Zhou DN, Kou ZK, Zhang ZS, Wang JP, et al. (2012) Effects of Tillage and Nitrogen Fertilizers on CH4 and CO2 Emissions and Soil Organic Carbon in Paddy Fields of Central China. Plos One 7: e34642.
3. Guo JP, Zhou CD (2007) Greenhouse gas emissions and mitigation measures in Chinese agroecosystems. Agricultural and Forest Meteorology 142: 270–277.
4. Rodhe H (1990) A Comparison of the Contribution of Various Gases to the Greenhouse-Effect. Science 248: 1217–1219.
5. Xu M, Qi Y (2001) Soil-surface CO2 efflux and its spatial and temporal variations in a young ponderosa pine plantation in northern California. Global Change Biology 7: 667–677.
6. Saito M, Miyata A, Nagai H, Yamada T (2005) Seasonal variation of carbon dioxide exchange in rice paddy field in Japan. Agricultural and Forest Meteorology 135: 93–109.
7. Liu Y, Li SQ, Yang SJ, Hu W, Chen XP (2010) Diurnal and seasonal soil CO2 flux patterns in spring maize fields on the Loess Plateau, China. Acta Agriculturae Scandinavica Section B-Soil and Plant Science 60: 245–255.
8. Maier M, Schack-Kirchner H, Hildebrand EE, Schindler D (2011) Soil CO2 efflux vs. soil respiration: Implications for flux models. Agricultural and Forest Meteorology 151: 1723–1730.

9. Yu XX, Zha TS, Pang Z, Wu B, Wang XP, et al. (2011) Response of Soil Respiration to Soil Temperature and Moisture in a 50-Year-Old Oriental Arborvitae Plantation in China. Plos One 6: e28397.
10. Li CF, Kou ZK, Yang JH, Cai ML, Wang JP, et al. (2010) Soil CO2 fluxes from direct seeding rice fields under two tillage practices in central China. Atmospheric Environment 44: 2696–2704.
11. FAO (2010) FAOSTATS. Food and Agriculture Organization, Rome, Italy. http://faostat.fao.org/default.aspx (Accessed July, 2010).
12. Solomon S, Qin D, Manning M, Chen Z, Marquis M, et al. (eds.) (2007) Contribution of Working Group I to the Fourth Assessment Report of the Intergovernmental Panel on Climate Change. Cambridge, UK: Cambridge University Press.
13. Lee CH, Do Park K, Jung KY, Ali MA, Lee D, et al. (2010) Effect of Chinese milk vetch (Astragalus sinicus L.) as a green manure on rice productivity and methane emission in paddy soil. Agriculture Ecosystems & Environment 138: 343–347.
14. Ruddiman WF, Thomson JS (2001) The case for human causes of increased atmospheric CH4. Quaternary Science Reviews 20: 1769–1777.
15. Fuller DQ, van Etten J, Manning K, Castillo C, Kingwell-Banham E, et al. (2011) The contribution of rice agriculture and livestock pastoralism to prehistoric methane levels: An archaeological assessment. Holocene 21: 743–759.

16. Miyata A, Leuning R, Denmead OT, Kim J, Harazono Y (2000) Carbon dioxide and methane fluxes from an intermittently flooded paddy field. Agricultural and Forest Meteorology 102: 287–303.

17. Iqbal J, Hu RG, Lin S, Hatano R, Feng ML, et al. (2009) CO_2 emission in a subtropical red paddy soil (Ultisol) as affected by straw and N-fertilizer applications: A case study in Southern China. Agriculture Ecosystems & Environment 131: 292–302.

18. Wassmann R, Papen H, Rennenberg H (1993) Methane Emission from Rice Paddies and Possible Mitigation Strategies. Chemosphere 26: 201–217.

19. Tyagi L, Kumari B, Singh SN (2010) Water management - A tool for methane mitigation from irrigated paddy fields. Science of the Total Environment 408: 1085–1090.

20. Li CL, Zhang JD, Hou ZQ (eds.) (1993) Techniques for high yield cultivations of several main crops. China Scientific and Technological Publishing House Beijing. (In Chinese).

21. Cheng YQ, Yang LZ, Cao ZH, Ci E, Yin SX (2009) Chronosequential changes of selected pedogenic properties in paddy soils as compared with non-paddy soils. Geoderma 151: 31–41.

22. Kogel-Knabner I, Amelung W, Cao ZH, Fiedler S, Frenzel P, et al. (2010) Biogeochemistry of paddy soils. Geoderma 157: 1–14.

23. Campbell CS, Heilman JL, McInnes KJ, Wilson LT, Medley JC, et al. (2001) Diel and seasonal variation in CO2 flux of irrigated rice. Agricultural and Forest Meteorology 108: 15–27.

24. Maier M, Schack-Kirchner H, Hildebrand EE, Holst J (2010) Pore-space CO_2 dynamics in a deep, well-aerated soil. European Journal of Soil Science 61: 877–887.

25. Itoh M, Sudo S, Mori S, Saito H, Yoshida T, et al. (2011) Mitigation of methane emissions from paddy fields by prolonging midseason drainage. Agriculture Ecosystems & Environment 141: 359–372.

26. Cai ZC, Tsuruta H, Rong XM, Xu H, Yuan ZP (2001) CH4 emissions from rice paddies managed according to farmer's practice in Hunan, China. Biogeochemistry 56: 75–91.

27. Minamikawa K, Sakai N (2006) The practical use of water management based on soil redox potential for decreasing methane emission from a paddy field in Japan. Agriculture Ecosystems & Environment 116: 181–188.

28. Bronson KF, Neue HU, Singh U, Abao EB (1997) Automated chamber measurements of methane and nitrous oxide flux in a flooded rice soil.1. Residue, nitrogen, and water management. Soil Science Society of America Journal 61: 981–987.

29. Cai ZC, Tsuruta H, Gao M, Xu H, Wei CF (2003) Options for mitigating methane emission from a permanently flooded rice field. Global Change Biology 9: 37–45.

30. Cao GM, Tang YH, Mo WH, Wang YA, Li YN, et al. (2004) Grazing intensity alters soil respiration in an alpine meadow on the Tibetan plateau. Soil Biology & Biochemistry 36: 237–243.

31. Han GX, Zhou GS, Xu ZZ, Yang Y, Liu JL, et al. (2007) Biotic and abiotic factors controlling the spatial and temporal variation of soil respiration in an agricultural ecosystem. Soil Biology & Biochemistry 39: 418–425.

32. Lee X, Wu HJ, Sigler J, Oishi C, Siccama T (2004) Rapid and transient response of soil respiration to rain. Global Change Biology 10: 1017–1026.

33. Chen D, Molina JAE, Clapp CE, Venterea RT, Palazzo AJ (2005) Corn root influence on automated measurement of soil carbon dioxide concentrations. Soil Science 170: 779–787.

34. Birch HF (1964) Mineralisation of plant nitrogen following alternate wet and dry conditions. Plant Soil 20: 43–49.

35. Unger S, Maguas C, Pereira JS, David TS, Werner C (2010) The influence of precipitation pulses on soil respiration - Assessing the "Birch effect" by stable carbon isotopes. Soil Biology & Biochemistry 42: 1800–1810.

36. Inglima I, Alberti G, Bertolini T, Vaccari FP, Gioli B, et al. (2009) Precipitation pulses enhance respiration of Mediterranean ecosystems: the balance between organic and inorganic components of increased soil CO2 efflux. Global Change Biology 15: 1289–1301.

37. Denef K, Six J, Bossuyt H, Frey SD, Elliott ET, et al. (2001) Influence of dry-wet cycles on the interrelationship between aggregate, particulate organic matter, and microbial community dynamics. Soil Biology & Biochemistry 33: 1599–1611.

38. Chang SC, Tseng KH, Hsia YJ, Wang CP, Wu JT (2008) Soil respiration in a subtropical montane cloud forest in Taiwan. Agricultural and Forest Meteorology 148: 788–798.

Permissions

List of Contributors

Yunbin Qin, Zhongbao Xin, Xinxiao Yu and Yuling Xiao
Institute of Soil and Water Conservation, Beijing Forestry University, Beijing, China

Roberto C. Izaurralde, Allison M. Thomson and Aaron G. Rappaport
Joint Global Change Research Institute, Pacific Northwest National Laboratory and University of Maryland, College Park, Maryland, United States of America

Charles W. Rice
Kansas State University, Department of Agronomy, Manhattan, Kansas, United States of America

Lucian Wielopolski and Sudeep Mitra
Brookhaven National Laboratory, Department of Environmental Sciences, Upton, New York, United States of America

Michael H. Ebinger and Ronny Harris
Los Alamos National Laboratory, Los Alamos, New Mexico, United States of America

James B. Reeves, III and Barry Francis
EMBUL, ARS, USDA, Beltsville, Maryland, United States of America

Jorge D. Etchevers
Rappaport and Associates, c/o Joint Global Change Research Institute, College Park, Maryland, United States of America

Kenneth D. Sayre
Soil Fertility Laboratory, Natural Resources Institute, Colegio de Postgraduados, Carretera México-Texcoco, México

Bram Govaerts
CIMMYT, Km. 45, Carretera México-Veracruz, Texcoco, México, México

Gregory W. McCarty
HRSL, ARS, USDA, BARC West, Beltsville, Maryland, United States of America

Xiao-Yan Chen
College of Resources and Environment/Key Laboratory of Eco-environment in Three Gorges Region (Ministry of Education), Southwest University, Chongqing, China

State Key Laboratory of Soil Erosion and Dryland Farming on the Loess Plateau, Institute of Soil and Water Conservation, CAS and MWR, Yangling, China

Yu Zhao, Bin Mo and Hong-Xing Mi
College of Resources and Environment/Key Laboratory of Eco-environment in Three Gorges Region (Ministry of Education), Southwest University, Chongqing, China

Tao Ren
College of Resources and Environmental Science, China Agricultural University, Beijing, China
College of Resources and Environment, Huazhong Agricultural University, Wuhan, China

Jingguo Wang, Qing Chen and Fusuo Zhang
College of Resources and Environmental Science, China Agricultural University, Beijing, China

Shuchang Lu
Department of Agronomy, Tianjin Agricultural University, Tianjin, China

Jirong Wu, Mingzheng Yu, Jianhong Xu, Juan Du, Fang Ji, Fei Dong and Jianrong Shi
Institute of Food Safety and Detection, Jiangsu Academy of Agricultural Sciences, Nanjing, China
Key Lab of Food Quality and Safety of Jiangsu Province—State Key
Laboratory Breeding Base, Nanjing, China
Jiangsu Center for GMO evaluation and detection, Nanjing, China

Xinhai Li
Institute of Crop Sciences, Chinese Academy of Agricultural Sciences, Beijing, China

Felicity V. Crotty, Rhun Fychan, Vince J. Theobald, Ruth Sanderson and David Christina L. Marley
Institute of Biological, Environmental and Rural Sciences, Aberystwyth University, Gogerddan, Aberystwyth, United Kingdom

R. Chadwick
Environment Centre Wales, School of Environment, Natural Resources and Geography, Bangor University, Bangor, Gwynedd, United Kingdom

Jay Ram Lamichhane, Alfredo Fabi and Leonardo Varvaro
Department of Science and Technology for Agriculture, Forestry, Nature and Energy (DAFNE), Tuscia University, Viterbo, Italy
Hazelnut Research Center, Viterbo, Italy

Roberto Ridolfi
Department of Science and Technology for Agriculture, Forestry, Nature and Energy (DAFNE), Tuscia University, Viterbo, Italy

Fanqiao Meng, Xiangping Sun and Wenliang Wu
College of Resources and Environmental Sciences, China Agricultural University, Beijing, China

Jørgen E. Olesen
Department of Agroecology and Environment, Faculty of Agricultural Sciences, Aarhus University, Tjele, Denmark

Ilja Sonnemann, Stefan Hempel, Maria Beutel, Nicola Hanauer and Susanne Wurst
Freie Universitaet Berlin, Dahlem Centre of Plant Sciences, Berlin, Germany

Stefan Reidinger
University of York, Department of Biology, York, United Kingdom

Genxin Song, Jing Zhang and Ke Wang
Institute of Agricultural Remote Sensing and Information Technique, Zhejiang University, Hangzhou, Zhejiang, China and Ministry of Education Key Laboratory of Environmental Remediation, Ecological and Health, Zhejiang University, Hangzhou, Zhejiang, China

Richard G. Smith, Lesley W. Atwood and Nicholas D. Warren
Department of Natural Resources and the Environment, University of New Hampshire, Durham, New Hampshire, United States of America

Enke Liu, Changrong Yan, Xurong Mei and Yanqing Zhang
Institute of Environment and Sustainable Development in Agriculture, Chinese Academy of Agricultural Sciences, Beijing, China
Key Laboratory of Dryland Farming g Agriculture, Ministry of Agriculture of the People's Republic of China (MOA), Beijing, China

Tinglu Fan
Dryland Agricultural Institute, Gansu Academy of Agricultural Sciences, Lanzhou, Gansu, China

Bo Ma
State Key Laboratory of Soil Erosion and Dryland Farming on Loess Plateau, Institute of Soil and Water Conservation, Northwest A&F University, Yangling, Shaanxi Province, China
College of Resources and Environment, Northwest A&F University, Yangling, Shaanxi Province, China

Xiaoling Yu and Faqi Wu
College of Resources and Environment, Northwest A&F University, Yangling, Shaanxi Province, China

Fan Ma
Institute of Desertification Control, Ningxia Academy of Agriculture and Forestry Science, Yinchuan, Ningxia Hui Autonomous Region, China

Zhanbin Li
State Key Laboratory of Soil Erosion and Dryland Farming on Loess Plateau, Institute of Soil and Water Conservation, Northwest A&F University, Yangling, Shaanxi Province, China

Shan-e-Ahmed Raza
Department of Computer Science, University of Warwick, Coventry, United Kingdom

Hazel K. Smith and Gail Taylor
Centre for Biological Sciences, Life Sciences, University of Southampton, Southampton, United Kingdom

Graham J. J. Clarkson
Vitacress Salads Ltd., Lower Link Farm, St Mary Bourne, Andover, United Kingdom

Andrew J. Thompson
Soil and Agri-Food Institute, School of Applied Sciences, Cranfield University, Bedford, United Kingdom

John Clarkson
School of Life Sciences, University of Warwick, Wellsbourne, United Kingdom

Nasir M. Rajpoot
Department of Computer Science, University of Warwick, Coventry, United Kingdom
Department of Computer Science and Engineering, Qatar University, Doha, Qatar

Yuying Wang, Chunsheng Hu, Hua Ming, Wenxu Dong, Yuming Zhang and Xiaoxin Li
Key Laboratory of Agricultural Water Resources, Center for Agricultural Resources Research, Institute of Genetics and Developmental Biology, Chinese Academy of Sciences, Shijiazhuang, Hebei, China
Oene Oenema
Department of Soil Quality, Wageningen University, Alterra, Wageningen, The Netherlands

Douglas A. Schaefer
Key Lab of Tropical Forest Ecology, Xishuangbanna Tropical Botanical Garden, Chinese Academy of Sciences, Menglun, Yunnan, China

Fazhu Zhao, Gaihe Yang, Xinhui Han, Yongzhong Feng and Guangxin Ren
College of Agronomy, Northwest A&F University, Yangling, Shaanxi, China and The Research Center of Recycle Agricultural Engineering and Technology of Shaanxi Province, Yangling, Shaanxi, China

Liang Wu, Xinping Chen, Zhenling Cui, Weifeng Zhang and Fusuo Zhang
Center for Resources, Environment and Food Security, China Agricultural University, Beijing, People's Republic of China

Jin Wu, Yanguo Teng and Xudong Jiao
College of Water Science, Beijing Normal University, Beijing, China

Sijin Lu and Yeyao Wang
China National Environmental Monitoring Center, Beijing, China

Yangquanwei Zhong and Zhouping Shangguan
State Key Laboratory of Soil Erosion and Dryland Farming on the Loess Plateau, Northwest A & F University, Yangling, Shaanxi, P.R. China

Dongmei He and Honghua Ruan
Faculty of Forest Resources and Environmental Science, and Key Laboratory of Forestry and Ecological Engineering of Jiangsu Province, Nanjing Forestry University, Nanjing, Jiangsu, China

Chunhua Zhang
Department of Geography and Geology, Algoma University, Sault Ste. Marie, Ontario, Canada

Dan Walters and John M. Kovacs
Department of Geography, Nipissing University, North Bay, Ontario, Canada

Angela Joy Eykelbosh
Institute for Resources, Environment, and Sustainability, University of British Columbia (UBC), Vancouver, British Columbia, Canada

Mark S. Johnson
Institute for Resources, Environment, and Sustainability, University of British Columbia (UBC), Vancouver, British Columbia, Canada
Department of Earth, Ocean, and Atmospheric Sciences, University of British Columbia (UBC), Vancouver, British Columbia, Canada

Edmar Santos de Queiroz and Eduardo Guimarães Couto
Faculdade de Agronomia, Medicina Veterinária e Zootecnia (FAMEV), Universidade Federal de Mato Grosso, Cuiabaá, Mato Grosso, Brazil

Higo JoséDalmagro
Instituto de Física, Universidade Federal de Mato Grosso, Cuiabá, Mato Grosso, Brazil

Maria Julia de Lima Brossi, Lucas William Mendes, Amanda Barbosa Lima and Siu Mui Tsai
Cellular and Molecular Biology Laboratory, Center for Nuclear Energy in Agriculture, University of São Paulo, Piracicaba, SP, Brazil

Mariana Gomes Germano
Brazilian Agricultural Research Corporation, Embrapa Soybean, Londrina, PR, Brazil

Yi Liu, Kai-yuan Wan, Yong Tao, Zhi-guo Li, Guo-shi Zhang and Fang Chen
Laboratory of Aquatic Botany and Watershed Ecology, Wuhan Botanical Garden, Chinese Academy of Sciences China, Wuhan, China

Shuang-lai Li
Institute of Plant Protection and Soil Fertilizer, Hubei Academy of Agricultural Sciences, Wuhan, China

Index

www.ingramcontent.com/pod-product-compliance
Lightning Source LLC
Chambersburg PA
CBHW080247230326
41458CB00097B/4065